Advances in Cryogenic Engineering – Materials: Proceedings of the International Cryogenic Materials Conference (ICMC 2017)

IOP Conference Series: Materials Science and Engineering
Volume 279

Madison, Wisconsin, USA
9 - 13 July 2017

Editor:

U. (Balu) Balachandran

ISBN: 978-1-5108-5533-5
ISSN: 1757-8981

Printed from e-media with permission by:

Curran Associates, Inc.
57 Morehouse Lane
Red Hook, NY 12571

Some format issues inherent in the e-media version may also appear in this print version.

Printed by Curran Associates, Inc. (2018)

For permission requests, please contact the Institute of Physics at the address below.

Institute of Physics
Dirac House, Temple Back
Bristol BS1 6BE UK

Phone: 44 1 17 929 7481
Fax: 44 1 17 920 0979

techtracking@iop.org

Additional copies of this publication are available from:

Curran Associates, Inc.
57 Morehouse Lane
Red Hook, NY 12571 USA
Phone: 845-758-0400
Fax: 845-758-2633
Email: curran@proceedings.com
Web: www.proceedings.com

Table of contents

Volume 279

Advances in Cryogenic Engineering - Materials: Proceedings of the International Cryogenic Materials Conference (ICMC) 2017

(previous edition: ICMC-2015)

9–13 July 2017, Madison, Wisconsin, USA

Accepted papers received: 27 November 2017
Published online: 30 December 2017

Preface

Papers

Cryogenic Properties of Metals and Non-Metals

Cryogenic Power Electronics and Application Technologies

Nb3Sn Superconductor

BSCCO, MgB2, Pnictide and YBCO Superconductors

Flux Pinning, Stability, and AC Loss

HTS and LTS Cables

IOP Conf. Series: Materials Science and Engineering **279** (2017) 011001 doi:10.1088/1757-899X/279/1/011001

Preface

The 2017 joint Cryogenic Engineering and International Cryogenic Materials Conferences were held from July 9 through July 13 at the Monona Terrace Community and Conference Center in Madison, Wisconsin. As at past conferences, the international scope of these meetings was strongly maintained with 28 countries being represented by 686 attendees who gathered to enjoy the joint technical programs, industrial exhibits, special events.

The program for the joint conferences included a total of 467 presentations in the plenary, oral, and poster sessions. Four plenary talks gave interesting in-depth overviews and discussions on the topics of NASA perspective on electric propulsion technologies for large commercial aircraft, liquid air energy storage, the pursuit of low cost access to space, and what productization of superconducting electronics requires of the cryocooler community. Contributed papers covered a wide range of topics including many aspects in advances in cryogenics and superconductors, along with their applications. In total, 261 papers were submitted for publication of which 237 are published in these proceedings.

The CEC/ICMC Cryo Industrial Expo displayed the products and services of 54 industrial exhibitors and provided a congenial venue for a reception and refreshments throughout the week as well as the conference poster sessions.

The conference venue in Madison had many charms, and offered a variety of social activities in a beautiful setting. Prior to the Conference kicking off with the Welcome Reception Sunday evening, participants had the opportunity to attend Madison's annual Art Fair on the Square and the Wisconsin Brewing Company's Third Annual "Depth Charge" event. The Welcome Reception featured the University of Wisconsin Marching Band who played entertaining songs that inspired some of the CEC/ICMC board members to join along in dancing. During the week, participants enjoyed the Cryo Expo Exhibitor's Reception on Monday, the Conference Banquet Event on Tuesday and the popular Fun Run/Walk and evening Concert on the Square in downtown Madison on Wednesday. Beautiful and entertaining music was provided for free, and many excellent food and beverage vendors were available in the local areas surrounding the square. The venue of Madison was large enough to provide many dining and entertaining comforts, and small enough to allow walking to most hotels and area attractions.

Conference Chairs for 2017 were John Pfotenhauer, University of Wisconsin-Madison, for CEC and Timothy Haugan from U.S. Air Force Research Laboratory for ICMC. Program Chairs were Jennifer Marquardt from Ball Aerospace & Technologies Corp. for CEC and Charles Rong from U.S. Army Research Laboratory for ICMC, assisted by the CEC Program Vice Chair, Luisa Chiesa from Tufts University. The Exhibit Chair was Richard Dausman of Cryomech, Inc. who also served as Publicity Chair. Finally, Awards Chairs were Peter Kittel, Consultant, for CEC and Eric Hellstrom from the Applied Superconductivity Center/NHMFL/FSU for ICMC.

On July 9, ICMC presented its first short course: *Cryogenic/Superconducting – Materials, Power Electronics Engineering, and Applications*, a half-day course given by L. Cooley, L. Graber, M. Sumption, M. Majoros and K. T. Hartwig. Also on July 9, the Cryogenic Society of America presented four short courses: *Design of Optimal Helium Refrigeration and Liquefaction Systems*, a full day course given by V. Ganni and P. Knudsen, *Cryogenic Safety*, a half-day course by T. Peterson, J. Jurns and J. G. Weisend II, *Cryocoolers*, a half-day course by P. Spoor; and *Thermal Properties: Changes in the Cryogenic Region*, a half-day course by J. Pfotenhauer and F. Miller. The courses were well attended and catered to both seasoned professionals and newcomers to cryogenics in engineering and materials.

The able leadership of Paula Pair with the hardworking staff from Centennial Conferences provided outstanding conference management and operations. Thank you, Paula and staff, for a truly outstanding and memorable conference experience! Finally, to the attendees, authors, editors, and reviewers whose combined efforts have contributed to these proceedings, thanks very much to you all.

We look forward to the next CEC/ICMC in Hartford, Connecticut, July 21 – 25, 2019. Similar to Madison and other conference locations, Hartford is large enough to provide many dining and entertaining opportunities, and small enough to provide beautiful charms and allow easy access by walking to most hotels and area attractions. It is also close to many popular tourist destinations, such as New York City and the Boston, Massachusetts area.

John Pfotenhauer
2017 CEC Conference Chairman

Timothy Haugan
2017 ICMC Conference Chairman

IOP Conf. Series: Materials Science and Engineering **279** (2017) 011002 doi:10.1088/1757-899X/279/1/011002

ICMC Awards

ICMC LIFETIME ACHIEVEMENT AWARD 2017

Professor Kozo Osamura

ICMC has presented its Award for Lifetime Achievement biannually at the CEC/ICMC conference since its 30th anniversary conference in Keystone in 2005 to an individual to "recognize a lifetime's achievement in advancing the knowledge of cryogenic materials".

The seventh recipient of the ICMC Lifetime Achievement Award is Kozo Osamura, who follows Edward Collings in 2005, David Larbalestier in 2007, Kyoji Tachikawa in 2009, René Flükiger in 2011, Harold Weber in 2013, and Herbert Freyhardt in 2015.

Kozo Osamura completed his Doctor of Engineering in Physical Metallurgy at Kyoto University in 1970. He was appointed as a Research Associate at Kyoto University in 1970, and was promoted to Associate Professor in 1976 and Professor in 1985. While he held the rank of Research Associate, he was a guest scholar for a year at the Max Planck Institute for Metallurgy in Stuttgart, Germany. He served as director of Mesoscopic Materials Research Center at Kyoto University from 1992 to 1994. In 2005 Osamura became a Professor Emeritus at Kyoto University, and subsequently moved to the Research Institute for Applied Sciences as a chief researcher. In 2015 he became a director of the Institute.

Osamura has been a successful and prolific researcher and professor. He published more than 400 scientific papers, covering a wide variety of subjects in superconductivity, all of which are based on his strong expertize in metallurgy, X-ray and neutron scattering and diffraction, and the mechanical properties of materials. He combined his unique expertise in materials synthesis, synchrotron X-ray and neutron low-angle scattering and diffraction techniques to study both low- and high-temperature superconductors with an emphasis on properties related to practical applications. In the early days of optimizing low-temperature superconductors, he led efforts in Japan that studied the role of α-Ti precipitates as pinning centers in Nb-Ti, and on the impact of mechanical deformation and pinning centers on Nb_3Sn. He developed an excellent relationship with Japanese industry and his scientific understanding of these materials directly contributed to optimizing production of long lengths of Nb-Ti and Nb_3Sn conductors. With the discovery of HTS, he applied his expertise in phase transformations to study reaction pathways in the synthesis of pure HTS materials such as YBCO and Bi-2223. He constructed a phase diagram for the Y-Ba-Cu-O system and discovered a new route to synthesize Bi-2223. Most notably, his phase transformation studies on Bi-2223 contributed significantly to the present reaction route to form Bi-2223 using a precursor consisting of Bi-2212, alkaline earth cuprates, and plumbate phases.

Osamura has been a leader in the study of the mechanical properties of superconductors, which is essential because of the brittle nature of all superconductors, except Nb-Ti, which must be accounted for in the design of successful high-field magnets. He has made important contributions to our understanding of the mechanical properties of Nb_3Sn conductors. He was also one of the first to study the mechanical properties of Bi-2223 multifilamentary tapes and YBCO coated conductors. Unlike round wire cross sections, these HTS wires have very anisotropic flat tape forms, making cabling and winding without J_c degradation more challenging than for round wires, and requiring a detailed understanding of how the fracture propagates and degrades J_c. For example, Bi-2223 tape conductors consist of flake-like Bi-2223 oxide in a ductile and soft Ag matrix. He and his team clarified where the cracks initiate in Bi-2223 and then propagate to failure under various stresses. Recently he has been studying the lamination strengthening of Bi-2223 tapes, important for high field magnet applications, and found that the stress to 95% I_c retention can exceeded 510 MPa by using pre-tensioned lamination. For YBCO coated conductors he clarified that, since the thin monolayer of YBCO is deposited on the flat, oxide-buffered and strong metal substrates, the mechanical behaviors under tensile and/or bending stress are significantly different from any other of the superconductors. These and his other HTS studies have laid the foundation for the industrialization and application of superconducting wires and tapes.

Osamura has also played an important role in the international superconductivity community, helping to establish the International Electrotechnical Commission (IEC) TC 90 (Superconductivity) working group that prepares the international standards for superconducting materials and devices. He was active in the critical current measurement, tensile test, and general requirement working groups. He was also one of the original organizers of the biennial International Workshop on Mechanical Electromagnetic Properties of Composite Superconductors (MEMS), which is the most important workshop for discussion of the mechanical properties of superconductors that are important for practical applications.

The citation on the award plaque.

In recognition of his unique combination of materials synthesis with synchrotron X-ray and neutron diffraction techniques to understand and advance superconductors, and for his many contributions to developing international standards for superconducting materials and devices.

IOP Conf. Series: Materials Science and Engineering **279** (2017) 011002 doi:10.1088/1757-899X/279/1/011002

ICMC CRYOGENIC MATERIALS AWARD FOR EXCELLENCE 2017

Dr. Tengming Shen

The ICMC Cryogenic Materials Award for Excellence, is awarded annually at the ICMC Conference, subject to the nomination of an appropriate candidate, to an individual, who is under 40 years of age by the application deadline, to “recognize excellence in advancing the knowledge of cryogenic materials over recent years”. The 2017 award was given to Tengming Shen. Previous awardees were Fumitake Kametani in 2014 and Kazumasa Iida in 2015.

Tengming Shen received his BS and MS in Electrical Engineering at Southwestern Jiaotong University, studying superconducting maglev and materials physics of the newly discovered MgB_2 bulk materials. In 2006, he moved to the US to do his PhD studies at Florida State University in 2010, also in Electrical Engineering, with an emphasis on applied superconductivity. His PhD studies played an important role in understanding the underlying materials science to increase the critical current density of Bi-2212. Working with colleagues at the National High Magnetic Field Laboratory and using quench and advanced microscopy techniques, he investigated how the microstructure develops in Bi-2212 round wire through the complex heat treatment. He was instrumental in discovering that the porosity in as-drawn Bi-2212 powder agglomerated into bubbles during the heat treatment, which were the biggest current limiting mechanism in wires. This was a key step leading to developing the overpressure processing heat treatment that squeezes out these bubbles, making Bi-2212 a viable high-field magnet material.

After his PhD studies, Shen received a Peoples Fellowship and moved to Fermi National Accelerator Laboratory, where he studied magnet technology and accelerators. In 2012, he won an Early Career award from the Department of Energy for “Engineering High-field Superconducting Materials for Frontier Accelerator Technology”. He led a team to build and test prototype magnets using new high-field magnet materials, in particular Bi-2212, and to reveal many new aspects of high-field wire and magnet engineering. He investigated the failure mechanisms of superconducting wire when a superconducting magnet quenches and the dependence of the maximum allowable temperature during a quench on winding, stress, and strain conditions of superconducting magnets. He is currently at Lawrence Berkeley Laboratory where he works on magnet technologies for accelerators, nuclear physics and fusion energy, and medical applications.

IOP Conf. Series: Materials Science and Engineering **279** (2017) 011002 doi:10.1088/1757-899X/279/1/011002

ICMC BEST PAPER AWARD

The award for the best papers of the 2015 Conference, published in the *IOP Conference Series: Materials Science and Engineering, Vol. 102*, was presented at the 2017 Madison Conference. From the large number of papers nominated by the reviewers, the committee selected one paper for the award.

Best Superconducting Materials Paper

J Kvitkovic, R Hatwar, S V Pamidi, S Fleshler, and C Thieme

for their paper

"Temperature dependence of critical current and transport current losses of 4 mm YBCO coated condurctors manufactured using nonmagnetic substrate"
IOP Conference Series: Materials Science and Engineering, Vol. 102, 2015; 012033

IOP Conf. Series: Materials Science and Engineering **279** (2017) 011003 doi:10.1088/1757-899X/279/1/011003

2017 International Cryogenic Materials Conference Board of Directors

CONFERENCE AND PROGRAM

Michael Sumption, *President* – The Ohio State University
Timothy Haugan, *Conference Chair* – U.S. Air Force Research Laboratory
Charles Rong, *Program Chair* – U.S. Army Research Laboratory

BOARD OF DIRECTORS

U. (Balu) Balachandran – Argonne National Laboratory
David A. Cardwell – University of Cambridge, United Kingdom
Lance D. Cooley – Applied Superconductivity Center/NHMFL/FSU
David Evans – Advanced Cryogenic Materials Ltd, United Kingdom
Herbert C. Freyhardt – University of Houston / University of Göttingen (retired) and ZFW gGmbH Göttingen, Germany
Wilfried Goldacker – Karlsruhe Institute of Technology, Germany
K. Theodore Hartwig, Jr. – Texas A&M University
Timothy Haugan – U.S. Air Force Research Laboratory
Eric E. Hellstrom – Applied Superconductivity Center/NHMFL/FSU
Juan Knaster – Fusion For Energy, Spain
Yanwei Ma – Institute of Electrical Engineering, Chinese Academy of Science, China
Thomas H. Nicol – Fermi National Accelerator Laboratory
Richard P. Reed – Cryogenic Materials, Inc.
Charles Rong – U.S. Army Research Laboratory
Kenichi Sato – Sumitomo Electric Industries Ltd., Japan
Jun-ichi Shimoyama – University of Tokyo, Japan
Hyung-Seop Shin – Andong National University, Korea
Michael D. Sumption – The Ohio State University
Herman H. J. ten Kate – CERN, Switzerland
Marcel ter Brake – University of Twente, The Netherlands
Robert P. Walsh – National High Magnetic Field Laboratory/FSU

AWARDS COMMITTEE

David A. Cardwell – University of Cambridge, United Kingdom
Lance Cooley – Applied Superconductivity Center/NHMFL/FSU
K. Theodore Hartwig, Jr. – Texas A&M University
Eric E. Hellstrom, *Chair* – Applied Superconductivity Center/NHMFL/FSU
Michael Sumption – The Ohio State University
Robert Walsh – National High Magnetic Field Laboratory/FSU

ICMC Technical Editors

U. (Balu) Balachandran
Chief Technical Editor
Argonne National Laboratory
9700 S. Cass Avenue
Building #212
Argonne, IL 60439
USA
Email: balu@anl.gov

Emanuela Barzi
Fermi National Accelerator Laboratory
P.O. Box 500
Batavia, IL 60510
USA
Email: barzi@fnal.gov

David Cardwell
University of Cambridge
Department of Engineering
Trumpington Street
Cambridge, Cambridgeshire CB21PZ
United Kingdom
Email: dc135@cam.ac.uk

Lance Cooley
ASC / NHMFL / FSU
2031 E. Paul Dirac Drive
Tallahassee, FL 32310
USA
Email: ldcooley@asc.magnet.fsu.edu

Timothy Haugan
U.S. Air Force Research Laboratory
AFRL/RQQM, 1950 Fifth Street
Bldg. 18
Wright-Patterson AFB, OH 45433
USA
Email: timothy.haugan@us.af.mil

Eric Hellstrom
ASC / NHMFL / FSU
2031 E. Paul Dirac Drive
Tallahassee, FL 32310
USA
Email: hellstrom@asc.magnet.fsu.edu

Takanobu Kiss
Kyushu University
Department of Electrical Engineering
744 Motooka, Nish-ku
Fukuoka, 819-0395
Japan
Email: kiss@sc.kyushu-u.ac.jp

Peter Lee
ASC / NHMFL / FSU
2031 E. Paul Dirac Drive
Tallahassee, FL 32310
USA
Email: lee@asc.magnet.fsu.edu

Tatiana Prikhna
Institute for Superhard Materials of the National Academy of Sciences of Ukraine
2, Avtozavodskaya Str.
Kiev, 04074
Ukraine
Email: prikhna@ukr.net

Wolfgang Stautner
GE Global Research
One Research Circle
EP 113
Niskayuna, NY 12309
USA
Email: stautner@research.ge.com

Michael Sumption
The Ohio State University
Materials Science Department
394 Watts Hall
2041 College Road
Columbus, OH 43210
USA
Email: sumption.3@osu.edu

Robert Walsh
NHMFL / FSU
1800 E. Paul Dirac Drive
Tallahassee, FL 32310
USA
Email: walsh@magnet.fsu.edu

Judy Wu
University of Kansas
1251 Wescoe Hall Drive
1082 Malott Hall
Lawrence, KS 66045
USA
Email: jwu@ku.edu

Editorial Management
Annett Cady and Carrie Lian
Centennial Conferences
908 Main Street, Suite 350
Louisville, CO 80027
USA
Email: cecicmc@centennialconferences.com

IOP Conf. Series: Materials Science and Engineering **279** (2017) 011005 doi:10.1088/1757-899X/279/1/011005

Acknowledgments

The CEC and ICMC Boards wish to thank our supporter and sponsors who have contributed to the 2017 CEC/ICMC Conference.

SUPPORTERS

- **Fermilab**
- **U.S. Department of Energy, Office of Science**

SPONSORS

- **CAD Cut**
- **Cryomech, Inc.**
- **Demaco**
- **Scientific Instruments, Inc.**

Their contributions helped ensure the success of the 2017 Conference.

Producing the **Open Access** ***IOP Conference Series: Materials Science and Engineering (MSE)*** requires the dedicated effort of many individuals. The quality of the publication relies on the dedication of those who took time from their busy schedules to review and edit papers. The Boards wish to specifically acknowledge the efforts of the managing editor, Annett Cady and Carrie Lian of Centennial Conferences, and the chief technical editors, John Weisend (CEC) and Balu Balachandran (ICMC). Their contributions to ensure a quality publication in a timely manner are greatly appreciated. The chief technical editors are indebted to the individual technical sub-editors for their time spent with reviewers and authors to get the final manuscripts ready for publication. The Boards also wish to thank all students and post-doctoral researchers from a number of universities for their service with the initial format checking of the papers submitted for peer review.

CEC/ICMC 2017 was managed by Paula Pair and Centennial Conferences. Recognizing this is a tremendous team effort, the Boards would especially acknowledge, in addition to Paula, all her staff – Jim Berhalter, Annett Cady, Troy Christensen, Johanna Hild, Brion Jacobs, Andrew Kahl and Carrie Lian – for the fantastic job they did wonderfully.

Last, but not least, the Conference would not have been possible without the effort and dedication of the attendees in preparing their presentations and papers, traveling to Madison, Wisconsin, and showing the professionalism and enthusiasm that makes the CEC/ICMC such a great Conference time after time.

Fatigue and fracture of three austenitic stainless steels at cryogenic temperatures

D M McRae[1], S Balachandran[1] and R P Walsh[1]

[1]National High Magnetic Field Laboratory, 1800 E. Paul Dirac Dr., Tallahassee, FL 32310

dmcrae@magnet.fsu.edu

Abstract. For the past couple decades, 316LN stainless steel has remained the "go-to" alloy for structural components intended for cryogenic temperature service, partially because of its favorable mechanical properties, but also because of the data available in the literature for T = 4 K. In recent years, some interest has arisen to investigate and develop stronger and tougher alloys for cryogenic structural components, particularly for magnet systems like ITER. This study presents new 4 K fatigue crack growth rate (FCGR) and fracture toughness data for Nitronic® 50 and JK2LB stainless steels, compiles existing data for these alloys, and compares them with 316LN data found in literature. This study intends to further expand the existing cryogenic data set for these alloys, clarify key differences between them to better facilitate mechanical design, and potentially bolster further alloy development.

1. Introduction

316LN stainless steel is currently one of the most commonly selected structural materials used in superconducting magnet systems, thanks in part to its high strength, good fracture toughness and low magnetic permeability after fracture at cryogenic temperatures – qualities which are not present in many steels. Additionally, its excellent 4 K database helps make it a comfortable choice during the design process, as there exists a substantial amount of tensile, fracture toughness, and S-n fatigue data in the literature for the alloy in various conditions at cryogenic temperatures. However, some systems could benefit from different alloys with slightly better, or more specifically matched properties for a given design. ITER is one such project, the scale of which has also facilitated R&D of new alloys for its purposes. The tie plates that maintain pre-compression on the Central Solenoid (CS) modules require higher room temperature yield strength than 316LN can offer, so Nitronic 50 (F XM-19, Fe-21Cr-12Ni-5Mn) was investigated and accepted for the task [1]. Additionally, a brand new alloy, JK2LB, was developed specifically for use as CS conduit material, primarily to minimize the thermal contraction mismatch between the conduit and Nb_3Sn superconducting cables it encases [2]. These alloys have been tested extensively and reported on by several labs in recent years, but are rarely compared with one another in a common study. This investigation aims to report typical ranges of fatigue and fracture behavior of each alloy in various conditions to highlight key differences between them, and potentially aid with material down-selection for future projects at the very least. It is a compilation of some previously-published data as well as brand new data.

2. Materials and methods

Three alloys are investigated in a wide array of conditions: 316LN, Nitronic 50 (N50), and JK2LB. Since 316LN has been studied and improved upon for several decades, relevant data must be selected for a fruitful comparison with other modern alloys. Arguably, one of the last substantial developments on 316LN specifically for conduit alloys was in Reed's study on grain boundary sensitization after Nb_3Sn heat treatment due to Carbon content [3]. It suggested that Carbon content be kept at or below 0.01% for a 700 °C/100 hr heat treatment to retain adequate toughness. Though the decrease in toughness was not as drastic in specimens whose heat treatment final step was 625 °C/75 hr, which is closer to most Nb_3Sn heat treatments today, only 316LN with Carbon content near 0.02% or less was chosen for the present study. Table I tabulates chemical composition of most materials reported here. An ideal academic comparison would normalize as many variables as possible (e.g. all specimens annealed, same orientation, deliberate grain size selection, etc.), but the data generated recently has been application-driven, in which clients need to confirm each materials' properties in the exact condition they are to be employed in the magnet structure. Thus, several conditions are reported here.

Table 1 - Chemical composition of available materials.

Mat'l Comp #	Material Description	Composition (wt%), Fe balance											Pub Ref
		C	Mn	P	S	Si	Ni	Cr	Mo	V	N	B	
1	JK2LB Conduit (Hamada 2008)	0.013	21.80	0.007	0.002	-----	9.25	12.80	0.98	-----	0.12	0.0036	12
2	JK2LB Conduit (Walsh 2009)	0.025	21.42	-----	0.002	0.41	8.43	11.93	0.78	-----	0.12	0.0013	13
3	JK2LB Conduit (Walsh 2015)	0.012	21.55	0.007	0.002	0.26	-----	12.72	0.98	-----	0.12	0.0034	14
4	NBS N50 (Read 1978)	0.041	4.96	0.020	0.015	0.49	12.37	21.15	2.17	0.15	0.31	-----	16
5	N50 Forgings, Slab & Head,Vendor A (2013)	0.035	5.18	0.024	0.004	0.27	12.09	21.16	2.12	0.14	0.28	0.0020	17,18
6	N50 Forgings, Slab & Head, Vendor B (2014)	0.026	5.29	0.022	0.002	0.33	13.04	20.97	2.00	-----	0.34	-----	18
7	N50 Plate (2017)	0.040	4.92	0.021	<0.001	0.34	11.62	20.96	2.07	0.18	0.32	-----	new
8	N50 Sheet (2017)	0.017	4.47	0.018	0.001	0.51	12.09	20.78	2.25	0.19	0.34	-----	new
9	N50 Forging Vendor C, Batch 1 (2017)	0.020	5.40	0.016	0.001	0.23	12.60	21.00	2.17	0.10	0.30	-----	new
10	N50 Forging Vendor C, Batch 2 (2017)	0.016	4.81	0.020	0.001	0.45	12.20	21.36	2.26	0.18	0.28	0.0014	new
11	N50 Forging Vendor D (2017)	0.024	4.80	0.020	0.001	0.34	12.18	21.30	2.18	0.15	0.28	0.0016	new
12	316LN, Alloy B (Simon 1988)	0.006-0.010	1.40	0.002	0.0002-0.001	0.40	11.40	18.50	2.50	-----	0.16	-----	19
13	316LN, Alloy E (Simon 1988)	0.016	1.58	0.021	0.021	0.48	13.90	17.40	2.50	-----	0.17	-----	19
14	316LN, Alloy H (Simon 1988)	0.016	1.10	0.020	0.015	0.54	10.50	18.00	2.10	-----	0.14-0.17	-----	19
15	316LN Mod (Walsh 1996)	0.010	1.53	0.011	0.005	0.26	13.75	16.91	2.32	-----	0.21	-----	5
16	316LN 80mm plate (Nyilas 2002)	0.020	1.58	0.027	0.0008	0.49	11.89	17.18	2.53	-----	0.16	-----	20
17	316LN Valinox jacket (Nyilas 2002)	0.012	1.64	0.025	-----	0.24	12.00	17.20	2.43	-----	0.18	-----	20
18	316LN Plate (Sa 2006)	0.022	1.14	-----	-----	0.69	10.10	17.70	2.01	-----	0.13	-----	21
19	316LN Conduit (Walsh 2006)	0.012	1.58	-----	-----	0.36	13.07	17.60	2.23	-----	0.15	-----	7

All of the 4 K tensile and fracture toughness tests performed at NHMFL are conducted on servo-hydraulic test machines that feature a load reaction frame and cryostat that facilitate testing in liquid helium. Typically, a 100 kN capacity load cell is used for standard tensile specimens as well as all sizes of load-line compact tension (CT) specimens for FCGR and fracture toughness. Shepic-style clip-on extensometers are used to measure strain in tensile tests, and a clip-on crack opening displacement (COD) gage is used to monitor load-line displacement of CTs. Tensile tests are performed according to ASTM E1450, and FCGR and fracture toughness tests are performed according to procedures outlined in E647 and E1820 respectively. Fracture toughness analysis is performed with commercial software via J-integral method to obtain $K_{IC}(J)$ per E1820.

3. Test data and discussion

3.1. Fracture toughness and yield strength

Table II shows all materials compared here, with their tested condition, grain size, and selected mechanical properties. The composition number column corresponds to the one shown in Table I for these materials, where available. The new data presented here is denoted with "new" in the Reference column. All mechanical property values shown are averages of several individual test specimens.

Table 2 – 4 K Mechanical properties of all materials included in this study.

Mat'l Comp #	Material Description	Material Condition	Grain Size	Yield Strength MPa	Tensile Strength MPa	Elongation %	Area Reduction %	$K_{IC}(J)$ MPa√m	FCGR Paris Parameters da/dN=C*ΔK^m C	m	ΔK Range (MPa√m)	Pub Ref
1	JK2LB Conduit TL (2008)	AR(*)	-----	787	1322	58	55	254	-----	-----	-----	12
1	JK2LB Conduit TL (2008)	AC(†) + AG(‡)	-----	1006	1376	47	52	224	-----	-----	-----	12
2	JK2LB Conduit TL (2009)	AC	ASTM 3.0, 100 μm nom	1015	1391	40	45	154	6.08E-10	3.50	30 - 60	13
2	JK2LB Conduit TL (2009)	AC+AG	ASTM 3.0, 100 μm nom	1063	1397	38	38	167	6.44E-10	3.52	30 - 60	13
3	JK2LB Conduit LT (2015)	AC + CW(§) + AG	ASTM 5.0	1085	1458	41	32	357	2.087E-10	3.649	23 - 64	14
-----	JK2LB Conduit LT (2015)	AC + CW + AG	-----	1029	1423	42	-----	307	5.81E-10	3.33	25 - 100	15
-----	JK2LB Manifold Forgings TL (2017)	-----	ASTM 3.0, 111 μm	935	1358	44	28	276	-----	-----	-----	new
4	N50 TL (1978)	Annealed 1120°C, 2h, WQ	ASTM 7.0, 34 μm	1446	1893	9	33	111	3.810E-12	4.660	30 - 60	16
4	N50 TL (1978)	Annealed 1177°C, 1.5h, WQ	ASTM 3.0, 120 μm	1330	1848	16	38	176	-----	-----	-----	16
5	N50 Head Forging LT, Vendor A (2013)	Annealed 1038°C/12.25h	ASTM 1, 226 μm	1287	1695	30	36	200	-----	-----	-----	17
5	N50 Head Forging LS, Vendor A (2013)	Annealed 1038°C/12.25h	ASTM 1, 226 μm	1287	1695	30	36	191	3.313E-10	3.511	27 - 93	17
5	N50 Slab Forging LT, Vendor A (2013)	Annealed 1038°C/12.25h	ASTM 4.0	1326	1742	26	42	167	3.624E-10	3.493	26 - 78	17
5	N50 Head, mid, LS, Vendor A (2015)	Annealed 1038°C/12.25h	-----	1310	1666	-----	-----	250	9.470E-11	4.410	22 - 45	18
5	N50 Slab, mid, LS, Vendor A (2015)	Annealed 1038°C/12.25h	-----	1266	1743	-----	-----	226	1.310E-09	2.930	22 - 42	18
6	N50 Slab Forging LS, Vendor B (2014)	Annealed 1100°C/3h, WQ	ASTM 2.0-3.0	1342	1682	17	15	183	7.080E-11	3.193	22 - 65	new
6	N50 Head, top, LS, Vendor B (2015)	Annealed 1100°C/3h, WQ	ASTM 3.5-4.5	1150	1699	-----	-----	220	1.010E-10	3.620	23 - 40	18
6	N50 Head, bottom, LS, Vendor B (2015)	Annealed 1100°C/3h, WQ	ASTM 1.5-2.5	1140	1692	-----	-----	197	-----	-----	-----	18
6	N50 Slab, mid, LT, Vendor B (2015)	Annealed 1100°C/3h, WQ	-----	1201	1709	-----	-----	226	-----	-----	-----	18
6	N50 Slab, mid, LS, Vendor B (2015)	Annealed 1100°C/3h, WQ	-----	1201	1709	-----	-----	243	-----	-----	-----	18
7	N50 Plate LT (2017)	Hot rolled plate, annealed	-----	1298	1960	27	32	118	1.33E-09	3.312	27 - 61	new
7	N50 Plate TL (2017)	Hot rolled plate, annealed	-----	1592	1925	13.5(**)	12.6(**)	109	6.97E-10	3.514	30 - 55	new
7	N50 Plate, thin TL (2017)	Hot rolled plate, annealed	-----	1592	1925	13.5(**)	12.6(**)	147	-----	-----	-----	new
8	N50 Sheet LT (2017)	Cold rolled, annealed 1066°C	ASTM 9.0	1528	2015	23	25	181	9.50E-10	3.28	27 - 64	new
8	N50 Sheet TL (2017)	Cold rolled, annealed 1066°C	ASTM 9.0	1622	1952	22	32	155	2.63E-10	3.596	28 - 62	new
9	N50 MTM Forging Batch 1 LT (2017)	Forged, Annealed	ASTM 3-4	1145	1647	38	45	180	-----	-----	-----	new
9	N50 Forging Batch 1 LS, Vendor C (2017)	Forged, Annealed	ASTM 3-4	1145	1647	38	45	168	-----	-----	-----	new
10	N50 Forging Batch 2 LT, Vendor C (2017)	Forged, Annealed	-----	1165	1643	27	24	131	-----	-----	-----	new
11	N50 Forging LT, Vendor D (2017)	"As-cast"	ASTM 1.5-2.5	1119	1703	31	30	149	7.15E-12	4.534	30 - 52	new
11	N50 Forging TL, Vendor D (2017)	"As-cast"	ASTM 1.5-2.5	1317	1698	33	20	126	1.36E-10	3.761	30 - 57	new
12	316LN Alloy B, coarse grain (1988)	-----	110 μm	921	-----	-----	-----	327	-----	-----	-----	19
12	316LN Alloy B, fine grain (1988)	-----	13 μm	1086	-----	-----	-----	216	-----	-----	-----	19
13	316LN Alloy E (1988)	-----	105 μm	1006	-----	-----	-----	207	-----	-----	-----	19
14	316LN Alloy H (1988)	-----	72 μm	861	-----	-----	-----	185	-----	-----	-----	19
15	316LN Mod (1996)	20% CW	ASTM 8.5, 17 μm	1284	1648	28	-----	159	-----	-----	-----	5
15	316LN Mod (1996)	20% CW + 340°C/48h + 625°C/70h	ASTM 8.5, 17 μm	1326	1677	31	-----	133	-----	-----	-----	5
15	316LN Mod (1996)	20% CW + 700°C/100h	ASTM 8.5, 17 μm	1202	1586	26	-----	90	-----	-----	-----	5
16	316LN 80 mm plate TL (2002)	Wrought	120 μm	1017	1638	41	-----	236	6.96E-11	4.141	20 - 40	20
16	316LN 80 mm plate LT (2002)	Wrought	120 μm	997	1610	43	-----	233	6.96E-11	4.141	20 - 40	20
17	316LN Valinox jacket TL (Nyilas 2002)	-----	-----	-----	-----	-----	-----	-----	3.12E-08	2.413	17 - 40	20
18	KSTAR 316LN TL (2005)	Hot rolled plate, annealed	-----	862	1759	51	60	180	-----	-----	-----	21
19	316LN Conduit TL (2006)	20% CW, AR	-----	-----	-----	-----	-----	170	4.00E-09	2.87	22 - 78	7
19	316LN Conduit TL (2006)	20% CW, AG	-----	-----	-----	-----	-----	164	5.00E-09	2.84	18 - 69	7

* AR: As-Received
† AC: As-Compacted; ITER CS conduit undergoes transverse compaction before winding
‡ AG: Aged; subjected to simulated Nb_3Sn reaction heat treatment
§ CW: Cold-Worked; simulated winding strain applied
** Specimens failed on gage marks

A common graphical presentation for 4 K mechanical performance of these alloys is a $K_{IC}(J)$ vs yield strength plot. For comparison of materials on the plot it is good practice that the applied load direction be the same for both the yield strength and toughness values plotted. Test standards ASTM E399 and E8 provide the standard nomenclature for referencing the test specimen orientation. Figure 1 portrays all data chosen here, with appropriate respective loading directions. Several data now appear outside the NIST trend line for 300 series austenitic stainless steels [4] on the plot, but one should approach this comparison with caution. The data used to establish the trend lines are for annealed alloys that are J-tested in TL orientation to generate conservative design data. We suspect that specimens are likely to deviate from this trend when tested in either LT or LS directions. However, in the case of ITER's mechanical qualification testing, these values are needed in LT and LS orientation, as it is a direct measurement of the real-life loading direction, using the exact manufacturing processes that will be employed on the full scale magnet system. This makes it useful data primarily for ITER and any other entity interested in properties in the same orientation, but trickier to compare with existing 316LN values. Additionally, several fracture toughness values on the plot are obtained from thin CT specimens that violate ASTM thickness requirements. But, since a significant amount of characterization has been done for thin-wall conduit [5-8], they are data points that must be carefully integrated with the rest. Previous studies have been done to quantify the effect of CT thickness on measured $K_{IC}(J)$ at 4 K [8-10]. The largest variance observed was a 10% increase on a 12.7 mm thick specimen [9], but both other studies concluded the difference to be inappreciable within the error of the measurement [8, 10]. However, when testing the N50 plate for this report, we machined standard-thickness (12.7 mm) 0.5-CTs as well as 1.6 mm thick 0.5-CTs to assess the effect of specimen thickness study on this alloy; the average results were 147 MPa√m for the thin specimens, and 109 MPa√m for full thickness - a 35% increase. It is unclear whether the degree of this thickness dependence is different for each alloy, but it highlights the care that must be taken when comparing non-standard thin-specimen results with those from full-thickness specimens.

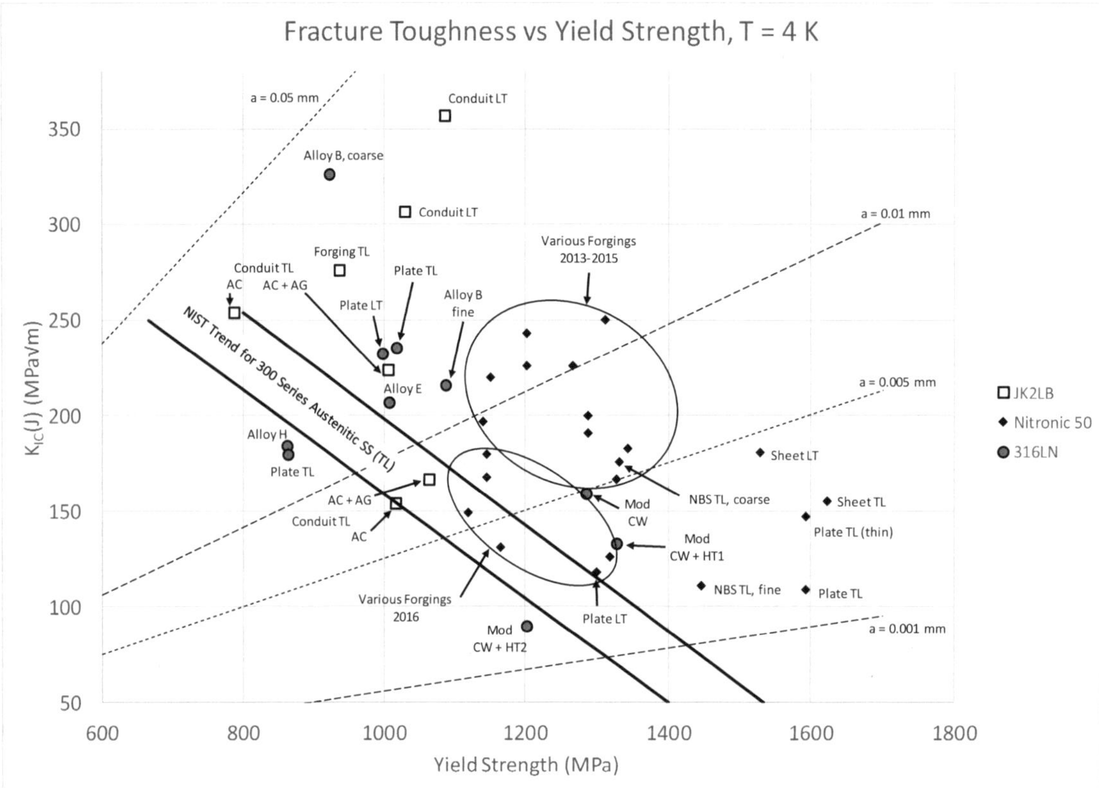

Figure 1 - Fracture toughness vs yield strength, T = 4 K.

The relationship between K_{IC} and yield strength in a given material is known to be inverse [11]. When comparing similar alloys on the plot, there are no singular factors present that stand out to explain the mechanisms that differentiate their properties relative to one another. Rather, they are the result of a combination of several variables - grain size, orientation, composition, manufacturing process, etc. Unfortunately, investigation into the effects of composition, microstructure, and other metallurgical variabilities was not included in the contracts that funded this work, so speculation on alloy optimization or improvement is outside the direct scope of this paper. However, reporting many of these variables in one common comparison alongside mechanical properties – as is the present intent – could make that investigation easier for the near future.

The three alloys do seem to populate broad common areas of this plot with one another. Most 316LN portrayed here does nearly follow the NIST trend, albeit with slightly higher strength and toughness than the upper band. JK2LB has consistently exhibited remarkably high toughness, in some cases exceeding maximum valid J-capacity of 0.5-CT specimens, thus requiring larger specimens. However, its strength has not yet achieved nearly that of some of the N50 investigated to-date. N50 generally has exceptionally high strength, often with lower toughness than that of JK2LB. Overall, these alloys combined appear to create a trend that surpasses the existing NIST trend, but additional testing could lend more confidence to this potential claim; for example, testing 2013-2015 N50 forgings in TL direction would help answer the question of orientation dependence, as would testing JK2LB from the same conduit and/or forging in two different directions.

3.2. Fatigue crack growth rate

Figure 2 shows FCGR curve fit data for all available materials here, separated by material – JK2LB, Nitronic 50, and 316LN at top left, top right, and bottom left, respectively. The graph at bottom right compares the range of crack growth data experienced in all three alloys. N50 in particular contains some dramatic outliers, though the bulk of data gathered here is grouped together and not dissimilar to 316LN and JK2LB. Those extreme values are from custom-specified thick forgings, received from different suppliers. However, other specimens from the same forgings behave similar to commercially-available material. There does not appear to be a characteristic crack growth rate dependence on crack plane orientation in most of these materials; the difference in da/dN curves between crack plane directions has often been observed to be on the same scale as variability between individual specimens of the same orientation. Thus, the high and low N50 curves are either a function of an exceptionally high degree of anisotropy or inhomogeneity within only those special forgings. None of these three alloys appears to have inherently higher or lower crack growth rate than the others.

3.3 Error Propagation Assessment

4 K mechanical properties testing is inherently challenging, and the test results exhibit data scatter due to both experimental procedure as well as material property variability. An experimental procedure error estimate is performed by evaluating the resolution and accuracy of the physical measurands that are used in calculating tensile and fracture toughness results. Using methods outlined in [22] and [23], which are based on standard error propagation calculation techniques, experimental procedure error is estimated to produce roughly 1-2% error in tensile test results, and can approach up to 4-6% for fracture toughness values presented here. These vary primarily with specimen size and equipment used to measure load, strain, load-line displacement, and specimen geometry. The typical data scatter observed in 4 K tensile and fracture toughness tests for a given material are quantified [13] with the standard deviations in a range closer to 10-20%, thus the additional spread is presumed to be attributed to variability within the material itself.

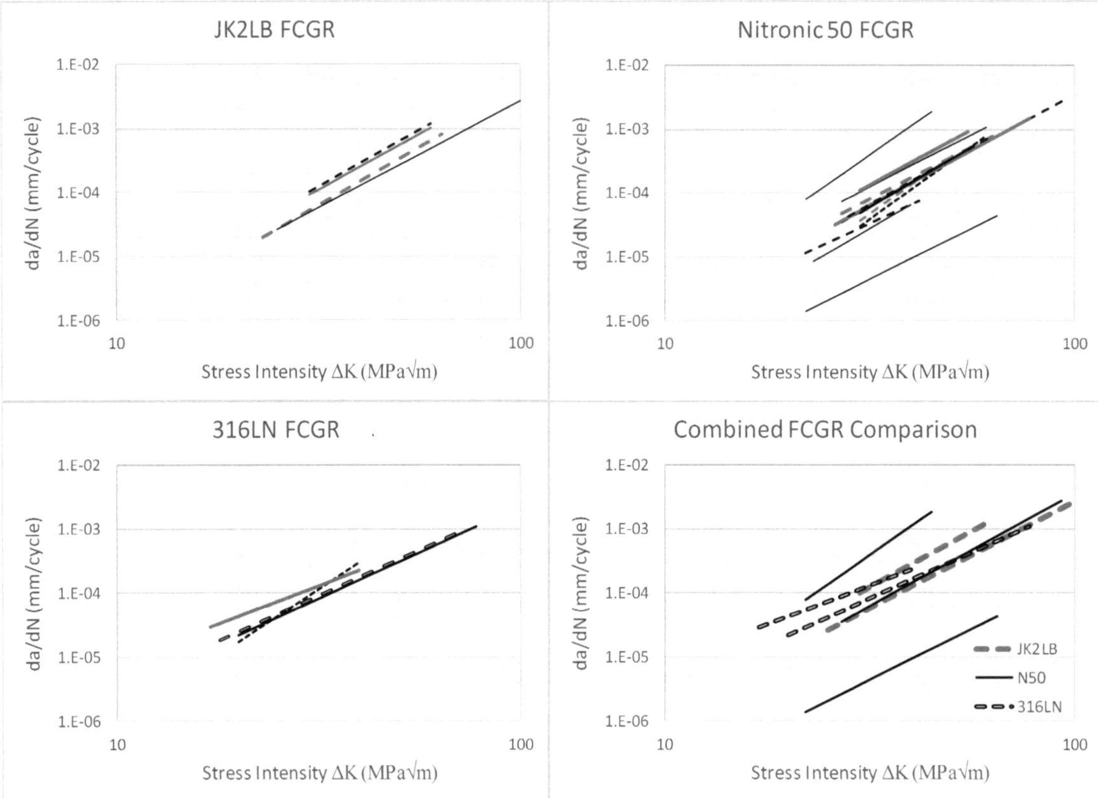

Figure 2 - 4 K fatigue crack growth rate comparison. Top left: JK2LB. Top right: Nitronic 50. Bottom left: 316LN. Bottom right: comparison of all 3 alloys.

4. Conclusions

4 K mechanical test data for three modern austenitic stainless steels has been compiled into one study for comparison. On the surface, they seem to represent an evolution of materials available for cryogenic structures, but more work is required for some concrete conclusions to be made. This report can serve as a launch pad for further alloy development - by organizing a considerable amount of research data into one source, potentially highlighting the next logical steps to take - but more immediately, it may serve as a useful tool during initial mechanical design decision-making.

5. References

[1] Freudenberg KD and Myatt RL 2011 ITER central solenoid support structure analysis 2011 IEEE/NPSS 24th Symp. on Fusion Eng.; DOI: 10.1109/SOFE.2011.6052298

[2] Nakajima H, Hamada K, Takano K, Okuno K and Fujitsuna N 2004 Development of low carbon and boron added 22Mn-13Cr-9Ni-1Mo-0.24N steel (JK2LB) for jacket which undergoes Nb3Sn heat treatment IEEE Trans. App. Supercon. 14, 2, pp. 1145-1148

[3] Reed RP, Walsh RP and McCowan CN 1992 Effects of Nb3Sn heat treatment on the strength and toughness of 316LN alloys with different carbon contents Adv. Cryo. Eng. Mat., 38, pp. 45-54; DOI: 10.1007/978-1-4757-9050-4_6

[4] Davis JR 1994 Stainless Steels, ASM International, p.502

[5] Walsh RP, Summers LT and Miller JR 1996 The 4 K tensile and fracture toughness properties of a modified 316LN conduit alloy Proc. of 16th CEC/ICMC Conf. Kitakyushu, Japan, Elsevier Science

[6] Walsh RP, Toplosky VJ, McRae DM, Han K and Martovetsky NN 2012 Cold work study on a 316LN modified alloy for the ITER TF coil conduit AIP Conf. Proc. 1435, 63; DOI: 10.1063/1.4712081

[7] Walsh RP, Toplosky VJ, Han K and Miller JR 2006 Effect of an aging heat treatment on the 4 K fracture and fatigue properties of 316LN and Haynes 242 AIP Conf. Proc., 824, 107; doi: 1063/1.2192340

[8] Walsh RP, Han K and Toplosky VJ 2008 Fracture toughness measurements and assessment of thin walled conduit alloys in a CICC application AIP Conf. Proc., 986, 68; DOI: 10.1063/1.2900398

[9] Yoshida Y, Nakajima H, Oshikiri M, Tobler RL and Shimamoto S 1988 Mechanical tests of large specimens at 4 K: facilities and results Adv. Cryo. Eng. Mat., 34, pp. 225-232

[10] Shindo Y, Horiguchi K, Kobori T, Nakajima H and Tsuji H 1996 Effects of specimen thickness and side-groove on fracture toughness of JN1 austenitic stainless steel rolled plate at liquid helium temperature Adv. Cryo. Eng. Mat. 42, pp. 345-352

[11] Ritchie RO 2011 The conflicts between strength and toughness Nature Materials, 10; DOI: 10.1038/NMAT3115

[12] Hamada K, Nakajima H, Matsui K, Kawano K, Takano K, Tsutsumi F, Okuno K, Teshima O and Soejima K 2001 Development of jacketing technologies for ITER CS and TF conductor AIP Conf. Proceedings 986, 76; DOI: 10.1063/1.2900399

[13] Walsh RP, Han K, Toplosky VJ, Martovetsky NN, Mann Jr TL and Miller JR 2010 Mechanical properties of modified JK2LB for Nb3Sn CICC applications AIP Conf. Proceedings 1219, 17; DOI: 10.1063/1.3402299

[14] Walsh RP, McRae DM, Han K and Martovetsky NN 2015 Tensile and fatigue qualification testing of ITER-CS conduit alloy JK2LB IOP Conf. Series: Mat. Sci. and Eng. 102, 1, 012002; DOI: 10.1088/1757-899X/102/1/012002

[15] Saito T, Kawano K, Yamazaki T, Ozeki H, Isono T, Hamada K, Devred A and Vostner A 2015 Mechanical properties of high manganese austenitic stainless steel JK2LB for ITER central solenoid jacket material Physics Procedia, 67, pp. 1016-21; DOI: 10.1016/j.phpro.2015.06.193

[16] Read DT and Reed RP 1978 Toughness, fatigue crack growth, and tensile properties of three nitrogen-strengthened stainless steels at cryogenic temperatures NBSIR 78-884, pp. 93-154, National Bureau of Standards, Boulder, CO

[17] McRae DM, Walsh RP, Dalder ENC, Litherland S, Trosen M and Kuhlmann DJ 2014 Fatigue and fracture properties of a super-austenitic stainless steel and 295 K and 4 K AIP Conf. Proceedings 1574, 59; DOI: 10.1063/1.4860605

[18] Langeslag SAE, Sgobba S, Libeyre P, Marcinek DJ and Zhang Z 2015 Extensive characterisation of advanced manufacturing solutions for the ITER central solenoid pre-compression system Fusion Eng. and Design, 98-99; DOI: 10.1016/j.fusengdes.2015.06.007

[19] Simon NJ and Reed RP 1988 Design of 316LN-type alloys NBSIR 88-3082, pp. 71-87, National Bureau of Standards, Boulder, CO

[20] Nyilas A, Portone A and Kiesel H 2002 European cryogenic material testing program for ITER coils and intercoil structures AIP Conf. Proc., 614, 123; DOI: 10.1063/1.1472534

[21] Sa JW, Kim HK, Choi CH, Kim HT, Hong KH, Park HK, Bak JS, Lee KW and Ha ET 2006 Mechanical characteristics of austenitic stainless steel 316LN weldments at cryogenic temperature Proc. 21st IEEE/NPS Symp. on Fusion Eng.; DOI: 10.1109/FUSION.2005.252881

[22] Gabauer W 2000 The determination of uncertainties in tensile testing *UNCERT Manual of codes of practice for the determination of uncertainties in mechanical tests on metallic materials* Code of practice No. 7 EU Contract SMT4-CT97-2165 Standards Measurement & Testing Programme ed Kandil FA et al (UK: National Physical Laboratory) ISBN 0-946754-41-1, Issue 1

[23] Plaza LM 2000 The determination of uncertainties in plane strain fracture toughness (K_{IC}) testing *UNCERT Manual of codes of practice for the determination of uncertainties in mechanical tests on metallic materials* Code of practice No. 3 EU Contract SMT4-CT97-2165 Standards Measurement & Testing Programme ed Kandil FA et al (UK: National Physical Laboratory) ISBN 0-946754-41-1, Issue 1

Acknowledgments
Research performed at NHMFL on these alloys has been funded by several grants from US Dept. of Energy via organizations within US-ITER and ITER-IO over a span of more than a decade. We value continued collaboration with our international colleagues, many whose work we cite here, and strive to maintain the semi-open-source nature of the ITER project, which facilitates faster development of these technologies for global benefit. We are also grateful to the NHMFL, Florida State University and the National Science Foundation for infrastructure and commitment of support.

Mechanical properties of AA5083 in different tempers at low temperatures

Chuanjun Huang[1], Zhixiong Wu[1], Rongjin Huang[1], Wei Wang[1], Laifeng Li[1, 2]

1 State Key Laboratory of Technologies in Space Cryogenic Propellants, Technical Institute of Physics and Chemistry, Chinese Academy of Sciences, Beijing, PR China

2 University of Chinese Academy of Sciences, Beijing, China

E-mail: wangwei@mail.ipc.ac.cn, laifengli@mail.ipc.ac.cn

Abstract. Aluminium alloy 5083 was chosen for use in the critical cryogenic applications of shipboard transportation of liquefied natural gas (LNG). In the present work, the tensile, Charpy impact, bend and fatigue crack propagation behaviours of aluminium alloy 5083 in temper O, H112, and H32 were investigated both at room and cryogenic temperatures.

Keywords: AA5083, tension, Charpy impact, fatigue crack growth rate

1. Introduction

The Al-Mg alloy 5083 (AA5083), which contains 4-4.9 wt% Mg, takes advantage of exceptional combination of economy of fabrication, weldability, and corrosion resistance. The AA5083 are a medium-strength, non-heat treatable wrought aluminium alloy and its strength increases with increasing Mg content. Since AA5083 contains more than 3wt% Mg, it is supersaturated at room temperature and the excess of Mg solute atoms tend to precipitate out as β-phase (Mg_2Al_3) particles located along the grain boundaries during prolonged aging at room temperature, or after shorter periods of exposure at slightly elevated temperatures in the range 65-200 °C [1]. Thus the AA5083 is suitable for utilization at room temperature and it is highly desired for being used in fast sea transportation for commercial and military applications, for example, military vehicles and high performance vessels [2, 3]. Moreover, AA5083 is found to demonstrate extremely high toughness both at room and cryogenic temperatures. In addition, its tensile strengths, elastic moduli and fatigue properties at cryogenic temperature are as high as, or even higher than those at room temperature, which has been attributed to its very low Peierls stresses, the inherent structure containing a small number of secondary precipitates, and a high notch yield ratio [4]. These properties make the AA5083 the most widely used for critical arctic and cryogenic applications such as bridges, marine structure, chemical processing equipment, superconducting machinery, and storage and transport tanks for cryogenic fluids.

The AA5083-O, which is strengthened by solid solution additions of Mg and used in a annealed condition, has been chosen for use in construction of large spherical tanks for shipboard transportation of liquefied natural gas (LNG) at service temperatures as low as 111 K [4, 5]. An interpretation of fatigue crack initiation and propagation behaviour of the AA5083-O at cryogenic temperature has been conducted to facilitate the design for leak-before-failure criterion for LNG tanks.

The temper designated in O state means in the annealed condition, which is applied to wrought aluminium alloys to obtain the lower strength temper but usually to increase subsequent workability. The AA5083-O has been investigated for LNG tanks and the AA5083 is also used in a temper designated H, i.e., strain hardened, which usually increases its strength by means of strain hardening. This temper may or may not have supplementary thermal treatments to produce some reduction in strength. The temper H1 represents strain hardened only, which is applied to AA5083 that has been strain hardened to obtain a desired level of strength but without a supplementary thermal treatment. The number following H1 indicates degree of strain hardening. The temper H3 represents strain hardened and stabilized, which is applied to AA5083 that has been strain hardened and then stabilized either by a low temperature thermal treatment or as a result of heat introduced during fabrication of the product. The stabilization treatment usually improves its ductility. The H3 temper is usually used only for those aluminium alloys that will gradually age soften at room temperature if they are not stabilized. The number following the H3 indicates the degree of strain hardening remaining after stabilization. In the present work, mechanical properties in terms of tensile, bend and Charpy impact at cryogenic temperatures of AA5083 in temper O, H112, and H32 were investigated. In addition, damage-tolerance fatigue life prediction methods are required for critical cryogenic components like LNG tanks, which is based on fatigue crack growth data, service loading conditions, and non-destructive inspection (NDI) results. Among these parameters, the fatigue crack propagation is mainly based on laboratory test data, thus the fatigue crack growth rates of the AA5083 in different tempers, i.e., O, H112, and H32, were investigated both at room and liquid nitrogen temperatures.

2. Experimental Details

AA5083 plates with a thickness of 30 mm in tempers O, H112, and H32 herein after referred as T-O, T-H112 and T-H32. The specimen geometries for tensile, fatigue crack growth rate (FCGR), Charpy impact, and bend testing are shown in Fig. 1(a-d). The dumb-bell shape tensile specimen has a gauge length, width, and thickness of 30 mm, 15 mm, and 2 mm respectively, which complies with ISO 6892-3. The compact tension specimen for test of FCGR, as shown in Fig. 1b, complies with ASTM E647 and ISO 12108, and the specimen has an initial ratio of crack length (a) to width (W) of 0.35. The V-notch specimen for Charpy impact testing, shown in Fig.1c, complies with ISO 148-1. The specimen geometry for three-point bend test, shown in Fig. 1d, complies with ISO 7438. Both the tensile and bend tests were conducted in liquid nitrogen in a displacement control on a 100 kN capacity electro-mechanical test machine (MTS model-SANS CMT5105S). The local deformation during tensile testing was monitored with a cryogenic-grade Epsilon extensometer with a gauge length of 25 mm. The mechanical tests at low temperatures were conducted with a homemade cryostat and the 77 K environment was obtained by immersing the test jig and specimen into liquid nitrogen in the cryostat. The Charpy impact test was conducted with an impact testing machine (model: MTS SANS XJ-300A) with an alternative hammer of 150 J energy. The Charpy impact test at 77 K was conducted with specimens immersed in liquid nitrogen for more than 15 min that were then transferred to the test machine within 5 s to keep the temperature, whereas the test at 150 K was conducted with specimens placed in 150 K environmental cryostat for more than 5 h and then were transferred to the test machine within 5 s. The fatigue crack propagation behaviour was investigated with a 120 kN capacity, servo-hydraulic dynamic test machine (Instron Model 8850). The crack mouth opening displacement (CMOD) was recorded with a cryogenic-grade Epsilon clip-on extensometer. The 77 K environment was obtained as that for both tensile and bend tests. The test was conducted with a constant-load controlled mode and with a stress ratio (R) of 0.1, i.e., a stress intensity factor K-increasing method, and the tests were conducted at a frequency of 20 Hz and 10 Hz at room temperature and 77 K respectively.

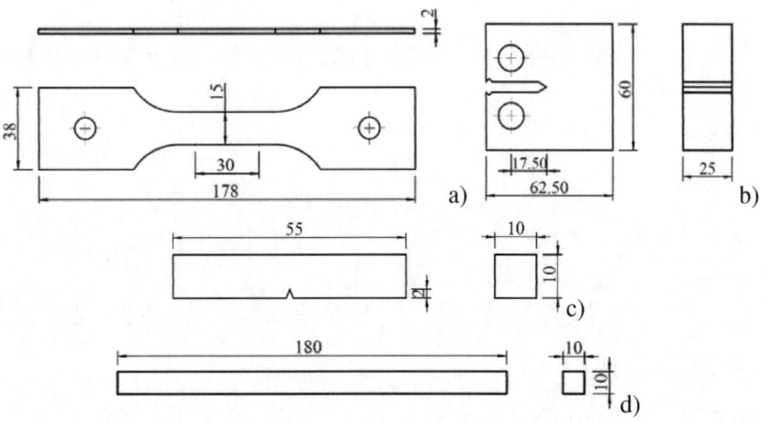

Figure 1. Specimen geometries for a) tensile, b) FCGR, c) Charpy impact, and d) bend testing.

3. Results and Discussion

3.1 Tensile properties

The cryogenic tensile properties of the AA5083 in different tempers are summarized in Table 1. It is found that the average 0.2% proof stress ($R_{p0.2}$) of the T-O, the T-H112, and the T-H32 at 77 K is 177.4 MPa, 175.4 MPa and 176.0 MPa, respectively. The Young's moduli of the T-O, T-H112 and T-H32 all demonstrate relatively large scatter but the average values are of a similar level. The average ultimate tensile strength (R_m) of the T-O, T-H112, and T-H32 at 77 K is 421.0 MPa, 417.0 MPa, and 419.8 MPa, respectively, whereas the average elongation at failure (A) of the T-O, T-H112, and T-H32 at 77 K are all around 45%. These results indicate that T-O, T-H112 and T-H32 fractured with a ductile behaviour in liquid nitrogen. Typical engineering stress-strain curve of AA5083 tested at 77 K is shown in Fig. 2, which confirms that the AA5083 undergoes a large plastic deformation before final fracture. Based on results of tensile properties including the 0.2% proof stress, ultimate tensile strength, and elongation at failure, the effect of O, H112, and H32 temper on cryogenic tensile strength and ductility at 77 K is slight.

Table 1 Tensile properties of T-O, T-H112, and T-H32 at 77 K

No.	$R_{p0.2}$ /MPa	R_m /MPa	E /GPa	A /%
T-O				
#-1	177	420	85	47
#-2	177	420	78	47
#-3	175	420	80	48
#-4	178	420	88	46
#-5	180	425	81	44
	177.4±1.82	421.0±2.24	82.4±4.04	46.4±1.52
T-H112				
#-1	174	415	75	43
#-2	176	420	72	47
#-3	176	415	88	45
#-4	178	420	89	48
#-5	173	415	89	47
	175.4±1.95	417.0±2.74	82.6±8.38	46.0±2.00
T-H32				
#-1	179	420	82	49
#-2	175	420	85	40
#-3	177	420	80	48
#-4	174	420	75	46
#-5	175	419	86	41
	176.0±2.0	419.8±0.45	81.6±4.39	44.8±4.09

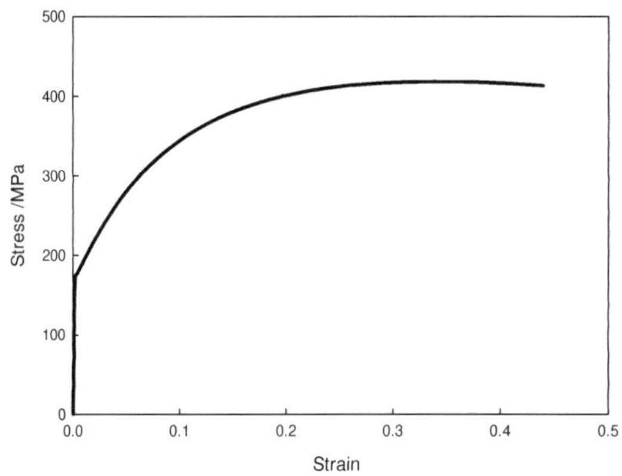

Figure 2. Tensile stress-engineering strain curve of T-O at 77 K.

3.2 Bend properties

The bend properties of T-O, T-H112, and T-H32 tested at 77 K are listed in Table 2. The average bend moduli of T-O, T-H112, and T-H32 are 76.0 GPa, 71.8 GPa, and 71.8 GPa. Considering the bend testing uncertainty, it is found that the temper has a negligible effect on the cryogenic bend modulus of AA5083. The average cryogenic bend strength of T-H32 is larger than that of T-O and T-H112, whereas T-O exhibits a slightly larger value than that of T-H112. A typical bending stress-strain curve of the AA5083 at 77 K is shown in Figure 3, which indicates a typical ductile behaviour. Moreover, serrated stress-strain behaviour was observed, as shown in Fig. 3.

Table 2 Bend properties of T-O, T-H112, and T-H32 at 77 K

	No.	Bend strength /MPa	Bend Modulus /GPa
T-O			
	#-1	695	81
	#-2	650	79
	#-3	675	77
	#-4	685	71
	#-5	645	72
		670±21.8	76.0±4.36
T-H112			
	#-1	650	72
	#-2	650	74
	#-3	640	69
	#-4	645	71
	#-5	660	73
		649±7.42	71.8±1.92
T-H32			
	#-1	715	70
	#-2	705	73
	#-3	700	72
	#-4	705	72
	#-5	700	72
		705±6.12	71.8±1.10

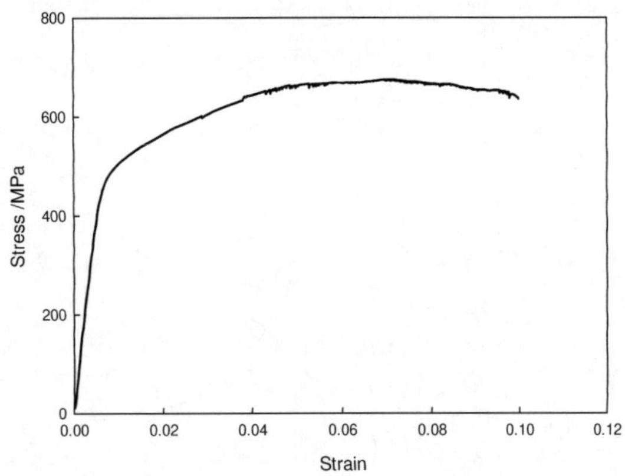

Figure 3. Bending stress-strain curve of T-O at 77 K.

3.3 Charpy impact properties

The Charpy impact energy can be described as notch toughness of a material, which is qualitative and comparative in nature. The Charpy impact energy of T-O, T-H112, and T-H32 test at 150 K and 77 K is listed in Table 3. At 150 K, T-O has Charpy impact energy of 24.7 J, which is the same as that of T-H112, whereas T-H32 demonstrates a slightly larger value, i.e., 26.7 J. When temperature decreased to 77 K, the average Charpy impact energy of all three tempers are around 21±1 J. Moreover, it is found that T-O, T-H112, and T-H32 exhibit significant reduction in Charpy impact energy when temperature decreases from 150 K to 77 K. It has been reported that the *J*-integral fracture toughness of T-O at 111 K demonstrated a slightly high value than that at 77 K, but it was dependent on specimen orientation [5]. The Charpy impact energy provides qualitative fracture toughness and the tendency with temperature change is consistent with that obtained by the *J*-integral test.

Table 3 Charpy impact energy of T-O, T-H112, and T-H32 at 150 and 77 K

No.	Temperature /K	Impact Energy /J
T-O		
#-1	150	25
#-2	150	25
#-3	150	24
		24.7±0.58
#-1	77	22
#-2	77	22
#-3	77	22
#-4	77	22
#-5	77	22
		22.0±0
T-H112		
#-1	150	25
#-2	150	24
#-3	150	25
		24.7±0.58
#-1	77	22
#-2	77	20
#-3	77	20
#-4	77	21
#-5	77	21
		20.8±0.84
T-H32		
#-1	150	26
#-2	150	27
#-3	150	27
		26.7±0.58
#-1	77	21

#-2	77	21
#-3	77	21
#-4	77	22
#-5	77	21
		21.2±0.45

3.4 Fatigue crack propagation behaviour

The fatigue crack growth rate (*da/dN*) as a function of stress intensity factor range (ΔK) of AA5083 in different tempers has been determined both at room and liquid nitrogen temperatures. Figure 4 provides the fatigue crack propagation behaviour of T-O, T-H112, and T-H32 tested both at room temperature and 77 K. As shown in Figure 4, the fatigue crack growth of AA5083 for all tempers occurs more slowly at low temperature than that at room temperature if small to moderate ΔK-values (10~23 MPa.m$^{1/2}$) are applicable, whereas faster crack growth was observed for larger ΔK-values with K_{max} close to plane stress fracture toughness K_c. This kind of behavior holds for most metallic materials and the crack growth rate was lower at small and moderate ΔK-values (i.e., the threshold (region I) and Paris (region II) region) has mainly been ascribed to the reduced cyclic plasticity at the crack tip at low temperature [6]. Tobler and his co-worker have attributed this to the presence of moisture in ambient air, which played a critical role in significant increase of the fatigue crack growth rate at room temperature compared to cryogenic temperature [5]. They also found that the fatigue crack growth rates of T-O at 111 K were nearly equivalent to those at 4 K. These findings indicated that moisture in air plays a crucial role in fatigue crack growth rate of AA5083, rather than the temperature effects per se. The increased fatigue crack growth behaviour for large ΔK-values at low temperature can be attributed to the reduced ductility of AA5083.

When investigating the effect of temper on fatigue crack propagation behaviour of AA5083 at 77 K, it is found that the FCGRs of T-H32 were higher than those of T-O, whereas T-H112 demonstrated the lowest FCGR.

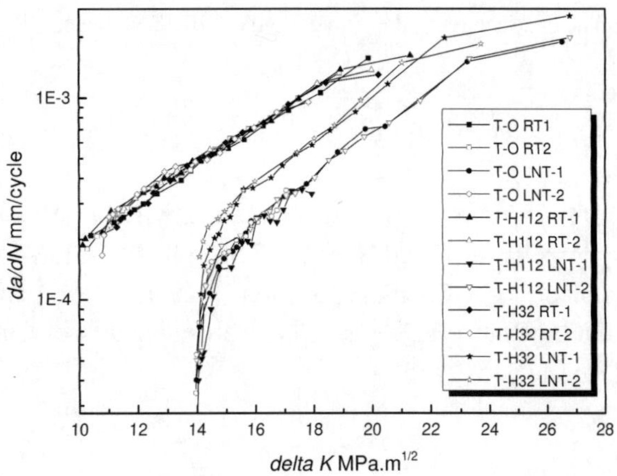

Figure 4. FCGRs for T-O, T-H112, and T-H32 at room temperature and 77 K.

IOP Conf. Series: Materials Science and Engineering **279** (2017) 012002 doi:10.1088/1757-899X/279/1/012002

4. Conclusion

In the present work, the mechanical properties of AA5083 in temper O, H112, and H32 were investigated both at room and low temperatures. It was found that the tensile properties including 0.2% proof stress, ultimate tensile stress and elongation at failure at 77 K depended slightly on the temper. For the bend properties at 77 K, it was observed that the temper H32 resulted in the highest bend strength whereas the temper H112 resulted in the lowest one. T-O exhibited the same Charpy impact energy value as T-H112 and both were slightly lower than that of T-H32 at 150 K. The Charpy impact energy of T-O, T-H112, and T-H32 was close at 77 K and all were lower than those at 150 K. The fatigue crack propagation of AA5083 in all three tempers occurred more slowly at low temperature than that at room temperature in the small to moderate ΔK-values range, whereas in the large ΔK-values range, the reverse condition was observed. Moreover, the FCGR of AA5083 at 77 K significantly depended on the temper condition.

References

1. R.Y. Chen, H.Y. Chu, C.C. Lai, and C.T. Wu, Effects of annealing temperature on the mechanical properties and sensitization of 5083-H116 aluminum alloy. *Proceedings of the Institution of Mechanical Engineers Part L-Journal of Materials-Design and Applications*, 2015. **229**: 339-46.
2. R.A. Sielski, The history of aluminum as a deckhouse materials. *Naval Engineers Journal*, 1987. **99**: 165-72.
3. A.P. Newbery, S.R. Nutt, and E.J. Lavernia, Multi-scale Al 5083 for military vehicles with improved performance. *JOM*, 2006. **58**: 56-61.
4. N.S. Raghavendran, M.E. Fourney, and F.V. Lawrence, Fatigue-fracture behavior of 5083-O aluminum magnesium alloy for LNG applications. *Engineering Fracture Mechanics*, 1988. **29**: 647-62.
5. R.L. Tobler and R.P. Reed, Fracture mechanics parameters for a 5083-O aluminum-alloy at low temperatures. *Journal of Engineering Materials and Technology-Transactions of the ASME*, 1977. **99**: 306-12.
6. J. Schijve, Fatigue of structures and materials. 2nd ed. 2009: Springer.

Acknowledgements

The authors thank financial support of the State Key Laboratory of Technologies in Space Cryogenic Propellants (Grant Nos. SKLTSCPQN201501 and SKLTSCP1701), the National Magnetic Confinement Fusion Science Program (Grant No. 2015GB121001), and the National Natural Science Foundation of China (Grant Nos. 51427806, 51401224, and 51577185).

Strength and electrical resistivity of heavily worked copper

J C Springs[1], Y T Kao[1], A Srivastava[2], Z S Levin[1], R E Barber[3], and K T Hartwig[1,2,3]

[1] Mechanical Engineering Department, Texas A&M University, College Station, TX 77843-3123, USA
[2] Materials Science and Engineering Department, Texas A&M University, College Station, TX 77843-3003, USA
[3] Shear Form, Inc. Bryan, Texas 77801, USA

Email: thartwig@tamu.edu

Abstract. Copper and its alloys are a topic of interest for various cryogenic conductor applications. Often, it is used as a supporting matrix for superconducting filaments requiring that it have good strength and high conductivity. One of the best methods to increase strength while preserving conductivity is work hardening. In this study, CDA101, CDA110, and C182 copper were processed by a severe plastic deformation (SPD) procedure called equal channel angular extrusion (ECAE). In this study we explore the relationships between the levels of plastic strain and annealing with tensile and hardness properties, grain size, and electrical resistivity. While C182 has the highest strength, it also has the lowest conductivity. CDA101 and CDA110 both retain over 95% of their conductivity in the fully worked state, while C182 has about 40% of the IACS value. Saturation of strength occurs around 3-4 ECAE passes. It is concluded that a lower amount of plastic strain via ECAE is best for creating a material with the highest combination of strength and conductivity, and is suitable for high strength high conductivity applications.

1. Introduction

In the relatively recent past, the needs of higher strength materials have prompted an increase into studies dealing with the manufacture of ultrafined-grained (UFG) and nanocrystalline (NC) metals [1]. Additionally, in cryogenics communities dealing with low temperature superconducting wires and electromagnets, electrical conductivity plays a most important role [2]. In electrical applications, copper and aluminum are the most commonly used metals due to their availability and high electrical conductivity. Pure copper is often the preferred choice because of strength and compatibility issues. Copper also has superior creep resistance when compared to aluminum, which decreases the possibility of wire failure. However, pure annealed copper does not have a particularly high strength, which would help for most superconducting wires. The problem is to find a method to increase the strength of copper while not decreasing its superior electrical conductivity. Methods to strength a material can include precipitation hardening, introduction of nanoparticles to metal matrix, heat treatment schedules, and plastic deformation. Some of these methods have the possibility of additionally increasing material conductivity as well [3-5].

One of the easiest methods to increase the strength of a material, while not heavily sacrificing electrical conductivity, is through severe plastic deformation (SPD). Some of the more common methods of severe plastic deformation include high pressure torsion, accumulative roll bonding, twist extrusion, and equal channel angular extrusion (ECAE). ECAE has a few advantages over the other SPD processes mentioned. First, ECAE is a repeatable process that does not change the initial dimensions of the work piece (billet), allowing for multiple extrusions, meaning you can accrue large amounts of strain. Second, there is not a geometric requirement, meaning you can extrude bars, cylinders, plates, or even tubes. Third, ECAE is easy to perform, and could be easily implemented in manufacturing facilities.

Equal channel angular extrusion can produce bulk ultrafined grained metals by subjecting a thin layer of material at the crossing planes of the die to simple shear. This process refines the grain structure, and introduces dislocations into the material, both of which increase strength. A schematic of the ECAE process is shown in Figure 1.

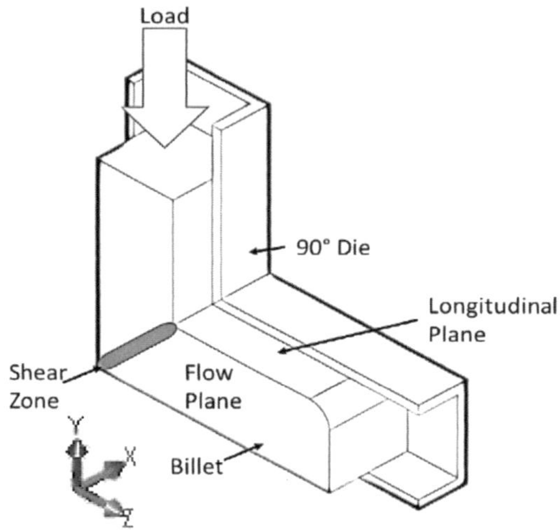

Figure 1: Schematic of ECAE deformation showing primary planes, directions, and the shear zone.

While many studies have been done with regard to strengthening copper and some of its alloys via SPD processes, relatively few have studied the impact ECAE has on the electrical resistivity. In previous works of SPD by ECAE, studies reported increases in hardness and strength, sometimes by up to as much as 300%. Torre et al. [6], reported tensile strengths as high as 455 MPa for copper extruded via route Bc (Routes are described later in Materials and Methods section). Similarly, Xue et al. [7] found an increase in strength to approximately 445 MPa for route Bc processed copper. Microhardness results are similar and multiple studies have obtained fully work hardened values for copper that fall in the ranges of 130-150 Vickers hardness [6-7]. Testing of copper alloys has been done by Wei et al. [5] and Ko et al. [8], who found that both alloys had significantly increased strength with only a small drop in resistivity following ECAE. Ko et al, concluded that after SPD processing and a proper heat treatment, both the mechanical and electrical properties of the Cu-0.5wt.% Cr alloy could be increased.

Other studies have focused more on the microstructure and recrystallization behavior of copper and its alloys. Torre et al. [6] and Etter et al. [9] have all reported grain sizes for ECAE processed copper in the ranges of 200 nm to 500 nm for routes Bc for both 4 and 8 pass samples. Additionally, for routes Bc the microstructure was found to be more equi-axed when comparted to other routes such route A. In almost all studies, the impact ECAE has on grain size is the most drastic for the first 2 passes, and then levels off with only a slight decrease after 4 passes.

Recrystallization behavior have been studied for copper with the recrystallization temperature falling between 200°C and 300°C depending on the total amount of accumulated strain [5]. Grain sizes were reported for recrystallized microstructures in the ranges of 0.75 μm to 3 μm for fully worked copper samples depending on the temperatures and length of heating [5]. Some studies have seen dynamic recrystallization occurring at room temperature after samples were subjected to rolling, whereas the purely ECAE processed microstructure apparently remains stable up to 150°C [10].

In this study oxygen free high conductivity (OHFC) copper (CDA101), commercially pure copper (CDA110), and a copper chromium alloy (C182), were processed via ECAE. Testing to determine tensile and yield strength, microhardness, recrystallization temperature, microstructure characteristics, and electrical resistivity were conducted. The testing results were used to determine if work hardening is indeed a suitable process to increase the strength of copper and its alloys, while maintaining the high conductivity needed in cryogenic applications.

2. Materials and Methods

The three different copper alloys studied, CDA101, CDA110, and C182, were purchased as 30 mm by 30 mm cross section bars. CDA101 with 99.99% Cu content was obtained from ThyssenKrupp Materials North America. CDA110 and C182 were purchased from New Southern Resistance Welding (NSRW). CDA110 possess 99.9% Cu content (silver counted as copper) and 0.04% oxygen. C182 contains 99.1% Cu content (Cu+Ag) and 0.6-1.2% Cr content.

The square cross section copper bars were first cut to 254 mm lengths and then recrystallized. The CDA101 and CDA110 bars were annealed at 350°C for one hour in a muffle furnace and then furnace cooled. The C182 was annealed at 990°C for 45 minutes and then quenched in room temperature water. Heat treatments were performed at 1 atm. Bars were then machined to 25.4 mm x 25.4 mm cross section.

Samples were ECAE processed though a square 90° sliding wall die at an extrusion speed of 2.54 mm/s at room temperature. The routes examined included routes A, B, Bc, and E, all processed to 8 passes. Route Bc was processed additionally to 16 passes. Between each pass, some flash on the bars was removed in order for the billet to fit back in the ECAE die. Table 1 summarizes the rotation angle between successive passes for each ECAE route. A more detailed description of the different routes can be found in Segal's paper "Material processing in simple shear" [11].

Table 1. Descriptions of ECAE routes.

Routes of ECAE	Route A	Route B	Route B_C	Route E
Odd passes-beginning at pass #3	0°	+90°	+90°	+180°
Even passes	0°	-90°	+90°	+90°

After the bars were processed by ECAE, 12.7×6.4×3.2 mm samples were cut from the flow plane via wire electric discharge machining (EDM) for heat treatment to determine recrystallization behaviour and to characterize the recrystallized microstructure. The CDA101 and CDA110 samples were subjected to temperatures ranging from 100°C to 500°C in a sand bath for 1 hour. The C182 samples were heat treated from 350°C to 500°C for 10 minutes. Recrystallization temperatures to determine recrystallized grain sizes in CDA101 Cu were 350°C, 275°C, 250°C, and 225°C for 15 minutes in a sand bath for 1 pass, 2 pass, 4 pass, and 8 pass samples respectively.

The 12.7×6.4×3.2 mm samples wire EDM cut from the CDA101 billet flow plane were mechanically ground starting from 320-grit emery paper to 1200-grit, followed by electrolytic polishing for microscopy. Electrolytic polishing was achieved using an electrolyte solution of 82.5 mL of 85% H_3PO_4 solution mixed with 17.5 mL of distilled water, applying a cell voltage of 2 V for 15 minutes. The solution was constantly stirred at low speed. Samples were cleaned with water then ethanol before imaging. Scanning electron microscope images were then taken with an accelerating voltage of 13 kV and a working distance of 10 mm. Additionally, Vickers hardness measurements were taken on the same sample to determine recrystallization behaviour.

Electrical resistivity measurements on the Cu alloy and CDA110 Cu were made using the four-point probe method on 3.2x3.2x50.8 mm samples using currents of 2.6 A at room temperature (298 K). The voltage drop was measured by a Keithley Model 181 nanovoltmeter and the resistivity calculated by equation 1 below. The distance between voltage taps was 25.4 mm, and the current leads being connected at the two ends.

$$\rho = \frac{VA}{IL} \tag{1}$$

where V and L stand for the voltage drop and the distance between the voltage taps, A is the cross section of the sample, and I is the current that passes through the sample. Resistivity ratios for the CDA101 material were determined by supplying a constant 1 A current and measuring the voltage drop across 2 mm diameter cylinders 25.4 mm long while immersed in liquid helium (4.2 K), liquid nitrogen (77 K), and an ice bath (273 K). The two voltage taps had a 12mm distance between them, with current leads again being at the two ends of the sample.

Microhardness measurements were taken using a Leco Microhardness Tester LM 300AT. The samples required the same polishing procedure as described previously. The samples were tested with a 300 g indenter load with a dwell time of 5 seconds. The hardnesses presented are an average of 10 separate measurements.

Dogbone samples were cut via wire EDM for tensile testing. The dogbones were 26 mm in total length with an 8 mm gauge length and 7 mm long tabs. The radius of the curve connecting the tab to the gauge length was 2 mm. The width of the tab was 7 mm and the width of the gauge length was 3 mm. The thickness of the samples was approximately 1.25 mm. The samples were tested using an MTS machine coupled with an 11 kN Interface 1010AF load cell and Epsilon Model 3442-008m-020-LHT strain gauge. The tests were run at 0.01mm/second until fracture. Three separate tests were done for each processing condition and the averages reported. Only CDA101 copper was tensile tested for this study.

3. Results

The average starting grain size for the CDA101 copper after the pre-processing anneal was found to be 29.5 μm. After a single ECAE pass the grain size decreased to 0.978 μm. With additional processing, grain size continued to reduce but much less drastically, coming to a near minimum with 4 pass route Bc at 0.412 μm. Recrystallized grain sizes follow a similar trend with route Bc having the smallest value ~1.3 to 1.9 μm, and the lower pass samples having the largest. Table 2 and Figure 2 summarize and show the changes in the microstructures after ECAE.

Table 2: Summary of as-worked grain sizes and recrystallized grain sizes for CDA101 copper

Route	Accumulated Strain	Grain Size (micron)	Standard Deviation (micron)	Recrystallized Grain Size (micron)	Standard Deviation (micron)
As Received	0	29.5	14.4	N/A	N/A
1A	1.1	0.98	0.67	6.3	4.5
2A	2.3	0.75	0.24	4.4	2.7
4A	4.6	0.58	0.20	2.2	1.2
4B	4.6	0.49	0.11	2.5	1.6
4Bc	4.6	0.41	0.13	1.9	0.8
4E	4.6	0.47	0.14	2.1	1.2
8Bc	9.2	0.42	0.16	1.4	0.7
8E	9.2	0.53	0.20	1.7	1.2
16Bc	18.5	0.42	0.13	1.3	0.7

Recrystallization based on Vickers hardness testing was studied for all three cases of copper for route 8Bc. CDA 110 and CDA101 showed a drop in hardness after heat treatment at 150°C. The C182 copper saw decreases in hardness starting after heat treatment at 400°C. Figure 3 gives examples of Vickers hardness for all the alloys at certain heat treatment temperatures.

Mechanical testing was performed for all routes for the CDA101 copper billets. Yield strength, tensile strength, and elongation to failure are presented in Table 3 along with total accumulated strain and microhardness. The lowest tensile strength is seen in the annealed as-received material, while the highest tensile strength of CDA101 Cu ECAE deformed samples comes from routes 4B, 4E and the three Bc routes. Generally, as the number of ECAE passes increase, the strength increases, but after 4 passes, the increases are small. Interestingly, route E gives a decrease in strength from 4 to 8 passes.

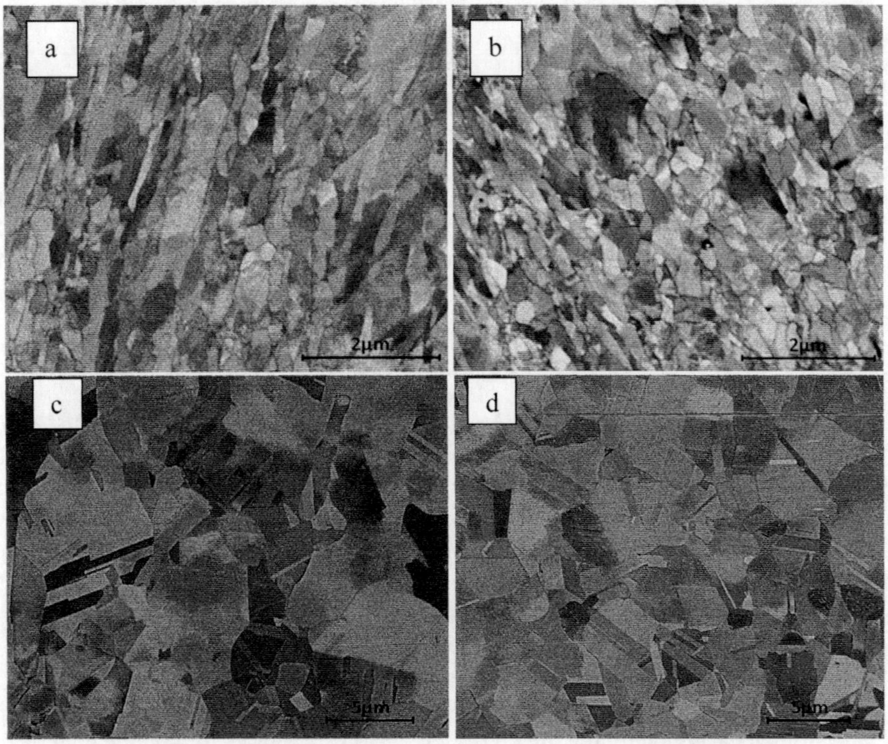

Figure 2: SEM images of as-worked 4A (a), as-worked 4Bc (b), recrystallized 4A at 250°C (c), and recrystallized 4Bc at 250°C (d)

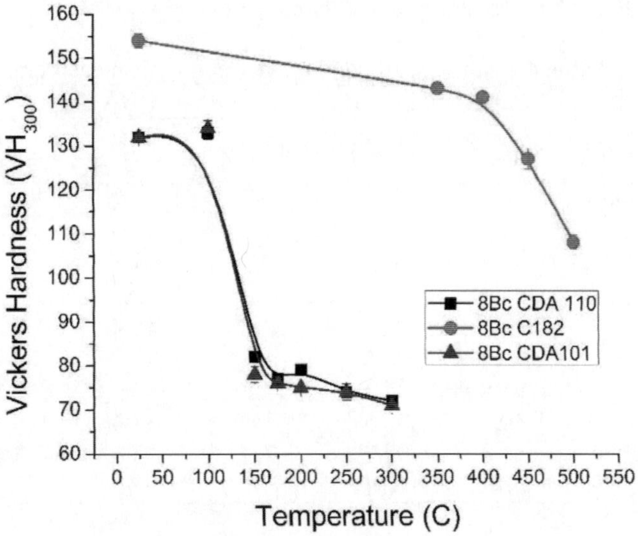

Figure 3: Vickers hardness as a function of temperature after 1 hour heat treatment for route 8Bc CDA101, CDA110, and C182 copper

Table 3: Summary of tensile strength and hardness data for CDA101 copper

Route	Accumulated Strain	Vickers hardness (VH_{300})	Yield Strength (MPa)	Tensile Strength (MPa)	Strain to Failure
AR	0	54 ± 1	181 ± 2	248 ± 2	0.38 ± .02
1A	1.1	126 ± 2	323 ± 5	349 ± 4	0.14 ± .01
2A	2.3	132 ± 5	364 ± 1	383 ± 4	0.14 ± .03
4A	4.6	137 ± 2	372 ± 1	397 ± 2	0.16 ± .01
4B	4.6	145 ± 4	399 ± 4	442 ± 4	0.16 ± .01
4E	4.6	143 ± 3	402 ± 1	438 ± 2	0.19 ± .01
4Bc	4.6	144 ± 2	383 ± 17	421 ± 6	0.16 ± .02
8Bc	9.2	141 ± 4	373 ± 3	437 ± 6	0.20 ± .01
8E	9.2	145 ± 2	382 ± 2	427 ± 1	0.15 ± .01
16Bc	18.5	136 ± 2	357 ± 3	438 ± 3	0.18 ± .01

Resistivity measurements for CDA101, CDA110 and C182 samples as well as resistivity ratios for the CDA101 copper case are presented in Table 4. The CDA101 and CDA110 copper samples had starting conductivity values matching almost exactly with the international annealed copper standard (IACS) value. After becoming fully worked, both CDA101 and CDA110 had conductivity that fell a maximum 4-5% below that of the IACS value. The C182 copper alloy had a starting conductivity of approximately 60% of the IACS value. After becoming fully worked, the C182 alloy had a final conductivity of approximately 40% of the IACS number.

The residual resistivity ratio (RRR) for the annealed CDA101 copper is 91. After one pass of ECAE, the RRR drops to just above 50. The RRR continually drops as more strain is accumulated within the material reaching a minimum 19.2 for route 16Bc. Table 4 gives a summary of conductivity, resistance ratio (RR), and RRR measurements for the CDA101, CDA110, and C182 copper materials. Figure 4 graphically represents the change in conductivity compared to the IACS value with the different routes as well as illustrates the correlation of tensile strength to RRR measurements on CDA101 for selected routes. The correlation is very linear, with R^2 values of just over 0.96, indicating that as you increase in tensile strength the RRR values decrease indicating a lower conductivity.

Table 4: Summary of conductivity and RRR measurements

	C182	CDA110	CDA101		
Processing	%IACS	%IACS	%IACS	RR (77K/4.2K)	RRR (273K/4.2K)
As Received	59.6	100.7	103.4	11.2	91.0
1A	41.2	97.0	98.2	7.40	51.6
2A	NM	NM	NM	6.05	36.4
2B	41.4	97.2	96.9	NM	NM
4A	NM	NM	NM	4.94	30.8
4Bc	40.4	96.5	95.6	4.14	24.5
4E	41.8	97.1	97.1	4.79	30.0
8Bc	39.1	96.5	95.4	3.64	21.1
8E	41.4	97.2	97.0	4.28	27.3
16Bc	NM	NM	NM	3.48	19.2

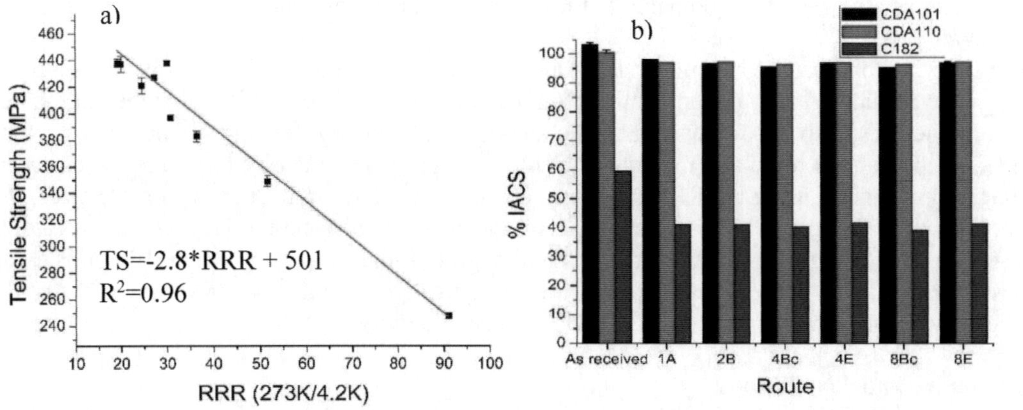

Figure 4: a) Tensile strength vs RRR for CDA101. b) Percent of international annealed copper standard (IACS) vs ECAE route for CDA101, CDA110, and C182 at 298 K

4. Discussion

The simple shear applied by ECAE across the intersecting planes of the die refined the microstructure and induced dislocations into the crystal structure. The SEM results shown in Figure 2 and grain size shown in Table 2 illustrate how the fully worked material has a much smaller grain size than the starting annealed material. With a smaller grain size and large number of dislocations, the material gains a much higher tensile strength and hardness. This is due to the numerous dislocations and grain boundaries inhibiting dislocation motion, meaning a larger force would need to be applied to deform the lattice. In the case of C182, the Cr precipitates inside the matrix also greatly increase hardness and strength due to the precipitates also blocking dislocation motion. It can be seen that the mechanical properties of copper are positively influenced by ECAE as seen by Table 3. As previously stated, through grain refinement and increasing the dislocation density in the material, dislocation motion is blocked. ECAE produces diminishing returns when looking at tensile strength increases. After ~4 passes, grain size and strength remain relatively constant. Fully worked samples of route Bc material show an increase in tensile strength of 78% with an increase in hardness of 166% when compared to the annealed starting material.

Another factor affected by ECAE processing is recrystallization. The required energy to start short range diffusion powered by elevated temperatures that leads to recrystallization has an inverse relationship to the amount of accumulated strain. The more strain and energy stored within the lattice, the easier the sample to recrystallize. The recrystallization temperature for route 8Bc is found to be 225°C as indicated in Figure 3. Additionally, for the higher applied strains, the recrystallization grain size is smaller compared to the lower strain samples. This is easily seen by Table 2. Route Bc, and E to some extent, have the lowest recrystallized grain size when compared to other samples with the same amount of strain. This could be due to the equiaxed nature of the grains. In the case of odd numbered passes, or routes that do not have full rotation, favourable sites for nucleation are more prevalent. However, if the grains stay equiaxed, the more common triple junction in the microstructure inhibits grain boundary migration, corresponding to a more stable structure [6-9,10]. Additionally, evidence of dynamic recrystallization can be seen by the decrease in strength after more processing in route E.

Resistivity testing indicates that as you increase the amount of cold work, the resistivity of the sample increases as shown in Table 4 and Figure 4. This is due to the scattering of electrons off numerous dislocations and grain boundaries. For the pure copper samples CDA101 and CDA110, the effect of extrusion on the conductivity is relatively minor. Fully worked samples retained, at a minimum, 95% of the conductivity compared to the IACS value. In comparison, the C182 alloy had a much lower conductivity when compared to its starting IACS value of 60%, only retaining 40% when fully worked. This is largely due to the alloy forming a solid solution which is particularly detrimental to easy electron flow through the material. There is still an effect due to increasing grain boundaries and dislocations, but it is relatively minor when compared to the effect from the chromium. Similar results were seen by Wang et al. [10].

When trying to select the best material for a high strength high conductivity application, a figure of merit can be considered in evaluating both strength and conductivity. By multiplying the microhardness values by their respective IACS percentage conductivity, and taking the largest value, we see that the best ECAE processing route and material is CDA101 copper processed by route 2B. When finding a similar figure of merit using the RRR measurements for the CDA101 copper, the tensile strength was multiplied by the RRR, and for this case route 1A is calculated to be superior. However, this is assuming that the importance of resistivity and strength are the same. If the need is for a material that is as strong as possible, and the 5% losses to conductivity are not critical, one of the 4 pass routes would most likely be chosen for high strength, yet still reasonable high conductivity.

5. Summary and Conclusions

In this study, different numbers of passes and routes of equal-channel angular extrusion (ECAE) were successfully applied to oxygen-free high conductivity (OFHC) copper (CDA101), commercially pure copper (CDA110), and copper chromium alloy (C182). The hardness, tensile properties, and electrical conductivity of the materials were measured. The main findings are:

- Near saturation of hardness and tensile strength occur at 3-4 ECAE passes.
- Obtaining a tensile strength of 500 MPa or greater could be possible with additional post processing techniques such as cold rolling
- Route Bc is a preferred method for creating the smallest grains possible in both as-worked and recrystallized conditions.
- Alloying, while good for increasing mechanical properties, causes a severe decrease in conductivity.
- A lower number of passes is best for creating a material with a high strength to resistivity ratio.

6. References

[1] Valiev R Z and Langdon T G 2006 *Principles of equal-channel angular pressing as a processing tool for grain refinement.* Progress in Materials Science **51** p 881-981

[2] Parrell J, Zhang Y, Field M, Meinesz M, Huang Y, Miao H, Hong S, Cheggour N, and Goodrich L 2009 *Internal Tin Nb3Sn Conductors Engineered for Fusion and Particle Accelerator Applications*. IEEE Transactions on Applied Superconductivity **19.3** p 2573-579

[3] O Umezawa, 2015 *Tensile properties and conductivities of a precipitation hardened and cold-rolled Cu-0.3Cr-0.1Zr alloy at cryogenic temperature* IOP Conf. Ser.: Mater. Sci. Eng. **102** 012006

[4] Zeon Han, Seung et al. 2015 *Design of Exceptionally Strong and Conductive Cu Alloys beyond the Conventional Speculation via the Interfacial Energy-Controlled Dispersion of Γ-Al2O3 Nanoparticles* Scientific Reports **5** 17364

[5] Wei K, Wei W, Wang F, Du Q, Alexandrov I, and Hu J 2011 *Microstructure, mechanical properties and electrical conductivity of industrial Cu–0.5%Cr alloy processed by severe plastic deformation* Materials Science and Engineering: A **528** p 1478-1484

[6] Dalla-Torre F, Lapovok R, Sandlin J, Thomson P, Davies C, and Pereloma E 2004 *Microstructures and properties of copper processed by equal channel angular extrusion for 1–16 passes* Acta Materialia **52** p 4818-4832

[7] Buet E, Dubois J, Olier P, Thilly L, Lecouturier F, and Renault P 2012 *Microstructure and texture of copper/niobium composites processed by ECAE* International Journal of Material Forming **3** p 1071-1074

[8] Ko Y, Namgung S, Lee B, and Shin D 2010 *Mechanical and electrical responses of nanostructured Cu–3wt%Ag alloy fabricated by ECAP and cold rolling* Journal of Alloys and Compounds **504** p S448-S451

[9] Etter A, Solas D, Baudin T, and Penelle R 2005 *Evolution of microstructure and texture during annealing of a copper processed by ECAE* Materials Science Forum **495-497** p 845-850

[10] Wang G, Wu S, Zuo L, Esling C, Wang Z, and Li G 2003 *Microstructure, texture, grain boundaries in recrystallization regions in pure Cu ECAE samples* Materials Science and Engineering: A **346** p 83-90

[11] V.M. Segal, 1995 *Material processing by simple shear* Materials Science and Engineering: A p 157-164

IOP Conf. Series: Materials Science and Engineering **279** (2017) 012004 doi:10.1088/1757-899X/279/1/012004

Temperature dependence on tensile properties of Cu-40mass%Fe dual phase alloy

N Koga[1], W Zhang[2], O Umezawa[1], V Tschan[3], J Sas[3], K P Weiss[3]
[1] Faculty of Engineering, Yokohama National University, 79-5 Tokiwadai, Hodogaya, Yokohama, 240-8501, Japan

[2] Graduate school of Engineering, Yokohama National University, 79-5 Tokiwadai, Hodogaya, Yokohama, 240-8501, Japan

[3] Institute for Technical Physics, Karlsruhe Institute of Technology, Postfach 3640, D-76021 Karlsruhe, Germany

E-mail: koga-norimitsu-sx@ynu.ac.jp

Abstract. The binary system of iron and copper shows low mutual solubility and cast Cu-Fe forms an iron (bcc) and copper (fcc) dual phase structure at room temperature. In this study, tensile properties, deformation and fracture behaviour of a rolled Cu-40mass%Fe alloy have been evaluated in order to reveal the temperature dependence on tensile properties in dual phase structures. The material formed a layer structure with ultra-fine grains of 1 μm in diameter. In both iron and copper grains, furthermore, many precipitates of copper or iron were revealed. The strength of this material increased at low temperatures, though the elongation was hardly changed, which suggests that fcc + bcc dual phase structure is effective to improve the tensile property at low temperature. Strain was inhomogeneously distributed at low temperature regardless of Cu and Fe region, and voids and cracks tended to form inside Cu layer. These results imply that the temperature dependence on tensile properties and deformation behaviour of each phase in dual phase structure is different from that of each single phase structure, and dual phase structure materials have a potential for becoming superior cryogenic structural materials.

1. Introduction

Mechanical properties of metallic materials depend on temperature, and the temperature dependence is closely related to the crystal structure. Bcc materials exhibit strong dependence and then their strength increases at lower temperature while the elongation is dramatically decreased. On the other hand, fcc materials are not too sensitive to temperature and show good ductility even at cryogenic temperature [1]. Although there are a lot of studies on the temperature dependence of mechanical properties in single phase materials, few studies on a dual phase structure have been done.

As an example, ferrite (bcc) + austenite (fcc) dual phase stainless steel shows good strength - elongation balance at low temperature [2]. However, the effects of temperature on mechanical properties for dual phase structure are unclear because metastable austenite transforms to martensite during deformation of this alloy. This transformation results in the higher work-hardening rate and elongation which is called the transformation induced plasticity (TRIP) effect [3]. Here, the binary system of iron and copper shows low mutual solubility in both matrix phases and cast Cu-Fe forms the iron (bcc) and copper (fcc) dual phase structure at room temperature. The TRIP effect does not occur in this alloy system. Tensile properties, deformation and fracture behavior of a cold rolled Cu-40mass%Fe alloy in each temperature had been evaluated in order to reveal the temperature dependence of tensile behavior in dual phase structure.

2. Experimental procedure

The test material was a commercial cold-rolled Cu-40mass%Fe alloy. The specimen was annealed at 1123 K for 1.8 ks followed by water cooling. Microstructure was observed by optical microscopy (OM) and scanning electron microscopy (SEM), and crystal orientation was analyzed by means of electron backscattered diffraction (EBSD). EBSD data were analyzed with the software program OIM™ analysis 7.0.1 for phase and crystal orientation. Strain distribution was analyzed by digital image correlation (DIC) calculating strain from the difference of the digital images between before and after deformation. DIC analysis was carried out for SEM images of before deformation and 5% tensile deformation at each temperature with the software program (VIC-2D) under 51 pixels in subset and 5 pixels in step. Tensile test specimens with 20 mm in gage length and 4mm in width were cut from plates along the rolling direction. Tensile testing was carried out at an initial strain rate of about 5.2×10^{-4} sec^{-1} at 293 K, 77 K and 8 K.

3. Results and discussion

3.1. Microstructure

Figure 1 shows OM (a) and SEM (b) images of the annealed material. There were many inclusions as shown in (a), which were determined to be iron oxide from the results of electron energy dispersive x-ray analysis. These inclusions formed during casting the specimen and were distributed regardless of phase. Cu and Fe phase have layer structure along rolling direction and many precipitates were observed inside Fe layer (b). Hardness of Cu and Fe layer were 84 HV and 134 HV, respectively. Fig.2 shows inverse pole figure along the rolling direction in Cu (a) and Fe (b) phase. Cu and Fe layers had fine grains which are smaller than 1 μm in diameter, and $\langle 111 \rangle_{Cu}$ // RD and $\langle 110 \rangle_{Fe}$ // RD texture. In addition, Fe and Cu precipitates as indicated by the white arrow in Fig. 2 were observed inside Cu and Fe layers. These precipitates might suppress grain growth during annealing and then fine grain structure was maintained. From above results, the material has Cu and Fe layer structure macroscopically and fine grains and many precipitates in each layer microscopically.

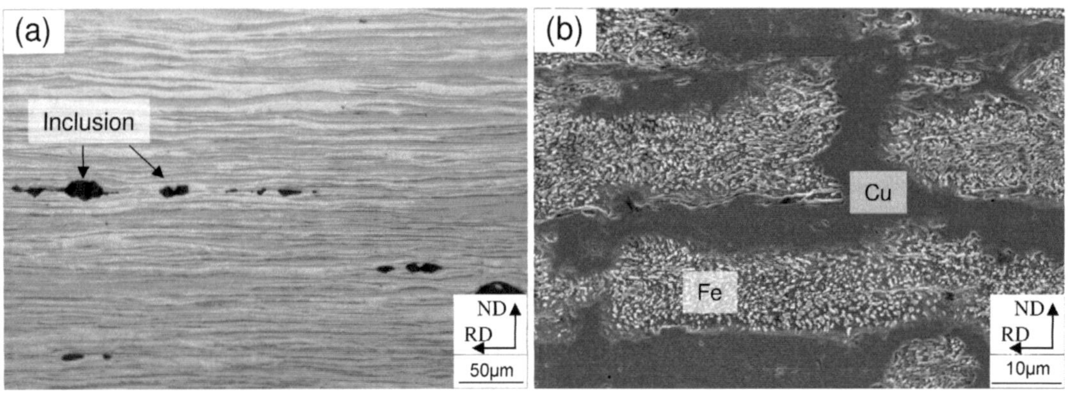

Fig. 1 OM (a) and SEM (b) image of annealed specimen.

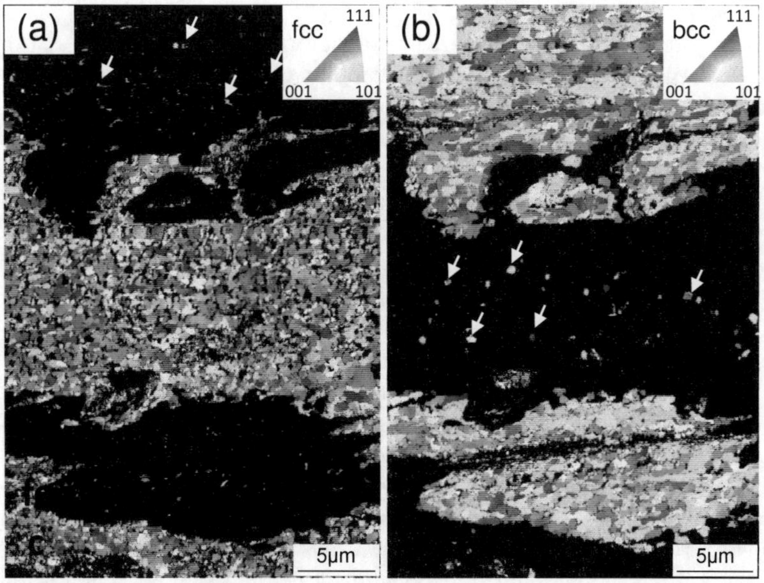

Fig. 2 Inverse pole figures of Cu (a) and Fe (b) phase in annealed specimen.

3.2. Tensile property

Figure 3 shows nominal stress - strain curves (a) and true stress or work hardening rate as a function of true strain (b) at 293 K, 77 K and 8 K. The work hardening rate had error at just after yield in tensile deformed at 293 K and 77 K because the flow stress was dramatically dropped after yield. Serration occurred by local rapid adiabatic deformation [4] appeared at 8 K, therefore, work-hardening rate at that temperature was calculated from the maximum stress at each serration neglecting other data. Serration may be formed by following mechanism: local rapid adiabatic deformation leads to rising the temperature within the local region and then applied stress dramatically decrease, after that, the stress continuously recovers with cooling the region and local rapid adiabatic deformation occurs again. Both yield and tensile stress continuously increased at lower temperature, and total elongation at 77 K is a little higher than the other test temperatures. The work - hardening rate also increased at lower temperature and uniform elongation is almost same among each temperature. From these results, it can be concluded that the strength - elongation balance in this alloy is improved at lower temperature. The Cu-Fe alloy has good temperature dependence on tensile properties than that of each phase single materials.

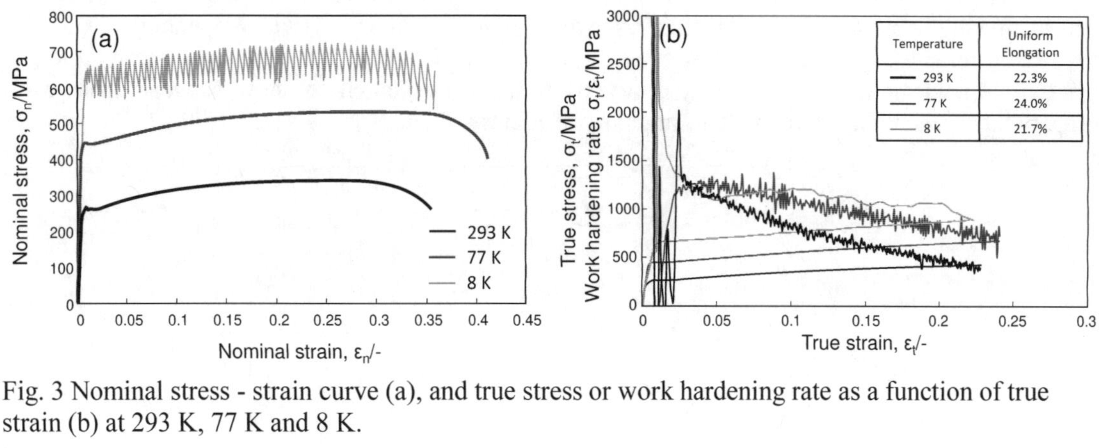

Fig. 3 Nominal stress - strain curve (a), and true stress or work hardening rate as a function of true strain (b) at 293 K, 77 K and 8 K.

IOP Conf. Series: Materials Science and Engineering **279** (2017) 012004 doi:10.1088/1757-899X/279/1/012004

3.3. Deformation and fracture behaviour

Figure 4 shows the ε_{xx} strain distribution in 5% tensile deformed material at 293 K (a), 77 K (b) and 8 K (c). Minimum and maximum strain in color bar are 0 and twice as much as average strain, respectively. The white dash line indicates Fe/Cu layer boundary. High strain tended to be distributed on Fe layer at 293 K as indicated by the black arrows in (a), implies that the Fe layer is more deformable than Cu layer at 293 K. Strain partitioning between Fe and Cu layer was low at 77 K (b) and 8 K (c); the average strain of Fe and Cu layer in tensile deformation at 8 K are 5.46% and 5.52%, respectively, means Fe and Cu layer were simultaneously deformed at these temperatures. On the other hand, strain distribution became inhomogeneous as lowering temperature, high strain region over twice as much as average strain and low strain region is approximately 0% at 8 K (c). In addition, strain was continuously distributed along 45 degrees which is the maximum shear stress direction regardless of Cu and Fe layers.

Figure 5 shows histograms of ε_{xx} strain obtained from the strain distribution (Fig. 4). Standard deviation corresponds to the width of the histogram and inhomogeneity of strain distribution in this analysis, it increases as lowering the temperature though the average strain is almost same among each temperature. This inhomogeneous strain distribution may be caused by local deformation of both phases; 48 slip systems in Fe phase: {110}<1-11>, {112}<11-1>, {123} <11-1> are restricted to 12 slip systems: {110} <1-11> at low temperature due to the temperature dependence of critical resolved shear stress in each slip system [5]; dislocation structure in Cu phase tends to be planer at low temperature due to the decrement of stacking fault energy as lowering the temperature. This inhomogeneous strain distribution may lead to high work-hardening rate and it is one of the reason for good strength - elongation balance at low temperature in this alloy.

Figure 6 shows SEM images at an uniform deformation (a) and a necking (b) regions in fractured specimen at 8 K. There were no voids or cracks in both phases at an uniform deformation region (a), while small voids or large crack as indicated by black arrows in (b) was observed within only Cu layer. Considering Fe and Cu layer were simultaneously deformed at 8 K, Fe layer has enough ductility even at 8 K which is under ductile to brittle transition temperature in polycrystalline Fe single phase material [6]. Fine grains or Cu precipitates may suppress the brittle fracture of Fe phase in this material, and the ductile deformation of Fe phase lead to good elongation at low temperature.

Figure 7 shows fracture surface at 293 K (a), 77 K (b) and 8 K (c). There were a lot of coarse voids generated from inclusions and micro voids formed regardless layers in every temperature. These coarse voids easily connected to each other through secondary voids which corresponds to micro voids in fracture surface, and then the specimen was broken. Therefore, this alloy has a possibility of improvement of elongation, especially local elongation, by removing these inclusions.

Temperature dependence on both tensile properties and deformation behaviour of each phase in dual phase structure materials disagree with that of each phase single materials. Fe (bcc) phase showed enough elongation even at low temperatures (Fig. 6). Cu (fcc) phase was simultaneously deformed with Fe phase (Fig. 4), which implies that Cu phase harden at lower temperatures. Although the reason for these unique properties of each phase at lower temperatures in dual phase structural material is unclear, dual phase structure materials have a potential to show superior mechanical property at cryogenic temperature than single phase materials.

IOP Conf. Series: Materials Science and Engineering **279** (2017) 012004 doi:10.1088/1757-899X/279/1/012004

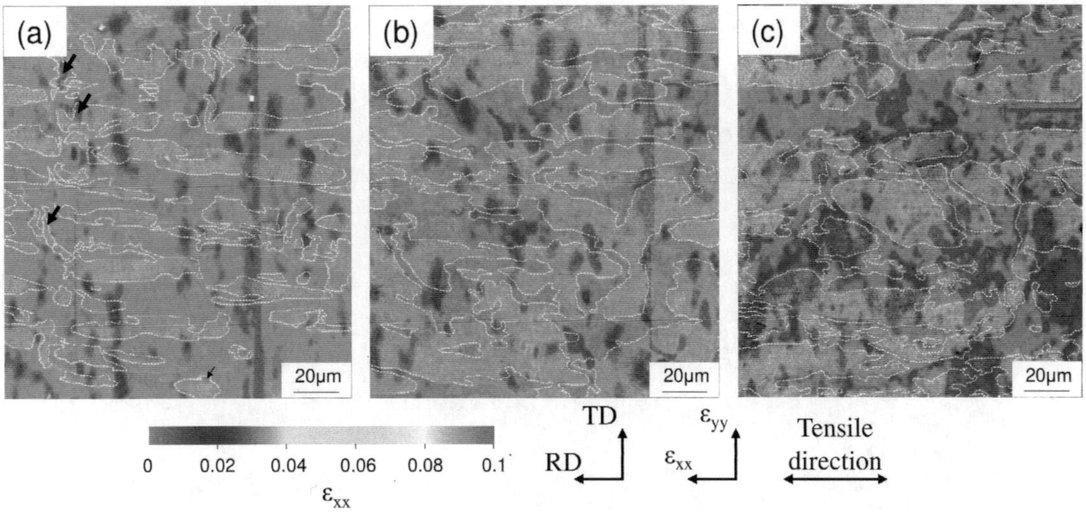

Fig. 4 ε_{xx} strain distribution in 5% tensile deformed at 293 K (a), 77 K (b) and 8 K (c).

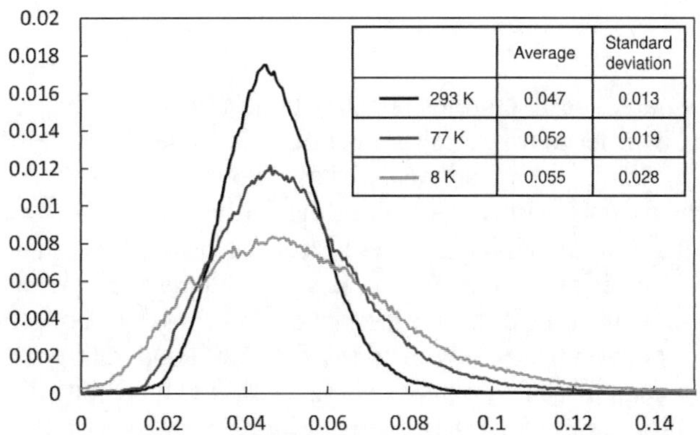

Fig. 5 Histograms of ε_{xx} strain in 5% tensile deformed at 293 K (a), 77 K (b) and 8 K (c).

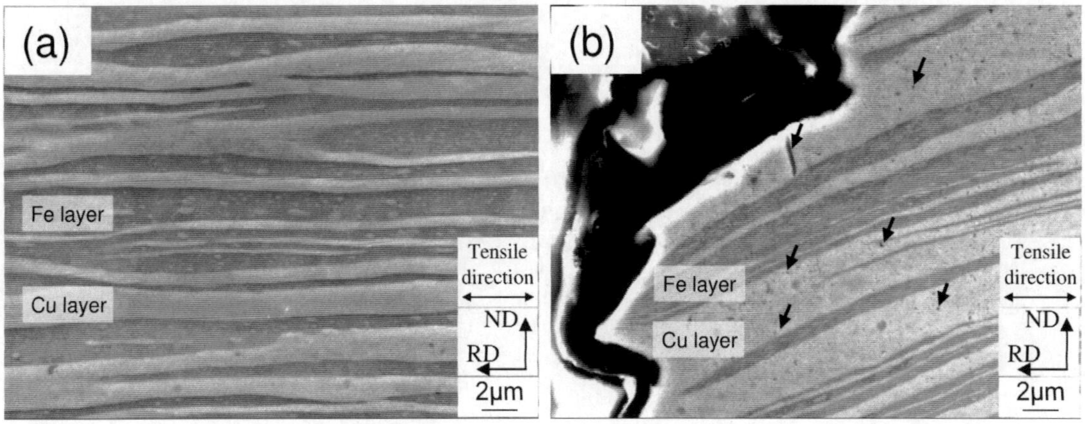

Fig. 6 SEM images at uniform deformation (a) and necking (b) region in fractured specimen at 8 K.

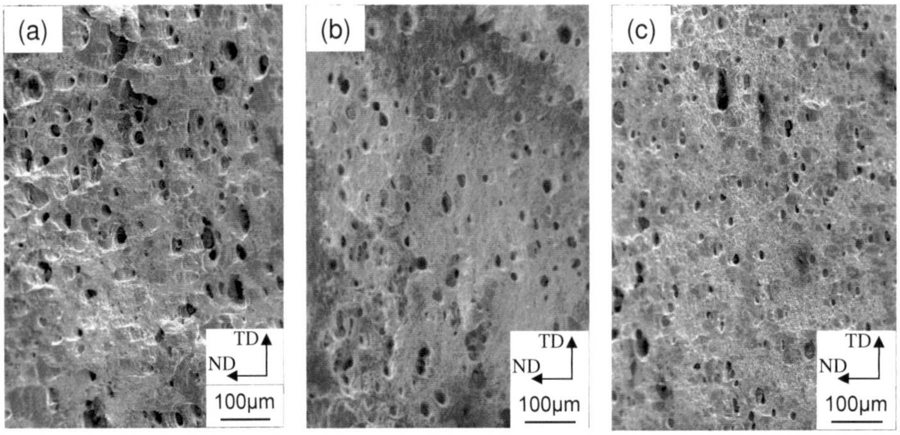

Fig. 7 SEM images of fracture surface at 293 K (a), 77 K (b) and 8 K (c).

4. Conclusions

Tensile properties, deformation and fracture behaviour of the rolled Cu-40mass%Fe alloy from 293 K to 8 K has been evaluated in order to reveal the temperature dependence on tensile properties in a dual phase structure material. Major results are summarized as follows:

The cold-rolled Cu-40%Fe alloy after annealing formed layer structure macroscopically and each layer had fine grains and precipitates. Yield stress, tensile stress and work-hardening rate continuously increased as lowering temperature though the elongation was hardly changed, which suggests that Cu-Fe alloy has the temperature dependence of strength in bcc structure and elongation in fcc structure. Strain was inhomogeneously distributed at low temperature regardless of Cu and Fe region. Voids and crack tended to generate inside Cu layer in fractured specimen at 8 K, implied that Fe layer has enough ductility even at 8 K. There were a lot of coarse voids generated from inclusions and micro voids formed regardless layers in every temperature on fracture surface. These results suggest that temperature dependence on tensile properties of dual phase structural materials is superior to that of each single phase materials and the materials have a potential to become good cryogenic structural materials.

5. References

[1] R. P. Reed, A. F. Clark 1983 *Materials at low temperature* (Ohio: American Society for Metals)
[2] K. Fukaura, H. Izumi, H. Kawabe *The Society of Materials Science* Japan **35** 603-609
[3] F. Zackay, E. R. Parker, D. Fahr, and R. Bush 1967 *Trans. ASM.,* **60** 252-259.
[4] T. Ogata, K. Ishikawa, O. Umezawa and T. Yuri 1988 *Cryogenics* **28** 511-515
[5] W. A. Spritzig, A. S. Keh 1970 *Acta Metallurgica* **18** 1021-1033
[6] H. Qiu, R. Ito, K. Hiraoka 2006 *Materials Science and Engineering A* **435-436** 648-652

IOP Conf. Series: Materials Science and Engineering **279** (2017) 012005 doi:10.1088/1757-899X/279/1/012005

Evaluation of the cryogenic mechanical properties of the insulation material for ITER Feeder superconducting joint

Zhixiong Wu[1] Rongjin Huang[1] ChuanJun Huang[1] Yanfang Yang[2] Xiongyi Huang[3] Laifeng Li[1,*]

[1] Key Laboratory of Cryogenics, Technical Institute of Physics and Chemistry, Chinese Academy of Sciences, Beijing 100190, China
[2]China Electric Power Research Institute, Beijing 100192, China
[3]Institute of Plasma Physics, Chinese Academy of Sciences, Hefei 230031, China

laifengli@mail.ipc.ac.cn

Abstract. The Glass-fiber reinforced plastic (GFRP) fabricated by the vacuum bag process was selected as the high voltage electrical insulation and mechanical support for the superconducting joints and the current leads for the ITER Feeder system. To evaluate the cryogenic mechanical properties of the GFRP, the mechanical properties such as the short beam strength (SBS), the tensile strength and the fatigue fracture strength after 30,000 cycles, were measured at 77K in this study. The results demonstrated that the GFRP met the design requirements of ITER.

1. Introduction

The ITER feeder system consists of 31 units, which was distributed around the Tokamak and connects the ITER magnet systems to the power supply [1-2]. The main function of the feeders is to convey the cryogenic supply and electrical power to the coils as well as house the instrumentation wiring. In the Feeder systems, hundreds of superconducting joints are used to assemble the segments of the superconducting busbars and to connect the busbars to the coil terminals, as shown in figure 1. In general, GFRP was selected as the insulation material and mechanical support material for the superconducting joints and the current leads of the ITER Feeder system due to its excellent mechanical, thermal and electrical properties [3-4]. During operational conditions, the joint insulation sustains stress from the large temperature gradient, the long term thermal cycle and the high electromagnetic force. To ensure the system reliability and performance, ITER established several acceptance criteria and expectations for the mechanical properties of the joint insulation and support material, including 77K static ultimate tensile strength (UTS) > 500 MPa, interlaminar shear strength >60 MPa, and residual UTS > 200 MPa after 30,000 cycles load for both the warp direction (0°) and the weft direction (90°). To evaluate the reliable application of GFRPs, their mechanical performance was characterized under static and dynamic load conditions at 77 K. In this paper, the results of the mechanical properties tests on GFRP fabricated by the vacuum bag process are described.

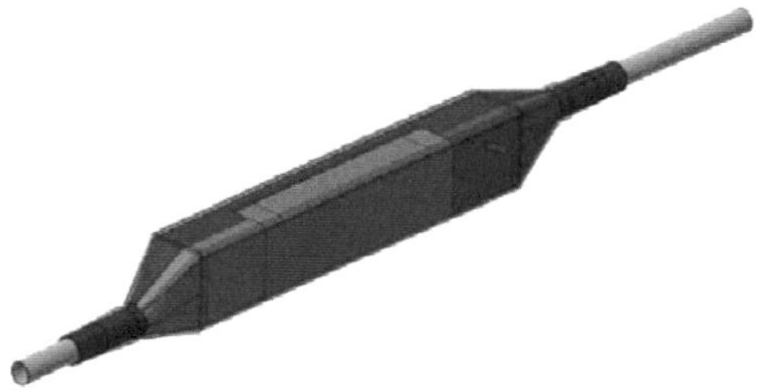

Figure 1. Schematic diagram of ITER Feeder superconducting joint.

2. Experimental

2.1. Sample preparation

The investigated GFRP laminate was fabricated using the vacuum bag process by Nanjing Fiberglass Research & Design Institute, which consists of a two-dimensional glass fabric and epoxy resin. All specimens were cut from the laminate, as shown in figure 2. Considering the anisotropic properties of the composite, the samples were cut in the warp direction (0°) and the weft direction (90^{0}) according to the ITER requirement. The outer dimensions and the diagrams of the samples are provided in table 1 and figure 2, respectively.

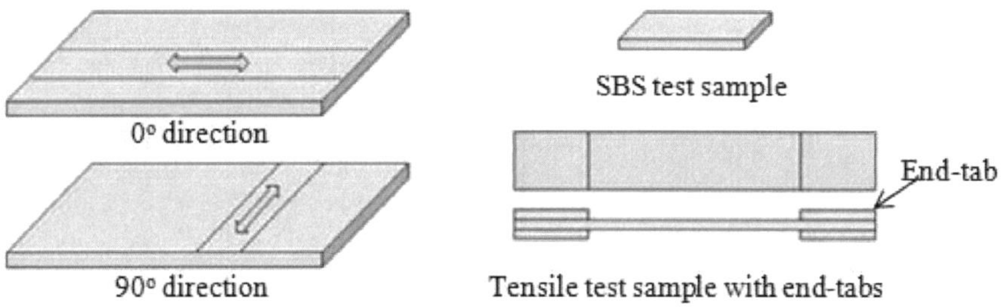

Figure 2. Schematic diagram of the test samples

Table 1. The outer dimensions of the samples.

Sample	Length (mm)	Width (mm)	Thickness (mm)
SBS test sample	15	5	2.5
Tensile test sample	250	25	4
Fatigue test sample	250	25	4

2.2. Test procedures

The short beam strength was assessed by the short-beam test according to the guidelines given in ASTM D2344. The short-beam test at 77 K was done using a cross-head speed of 1 mm/min with a SUNS UTM4204 universal testing equipped with a liquid nitrogen cryostat. The span-to-thickness ratio was 5:1 and the radius of the nose and supports was 3.0 mm. The SBS can be calculated using the following formula:

$$SBS = \frac{0.75P_m}{b \cdot h} \tag{1}$$

where P_m is the maximum load, N; b is the specimen width, mm; h is the specimen thickness, mm.

The UTS was measured according to the guidelines given in ASTM D3039. All tensile testing at 77K were carried out using an MTS-SANS 5000 universal testing with a cross-head speed of 2 mm/min. In order to prevent gripping damage, the samples were equipped with aluminum end-tabs of 2 mm thickness.

The tension–tension fatigue tests were performed at 77 K according to the standard ASTM D3479 in load control using a sinusoidal load function. The fatigue tests were run with different peak load of 200, 300 and 400MPa for 0° sample, and 100, 150 and 200MPa for 90° sample. All fatigue testing was carried out at a load frequency of 5Hz with a minimum-to-peak stress ratio of R= 0.1 using an Instron 8850 axial tension-torsion fatigue machine. The fatigue experiment was stopped manually after 30,000 load cycles. Then, the static UTS of the sample after the fatigue test was measured at 77 K.

3. Results and discussion

3.1. SBS test

The SBSs of the 0° sample and the 90° sample were measured at 77 K. The fracture mode of all tested samples was interlaminar shear fracture. Typical stress-deflection curves are shown in figure 3(a). It can be seen that both curves show a nearly linear elastic trend during the early stage of loading and rise gradually to the peak load, then drop suddenly, which reflects an interlaminar fracture. The SBS of 0° sample is higher than that of 90° sample. The results of SBS are summarized in table 2. The measured SBSs were 125.0±1.1 MPa for 0° sample and 104.0±4.3 MPa for 90° sample, which are almost 2 times than the interlaminar shear strength required by ITER. It indicates that all tested samples should be able to meet the ITER requirement although the test results are not a direct measurement of the shear strength.

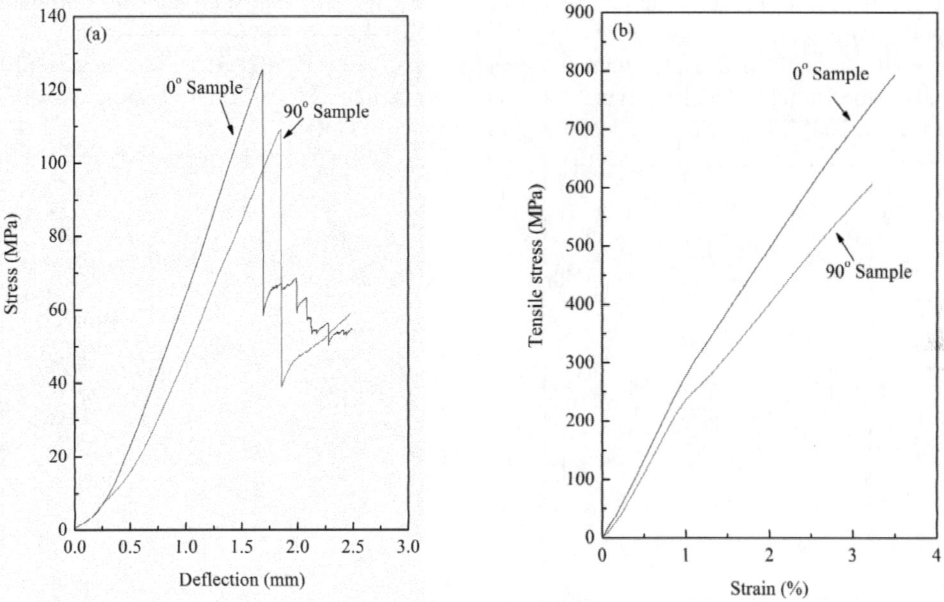

Figure 3. SBS test curves (a) and UTS test curves (b) of 0° sample and 90° sample.

3.2. Tensile test

Figure 3(b) shows the representative stress–strain curves obtained from the tensile test at 77K. As seen in figure 3(b), the slope of the stress–strain curve changed abruptly, presenting a knee behavior. The stress–strain behavior observed in this study is similar for all other woven-fabric glass/epoxy laminates [5-6]. The results of UTS for both directions are summarized in table 2. In 0° direction the UTS is 759±41.9 MPa, and in 90° direction 638±38.1 MPa. All test results are higher than 500 MPa which was required by ITER.

Table 2. The results of SBS and UTS measured at 77 K.

	SBS (MPa)		UTS (MPa)	
	0° direction	90° direction	0° direction	90° direction
1	126.6	106.7	711	638
2	125.4	98.8	770	606
3	124.6	109.2	801	665
4	124.9	104.7	719	685
5	123.5	100.5	794	595
Mean value	125.0±1.1	104.0±4.3	759±41.9	638±38.1
ITER Requirement	60[a]	60[a]	500	500

[a] Interlaminar shear strength

3.3. Fatigue test

In this study, the fatigue tests were run with different peak load of 200, 300 and 400 MPa for 0° sample, and 100, 150 and 200 MPa for 90° sample. The total load cycle was 30,000 for each sample. All samples were not damaged after 30,000 load cycles. Then, the residual UTS of these samples at 77K was tested and the stress-strain curves and results were shown in figure 4 and figure 5, respectively. As shown in figure 4, it is seen that the stress-strain curve for the composite obtained from the static tensile test is significantly affected by fatigue load. The knee behavior is becoming less and less obvious as the peak stress used in fatigue tests increases.

From figure 5, it is seen that the fatigue load degraded the UTS of the composites and the residual UTS significantly decreases with increasing peak stress used in fatigue tests. The residual UTS of the 0° sample after 30,000 load cycles with the peak stress of 400 MPa is 518 MPa, degraded by 31% compared to the initial UTS. However, note that the residual UTS of all test samples satisfied the residual UTS > 200MPa after 30,000 load cycles required by ITER.

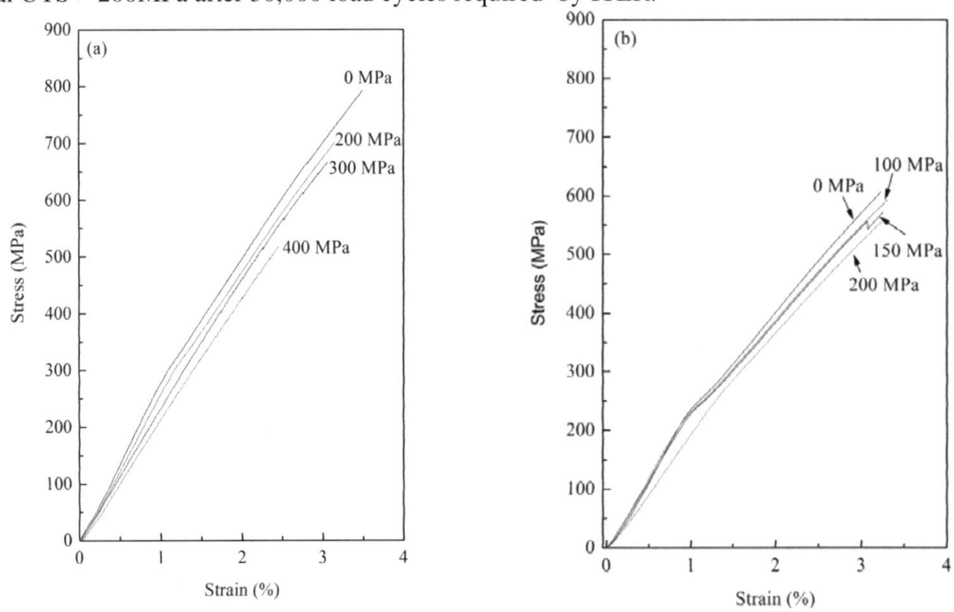

Figure 4. Tensile stress-strain curves of 0° sample (a) and 90° sample (b) after 30,000 cycles.

IOP Conf. Series: Materials Science and Engineering **279** (2017) 012005 doi:10.1088/1757-899X/279/1/012005

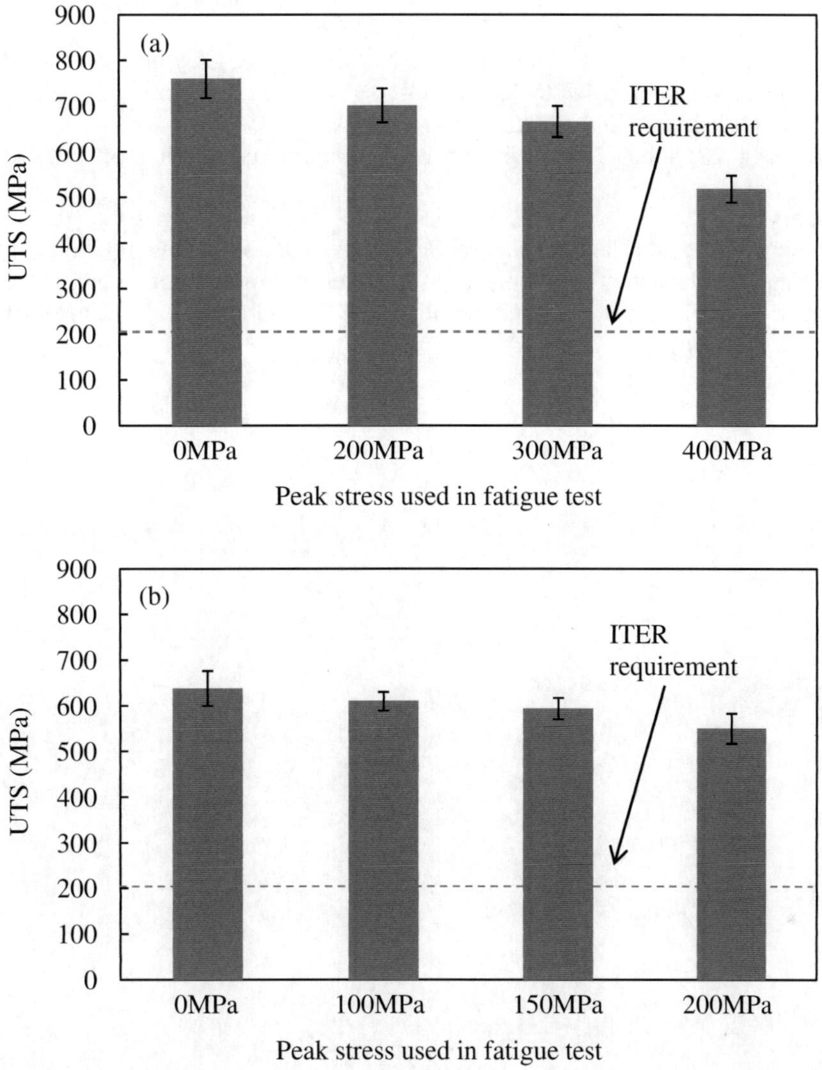

Figure 5. The results of UTS for the 0° sample (a) and the 90° sample (b) after fatigue test.

4. Conclusion

In the present study, the evaluation of the mechanical properties of the GFRP fabricated by vacuum bag process was performed. The SBS, the static UTS and the residual UTS after 30,000 cycles fatigue test of the composites for both directions satisfied the ITER requirements. In addition, the mechanical property in 0° direction is better than that in 90° direction. The fatigue load degraded the UTS of the composites and the residual UTS significantly decreases with increasing peak stress used in fatigue tests.

5. References

[1] Song Y, Lu K, Huang X, Chen Y, Zhou T, Liu S, Bauer P, Bi Y, Chen Y, Devred A, Ding K, Niu E, Gung C, Mitcell N, Shen G and Wang Z 2012 *IEEE Trans. Appl. Supercond.* **22** 4800404

[2] Ding K, Liu C, Bi Y, Feng H, Du Q, Zhou T, Lu K, Song Y and Xu L 2016 *IEEE Trans. Appl. Supercond.* **26** 9500604

[3] Huang X, Song Y, Zheng J, Yu X, Xiao W, Gao D, Tao Y, Wang C, Gung C, Clayton N, Niu E, Chen Y and Mitcell N 2013 *Fusion Eng. Des.* **88** 696
[4] Huang X, Yu X, Wang C, G Li, Song Y, Ku L, Niu E, Gung C, Clayton N and Su M 2014 *IEEE Trans. Appl. Supercond.* **24** 7700304
[5] Dzenis Y and Qian J. 2001 *Int. J. Solid Struct.* **38** 1831
[6] Takeda T, Takano S, Shindo Y and Narita F 2005 *Compos. Sci. Technol.* **65** 1691

Acknowledgments
This work was supported by the National Natural Science Foundation of China (Grant No.: 51407180, 51577185), the National Magnetic Confinement Fusion Science Program (Grant No.: 2015GB121001) and the Scientific Project of the State Grid Corporation of China (Grant No.: DG71-16-001).

Cryogenic electrical properties of irradiated cyanate ester/epoxy insulation for fusion magnets

X. Li[1,2], Z.X. Wu[1], J. Li[1,2], D. Xu[1], H.M. Liu[1], R.J. Huang[1,2], and L.F. Li[1,2]

[1]State Key Laboratory of Technologies in Space Cryogenic Propellants, Technical Institute of Physics and Chemistry, Chinese Academy of Sciences, Beijing 100190, P. R. China. [2]University of Chinese Academy of Sciences, Beijing 100049, P. R. China

lfli@mail.ipc.ac.cn

Abstract. The insulation materials used in high field fusion magnets require excellent mechanical properties, high electrical breakdown strength, good thermal conductivity and high radiation tolerance. Previous investigations showed that cyanate ester/epoxy (CE/EP) insulation material, a candidate insulation for fusion magnets, can maintain good mechanical performance at cryogenic temperature after 10 MGy irradiation and has a much longer pot life than traditional epoxy insulation material. In order to quantify the electrical properties of the CE/EP insulation material at low temperature, a cryogenic electrical property testing system cooled by a G-M cryocooler was developed for this study. An insulation material with 40% cyanate ester and 60% epoxy was subjected to ^{60}Co γ-ray irradiation in air at ambient temperature with a dose rate of 300 Gy/min, and total doses of 1 MGy, 5 MGy and 10 MGy. The electrical breakdown strength of this CE/EP insulation material was measured before and after irradiation. The results show that cryogenic temperature has a positive effect on the electrical breakdown strength of this composite, while the influence of ^{60}Co γ–ray irradiation is not obvious at 6.1 K.

1. Introduction

In order to maintain fusion for long periods of time, a high magnetic field is required in ITER to control the high temperature plasma, and it is crucial for superconducting magnets operating in cryogenic environment to maintain their superconductivity. Supercritical helium at 4 K is used to cool the ITER magnets and the superconducting magnet is surrounded by a large cryostat and an actively-cooled thermal shield with forced flow helium at 80 K [1]. At the same time, there are many high-energy neutrons generated by deuterium-tritium fusion reactions in the plasma, and the insulation materials covering magnetics are irradiated by these neutrons escaping from the plasma.

Insulation material is an important part of all superconducting magnets. It works as an insulator, supporter and protector, and needs to withstand a radiation environment in ITER. The insulation of the superconducting coils is affected by a combination of low temperature, irradiation and high voltage, as well as by mechanical stresses [2]. Thus, it is important to find insulation materials with excellent mechanical properties, high electrical breakdown strength, and high radiation resistance at low temperature [3, 4].

The electrical insulation system of the Toroidal Field (TF) magnet system requires an irradiation resistance of 10 MGy in ITER. Recent studies showed that cyanate ester/epoxy (CE/EP) insulation material has an excellent resistance against gamma ray and neutron irradiation, and thus it considered as a candidate insulation material for fusion magnets [5-9]. However, previous works for this kind of material were focused on its mechanical properties, while electrical properties tests have been seldom conducted. Therefore, it is essential for us to test the electrical properties of cyanate ester/epoxy (CE/EP) insulation material after irradiation and compare the differences when irradiated by different doses of gamma rays.

This work reports on the fabrication of a device to measure the electrical properties of insulation materials under vacuum at cryogenic temperature, in order to assess the material performance under ITER relevant conditions, and then addresses the breakdown strength of cyanate ester/epoxy at 6.1 K before and after irradiation. Moreover, the insulation breakdown strength for different irradiation doses was measured in this study.

2. Experiment

2.1. Material

The insulation material used in this work consists of two parts, the cyanate ester, which is 1,1-Bis(4-cyanatophenyl) ethane (DCBE, Primaset LECy from Lonza), and the epoxy resin, which is bisphenol-F (DGEBF, GY285 from Huntsman). Figure 1 shows the chemical structures of DGEBF and DCBE. The CE content is 40%, and the EP content is 60%. The catalyst is acetyl acetone cobalt(II) bought from Alfa Aesar. The reinforcement material is boron free glass fiber cloth (RW220-90, Sinoma Science and Technology Co., China) treated by a silane coupling agent. The fabric was 0.2 ± 0.022 mm thick with a count of 18 ± 1 threads/cm in the warp and 14 ± 1 threads/cm in the fill.

This material was fabricated using the vacuum pressure impregnation (VPI) technique. Firstly, cut and seal the woven glass fiber into a metal mould coated with releasing agent. Then, dry the metal mould and glass fiber at 100°C for 10 h. The 60 wt% epoxy resin and 40 wt% cyanate ester with 0.01 wt% $Co(acac)_2$ were mixed in a 250 ml beaker. Impregnating the fluid resin into the preheated mould followed by curing at 100°C for 6 h, 120°C for 4 h, and then 150°C for 16.5 h. ^{60}Co γ –ray was used to irradiate the specimens in air at ambient temperature with a dose rate of 300 Gy/min. The total doses were 1 MGy, 5 MGy and 10 MGy.

Previous studies showed that glass, resin and void content are important parameters that affect electrical properties of dielectric material [10-14]. The insulation tested in this study consists of resin, glass fiber and void, so the test methods of references [15], [16] and [17] are applicable to it. The fiber glass content was obtained using the method in reference [15], the resin content uses the method in reference [16], and the void content uses the method in reference [17]. The resin content was measured to be 45.2 volume %, the glass content was 53.0 volume %, and the void content was 0.8 volume %.

2.2 Apparatus

Figure 1 shows a sketch of the cryostat with a GM cryocooler, which was employed to measure the electrical breakdown strength and the breakdown characteristic of the insulation surface at low temperature. The GM cryocooler used in this equipment is an ESYCOOL KDE415SA with two stages. The GM cryocooler specifications are 35W@50K on the first stage, and 1.5W@4.2K on the second stage.

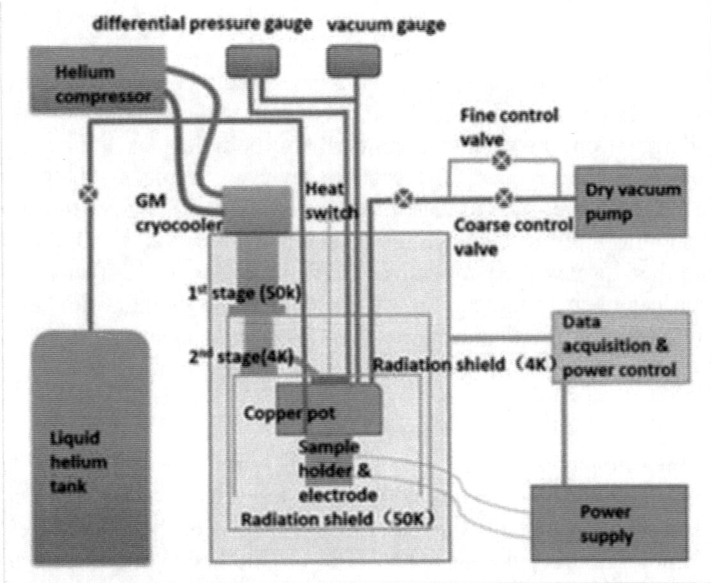

Figure 1. Sketch of cryostat with a GM cryocooler

This testing system consists of three parts: A G-M cryocooler and a cryogenic liquid helium pot, which cools down the whole system, an electrode system that is used to measure the breakdown strength of samples, and finally a thermal insulation system to maintain a cryogenic environment. The electrode system can be changed so that the system can adapt to different kinds of electrical property tests.

Equipment designed in this study are able to reach 4.2 K, however, due to the low thermal conductivity of CE/EP insulation material, hard radiation heat transfer at low temperature and almost no convective heat transfer at high-vacuum, samples tested in this paper were unable to reach 4.2 K, but would stay at 6.1 K [18]. Therefore, the data obtained are at 6.1 K.

The main goal was to investigate the breakdown strength of cyanate ester/epoxy before and after irradiation at 6.1 K. Therefore, the electrodes used in this equipment were especially designed for electrical breakdown strength testing. The diameters of the top electrode and the bottom electrode were 6 mm and 25 mm. An insulation bushing is used to press the samples to avoid cold-contraction of the insulation materials.

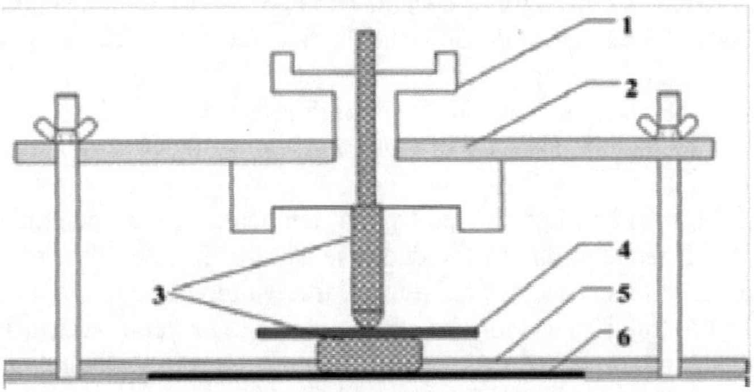

1-insulation bushing; 2-top support plate;
3-electrodes; 4-test sample;
5-lower support plate; 6-insulation layer

Figure 2. Structure of electrode system for electrical breakdown strength

A vacuum was used to reduce convective heat transfer and simulate the environment in ITER. This incorporated two thermal shields made of copper. The outer shield connects with the first stage cold head of the cryocooler, which can reach 50 K, and the inner shield connects with the second stage cold head, which can reach 4 K. The thermal shields were covered with ten layers of multi-layer insulation in order to reduce radiation from the vacuum chamber. A turbomolecular pump was used to create and maintain the vacuum inside the cryostat. The vacuum level was typically maintained below 10^{-5} Pa and monitored using a wide range vacuum gage mounted on the top flange of the cryostat.

The temperature monitoring and controlling system was based on a labview software platform, which can collect and then display the temperature of the samples and at other important places in the equipment. The electrical temperature controllers used in this system were Lakeshore Cryogenics model 340, which utilize a heater and thermometer coupled on an electrode. With this control system, the heating unit can be accurately controlled from 6.1 K to 300 K.

2.3 Test procedures

In this experiment, the sample was first sandwiched between the top electrode and the bottom electrode. The top electrode and bottom electrodes were connected to the high voltage and grounds, respectively. The samples were put on the bottom and the bolts by sides were adjusted until the electrode touch the sample. Second, air was exhausted from the vessel to create a vacuum environment. Next, the cryocooler was turned on until the temperature stayed constant, and then liquid helium was injected and a heater was turned on to keep the temperature. The high voltage DC power supply was then ramped up with a rate of 1 kV/s. Results were recorded when breakdown occurred. Each data represents an average, calculated from five measurements (samples showing significant deviation were rejected in the statistical calculations).

The breakdown voltage was divided by the thickness of the sample in order to obtain normalized values (in kV/mm) [19]. The thickness of the samples in this work is 0.5 mm. The electrical breakdown strength was calculated from the average voltage of five samples using the following formula:

$$E_b = \frac{U_b}{d} \tag{1}$$

E_b = electrical breakdown strength,
U_b = average breakdown voltage of five samples, and
d = thickness of the samples.

3. Results and discussion

In this paper, electric breakdown strength of cyanate ester/epoxy (CE/EP) insulation material at 6.1 K, 77 K (the temperature of liquid nitrogen) and 300 K (the room temperature) was studied. Table 1 shows that electrical breakdown strength of this composite improved obviously when the temperature decreases. Furthermore, the average breakdown strength at 6.1 K is 68.9 kV/mm, about 18% higher than that at 300 K. The average breakdown strength at 77 K is 63.8 kV/mm, about 9.4% higher than that at 300 K.

The increase of breakdown strength caused by low temperature may be explained as follows: the free volume of the CE/EP reduced due to the decrease of temperature. The molecular motion of the composite confined because less space of action if the free volume of the composite is reduced. There would not be enough molecular space for the charge carrier to accumulate easily when an external electrical field is applied, and much more polarized charge carriers in the composite would be released immediately, which helps the improved dielectric breakdown strength [20].

Table 1. Electrical breakdown strength of the CE/EP composite at different temperatures

Temperature (K)	6.1	77	300
Breakdown strength (kV/mm)	68.6	65.2	57.9
	69.2	63.9	55.3
	69.9	64.6	61.6
	67.5	61.3	59.8
	69.3	64.0	56.9
Average(kV/mm) (X±2 std. deviations)	68.9±1.6	63.8±2.6	58.3±4.4

Table 2 shows the electric breakdown strength of cyanate ester/epoxy (CE/EP) before and after irradiation by a total dose of 1 MGy, 5 MGy or 10 MGy at 6.1 K. Slightly higher values were found with the increase of dose range, but the overall effects are within the calculated standard deviation. That evidently indicates irradiation by ^{60}Co γ - ray effect slightly on electric breakdown strength of this composite. That is to say, the chemical structure of the CE/EP is scarcely changed by the irradiation up to 10 MGy. All results obtained by UV-Vis, FT-IR and mechanical properties tests in our previous study are consistent with this point [21].

Table 2. Electrical breakdown strength of the CE/EP composite at 6.1 K before and after irradiation

Radiation doses (MGy)	0	1	5	10
Breakdown strength (kV/mm)	68.6	75.5	72.1	73.8
	69.2	65.1	71.2	70.6
	69.9	69.5	76.1	71.2
	67.5	68.6	74.5	74.8
	69.3	72.3	72.6	73.6
Average (kV/mm) (X±2 std. deviations)	68.9±1.6	70.2±7	73.3±3.6	72.8±3.2

4. Conclusion

In this work, equipment was fabricated to measure electrical properties at a temperature range from 4.2 K to 300 K under vacuum. The electrical breakdown strength of cyanate ester/epoxy blend with 40% CE and 60% EP was measured at 6.1 K, 77 K and 300 K. At the same time, electrical breakdown strength of this composition at 6.1 K before and after irradiation was studied. In conclusion, cryogenic temperature has a positive effect on electrical breakdown strength of the composite, while the influence of ^{60}Co γ - ray irradiation is not evidential at 6.1 K.

References

[1] Vaghela H, Sarkar B, Bhattacharya R, et al. 2014 *American Institute of Physics* 872-879.
[2] Prokopec R, Humer K, Maix R K, et al. 2010 *Fusion Eng. Des.* **85(2)** 227-233.
[3] Hong J, Heo J I, Nam S, et al. 2013 *IEEE Trans. Appl. Supercond.* **23(3)** 7700604-7700604.
[4] Tuncer E, Sauers I, James D R, et al. 2009 *IEEE Trans. Appl. Supercond.* **19(3)** 2359-2362.
[5] K. Humer, R. Prokopec, R. K. Maix, H. Fillunger and H. W. Weber 2009 *Fusion Eng. Des.* **84**

(2) 960-963.
[6] R. Prokopec, K. Humer, R. K. Maix, H. Fillunger and H. W. Weber 2009 *Fusion Eng. Des.* **84 (7)** 1544-1547.
[7] Y. Tomi, F. Mishima, Y. Akiyama, Y. Izumi, A. Nishimura and S. Nishijima 2012 *IEEE Trans. Appl. Supercond.* **22 (3)** 7701104-7701104.
[8] Bittner-Rohrhofer K, Humer K, Fillunger H, et al. 2004 *Journal of Nuclear Materials* **329–333(1)** 1083-1087.
[9] Bittner-Rohrhofer K, Humer K, Wang Z D, et al. 2003 *Fusion Eng. Des.* **66(03)** 1209-1213.
[10] Wichmann M H G, Sumfleth J, Gojny F H, et al. 2006 *Engineering Fracture Mechanics* **73** 2346-2359.
[11] Tsotra P, Friedrich K 2004 *Synthetic Metals* **143(2)** 237-242.
[12] Shokrieh M M, Esmkhani M, Vahedi F, et al. 2013 *Iranian Polymer Journal* **22(10)** 721-727.
[13] Bakar A S A, Misnon M I, Ghodgaonkar D K, et al. 2004 *RF and Microwave Conference* **2004** 164-168.
[14] Ota W N, Amico S C, Satyanarayana K G 2005 *Composites Science & Technology* **65(6)** 873-881.
[15] ASTM D 792 Test Methods for Density and Specific Gravity (Relative Density) of Plastics by Displacement
[16] ASTM D 2584-11 Standard Test Method for Ignition Loss of Cured Reinforced Resins
[17] ASTM D 2734. Standard Test Method for void content of reinforced plastics
[18] Green M A 1998 *Advances in Cryogenic Engineering* 1313-1318.
[19] *Internal ABB report*
[20] Wan W, Yu D, He J, et al. 2008 *Journal of Applied Polymer Science* **107(2)** 1020-1028..
[21] Li J, Wu Z, Huang C, et al. 2014 *Fusion Eng. Des.* 89(12) 3112-3116.

Acknowledgments
This work was supported by the National Natural Science Foundation of China (Grant No.: 51407180, 51577185), the National Magnetic Confinement Fusion Science Program (Grant No.: 2015GB121001) and the Scientific Project of the State Grid Corporation of China (Grant No.: DG71-16-001).

Characterization of low temperature high voltage axial insulator breaks for the ITER cryogenic supply line

P Fernandez Pison[1,2], S Sgobba[1], I Aviles Santillana[1,2], S A E Langeslag[1], M Su[3], R Piccin[3], J Y Journeaux[3], A Laurenti[3] and W Pan[4]

[1] CERN, CH-1211 Genève, Switzerland
[2] University Carlos III of Madrid, Campus de Leganes, Av. Universidad 30, 28911 Madrid, Spain
[3] ITER Organization, Route de Vinon sur Verdon, CS 90 046 - 13067 St. Paul lez Durance cedex, France
[4] Institute of Plasma Physics, Chinese Academy of Sciences, Shushanhu 350, Hefei, 230031 Anhui, China

pilar.fernandez.pison@cern.ch

Abstract. Cable-in-conduit conductors of the ITER magnet system are directly cooled by supercritical helium. Insulation breaks are required in the liquid helium feed pipes to isolate the high voltage system of the magnet windings from the electrically grounded helium coolant supply line. They are submitted to high voltages and significant internal helium pressure and will experience mechanical forces resulting from differential thermal contraction and electro-mechanical loads. Insulation breaks consist essentially of stainless steel tubes overwrapped by an outer glass – fiber reinforced composite and bonded to an inner composite tube at each end of the stainless steel fittings. For some types of insulator breaks Glass – Kapton – Glass insulation layers are interleaved in the outer composite. Following an extensive mechanical testing campaign at cryogenic temperature combined with leak tightness tests, the present paper investigates through non-destructive and destructive techniques the physical and microstructural characteristics of the low temperature high voltage insulation breaks and of their individual components, thus allowing to correlate the structure and properties of the constituents to their overall performance. For all the tests performed, consistent and reproducible results were obtained within the range of the strict acceptance criteria defined for safe operation of the insulation breaks.

1. Introduction

New generation superconducting magnets for fusion projects, such as ITER, widely use coils based on technology of Cable-In-Conduit Conductors (CICCs) [1]. CICCs are designed based on various types of stainless steel jackets, densely filled with Nb_3Sn or NbTi superconducting strands, and subsequently compacted [2]. They are cooled by a forced flow of supercritical helium [3]. The magnet coils are directly connected to the helium supply pipes of the electrically grounded cryogenic distribution system.

In order to guarantee the electrical insulation between the helium supply pipes and the CICCs, Insulation Breaks (IBs) are used [4]. The IBs will be placed at the helium supply (inlets) and return points (outlets) of the different cryogenic circuits foreseen in the ITER magnet system. Depending on the operating environment of these cryogenic circuits, the IBs will withstand different potentials and temperatures. A High-Voltage Low-Temperature (HVLT) IB will operate at 4.5 K and could experience a voltage difference of up to 28 kV in case of failure [4]. These HVLT IBs consist of stainless steel end-fittings overwrapped by an outer glass – fiber reinforced composite and bonded to an inner composite tube at each end of the stainless steel fittings. For some types of insulator breaks Glass – Kapton – Glass (GKG) insulation layers are interleaved in the outer composite (figure 1).

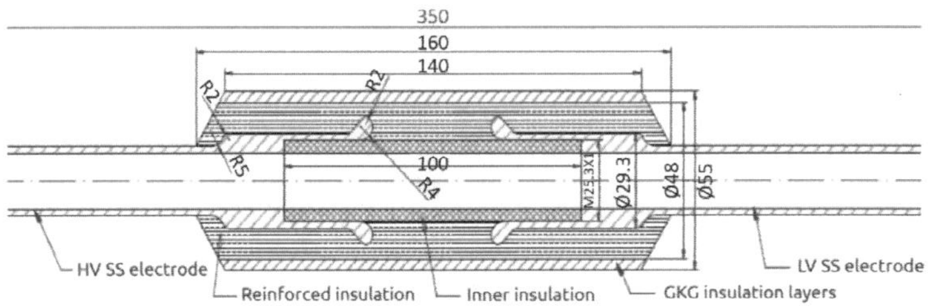

Figure 1. Schematic drawing of a HVLT IB (dimensions in mm), from [5].

In addition to the electrical potential and internal helium pressure that the IBs have to withstand at cryogenic temperature, they will also be submitted to a variety of mechanical forces resulting from differential thermal contraction and electro-magnetic loads [4]. Prior to the present assessment of the IBs, a characterization campaign aiming at testing their mechanical robustness was defined and executed. Mechanical testing at 77 K combined with leak tightness tests were carried out [5].

To further validate HVLT IBs against state of the art and specification requirements, an extensive assessment of the physical and microstructural characteristics of individual HVLT IBs and of their components have been performed. This comprehensive campaign addresses the material assessment of HVLT IBs for cryogenic use, which have already successfully passed prior qualification tests (mechanical, leak and high voltage tests). This campaign is aimed at confirming that IBs can withstand the highly demanding operating conditions in a cryogenic environment.

2. Components

Inspections were performed based on two identical 30 kV HVLT IBs (figure 1) identified as IBDH109 and IBDH251, respectively. Previous to the testing campaign reported here the two IBs were submitted to qualification tests. The latter included thermal cycling between room temperature (RT) and 77 K (100 thermal cycles), 4-point bend tests at RT, leak tightness tests, and high voltage electrical tests (partial discharge). Bearing this in mind, the IBs cannot be considered to be in a fully pristine state.

3. Experimental methods and results

An extensive characterization has been defined consisting of several reliability tests of individual IBs relevant for magnet operation. Destructive and non-destructive examinations included:

1. assessment of imperfections by visual and/or microoptical inspections in the axial and radial planes, and by die penetrant testing (PT);
2. measurement of the density of inner and outer ("reinforced") insulation composites (see figure 1), respectively;
3. assessment of the void fraction, fiber fraction and resin fraction of both insulation composites;
4. measurement of the fiber orientation;
5. assessment of the glass transition temperature (T_g) of the resin.

In order to evaluate the manufacturing quality of the inspected IBs for cryogenic use, they needed to be sectioned to access their internal structure. Firstly, sectioning along their axial center line was performed to obtain a flat finish and an exploitable surface for the analysis. One of the so obtained halves

IOP Conf. Series: Materials Science and Engineering **279** (2017) 012007 doi:10.1088/1757-899X/279/1/012007

was further sectioned in three portions, allowing one portion to be directly observed in a radial plane, and specimens from the inner and outer composites to be removed from the two other portions, respectively, for their individual assessment (figure 2).

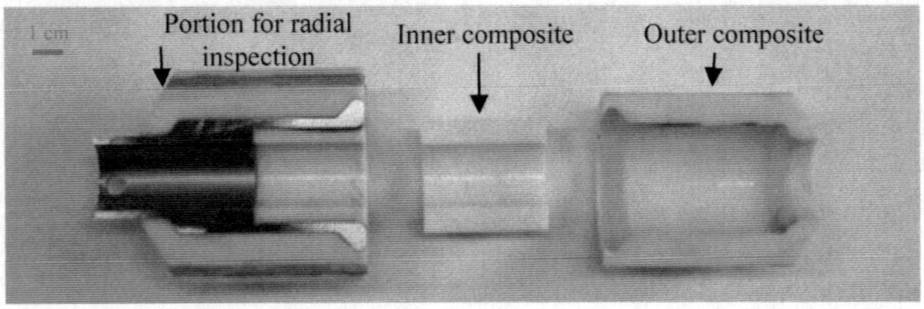

Figure 2. IBDH251 sectioned into three cross-sectional portions.

3.1. *Assessment of imperfections*

The destructive and non-destructive examinations applied for the assessment of imperfections are described hereafter.

3.1.1. Inspection in axial plane. The non-sectioned half of the individual IBs was microoptically inspected for any flaws in the structure (digital microscope VHX 1000E by KEYENCE). Since the cutting operation might have induced minor localized damage in the surface, such as slightly burnt regions in the composites, the surface was manually polished prior to the inspections. Figure 3 shows the state of IBDH251 prepared for the inspections.

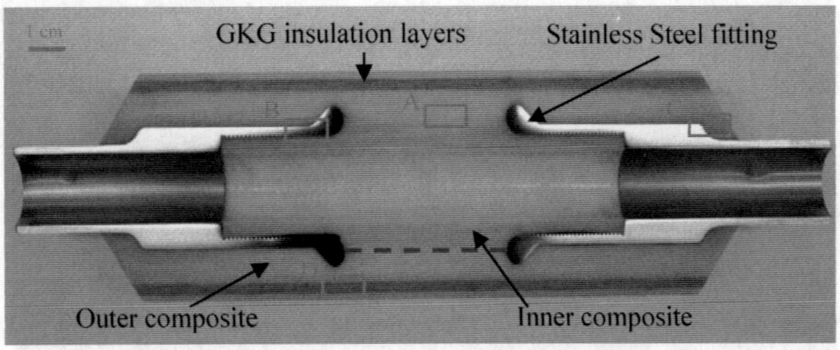

Figure 3. Longitudinal cut of IBDH251, showing the fields of view used for identifying specific features and their location. In fields of view A, B, C and D detailed observations were carried out.

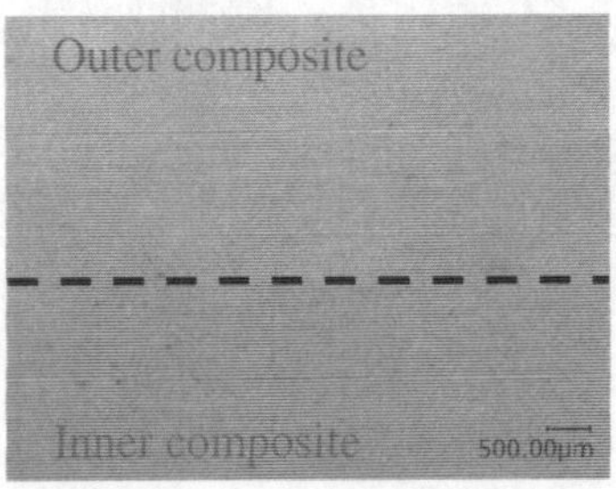

Figure 4. Interface between both composites; field of view A in figure 3.

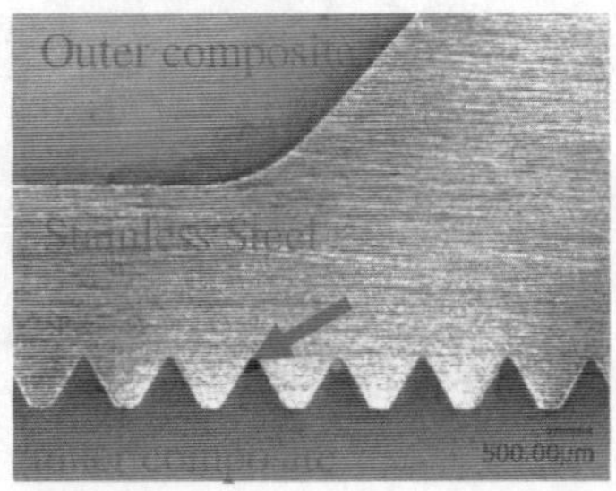

Figure 5. Void close to the stainless steel fitting; field of view B in figure 3.

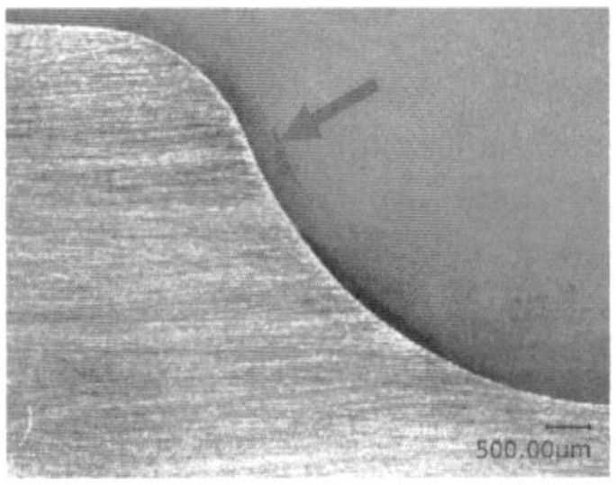

Figure 6. Cracks close to the stainless steel fitting; field of view C in figure 3.

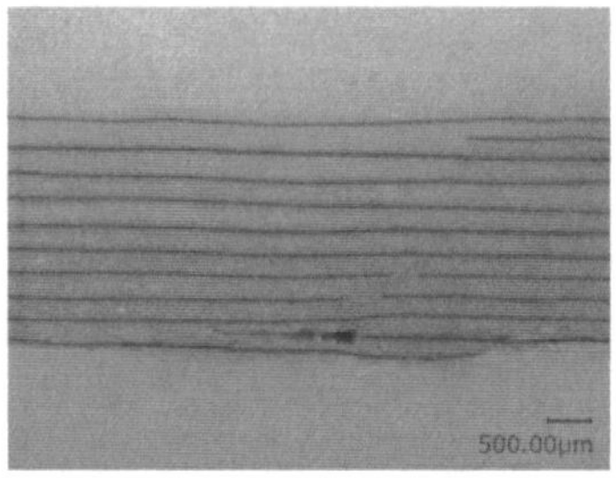

Figure 7. Voids in the GKG insulation layers; field of view D in figure 3.

For both IBs analyzed, a good continuity between the two composites was confirmed (figure 4, field A). No gross porosities were visible within the composites. However, in some regions around the stainless steel fittings, some voids and cracks were observed (figure 5, field B; and figure 6, field C). The GKG insulation layers showed some voids particularly concentrated towards the outer layer, more pronounced for IBDH251 than for IBDH109 (figure 7, field D).

3.1.2. Penetrant Testing (PT). Both halves of the individual IBs were examined by dye penetrant testing, Type II (visible dye) method A (water washable) according to ASTM E1417. PT was applied before performing further cross-sectional cuts. For both tested IBs, 20 minutes after the developer was applied (solvent-based white-powder penetrant developer), very pronounced linear indications were observed. They were mainly located at the stainless steel to composite interfaces and in the GKG insulation layers, which agree fairly well with the location of the imperfections observed in the axial inspection. The large amount of dye penetrant observed in some regions revealed deep discontinuities associated to the respective indications. Figure 8 shows the results obtained for IBDH109.

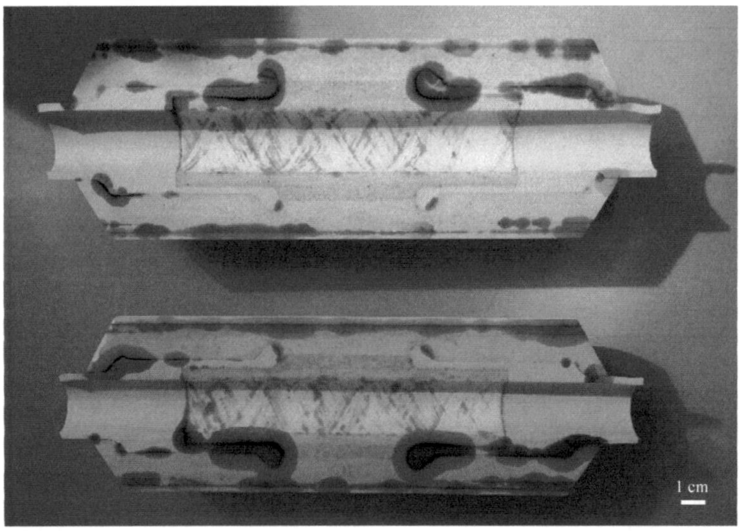

Figure 8. IBDH109 after application of developer. Dwell time: 20 minutes.

3.1.3. Inspection in radial plane. One cross-sectional portion of the individual IBs was used for the inspection. The radial plane was observed after manual polishing. For both IBs, small linear indications were observed in the stainless steel to composite interface. Long indications were found in the last layer of the outer GKG insulation (see figure 9 for IBDH251). Both cross sections were performed at the positions where major indications from PT performed in the axial plane were observed. Consistently, in these positions, the radial inspections have confirmed the presence and the extent of the imperfections already identified in the axial cut.

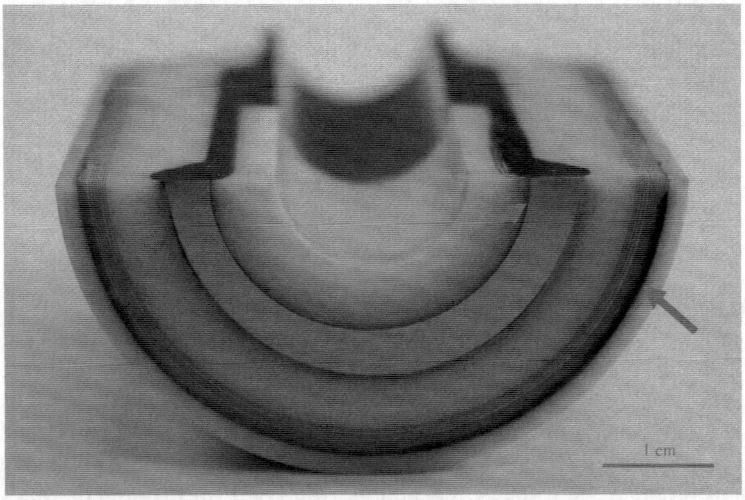

Figure 9. Radial inspection of IBDH251. The arrows show the indications already identified by PT on the axial cut, located in the stainless steel to composite interface and in the last layer of the GKG insulation layers, respectively.

3.2. Density

The density was measured in both inner and outer composites of the individual IBs, based on Archimedes' principle according to ASTM D792 (pure ethanol was used as the process fluid). Two specimens of approximately 5 g each, as required by the standard, for each composite were extracted from the two individual cross-sectional portions respectively (figure 2). The density was measured using a Sartorius analytical balance (accuracy class I). The specimens were measured three times for repeatability purposes. The results are shown in table 1 where similar densities for both IBs are observed.

Table 1. Summary results of densities for different components.

	IBDH109	IBDH251
Inner composite	1.88 g/cm^3 ± 0.01 g/cm^3	1.91 g/cm^3 ± 0.00 g/cm^3
Outer composite	1.86 g/cm^3 ± 0.01 g/cm^3	1.84 g/cm^3 ± 0.02 g/cm^3

3.3. Void, fiber and resin fraction

The assessment of the volume fraction of fiber, resin and void was performed according to ASTM D2584 (ignition loss method) on the same specimens used for the assessment of the density of the composites. Following weighing in an analytical balance the specimens were fired in a furnace at 550 ^{0}C for approximately 2 hours. Afterwards, they were cooled inside the furnace by natural convection and weighed again as soon as they reached RT. In order to confirm the complete removal of the organic part of the composite material, an infrared spectroscopy analysis was carried out for one of the specimens of the outer composite of IBDH109, both before and after heat treatment. A full decomposition of the resin after 2 hours at 550 ^{0}C was confirmed. The results of the volume fractions are shown in table 2. It can be noticed that the void volume fraction of the outer composite of IBDH251 is slightly higher than the inner composite of the same IB and for both composites of IBDH109.

Table 2. Summary results of volume fraction of fiber, resin and void for different components.

	IBDH109			IBDH251		
	Vol. Fiber	Vol. Resin	Vol. Void	Vol. Fiber	Vol. Resin	Vol. Void
Inner composite	48.6 % ± 0.1 %	49.7 % ± 0.4 %	1.7 % ± 0.1 %	50.6 % ± 0.1 %	47.5 % ± 0.2 %	1.9 % ± 0.1 %
Outer composite	46.9 % ± 0.1 %	51.3 % ± 0.1 %	1.8 % ± 0.0 %	45.6 % ± 1.5 %	52.2 % ± 1.5 %	2.2 % ± 0.0 %

3.4. Fiber orientation

In order to estimate the fiber angle in the composites of the individual IBs, the specimens previously used for density and void fraction assessment were observed after combustion. The fired specimens were optically observed using a digital microscope (VHX 1000E by KEYENCE). The fiber angle was measured from the images using microscope specific image analysis software. It is important to note that only the fiber angle from the near surface layers visually accessible could be measured due to the fact that removing external layers of fibers to examine underlying fibers could modify the orientation of the latter, introducing errors. However, specimens for observation from both inner and outer diameter were prepared in order to investigate a possible change of the angle across the thickness. All of the composites show a braid pattern (figure 10). The braid angles measured with respect to the main axis (figure 11) are shown in table 3, similar values for each composite of both IBs were obtained.

Figure 10. Inner composite of IBDH109 after firing.

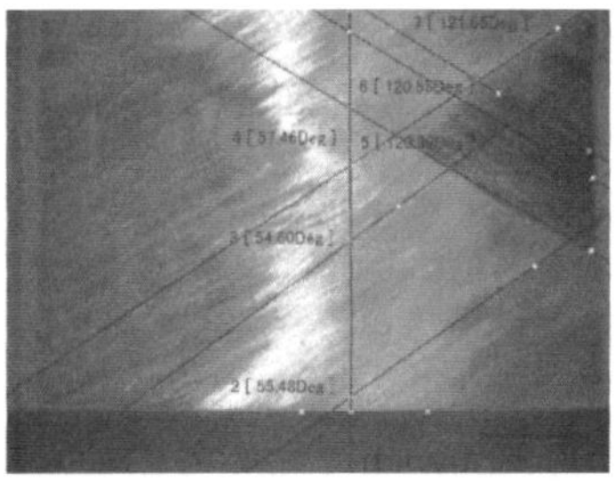

Figure 11. Fiber angle measurement for the inner composite of IBDH109.

Table 3. Summary results of braid angles, referred to the main axis, for different components.

	IBDH109	IBDH251
Inner composite	± 60.6° ± 3.3°	± 60.8° ± 2.2°
Outer composite	± 70.4° ± 3.1°	± 73.5° ± 4.4°

3.5. Glass transition temperature (T_g)

The evaluation of T_g of each of the two composites was performed according to IEC 61006 via Differential Scanning Calorimetry (DSC) using a SETARAM 131 calorimeter. For the measurement, 20 mg powder specimens from the two individual cross-sectional portions (figure 2), respectively, were obtained by ball milling and were warmed up with a ramp of 10 ^{0}C/min to a temperature of 200 ^{0}C under nitrogen. Measurement was repeated twice for each of the specimens, with an intermediate cool down (50 mL/min of pure nitrogen) to RT by forced convention. A third DSC analysis was carried out after the same intermediate cool down for the outer composite of IBDH109, in order to confirm the stabilization of T_g. The results for each of the composites for both IBs are displayed in table 4. An increment of T_g between the first and the second heating cycles is observed for both composites of both IBs. All specimens reach the same T_g (132 ° ± 2 °) after the second heat cycle. The third DSC analysis showed no evolution in T_g.

Table 4. Summary results of T_g for different components.

	IBDH109			IBDH251		
	T_{g1}	T_{g2}	T_{g3}	T_{g1}	T_{g2}	T_{g3}
Inner composite	124 °C	131 °C	n.a.	122 °C	134 °C	n.a.
Outer composite	114 °C	129 °C	130 °C	122 °C	135 °C	n.a.

4. Discussion

In both axial and radial planes several cracks and discontinuities were observed (figure 5 to figure 9), being of a considerable depth for most of them as revealed by PT. This confirms that although the observations were carried out in a single cross section per direction (axial and radial), the imperfections extend within the bulk. The most critical defects for both individual IBs were observed around the stainless steel fittings, towards the interface between them and the composites. Due to their axial orientation and their location, these discontinuities seem to have been induced by the hundred thermal cycles to liquid nitrogen to which the IBs were submitted prior to destructive examination. They are likely a consequence of high stresses induced by the difference in thermal expansion coefficient between the dissimilar materials. However, these cracks do not form a continuous network along the interfaces, and therefore they do not impair the strength and the leak tightness of the assembly [5]. Another critical region where imperfections were observed is the GKG insulation layers, which showed less soundness for IBDH251 compared to IBDH109. In general these imperfections present an ellipsoidal shape and did not evolve into long imperfections. They are generally small and their geometry is more favorable than elongated defects such as sharp cracks in terms of stress concentration.

The repeatability of the density results for the same IB is high, and in addition, similar densities were obtained for both IBs (table 1), meaning that during the fabrication, a stable process was achieved leading to homogeneous and dense composites.

Regarding the volume fraction of the different IBs for both composites, it is important to note that the void volume fraction of the outer composite of IBDH251 is slightly higher than the inner composite of the same IB and for both composites of IBDH109 (table 2). Moreover, the orientation of the fibers resulted in similar values for each composite of both IBs confirming repeatability and robustness of fabrication (table 3).

Based on the DSC results, an increment of T_g between the first and the second heating cycles was observed for both composites of both IBs (table 4). After the second heat cycle, all specimens reached the same T_g (132 °C ± 2 °C). The third DSC analysis showed no evolution in T_g after an additional heat cycle, confirming the stability of the obtained value. Based on these results, an incomplete polymerization of all studied composites is evident, which could be due to the high volume of the composite to be cured together with an expected low thermal conductivity of this type of material.

However, since both IBs successfully passed all tests included in the validation protocol, the impact of this fact is small. Moreover, an additional polymerization step will be carried out during assembly by an additional thermal cycle at 80 °C.

The obtained results, together with previous tests confirming that IBs are leak tight against severe tensile loads at cryogenic temperature, represent a sufficient basis of qualification for their operation.

5. Conclusion

An in-depth characterization was successfully performed for the assessment of the physical properties and microstructure of HVLT IBs and of their individual components. Two individual IBs were assessed through non-destructive and destructive techniques.

Imperfections were observed by an axial and a radial inspection of the IBs and confirmed by PT. However, they proved not to impair the leak tightness within the range of design stresses at cryogenic temperature [5]. The properties of the bulk composites were assessed by measuring their respective density and calculating the volume fraction of fiber, resin, and void individually. Furthermore, the fiber orientation was measured through optical techniques, and T_g was determined via DSC. For all the tests performed, consistent and reproducible results were obtained.

To the extent of the investigations performed, the results are within the range of the strict acceptance criteria defined for safe operation of the IBs, therefore a reliable overall performance of the analyzed HVLT IBs is expected on this basis.

The views and opinions expressed herein do not necessarily reflect those of the ITER Organization.

References

[1] Mitchell N, Devred A, Libeyre P, Lim B and Savary F 2012 The ITER Magnets: Design and Construction Statud *IEEE Trans. Appl. Supercond.* **22** no. 3 4200809

[2] Turck B, Bessette D and Ciazynski D 1993 Design methods and actual performances of conductors for the superconducting coils of Tokamaks *Fusion Eng. Des.* **2** 667-70

[3] Bottura L and Luongo C 1999 Superconductors, stability in forced flow *Wiley Encyclopedia of Electrical and Electronics Engineering*

[4] Rodriguez-Mateos F, Evans D, Devred A, Laurenti A and Mitchell N 2014 Essential Design, Construction and Test Elements of ITER Axial Insulation Breaks *AIP Conf. Proc.* **1574** no.1 117-24

[5] Langeslag S A E, Rodriguez Castro E, Aviles Santillana I, Sgobba S and Foussat A 2015 Design of load-to-failure tests of high-voltage insulation breaks for ITER's cryogenic network *IOP Conf. Ser.: Mater. Sci. Eng.* **102** no.1, 012009

Acknowledgements

The cryogenic cycling, mechanical and electrical measurements and leak tightness tests were carried out by the Magnet Infrastructure Facilities for ITER (MIFI) under the supervision of the ITER Organization; T_g was measured by A. Rivière at the chemistry laboratory of CERN (TE/VSC/SCC); PT was performed by A.M. Piguiet, a level 2 certified operator (COFREND ID n° 012084) of the Materials and Metrology section of CERN (EN/MME/MM).

Evaluation of the Fretting Resistance of the High Voltage Insulation on the ITER Magnet Feeder Busbars

N Clayton[1], **M Crouchen**[2], **D Evans**[1], **C-Y Gung**[1], **M Su**[1], **A Devred**[1] and **R. Piccin**[1]

[1] ITER Organization, Route de Vinon-sur-Verdon, CS 90 046, 13067 St. Paul Lez Durance Cedex, France,
[2]Rockwood Composites Ltd, 8 Venture Court, Bradley Lane, Newton Abbot, Devon TQ12 1NB, U.K.

David.Evans@iter.org

Abstract. The high voltage (HV) insulation on the ITER magnet feeder superconducting busbars and current leads will be prepared from S-glass fabric, pre-impregnated with an epoxy resin, which is interleaved with polyimide film and wrapped onto the components and cured during feeder manufacture. The insulation architecture consists of nine half-lapped layers of glass/Kapton, which is then enveloped in a ground-screen, and two further half-lapped layers of glass pre-preg for mechanical protection. The integrity of the HV insulation is critical in order to inhibit electrical arcs within the feeders. The insulation over the entire length of the HV components (bus bar, current leads and joints) must provide a level of voltage isolation of 30 kV. In operation, the insulation on ITER busbars will be subjected to high mechanical loads, arising from Lorentz forces, and in addition will be subjected to fretting erosion against stainless steel clamps, as the pulsed nature of some magnets results in longitudinal movement of the busbar. This work was aimed at assessing the wear on, and the changes in, the electrical properties of the insulation when subjected to typical ITER operating conditions. High voltage tests demonstrated that the electrical isolation of the insulation was intact after the fretting test.

1. Introduction

After installation, the insulated bus bars will be supported by a series of stainless steel supports (clamps). There is a nominal 2 mm radial clearance between the insulated busbar and the clamp, and during machine cool-down the approximately 30 m long bus bars will slide within these clamps by up to 50 mm; the machine design is for 100 cool-downs.

During operation of the ITER device, there will be large Lorentz forces on the bus bars that are reacted through the clamps. Many of the ITER magnets are pulsed and during machine operation the bus bars will slide up to 5 mm during each of the 30,000 lifetime pulses.

To assess the overall effects of the busbar insulation sliding within a close fitting stainless steel clamp, a test assembly and test specimen were devised and manufactured to produce a configuration that closely matched that of an operational insulated bus bar.

The system has to apply a constant load, irrespective of wear on the insulation, since in practice the Lorentz force of up to 15 kN per clamp [1] will ensure that the busbar remains firmly in contact with one surface of the clamps. The test assembly must ensure that the applied load is maintained when the system is cooled to 77 K and remains constant while cycled through ±5mm for up to twice the ITER lifetime of 30,000 cycles.

After cycling in this way, specimens were subjected to high voltage and partial discharge testing to assess the integrity of the insulation.

Test specimens that are used to evaluate the response of the insulation to these demanding conditions must be manufactured using identical materials and process conditions to those used for the feeder busbar. A clamp system that exactly replicates that of the ITER design forms the other component of the test set-up.

2. Materials

A stainless steel bar was machined to a diameter of 43 mm and was used as a 'dummy' busbar. The insulation was based on a plain weave, S-glass fabric of approximately 0.25 mm thickness pre-impregnated with ~ 40 weight % epoxy resin (Gurit SE84LV) [2] that has been selected, tested and described in more detail by Clayton[3].

Pre-impregnated glass tapes 25 mm wide were combined with 20 mm wide Kapton HN tape [4]. In this development stage of the project, only limited quantities of material were required for testing and therefore manual techniques were used to combine the glass and Kapton tapes; the method of combining these tapes is also described fully in [3].

The 316L stainless steel bar was grit blasted to produce a surface roughness of Ra = 4 microns, detergent cleaned and its cleanliness confirmed using the "water break test' as described in [5].

3. Test Assembly

The principles of the fretting rig are illustrated in Figure 1. The bus-bar clamps are located between two side plates, one clamp is firmly bolted to each side plate and the other clamp floats within guide rails. A pivot pin runs between the side plates, located at the bottom of the assembly to provide a fulcrum for a "loading bar".

External to the cryostat, a 'lever arm' is attached to the loading bar with the required mass at the end of this arm to produce the necessary balancing moment. The distance between the fulcrum and the point on the floating clamp where the force is applied is 50 mm. Since the force required is 15'000 N, the moment required (15'000 x 50) is 750,000 Nmm.

To balance this moment, with a bar length of 1051 mm, a total mass 713.6 N (72.74 kg) is necessary. The mass at the end of the bar, including the hanger for additional weights, is 2.74 kg and therefore an additional 70 kg was required at the end of the lever arm to provide the necessary load on the clamp.

To off-set the effects of a large load on one side of the system, a balancing arm of the same length and with the same added mass is attached to the opposite side.

As wear on the insulation progressed, the angle of the lever arm may change and this can be compensated for by adjustment of a compensation "wedge". In practice, the wear was not found to be sufficient to warrant any changes in the wedge position.

A shell was added to the bottom of the assembly in Figure 1 to form a dewar to contain the liquid nitrogen (LN2). In this way, the fretting specimen and clamps were immersed in LN2 for the fretting test; however the arrangement allowed the specimen to be readily warmed to room temperature for inspection.

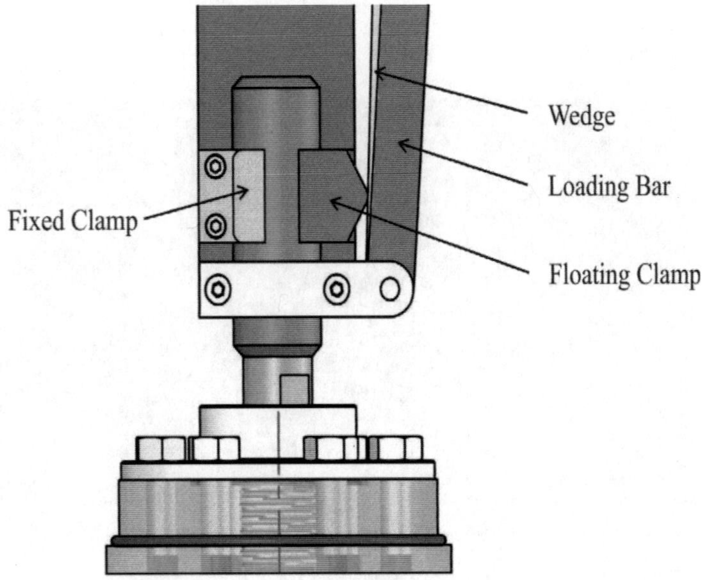

Figure 1. Diagrammatic view of the specimen, clamps and lever arm loading system

4. Manufacture of Fretting Specimen and Clamps

A total of 4 fretting specimens were prepared and tested. The first two specimens were identical, and their overall insulation architecture was applied to the stainless steel bar in multiple stages, with a cure between each stage as an aid to minimizing trapped voids and 'wrinkling' of the Kapton tape. A vacuum bag was used to apply compaction pressure on the insulation during cure. The first stage consisted of wrapping one layer of half-lapped (i.e. two thicknesses) of pre-impregnated (pre-preg) glass tape on the steel bar, followed by 4 half-lapped layers of combined glass / Kapton (GK) tape, wound with the Kapton facing outwards. On completion of these layers, the insulation was covered with 'peel ply' and cured. Stage 2 was to apply a further 5 half-lapped glass / Kapton layers, followed by two half-lapped layers of pre-preg, then cure again. The moulding with all the GK layers was typically 5.5 mm thick.

It was found that the use of the vacuum bag led to small wrinkles in the surface of the insulation. In order to avoid these, the last two fretting specimens were cured using trapped silicone rubber to apply the compaction pressure instead of a vacuum bag. This technique is fully described in [3] but in essence uses the high thermal expansion of silicone rubber to apply a compaction pressure to the insulation. The process consisted of wrapping two layers of 3 mm thick, 25 mm wide, extruded silicone rubber tape, trapezoidal in cross-section, around the structure and over-wrapping with five layers of glass tape. When heated to the cure temperature of the pre-preg, the glass tape restrains the expansion of the rubber and so a pressure is applied to the composite insulation.

Specimens #3 and #4 were also built with a ground screen located on top of the GK insulation, but underneath the uppermost two half-lapped layers of pre-preg. The selected ground screen consisted of a nonwoven copper and nickel coated carbon fibre veil [6]. This material is porous to resin, and so can be well bonded into the insulation structure. In order to accommodate the veil, a third manufacturing stage was added which involved wrapping a layer of pre-preg on top of the 9 GK layers, then applying the veil (Figure 2), followed by two half-lapped layers of pre-preg for mechanical protection, and then the final cure.

Electrical connection to the veil was made using a thin rectangle of copper foil (15 mm by 60 mm), to which a wire had been soldered; the copper foil was laid directly on the veil and overwound by the two half-lapped layers of pre-preg.

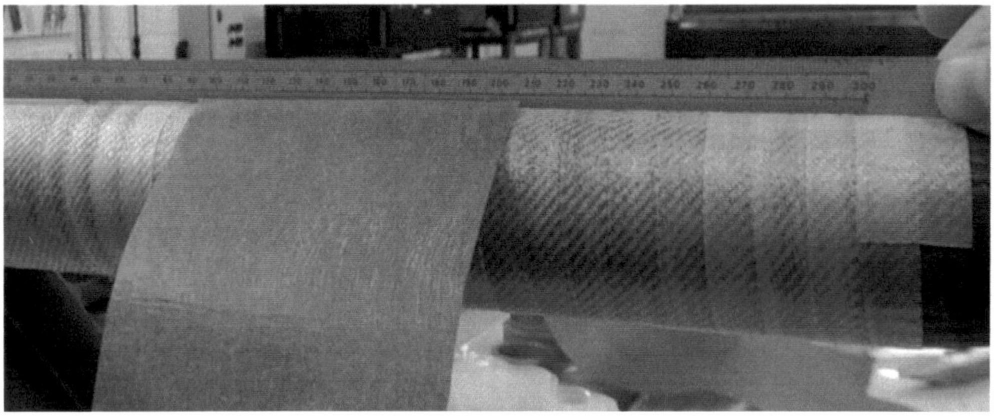

Figure 2 Applying the 'veil' on top of the pre-preg layer

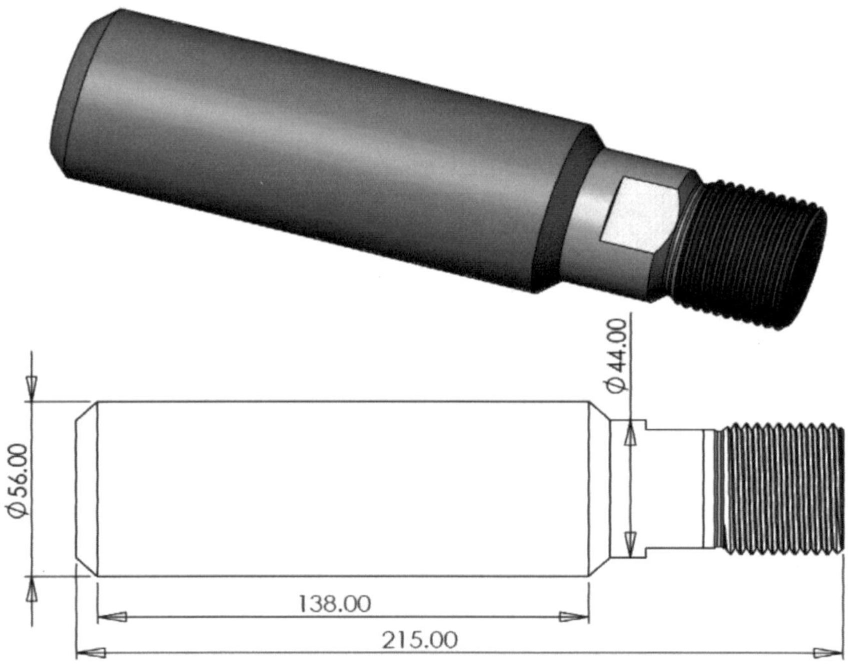

Figure 3. Test specimen machined from the insulated mandrel

The stainless steel bars were 400 mm in length in order to provide a realistic length for wrapping insulation, and from this length, one specimen was machined. The insulation length was machined to

IOP Conf. Series: Materials Science and Engineering **279** (2017) 012008 doi:10.1088/1757-899X/279/1/012008

215 mm long, and the threaded portion, for holding the specimen in the test machine, was an M42 x 3 thread, 40 mm in length and 'flats' machined at the opposite end were used for tightening the specimen onto the test machine (Figure 3).

Clamps were machined from 316 L stainless steel, and prepared with a surface roughness of Ra = 1.6 microns. As in the feeders, the clamps were 50 mm long, and were prepared with an R5 radii on their leading and trailing edges. New clamps were used for each fretting test, and the first three fretting specimens were tested with identical clamps. The 4th fretting test employed clamps with an 8 mm wide "flat" region along the centre line, such that the fretting specimen was in contact with this flat along its length. This final clamp design is employed in some locations of the feeder in order to accommodate increased displacements of the busbar.

5. Fretting Test Procedures

All mechanical tests were carried out on an Instron, model 8801 by Composite Test & Evaluation Ltd, Honiton, Devon, UK. The bottom of the specimen was fixed to the moving actuator while the clamps were attached to the test assembly, that in turn was connected to the load weighing system.

Initial trials were conducted at a number of frequencies and for practical reasons, 1Hz was selected as the most useable frequency and the amplitude was ±5 mm about a zero point.

A thermocouple was used to ensure that the friction did not cause the specimen to warm up in the LN2 dewar. The diameter of the test specimen was measured in a number of positions around the circumference at room temperature using a hand held micrometer and the mean value calculated. Measurements were made at a height on the specimen that corresponded to a position at the centre point of the clamps. The fretting test procedure was to perform a number of fretting cycles on the specimen at 77 K, and then warm it up to inspect and measure any wear on the insulation. The specimen was then re-cooled to 77 K, and the fretting test continued. The first specimen was inspected and measured after 500, 1000, 5000, 10000 cycles, and then every 10000 cycles up to 60000 cycles. The subsequent specimens were inspected less regularly, but at each inspection it was noted that the test assembly and specimen were coated with a fine black dust.

Each specimen took around 3 full days to test.

6. Results

Results are considered under three separate headings:

(a) Wear on insulation
(b) Electrical Characteristics
(c) Calculation of Friction Factor

6.1 Wear on Insulation

The central result from this study is that the specimens suffered wear to the insulation underneath the clamps, but did not lose their overall mechanical integrity. Figure 4 shows the wear on the insulation as a function of fretting cycle, measured for the 4 specimens. As shown in Figure 1, the test set-up incorporates two clamps, and so there are two wear zones on each specimen. The data in Figure 4 shows the average of these two wear zones, and as such corresponds to an average radial wear.

Initially, all 4 specimens displayed accelerated rates of wear, which then stabilised after a few thousand cycles. This may have been due to any outstanding features on the insulation surface, such as small irregularities, being ground away early on in the test. It is interesting to note that Specimens #1 and #2 display significantly different rates of wear despite being ostensibly identical. Figure 5 shows Specimen #2 and its moveable clamp at the end of the fretting test, and a significant ridge can be seen in the insulation that extended into the wear zone. It is postulated that this ridge acted as a stress concentration region and resulted in additional dust formation that accelerated the wear on this specimen compared Specimen #1, which did not have such a prominent feature.

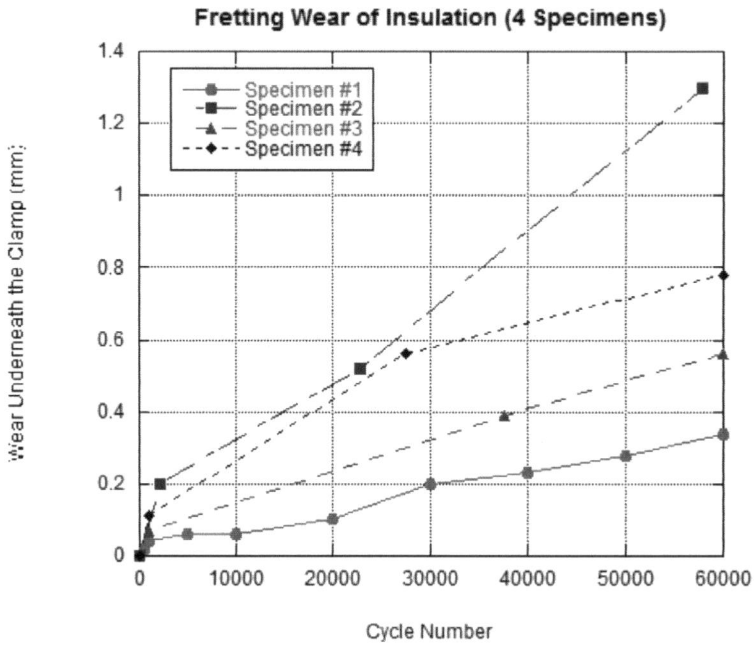

Figure 4. Average radial wear on the insulation as function of fretting cycle number (all 4 specimens)

Figure 5. (Left) Photograph of Specimen #2 after the fretting test had been completed, showing a significant ridge in the insulation around the wear zone. (Right) Photograph of the moveable clamp after the fretting cycles showing the polishing action on the contact surface.

IOP Conf. Series: Materials Science and Engineering **279** (2017) 012008 doi:10.1088/1757-899X/279/1/012008

6.2 Electrical Characteristics

Two specimens (#3 and #4) were evaluated after mechanical testing to examine their electrical characteristics and confirm (or otherwise) that they still met the ITER requirements. These specimens were fitted with integrated ground-screens which facilitated these tests. The insulation resistance was measured between the steel thread and the veil ground-screen using a Megger, model MIT 1525 at 1 kV DC for 1 minute. The insulation resistance of specimen #3 was found to be 2 x 10^9 Ohms (2 TΩ) and specimen #4 was found to be 1.8 x 10^9 Ohms (1.8 TΩ), significantly exceeding the ITER requirements.

The specimen geometry did not permit a breakdown test to be performed due to the proximity of the edges of the ground-screen to the high voltage potential on the specimen, which resulted in surfaces discharges. Partial discharge measurements were made to characterise the insulation, which were limited to 10 kVrms for a period of 30 minutes. The partial discharge measurements showed large surface discharges (>2 nC) which masked the signal generated by internal voids. Nonetheless, the partial discharge signal did not increase over the duration of the test, indicating that the electrical performance of the insulation was maintained.

6.3 Calculation of Coefficient of Friction

The experiment permitted a rough estimate of the dynamic coefficient of friction to be made. With reference to Figure 6, the coefficient of friction is defined as μ = Ff/Fn, where μ is coefficient of friction, Ff is frictional force, and Fn is normal force. In the experimental set-up, the specimen is fretted against two clamps, each applying 15 kN of normal force, and the friction force is measured directly by the Instron testing machine. Given that the profile of the clamps is identical, it is assumed that the friction from each one is the same; therefore the coefficient of friction underneath one clamp may be estimated by considering half the measured frictional force.

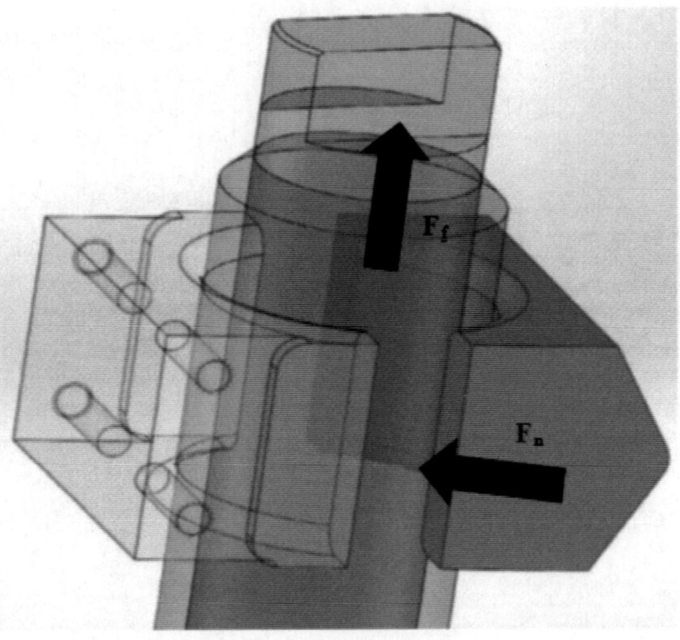

Figure 6. Representation of Friction Force (Ff) and Normal Force (Fn)

IOP Conf. Series: Materials Science and Engineering **279** (2017) 012008 doi:10.1088/1757-899X/279/1/012008

It was found that the test results from all four specimens could not be readily interpreted. Careful tests showed that the up and down forces in a fretting cycle were not symmetric, which might be caused by the presence of a residual force as the specimen direction reverses. This indicates that this experimental set-up cannot be used to reliably determine the friction factor. Nonetheless, as an approximation, a friction factor in the range of 0.2~0.4 was deduced.

7. Conclusions

A series of fretting tests has been performed on insulation specimens representing the high voltage insulation on the ITER feeder busbars. It was found that the specimens suffered wear to the insulation underneath the clamps, but did not lose their overall mechanical integrity. The amount of wear varied between the specimens, and in all but one was confined to the 'capping' layers of pre-preg which are designed to protect the ground-screen.

The reason for the increased wear on the 4th specimen may be the presence of a crease in the surface of the insulation which led to a concentration of the load in this region.

Electrical tests on two of the specimens fitted with integrated ground-screens revealed that the high voltage insulation was intact electrically after the fretting test. It can be concluded that as long as significant ridges in the insulation are avoided, wear due to fretting can be confined to the pre-preg capping layers whilst maintaining the integrity of the underlying high voltage insulation.

The friction force generated during low temperature cycling has been used to derive an approximation for the coefficient of friction in the range of 0.2~0.4.

8. Disclaimer

The views and opinions expressed herein do not necessarily reflect those of the ITER Organization.

References

[1] Private Communication – "Summary of TF ICF Analysis", ITER IDM MBYENM v2.4.
[2] Gurit Technical Data Sheet on SE 84LV pre-preg
[3] N. Clayton, M. Crouchen, A. Devred, D. Evans, C.-Y. Gung and I. Lathwell Cryogenics 83 (2017) 64-70
[4] Kapton HN – DuPont Technical Data Sheet
[5] Standard Test Method for Hydrophobic Surface Films by the Water-Break Test. ASTM F22-02 (Reapproved 2007)
[6] Technical Fibre Products Ltd. Optiveil, Part number: 20444A 34g/m2 Cu-Ni-Carbon veil

IOP Conf. Series: Materials Science and Engineering **279** (2017) 012009 doi:10.1088/1757-899X/279/1/012009

Degradation Free Vacuum Epoxy Impregnated short REBCO Undulator Magnets

I Kesgin[1,3], Q Hasse[1], Y Ivanyushenkov[1] and U Welp[2]

[1]Advanced Photon Source, Argonne National Laboratory, 9700 South Cass Avenue, Lemont, IL 60439, USA
[2]Materials Science Division, Argonne National Laboratory, 9700 South Cass Avenue, Lemont, IL 60439, USA

E-mail: ikesgin@aps.anl.gov

Abstract. The source of high–brilliance, hard x-ray radiation in light sources and free electron lasers are undulator insertion devices. Recently, REBCO-based superconductors have been demonstrated to be a potential candidate to enhance the performance of the undulator insertion devices. Epoxy impregnation remained as one of the steps which is crucial to fully realize this technology. Epoxy impregnation of magnets is required because it prevents the motion of the wire and provides conduction cooling to the winding layers. Up until now, most of the impregnated REBCO coils/magnets showed some degree of degradation after cold cycles. We developed a vacuum impregnation technique that does not degrade the performance of the short REBCO prototype undulator magnets after cold cycles. The results showed that in order to prevent degradation, a bumper layer between the magnet winding stacks and the epoxy/powder mixture is required. Microstructure images of vacuum impregnated coil packs showed homogenously-spaced REBCO winding layers and very thin epoxy fillings between the layers, which is important for the overall performance of the device.

1. Introduction

REBCO (RE = rare earth, barium copper oxide) coated conductors (CCs) are becoming a promising candidate to enhance the performance of an undulator's magnetic structure [1-3]. The main advantage of using REBCO CCs in undulator structures, compared to its counterparts, is that it does not require a high temperature reaction process (Nb_3Sn ~650 °C for ~48 hrs). Typically, the wire is reacted after winding the magnet due to its brittle nature in superconducting phase. This high temperature reaction process poses additional magnet design challenges. In contrast, REBCO CCs are used as received without any post-processing. Avoiding this extreme reaction temperature is a significant convenience considering the possible deformation of the stringent tolerances required by the field quality. Most of the technical questions regarding the winding, insulating, and measuring of the undulator magnet have been addressed in our previous works [1-3]. However, the epoxy impregnation of the REBCO magnet systems remains a challenge. Epoxy impregnation is especially necessary for conduction-cooled magnet systems such as undulators. It serves as a mechanical support structure, secures the tape conductor and impedes the conductor motion that could, otherwise, result in a catastrophic failure. While the epoxy impregnated magnets are mechanically more robust and uniformly-cooled compared to non-epoxy

[3] To whom all correspondence should be addressed

impregnated magnets, it has been reported that most of the epoxied coils or magnets demonstrated some degree of degradation [4-6]. This behaviour is mainly attributed to the weak c-axis or transverse direction delamination strength, which is possibly a result of in-plane (ab plane oriented) defect structures [7]. Thermal stresses potentially originating from the different thermal expansion coefficients among different components in a magnet system could trigger this type of delamination. We also recently found that cracks in the epoxy is a major contributor to the performance degradation [6] in REBCO coils. Dry-wound or soft impregnated coils, due to the weak bounding between soft impregnation material and REBCO tape, did not show any degradation, suggesting that the epoxy component has to be compatible with the REBCO tape as the epoxy is strongly bound to the REBCO tape. Typically, epoxy resins have a much larger thermal expansion coefficient than REBCO tapes, and have to be altered to reduce that difference. In the past, we have added different types of powders to make the epoxy component more compatible with the REBCO tape. Regardless of the powder type and size, all the impregnated coils have been degraded [6]. Figure 1 presents the percentage of the critical current, I_c, degradation with respect to different powder types and sizes along with respective weight percentage of the powder content. Epoxy with diamond and alumina fillers offer the least performance degradation although alumina degradation is slightly higher. *n*-value degradations were also provided in the same figure. We also found that reducing the epoxy/powder mixture thickness can also eliminate performance degradation since the impact of the thin epoxy to the REBCO tape is not as severe as the thick epoxy. Keeping the epoxy thickness under control is very difficult in an intricate undulator magnet geometry (Figure 3 and 7). A different procedure is therefore required to prevent the degradation. We developed a vacuum impregnation technique that does not degrade the performance of the REBCO undulator prototype magnets after cold cycles. We investigated wet-wound and vacuum impregnated short prototype undulator magnets. The results showed that the most effective way of preventing the degradation is that of using a bumper layer between the magnet winding stacks and the epoxy. Microstructure images of vacuum impregnated coil packs showed uniformly spaced REBCO winding layers and very thin epoxy fillings between the layers, which is important for the overall performance of the device.

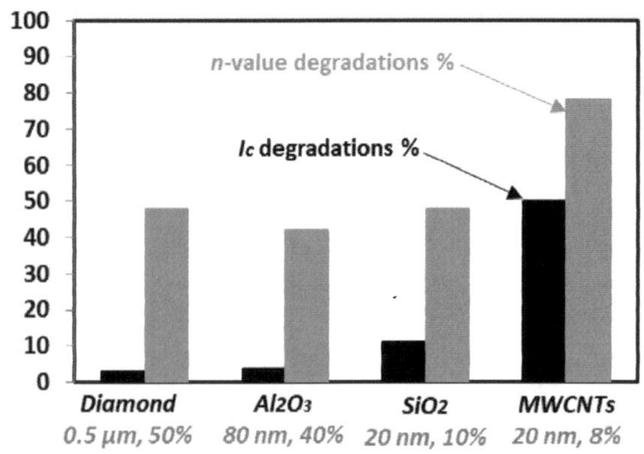

Figure 1. Performance degradations on epoxy impregnated coils with respect to filler types. The weight ratios, filler types, and sizes are indicated in the horizontal axis. The filler ratio is adjusted to keep each epoxy component at the same viscosity level. Epoxy with diamond and alumina fillers offer the least performance degradation.

2. Experimental

2.1. Conductor and impregnation details

The REBCO coated conductors used in this work were fabricated by SuperPower Inc (SP). The conductor architecture consists of buffer layers of alumina, yttria, MgO, and $LaMnO_3$ (LMO), deposited on a non-magnetic electro-polished Hastelloy substrate followed by deposition of ~1 μm thick REBCO by metal organic chemical vapour deposition (MOCVD). The superconducting film was capped by a ~2 μm annealed Ag stabilizer and a 20 μm electroplated copper on both sides. We used Araldite epoxy and Aradur hardener with a mixing ratio of 10:1 in the epoxy impregnation of the prototype undulator magnets. The advantage of using araldite epoxy systems has been discussed in [6]. In order to reduce the CTE, we mixed the epoxy with diamond and Al_2O_3 powders. The bare epoxy has extremely low fracture toughness and it is susceptible to crack formation during thermal cycles. After mixing the epoxy and powders, the mixture was degassed in a vacuum and the air were removed as much as possible. The magnets were then impregnated after adding the Aradur hardener into the mixture.

Table 1. Details of the short prototype undulator magnets prepared for epoxy evaluations.

	Number of racetrack coils	Wet-would (WW) or vacuum impregnated (VI)	Number of layers in a coil	Powder, weight ratio [%], size [μm]
Magnet 1	1	WW	25	Diamond, 40, 0.1
Magnet 2	3	VI	30	Diamond, 20, 0.1
Magnet 3	3	VI	50	Al_2O_3, 15, 0.08

2.2. Prototype undulator windings and epoxying
We fabricated three prototype undulator magnets – two for the vacuum impregnation and one for the wet-winding (see Table 1 for details). Prototype undulator magnet cores for the vacuum impregnation were 3D printed and the wet-wound magnet core was fabricated from a low carbon steel (LCS) (Figure 2b). The width and depth of the winding grooves are 4 and 6 mm, respectively. We use a special scheme to continuously wind the magnets, and details of this are provided in [2]. The magnet structures have racetrack coils with alternating current directions (Figure 7b), thus creating a periodic on-axis magnetic field pattern that is necessary for producing x-ray beam. For winding the magnets, a home-made winder was used. The tape motion and tension were provided by a stepper and a torque motor, respectively. Intermediate spools, as shown in Figure 2a, are inserted into the winding system during the wet winding to immerse the tape into the epoxy/powder mixture. Excess epoxy is removed by spring loaded metal plates. Higher filler ratios can be afforded by this method and the Araldite epoxy mixed with 40 wt. % diamond powders to wet wind the magnets. A picture of the wet-wound coil pack has been provided in Figure 2b.

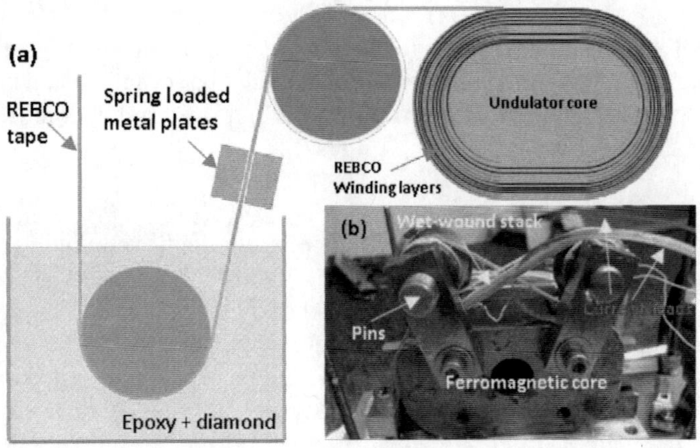

Figure 2. Wet winding schematic (a) and the wet-wound magnet (b). The excess epoxy is removed by the metal plates. The load on the plates are adjusted via springs.

Vacuum impregnation steps are presented in Figure 3. First, the prototype undulator magnet is dry-wound (Figure 3a) and characterized at 77 K. Since the epoxy thickness is not easily controllable in the vacuum impregnation due to the complex shape of the undulator magnet, the magnet winding stacks were decoupled by locating a bumper layer between the epoxy mixture and the REBCO layers (Figure 3b) from the winding stack before the vacuum impregnation. The bumper layer used was a dummy sacrificial REBCO tape in the first prototype and sticky silver tapes in the second prototype. The sticky Ag tape was chosen in the second one because it was easier to apply. This bumper layer (sacrificial REBCO tape or sticky Ag tape) is used to prevent the damage on the REBCO tape by decoupling them from the epoxy. Any kind of crack propagation is absorbed by this outermost bumper layer as illustrated in Figure 3b and the magnet windings are protected. The winding layers also supported by fiberglass sleeves which acts as an extra reinforcement material. An epoxy tight 3D-printed mold was used during the vacuum impregnation (Figure 3c). The pictures of the prototype undulator after disassembling the mold have been presented in Figure 3d. The current leads are visible in the figure. The electron beam side of the undulator magnet has been shown in Figure 3e, revealing the winding stacks that are at the same level with the plastic poles on this side.

2.3. Measurements

A conventional four-probe method was used for the electrical characterization of the magnets. The magnet voltage was recorded at a 24-bit Dynamic Signal Analyser (DSA) module in a PXI chassis. A calibrated current transducer was used to measure the current. All magnets were slowly cooled (~4 hours) and tested at liquid nitrogen temperature (~77K). In addition, repetitive fast cooling cycles were performed. In these fast cooling tests, the magnets were essentially dunked into LN_2 very quickly, followed by a performance measurement.

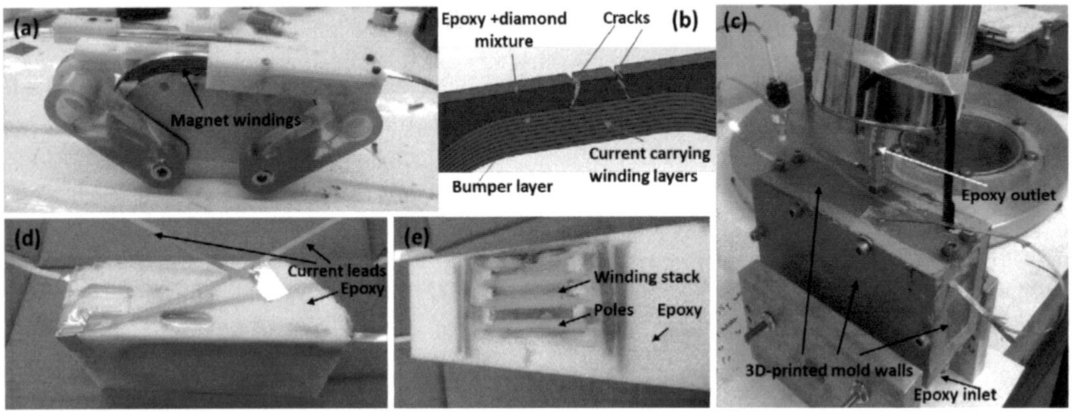

Figure 3. Prototype undulator magnet with a bumper layer (a), a 3D sketch of the bumper layer located between winding stacks and the epoxy/diamond mixture (b) the 3D printed vacuum impregnation mold (c). Magnet after vacuum impregnation and disassembling it from the mold (d) and electron beam side showing the magnet stacks (e).

3. Results and Discussions

Each prototype magnet was tested before and after application of the epoxy through wet-winding or vacuum impregnation, and had its performance evaluated. In Figures 4, 5 and 6, the markers are the measurements and the solid lines are fits to a power–law, $V=V_c(I/Ic)^n$, where V_c is the critical voltage. The power–law fits describe our measurements well. In the interest of simplicity, because the inductive voltage is not the interest here, it was subtracted from the total voltage. This subtraction is evident from Figure 6, as the voltage starts from a negative value (~0.25 mV). Each magnet was wound with a single long tape; however, we used different tapes from different batches to wind different prototypes and, therefore, there are no I_c correlations among the prototype magnets. For example, the I_c of the wet-wound magnet is low even though it has the lowest number of winding layers compared to other magnets because the received I_c of the tape is also low.

3.1. Wet-winding

The thickness of the epoxy can be kept relatively small in the wet-winding with the fixture described (Figure 2a). Two coil packs were first dry-wound with 25 and 5 layers with a transition piece used to continuously wind the magnet. Current and voltage (*IV*) curves of the dry-wound coil were first measured at 77 K, and the magnet was unwound onto a spool. Then the same unwound tape was used to wet-wind the magnet. After a slow cool down, the I_c was measured. Exactly the same I_c, 37.6 A at 1μV/cm criteria value was obtained before and after the wet-winding indicating that there is no performance degradation (Figure 4). In Figure 4, 'BWW' stands for 'before wet-winding' and 'AWW' stands for 'after wet-winding.' Also in the figure, 'SC' and 'FC' stand for slow-cooled and fast-cooled, respectively. Slow cooling denotes that each coil was first exposed to cold nitrogen gas and then slowly immersed into the LN_2 bath. The cool down time for each coil was ~4 hours. Fast cooling denotes that the coil was suddenly immersed into the LN_2. The degradation due to the epoxy impregnation is unpredictable and even a low level of or no I_c degradation does not guarantee that repeated cooling cycles do not result in further degradation. It is therefore important to cool the magnet repetitively and to measure the I_c performance. In this wet-wound magnet, we repeatedly cooled the magnet down to 77 K – the same as the first cool down for the following four consecutive days, and each time exactly the same I_c value was obtained. Furthermore, shock tests were performed to determine if the fast cooling would result in any performance degradation. Even essentially dunking the magnet into the LN_2 did not result in any I_c degradation. n-values were not also affected by the first and repetitive cool downs and the same value was obtained after each cool down. These results indicate that wet-winding and a diamond powder addition eliminate the performance degradation in the short REBCO prototype undulator magnet.

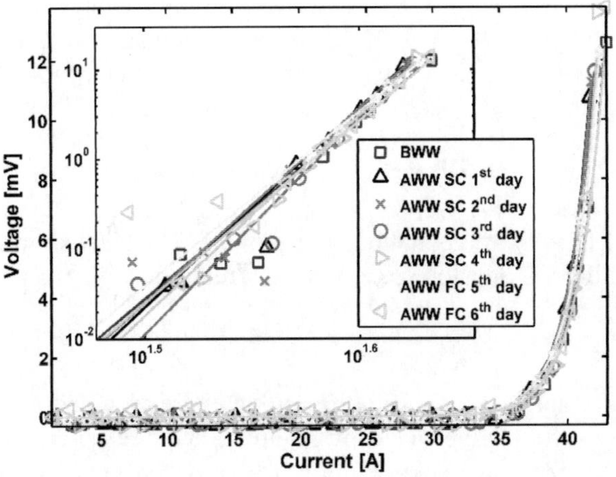

Figure 4. *IV* curves of the wet-wound prototype magnet obtained at 77 K. 'BWW' stands for 'before wet-windings' and 'AWW' stands for 'after wet-winding'. Markers are the measurements and solid lines are the power-law fits. Respective n-values starting from the first legend to last as follows: 21, 22.3, 21.7, 23, 22.62, 22.4 and 20.3. I_c is 37.6 A at 1 μV/cm criterion. The inset shows the log-log scale transition region

3.2. Vacuum impregnation

Two short prototype magnets were prepared for the vacuum impregnations. The magnets were dry-wound and tested at ~77 K. These magnets were then vacuum impregnated using Araldite and Aradur epoxy systems mixed with 20 wt. % diamond powder in the first short prototype and 15 wt. % Al_2O_3 powder in the second short prototype. The reason these two powders were chosen is because these powders showed the least performance degradation in our previous study [6]. In Figures 5 and 6, 'BI' stands for 'before impregnation' and 'AI' stands for 'after impregnation.' Again, the markers in the figures are the measurements and the solid lines are fits to the power–law.

The *IV* curves for the first vacuum impregnated magnet epoxy filled with diamond powders are provided in Figure 5. The I_c values and n-values are almost the same before and after the vacuum impregnation. Even after two-day shock tests, the I_c values are unchanged. The small fluctuations in the n-values are within the tolerance of the measurement system. These results prove that the bumper layer effectively absorbs the energy produced by the cracks and does not transfer it to the REBCO tapes. In addition, the bumper layer also prevents any possible delamination as it physically detaches the REBCO conductors from the surrounding epoxy compound. The second prototype coil was impregnated using Al_2O_3 fillers. The results showed some fluctuations – slightly more than the first one (diamond loaded epoxy used) – in the I_c and n-values (Figure 6). The n-value increased from 16 to 17.4 after one slow and two fast cool downs, which is still slightly lower than the value before the vacuum impregnation. I_c values of the magnet after the epoxy impregnation increased about 1.5 A. These fluctuations could be

IOP Conf. Series: Materials Science and Engineering **279** (2017) 012009 doi:10.1088/1757-899X/279/1/012009

due to the noise or very small degradation resulted using the alumina fillers. The insets in Figures 5 and 6 are the log-log scale of the transition regions.

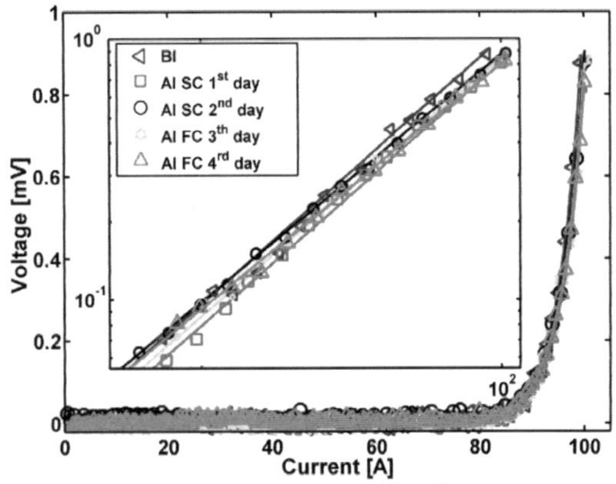

Figure 5. *IV* curves of first vacuum impregnated prototype undulator magnet at 77K. Diamond powders used as a filler. 'BI' stands for 'before impregnation' and 'AI' stands for 'after impregnation.' Markers are the measurements and solid lines are the power-law fits. Respective n-values starting from the first legend to last as follows: 22.3, 22, 20.8, 21.7 and 20.6. I_c is about 100.2 A and the same after each additional cold cycle. The inset shows the log-log scale transition region.

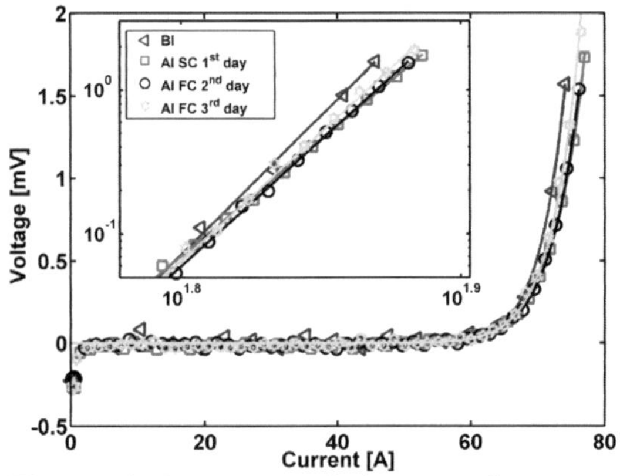

Figure 6. *IV* curves of the second vacuum impregnated prototype undulator magnet at 77K. Al_2O_3 powders used as a filler. Markers are the measurements and solid lines are the power-law fits. Respective n-values starting from the first legend to last as follows: 18.4, 16, 16.1, 17.4. I_c is 74 and 75.5 A before and after impregnation, respectively. The inset shows the log-log scale transition region.

Microstructures of the vacuum impregnated and wet-wound short prototypes are provided in Figure 7. A sketch of the undulator magnet and cross-section are provided in Figure 7a and b respectively. The polarity of the current is displayed inside the winding stacks. Figure 7c is a cross sectional optical microscope image of the vacuum impregnated magnet, showing the REBCO tapes and epoxy between the layers. The layers are almost equally-spaced and there is very little epoxy build-up between the layers. Figure 7d is an SEM image of the cross-section of the vacuum impregnated magnet. Cu stabilizer, epoxy, Ag protective, Hastelloy substrate, and REBCO superconducting layers are visible from the figure. The thickness of the epoxy layer is about 5 μm. Figure 7e is the cross-section of the wet-wound coil. The epoxy thickness between the different layers is highly irregular and changes drastically. The average thickness of the epoxy layer between the REBCO layers in wet wound magnet is about 10 times larger than the vacuum impregnated magnet (Figure 7f). One important difference between the vacuum impregnated and wet-wound magnet is that the vacuum impregnated magnet does not have any diamond powder between the REBCO layers. This can easily be seen from the SEM image (Figure 7d). The epoxied region does not have any contrast variation, which is an indication of the diamond powder in the epoxy. This is because such regions are so tiny and diamond powders accumulate, act as a filter, and do not allow the powders to pass. Only epoxy can squeeze into those places and access to those remote portions of the magnet. However, it is evident from Figure 7f that diamond powders are presented in the wet-wound magnet. The diamond powder is agglomerated inside the epoxy matrix (Figure 7f). Such agglomerations could reduce the effectiveness of the powder fillers and promote formations of the micro-cracks. The source of these agglomerations are not known at this time, but could be due to the poor mixing of the epoxy and powders. Perhaps more sophisticated powder mixing methods need to be employed in order to fully disperse the nano-powder inside the epoxy matrix.

IOP Conf. Series: Materials Science and Engineering **279** (2017) 012009 doi:10.1088/1757-899X/279/1/012009

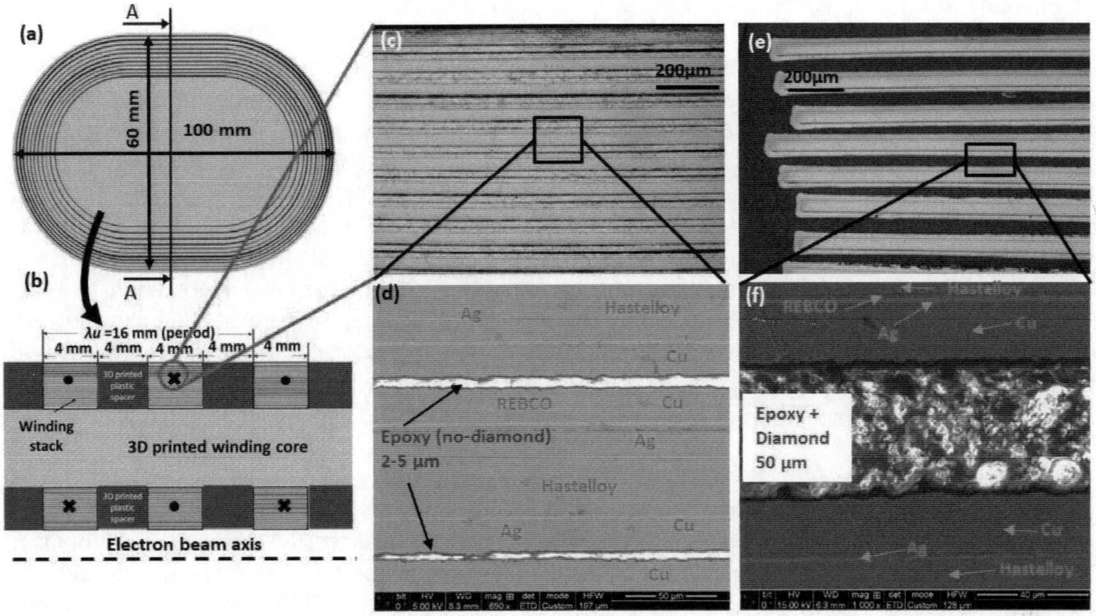

Figure 7. Cross-section SEM images of the vacuum impregnated and wet-wound coils. Sketch of the vacuum impregnated magnet showing dimensions of the prototype magnets with a period of 16 mm (a) and the cross-section (b). Optical microscope image after cross-sectional polishing (c) and an SEM image showing all the layers and the ~5 microns thick epoxy layer. An optical microscope image of the cross-section of the wet-wound magnet (e) and a SEM image providing more details about the thickness of the epoxy layer in the wet-winding technique.

The epoxy thicknesses along the width of the winding stack after the wet-winding at four different positions are provided in Figure 8. It seems that the thickness not only drastically varies between different layers, but also in the each individual layer due to the inclined tape winding. This happens as a result of the non-uniform epoxy build-up in different regions during the wet-winding. Such a thickness variation can be more than 15 μm and accumulated thickness irregularity can be detrimental to the field quality.

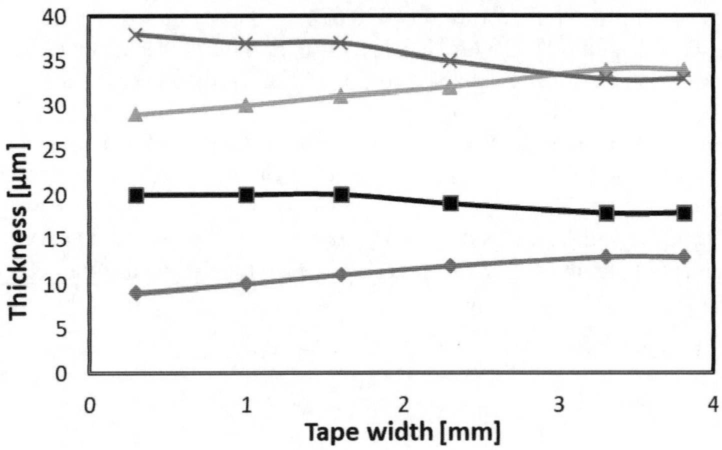

Figure 8. Epoxy layer thickness along the width of the tape at four different gaps.

IOP Conf. Series: Materials Science and Engineering **279** (2017) 012009 doi:10.1088/1757-899X/279/1/012009

4. Conclusions

We evaluated vacuum impregnation and wet-winding techniques using short prototype undulator magnets and both approaches offered no degradation to the prototype performances. The wet-winding process has been found to be time consuming and difficult to implement, especially for the long undulator magnets. Introduction of a bumper layer has eliminated the performance degradations in the vacuum impregnated magnet when diamond fillers used. Vacuum impregnation is a relatively quick process, although preparation might take a longer amount of time. The actual epoxy drawing time is short: An hour at most for a 1.5 m undulator magnet. In addition, additional attention during the process is not required as wet-winding requires constant surveillance – guiding the tape, and making sure that the thick epoxy build-ups are avoided. Finally, the microstructures that were evaluated showed very thin, homogenously spaced epoxy layer between the REBCO layers, which is crucial for the overall performance. For these given reasons, the vacuum impregnation technique with a bumper layer is suitable for vacuum impregnations of the REBCO undulator coils.

5. Acknowledgments

The authors thank Kurt Boerste from the Accelerator Systems Division at Argonne National Laboratory for the technical help. IK acknowledges support from Laboratory Directed Research and Development (LDRD), and funding from Argonne National Laboratory, provided by the Director, Office of Science of the US Department of Energy under Contract No. DE-AC02-06CH11357 for purchase of HTS wires, magnet fabrications, winding and testing, the preparation of publications and presentations. UW acknowledges the support from the US Department of Energy (DOE), Office of Basic Energy Sciences, as part of the Center for Emergent Superconductivity Energy Frontier Research Center.

6. References

[1] Kesgin I, Doose C L, Kasa M T, Ivanyushenkov Y, Zhang Y, Knoll A, Brownsey P, Hazelton D and Welp U 2016 Feasibility and electromagnetic analysis of a REBCO superconducting undulator *Supercond. Sci. Technol.* **29** 055001

[2] Kesgin I, Kasa M, Ivanyushenkov Y and Welp U 2017 High-temperature superconducting undulator magnets *Supercond. Sci. Technol.* **30** 04LT1

[3] Kesgin I, Doose C L, Kasa M T, Ivanyushenkov Y and Welp U 2015 Design of a REBCO thin film superconducting undulator *IOP Conf. Ser. Mater. Sci. Eng.* **101** 012053

[4] Takematsu T, Hu R, Takao T, Yanagisawa Y, Nakagome H, Uglietti D, Kiyoshi T, Takahashi M and Maeda H 2010 Degradation of the performance of a YBCO-coated conductor double pancake coil due to epoxy impregnation *Physica C* **470** 674-7

[5] Yanagisawa Y, Nakagome H, Takematsu T, Takao T, Sato N, Takahashi M and Maeda H 2011 Remarkable weakness against cleavage stress for YBCO-coated conductors and its effect on the YBCO coil performance *Physica C* **471** 480-5

[6] Kesgin I, Hasse Q, Ivanyushenkov Y and Welp U 2017 Performance of 2G-HTS REBCO Undulator Coils Impregnated Epoxies Mixed With Different Fillers *IEEE Trans. Appl. Supercond.* **27** 1-4

[7] Kesgin I, Khatri N, Liu Y, Delgado L, Galstyan E and Selvamanickam V 2016 Influence of superconductor film composition on adhesion strength of coated conductors *Supercond. Sci. Technol.* **29** 015003

IOP Conf. Series: Materials Science and Engineering **279** (2017) 012010 doi:10.1088/1757-899X/279/1/012010

Efficacy of Tantalum Tungsten Alloys for Diffusion Barrier Applications

D B Smathers and P R Aimone

H. C. Starck, Inc. 45 Industrial Place, Newton, MA 02461 USA

E-mail: david.smathers@hcstarck.com, paul.aimone@hcstarck.com

Abstract. Traditionally either Niobium, Tantalum or a combination of both have been used as diffusion barriers in Nb_3Sn Multi-filament wire. Vanadium has also been used successfully but the ultimate RRR of the copper is limited unless an external shell of Niobium is included. Niobium is preferred over Tantalum when alternating current losses are not an issue as the Niobium will react to form Nb_3Sn. Pure Tantalum tends to deform irregularly requiring extra starting thickness to ensure good barrier qualities. Our evaluations showed Tantalum lightly alloyed with 3 wt% Tungsten is compatible with the wire drawing process while deforming as well as or better than pure Niobium. Ta3wt%W has been processed as a single barrier and as a distributed barrier to fine dimensions. In addition, the higher modulus and strength of the Tantalum Tungsten alloy improves the overall tensile properties of the wire.

1. Introduction

The progression of technology from discovery to multi-filamentary wire was very rapid for Nb_3Sn. The history of Nb_3Sn and the need for finely divided filaments is very nicely summarized in two recent theses[1,2]. The brittle nature of Nb_3Sn required the clever invention of creating the multi-filamentary structure first and then converting ductile niobium into Nb_3Sn by solid state diffusion of tin through a copper matrix. The quality of the Nb_3Sn filaments can be very good, especially when very high fractions of niobium can be incorporated and an excess of tin provided. A need arose at the start to prevent the excess tin from contaminating the parallel, stabilizing copper conductor. A barrier between the highly active tin and the high purity copper was needed to prevent diffusion of tin through the barrier layer during the reaction heat treatment. Tantalum was an early choice and theoretically more ideally suited for the purpose than niobium.[3,4] Tantalum has a lower critical temperature than niobium and any reaction between tantalum and tin to form Ta_3Sn also results is a low critical temperature. Tin diffusion through tantalum is slower than through niobium due to the differences in melting temperature. However, tantalum did not co-process as well as pure niobium. Some solutions were found to improve the formability of pure tantalum and it is primarily incorporated when low a.c. losses are required in the reacted conductor[5].

One of the issues often seen with pure tantalum is wall thickness instability of the barrier layer[6,7]. The consequence of this instability is a tendency to use thicker tantalum in the barrier layer to be sure a minimum thickness is available in all locations. The excess tantalum is neither stabilizer nor superconductor and adds undesired cost to the wire. Complicated or expensive alternatives include pure vanadium [8], a combination of vanadium and niobium [9], a combination of tantalum and niobium [10] or a Ta40Nb alloy[11].

In this work, we study the potential of using tantalum lightly alloyed with 3 weight percent tungsten (Ta3W) as a diffusion barrier material in place of pure tantalum. We initially studied this alloy as a wire reinforcement material [12,13]and found the stability of the copper-tantalum interface is superior for the Ta3W alloy. The improved interfacial stability reduces the need for excess thickness resulting in an improved cost model. The primary benefit was higher wire mechanical properties with the Ta3W alloy as either a parallel filament or a diffusion barrier. Tubular Ta3W was successfully co-processed with a multi-filamentary copper niobium and copper niobium tin composite and tested for the efficacy of the diffusion barrier properties. Since Ta3W is a commercial alloy in regular production the comparative cost to pure tantalum is quite good. H. C. Starck produces and sells this alloy under the trade mark NRC76™.

2. Wire Construction

The wire process used to make test specimens was previously reported[12]. The process was improved by increasing the copper to niobium local ratio and improving the bonding process using low temperature heat treatments and limiting consolidation to drawing[13]. The earlier trials were consolidated by swaging the assembly full length.

To test the process capability of a tantalum tungsten diffusion barrier, a seam welded Ta3W tube (12.7mm diameter by 0.38mm wall) was added to a second stage assembly between a 19 x 19 (2.29mm f-f) stack and a "stabilizer" copper tube (16.5mm x 1.5mm wall). This assembly was cold drawn to 0.72mm as a mono-barrier assembly, figure 1 (a). Some of this construction was converted to 3.05mm hexagonal element from 3.66mm diameter wire and re-stacked as an 18 element in a similar pure copper tube (18.3nn outer diameter by 1.5mm wall). The two stage assembly was cold drawn to 0.72mm as a distributed barrier composite. These wires only evaluated the ability of the Ta3W alloy to co-process with the copper-niobium as no tin elements were incorporated. An unexpected result was the survivability of the seam weld microstructure through ~500:1 area reduction. Figure 1 shows the wire cross section of the mono barrier assembly and distributed barrier assemblies at 0.72mm diameter. While the mono-barrier wire shows a continuous barrier, figure 1a, the distributed barrier shows some failure, usually near the weld area. There was no filler between the 19 element hexagonal geometry and the round tube. The Ta3W conformed to the stack but this extra deformation may have introduced some unintentional strain into the elements leading to barrier failure, figure 1 (b).

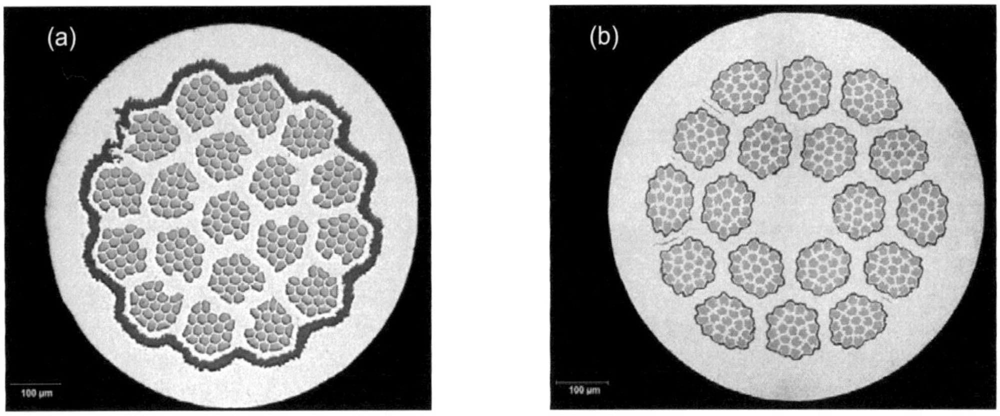

Figure 1. (a) Mono barrier wire with copper-niobium elements and (b) distributed barrier version wires at 0.72mm diameter. Optical micrographs.

The success of the early experiment lead to the desire to test the Ta3W alloy as a real diffusion barrier. Tin-titanium elements were provided by Luvata [15]. The only wire assembly method available for this experiment was stacking hexagonal elements in round tubes. Providing a large tin core to a filament array is often done by gun drilling an extruded assembly. For this experiment, all the hexagonal elements needed to be the same diameter. A 37 element stack was chosen with the 7 element core being copper clad tin-titanium rods. This non-optimum choice would lead to issues in the wire drawing. To make the 37 element stack more circular, the 6 peak elements were drawn to a smaller diameter. 31 elements (7 tin-titanium and 24 copper-niobium) were drawn to 1.57 mm face to face hexagonal and 6 elements (copper niobium) to 1.29 mm face to face. In this assembly, an overlapped rather than welded tube, 0.38mm thick by 12.7mm diameter, was used. Figure 2 shows the mono-barrier version at 0.72mm diameter. The distributed barrier version was not drawn below 2.6mm diameter due to excessive breakage. The breakage was a result of stacking faults in the assembly rather than the diffusion barrier material. During the assembly process, some of the elements shifted from their ideal position leading to the 7 tin cores being inadequately contained, figure 2.

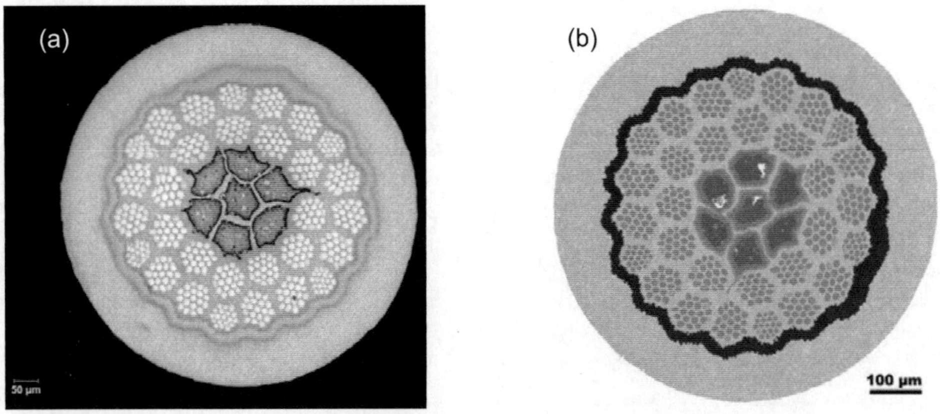

Figure 2. Mono barrier wire with copper-niobium and tin-titanium elements at 0.72mm diameter. (a) As drawn, optical image. (b) 210°C/72hr, SEM image [inverted contrast]

The 0.72mm mono-barrier was drawn with sufficient length to run a reaction heat treatment to test the barrier quality. At 0.72mm, the Ta3W barrier thickness is ~0.02 mm and the niobium filaments are ~0.015 mm. The niobium filament size is much larger than would be appropriate for a true Nb-Sn conductor. The goal of the experiment is to test fabrication and efficacy, not make a competitive superconductor wire.

3. Results

3.1 Wire Processing

The general fabrication of the wires went very well. The experimental design choices that led to issues at small diameters can be easily overcome by more experienced practitioners. The key parameter of interest was whether the Ta3W co-processes with the pure copper and niobium elements. The wall thickness and area uniformity of the Ta3W material stayed consistent after the wire was fully dense. The Cu-Ta3W interface is much more uniform than a conventional Cu-Ta interface.

Even though the Ta3W hardness is high compared to the niobium and especially copper, it still processes consistently. At low strains, typical of wire drawing passes, the small difference in flow stress between tantalum and Ta3W suggests they process equivalently[14]. The welded tube survived a

500:1 area reduction. The overlapped tube design in the wire that incorporated the tin elements also processed well through 500:1 area reduction. The overlap introduced some sharp edges and the wall thickness is doubled leading to local stiffening that could impact the internal stresses during wire fabrication. A slightly worked weld could make the welded tube design much more practical.

3.2 Mechanical Properties

The increased mechanical strength of Nb-Cu wires with the substitution of Ta3W has been previously shown[12]. In the previous work the strengths were shown with a reported 700°C heat treatment. A correction was submitted to properly report the heat treatment as 8 hours at 250°C. When the wires were heat treated for 3 hours at 700°C, the wire without tantalum reinforcement softens much more substantially while the wire with Ta3W barriers maintains more of its strength[13], figure 3. This is true when the reinforcement Ta3W element is a large filament, a tube or diffusion barrier, whether single or distributed.

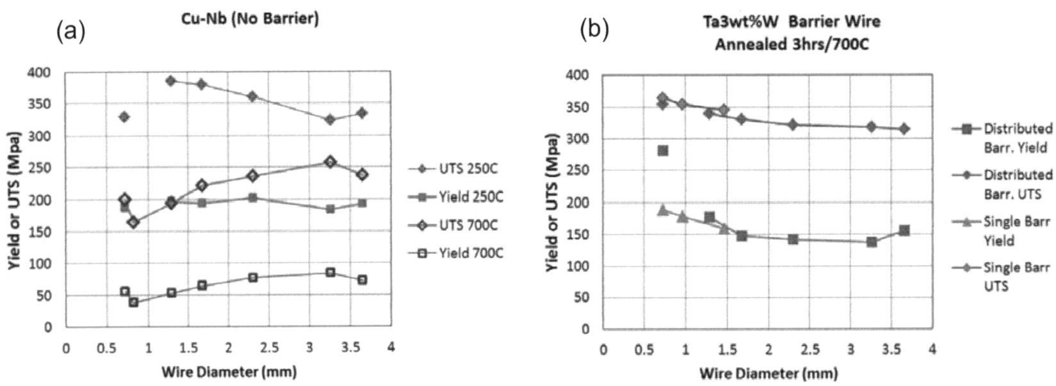

Figure 3. Mechanical properties for Cu-Nb multi-filamentary wires without (a) and with (b) Ta3W elements. A wire with a Ta3W element maintains more strength at high reaction temperature as compared to a wire without a Ta3W element. Heat treated 8 hours at 250°C or 3 hours at 700°C.

3.3 Diffusion Heat Treatment and RRR

Heat treatment and RRR measurements were performed at the Applied Superconductivity Center at Florida State University. Two heat treatments were run on the wire prior to testing the residual resistivity ratio (RRR) to determine the value before and after tin had the potential to penetrate the barrier. The first heat treatment was at 210°C for 72 hours and was meant to simply anneal the copper without stressing the diffusion barrier since only short range solid state diffusion was possible. The second heat treatment was representative of a real superconductor wire, 210°C for 72 hours followed by 400°C for 48 hours and then 640°C for 48 hours. The RRR was measured using a 4-point probe with a current of 0.5 amp. The reacted sample was observed to begin the superconducting transition at about 17.2 K. The resistance values were compared at 300K and 19K. After the 210°C/72 hour treatment, the wire RRR was 113 +/- 3. Following the reaction heat treatment, the RRR increased to 247 +/- 7. The wire geometry may be playing a role. The un-reacted wire has more copper to participate in the room temperature conductivity. When the mean free path increases as the temperature is reduced the copper between the niobium filaments may contribute less to the conductivity. The fine filament structure effectively alters the amount of copper in the measurement between 300K and 19K. At 19 K, the wire resistance is about the same but at 300K, the conductivity of the un-reacted wire is lower than the reacted wire. As can be seen in figure 2, the grain size of the stabilizing copper is much smaller than after the full reaction heat treatment, figure 4. Nevertheless, the Ta3W barrier has withstood the tin activity and preserved the copper purity.

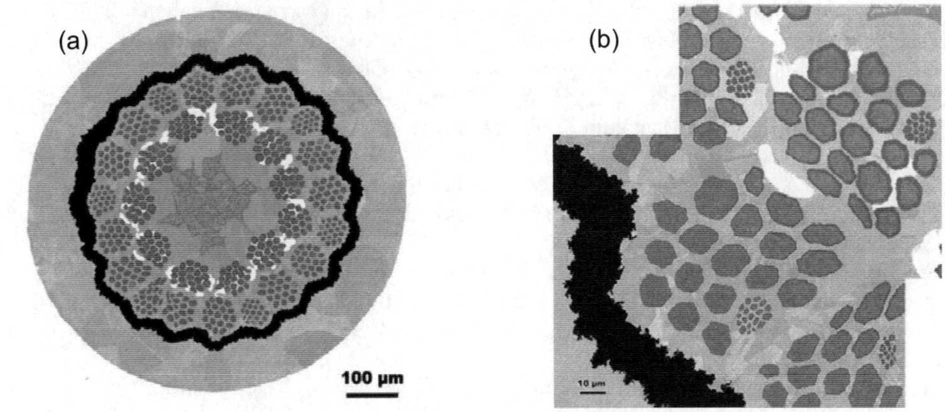

Figure 4. Scanning electron microscope images of reacted 0.72mm diameter wire. Full cross section (a) and higher magnification image (b) shows the bronze formation near the barrier and reaction of the smaller niobium filaments. The outer pure copper shell has a RRR of 247 +/-7 and large grains. Voids appear white due to inverting the SEM image contrast to compare more easily with the optical images. The large niobium filaments show little reaction near the barrier but the small niobium filaments in the element keys are almost completely reacted.

4. Discussion

Ta3W has great potential to replace pure Ta as a single layer diffusion barrier option due to its better interfacial stability during wire drawing and the added strength and temperature stability provided in the reacted wire product. A model wire incorporating a singleTa3W diffusion barrier has been processed with reactive tin cores through more than 500:1 area reduction and survived a standard reaction heat treatment while preserving the purity of the outer copper stabilizer. Experienced practitioners can incorporate the benefits of Ta3W diffusion barriers into future low a.c. loss wire designs. The high survivability of a weld microstructure during co-processing could also make a continuous tube forming concept[16] more productive.

Acknowledgments
Measurements at Florida State University were supported by the Department of Energy under Award Number DE-SC0012083 and were performed at the National High Magnetic Field Laboratory, which is supported by National Science Foundation Cooperative Agreement No. DMR-1157490 and the State of Florida. Many thanks to William Starch and Brandon Chew for assistance in wire drawing, Christopher Segal for RRR measurements, Shreyas Balachandran and Peter Lee for analysis and discussion.
Optical microscopy was performed under contract with Paul Danielson at eImaging in Bend Oregon. Special thanks to Tae Pyon at Luvata Special Products for the supply of tin-titanium rod.

References

[1] Sanabria C, 2017 *A new understanding of the heat treatment of Nb-Sn superconducting wires,* PhD thesis, University of Florida-Tallahasee

[2] Godeke A, 2005 *Performance boundaries in Nb3Sn superconductors*, PhD thesis, University of Twente, Enschede, The Netherlands

IOP Conf. Series: Materials Science and Engineering **279** (2017) 012010 doi:10.1088/1757-899X/279/1/012010

[3] McDougall I, 1975 Method of fabricating a composite intermetallic-type superconductor, US Patent 3,876,473

[4] Young M and Larbalestier D, 1981 Wrapped tantalum diffusion barrier, US Patent 4,285,740

[5] Nijhuis A, Knoopers H and ten Kate H, 1994 The influence of the diffusion barrier on the AC loss of Nb3Sn superconductors, *Cryogenics* 34 (1), 547-550

[6] Hartwig K, Balachnadran S, Mezyenski R, Seymore N, Robinson J and Barber R, 2014 Fabrication of fine-grain tantalum diffusion barrier tube for Nb3Sn conductors, *Adv. in Cryogenic Eng.*, vol **60** 1574: 204-210

[7] Balachandran S, Hartwig K, Baars D, Mathaudhu S, Bieler T, Pyon T and Barber R, 2009 Fabrication of tantalum sheet for superconductor diffusion barriers, *IEEE Trans on Appl Superconductivity* **19** (3) 2606-2609

[8] Smathers D, O'Larey P, Siddall M and Peterson R, Characterization of vanadium diffusion barriers in Nb-Sn composite wires, 1987 *IEEE Trans. Mag*, Vol **MAG-23**, no. 2, 1347-1350

[9] Smathers D, O'Larey P, Siddall M and McDonald W, 1988 Status of the superconductor development program at Teledyne Wah Chang Albany, *Adv. in Cryogenic Eng.*, vol **34** 515-522

[10] Hartwig K, Balachandran S, Barber R, Pyon T and Griffin R, 2011 Design aspects of dual Nb-Ta sheet diffusion barriers for Nb3Sn conductors, *IEEE Trans. Appl. Superconductivity*, **21**(3): 2563-2566 Part 3

[11] McKinnell J, O'Larey P, Jablonski P and Siddall M, 1997 Tantalum-niobium diffusion barriers for superconducting Nb3Sn wires in fusion applications, *Adv. in Cryogenic Matls*, Vol **42**, 1415-1422

[12] Smathers D, Aimone P, 2017 Comparison of Ta3wt%W and Pure Ta as a Wire Reinforcement Element in a Multifilament Composite, *IEEE Trans. on Appl. Superconductivity* Vol **27** (4)

[13] Smathers D, Aimone P, 2017 Replacement of Ta Elements with Ta-W Elements in Nb3Sn Superconductors, to be published in: *Proc. 19th Plansee Seminar* (eds. L.S. Sigl et al., Reutte 2017).

[14] Weiss K, Nyilas A, Thoener M and Seeber B, Proof Strength behaviour of Nb3Sn bronze route wires between 300 and 4K, *Adv. in Cryogenic Eng.* Vol **43**: Transactions of the international Cryogenic Materials Conference Volume AIP Conference Proceedings 986: 293 2008

[15] Pyon T, Luvata Special Products, Waterbury, CT, private communication 2017.

[16] Fu X, Gregory E, McIntyre P, Smith D and Tomsic M, 2004, Continuous tube forming of the Ta diffusion barrier and Cu stabilizer on Internal-Nb3Sn subelement, *AIP Conf. Proc.* **711**, 382

IOP Conf. Series: Materials Science and Engineering **279** (2017) 012011 doi:10.1088/1757-899X/279/1/012011

Cryogenic power electronics at megawatt-scale using a new type of press-pack IGBT

L Graber[1], M Saeedifard[1], M J Mauger[1], Q Yang[1], C Park[1], T Niebur[1], S V Pamidi[2,3], S Steinhoff[4]

[1]School of Electrical and Computer Engineering, Georgia Institute of Technology, Atlanta, USA

[2]Department of Electrical and Computer Engineering, FAMU-FSU College of Engineering, Florida State University, Tallahassee, USA

[3]Center for Advanced Power Systems, Florida State University, Tallahassee, USA

[4]IXYS UK Westcode Ltd., Chippenham, UK

lukas.graber@ece.gatech.edu

Abstract. Electrical characteristics of a press-pack IGBT stack have been investigated experimentally in both static and dynamic tests at room temperature and at 77 K in an effort to identify a megawatt-scale class semiconductor that performs well electrically, thermally, and mechanically. It has been observed that the forward voltage drop and the turn-off time are both noticeably reduced at cryogenic temperatures, suggesting significant improvement in the device operating area. Furthermore, the stack was subjected to repeated thermal cycles between room temperature and 77 K without observing any degradation. A circuit simulation of a 200-kW boost converter based on the characterized press-back IGBT stack exhibited a reduction in switching and conduction losses at 77 K by 60% and 24%, respectively compared to the values at room temperature. It is expected that the type of press-pack IGBT stacks tested will be key elements to build high power dense, megawatt-scale cryogenic converters for future all-electric ships and electric airplanes.

1. Introduction

It may appear counterintuitive to operate a device that produces large amounts of heat, such as power electronic devices (PEDs), at cryogenic temperatures. The motivation behind developing cryogenic PEDs is to increase the power density, system wide efficiency, and enhance the reliability of the PEDs. The electric power distribution system on future all-electric ships and aircrafts is expected to operate at cryogenic temperatures to allow the use of high power dense superconducting motors, generators, and cables [1, 2]. For efficiency reasons, the power distribution system is likely to be based on medium voltage direct current (MVDC) topology. PEDs are required between AC devices like generators and motors and the DC bus (Fig. 1 (a)). The traditional warm PEDs would introduce a substantial heat load to the cryogenic system and require specialized interface components between the conductors at cryogenic temperature and the conductors at room temperature (RT) [3].

A closed loop cryogenic cooling system with a centrally located cryorefrigerator would be the most versatile and efficient system to serve multiple devices with diverse cryogenic operating temperatures of 30-70 K. In such a system, lowest temperature cryogenic fluid will probably be tapped for devices that benefit from operating at lowest possible temperatures. High temperature superconducting (HTS) generators and motors are examples of devices that are likely to operate at the low end of the available temperature window. HTS cables possess substantial power densities even in the warm end of the temperature window of 70 K. Based on specific application and system design, it might still be beneficial to operate HTS cables at 50-60 K to reap the benefits of high power densities or to reduce the amount of superconductor needed. (50 K in Fig. 1(b)). The cable terminations that serve as interfaces between the cable and the rest of the power system typically have a significant temperature gradient from the cable operating temperature of 50 K to the outer end of the current lead/bushing as high as 80 K. This would allow the PEDs that are in between the HTS cable system and the other devices to operate at 80 K. Since there is no liquid cryogen with such a temperature range, the proposed cooling loop would need to be based on helium gas, hydrogen gas, or sealed liquid nitrogen bath that is conductively cooled by the circulation loop.

Operating PEDs at cryogenic temperature can have other advantages besides reducing the heat influx into the cryogenic system: The reduction of losses of many of the components in the PEDs, the high heat transfer capabilities, and the high dielectric strength of the cryogenic fluid are expected to substantially increase the power density. Furthermore, the lower component temperature and reduced temperature cycles are expected to further increase the reliability [4].

In certain applications such as electric aircraft, the cryogen could also act as the fuel [2]. In such a case, the system could be operated as "open-loop" with the liquid cryogenic fuel (e.g. liquid hydrogen, liquid methane, or liquid natural gas) being pumped from the fuel tank to the power system devices and from there to the gas turbines or fuel cells.

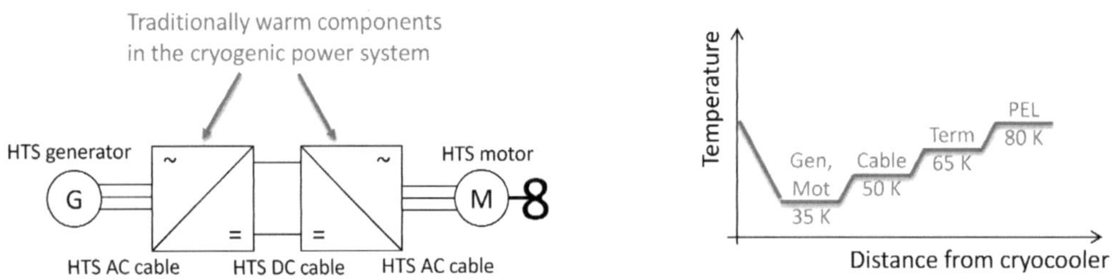

(a) (b)

Fig. 1: Block diagram showing the power electronic devices between a HTS generator and a HTS motor (a) and the temperature profile of a proposed cryogenic cooling loop (b).

2. Power Electronic Components at Cryogenic Temperature

The power electronic components of PEDs include semiconductors, capacitors, inductors, transformers, resistors, and varistors. Other components, which are not in the high power path, are not discussed here since they are considered to stay at room temperature (RT). These include the controller, parts of the current and voltage transducers, as well as the gate drivers. Current transducers might be based on transformer principle (AC only) or compensation current probes like Rogowski coils or Hall probe based transducers (mixed AC-DC) with a cryogenic primary and a RT secondary. The design of the converter will include a well-insulated cryogenic section (cryostat) and a RT section.

2.1. Semiconductors

Power semiconductors are at the heart of any power electronic converter and their operation and characterization at cryogenic temperatures have been the main focus of the ongoing research effort on cryogenic power electronics for the past 20 years [3]. In general, many silicon-based power devices show significant static and dynamic performance improvements at cryogenic temperatures, along with an increased reliability due to lower junction temperatures and an increased integration density thanks to the higher thermal conductivity of packaging material and silicon. The reduced heat capacity at cryogenic temperatures will need to be taken into account, though.

Bipolar controllable devices such as power bipolar junction transistors (BJTs) and thyristors have been among the first power devices to be studied at cryogenic temperature. Thanks to the reduced carrier lifetime at cryogenic temperatures, an order of magnitude reduction in the turn-off time with a decrease in temperature from 300 K to 77 K has been reported for power BJTs [5]. An increase in the open base collector-emitter breakdown voltage by about 20% has also been observed. However, because of a reduced emitter injection efficiency and a decreased electron diffusion length in the base, a large reduction in the current gain by more than an order of magnitude is observed at cryogenic temperatures, which causes a serious problem in controlling these devices. Furthermore, a slight increase in the forward voltage drop is observed because of carrier freeze-out and increased emitter current crowding. Although an optimization of the BJT structure would be possible to overcome the current gain limitation at cryogenic temperatures as proposed by Singh *et al.* [5], the operation of commercially available power bipolar devices at these temperatures is not practical.

Field effect transistors such as MOSFETs do not suffer from the reduced controllability at cryogenic temperatures and display a lower on-state voltage drop and better switching characteristics [6–8]. However, the currently available voltage and current ratings in commercial devices do not meet the power level requirements of the envisioned applications that could benefit from power converters operating at cryogenic temperatures. Initial experiments with silicon carbide (SiC) based devices, which are expected to enable higher voltage and power ratings than their silicon counterparts, did not perform well at cryogenic temperature [9]. On the contrary, emerging AlGaN/GaN HFETs are promising for high power cryogenic applications and will need to be considered for detailed studied as they become commercially available [10].

IGBTs rely on the same control principle as FETs, therefore offering full controllability at cryogenic temperatures, are available in the medium power range with devices rated up to 6.5 kV and 3 kA, and as such constitute the most promising candidate for the target applications. Several studies [11–15] have shown a large reduction in the tail current owing to reduced carrier lifetime at cryogenic temperatures, which could be a motivation to increase the switching frequency and/or the current rating. A decrease of the on-state voltage has also been observed thanks to the large increase in the inversion layer carrier mobility that overbalances the increase of the knee forward voltage intrinsic PN junction due to carrier freeze-out. As for any FET device, an increase in the gate threshold voltage due to carrier freeze out at lower temperature has been reported [11–15].

Although low power IGBT devices have been extensively studied, no experimental results on medium power IGBTs are available. Furthermore, very little attention has been paid to the device packaging and the mechanical constraints arising from the coefficient of thermal expansion (CTE) mismatch of materials, potentially damaging the device during the thermal cycles. In a conventional power module, several potential weaknesses for cryogenic operation, including repeated thermal cycling between room temperature and cryogenic temperatures, can be anticipated: the solder joints between the chips and substrates; the joint between the substrates and the baseplate; the busbar connections to the substrates (soldered or welded); the interface between the gel (or other potting materials) and the chip termination coatings; and, maybe most critical, the interaction of the gel (and other potting materials) with the wire bonds. Most promising instead seem to be pure pressure contact devices with packages that are based on metals and ceramics only, and which construction avoids soldered joints, wire bonds and gel potting altogether. Press-pack devices such as high power diodes and thyristors fall into this category. However, only recently a manufacturer (IXYS UK Westcode) started packaging IGBTs in traditional press-pack modules for megawatt-scale, high reliability applications [16, 17]. These devices combine the most suitable semiconductor type with the most suitable package, making them a good choice for megawatt-scale cryogenic power applications.

2.2. Inductors and Transformers
The design of power electronic devices typically include inductors and sometimes transformers. These typically consist of copper or aluminum windings on a magnetically soft core. The core increases the inductance of the inductors and improves the coupling between the windings in transformers. An ideal core has high magnetic permeability over the desired frequency range, a high saturation flux density, low hysteresis, and low electric conductivity. The low conductivity is important to minimize eddy currents that cause losses in the core.

There are two common materials used for the core: laminated transformer steel and compacted metal powder. Laminated steel cores are common in high power devices, which typically operate at switching frequencies below 15 kHz. Modern grain oriented silicon steel is an anisotropic material, which offers high permeability at comparatively low electrical conductivity. In certain designs, they incorporate an air gap, which avoids core saturation in certain types of converters such as flyback converters, which use the core to store energy between switching cycles. Powder cores are more often used in lower power devices, which operate at higher switching frequencies typically above 100 kHz. Powder cores offer high magnetic permeability, a so-called distributed air gap, and low conductivity, minimizing eddy currents without the need for lamination. There is a wide range of sintered magnetic materials used for the fabrication of powder cores including iron, carbonyl iron, nickel-iron, and molypermalloy.

Neither of these core types are expected to be suitable for cryogenic use. Transformer steel shows a substantial increase in conductivity, increasing eddy currents and core losses. No literature could be found on voluminous powder cores at cryogenic temperature but the authors would expect them to exhibit issues with material embrittlement and eventually cracking. The development of new core materials for cryogenic use might be a valuable research task.

Two solutions are proposed:

1. Keep the core at RT and operate only the winding(s) at cryogenic temperatures. This requires a non-magnetic cryostat that insulates the cold winding(s) from the warm core.
2. Design of core-less ("air core") inductors and transformers. This might result in substantially more voluminous solutions. However, the higher permissible current density in the copper or aluminum windings operating at cryogenic temperatures might compensate partially for the increase in volume. The current density could potentially be further increased if superconducting windings would be used.

2.3. Capacitors and Varistors
Capacitors and varistors are problematic to operate at cryogenic temperature. There are several different capacitor technologies, typically named after the dielectric material. The dominating technology for large size DC link capacitors is electrolytic, which does not operate at temperatures below the freezing point of the electrolyte. Other technologies include polymer film capacitors, ceramic capacitors, tantalum and niobium capacitors, and vacuum capacitors. As the temperature drops, the metallized layers/foils as well as the polymer foil shrinks according to their respective material properties CTE). A mismatch in CTE can lead to delamination and damage of the capacitor. There are some reports of encouraging results of capacitors at cryogenic temperature [18–20]. However, these studies focused on small volume capacitors only. It is expected that the mismatch in CTE becomes a bigger problem for larger size capacitors.

Two solutions are currently being investigated: The use of capacitors at RT as well as focusing exclusively on converter technologies that work without large size capacitors such as current source converters and direct converters, for instance matrix converters. However, even those topologies typically require (smaller) capacitors in the filter and snubber circuits.

Varistors share the difficulties that capacitors have with the large interface area between two materials of dissimilar CTE. However, since they are expected to absorb surge energy by turning it into heat, varistors also suffer from the reduced heat capacity of solid materials at cryogenic temperatures [21].

2.4. *Interconnects*

Power electronic devices need to interconnect semiconductors, capacitors, inductors, current sensors, voltage sensors, and circuit breakers in a way that is most compact and often at lowest inductance. This is typically achieved by bus plates that are stacked and insulated against each other by plates often made of glass-fiber reinforced resins. The bus plates are typically made of copper or aluminum. The resistivity of these materials drop substantially at cryogenic temperature, reducing the losses and/or reduce the necessary plate thickness. The resistivity of (pure) aluminum and copper are comparable at 4.96×10^{-10} Ωm, and 5.01×10^{-10} Ωm, respectively at 50 K so that using the lighter aluminum could be an attractive solution [22]. At temperatures below approximately 50 K, aluminum features a higher conductivity than copper. However, the typically increased contact resistance of aluminum needs to be taken into account and might require electroplating of the interconnects. It might also be possible to develop superconducting interconnects to further increase the power density and efficiency.

2.5. *Cryogenic Media*

There are both liquid and gaseous cryogenic media that would be suitable for PEDs. Liquid media generally have the advantage of higher dielectric strength and improved heat transfer, which would allow to increase the volumetric power density. In case of a liquid cryogenic medium, the losses in the PED will evaporate some of the liquid cryogen. This mass flow rate $\dot{m}_{evap}$ formed can be calculated:

$$\dot{m}_{evap} = \frac{P_{Loss}}{\Delta H_{vap}} \qquad (1)$$

where, P_{Loss} are the losses in the converter and ΔH_{vap} is latent heat of vaporization of the cryogen, which is approximately 199.2 J/g and 510.8 J/g for LN_2 and LCH_4, respectively [23]. For each kilowatt of losses, 5.02 g/s of LN_2, or 1.96 g/s of LCH_4 would be evaporated. Depending on the application, the vapor can be condensed back into a liquid, used as a gaseous medium or fuel, or vented.

3. Static and Dynamic Characterization of a Press-Pack IGBT Stack in LN_2

3.1. *Static Characterization*

In an effort to identify a power semiconductor that performs well electrically, thermally, and mechanically and that could be used in a megawatt-scale class cryogenic power converter, a medium voltage IGBT stack from IXYS/WESTCODE has been characterized at 77K in LN_2. The stack of IGBTs consists of two identical IXYS T0160NB45A modules, rated for 4500 V and 160 A (Fig. 2 upper left). The modules include an IGBT and an antiparallel diode as needed for voltage source converter applications. The series include models with ratings of up to 6500 V and up to 3000 A. All IGBTs of this series are of ceramic press-pack type, which is foreseen as the most promising packaging type for cryogenic applications. In this experiment, the top flange was ground off to allow LN_2 to flow into the package and cool the IGBT and diode dies. LN_2 also serves as the dielectric and provides sufficient dielectric strength well in excess of the electric fields expected in a 4500 V rated device. The stack does not include dedicated heatsinks. However, the leads consisting of copper plates are designed to help dissipate the heat generated.

Only one of the two IGBTs was used for the static experiment. A 5 V and 900 A rated DC power supply was used in current control mode to provide a current of up to 160 A. Since the stack did not include any particular measures for thermal control (heat sinks etc.), each test was only conducted for a few seconds to avoid excessive heating. The gate voltage was provided by a second power supply, set to voltages of $V_G = \{11, 13, 15, 17, 20\}$ V. For each of the gate voltages, the V_{CE} voltage drop was measured. The experiment was repeated in ambient air at room temperature. The results indicated a substantial reduction of V_{CE} at cryogenic temperature (Fig. 3). This could result in a reduction of conduction losses of approximately 5-20%.

IOP Conf. Series: Materials Science and Engineering **279** (2017) 012011 doi:10.1088/1757-899X/279/1/012011

It should be noted that the actual operating temperatures will deviate from 77 K and RT respectively. For standard converter applications, the thermal design is typically designed to maintain a maximum temperature at full power substantially above RT but well below the rated junction temperature of 125°C. For cryogenic applications, the thermal design could either be optimized to operate at a temperature just slightly above the boiling point of the cryogen using its enthalpy of evaporation to maintain the temperature over a wide range of power. Alternatively, the temperature could be substantially above the boiling point, which would displace the liquid inside the package by gas. Under these conditions, both the heat transfer and dielectric strength would be reduced and would need to be carefully managed.

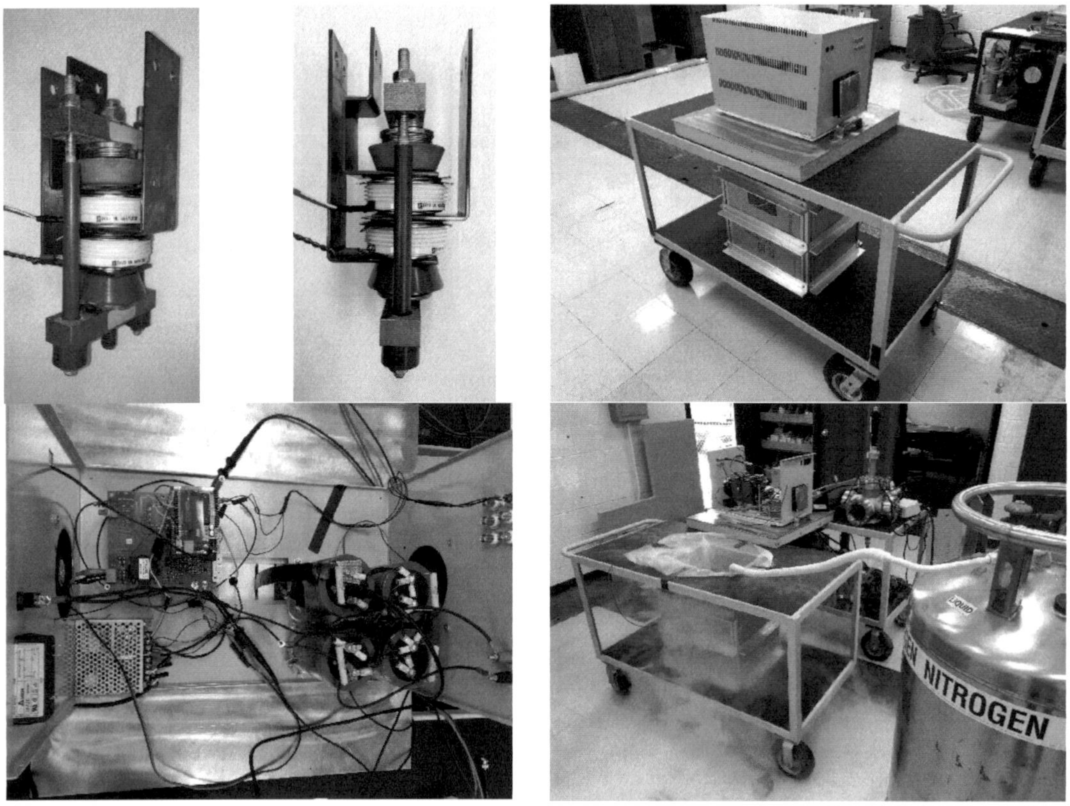

Fig. 2: Picture of the IGBT stack (top left), cart with integrated cryostat (top right), DC link and controller (bottom left), and filling the cryostat with LN_2 (bottom right)

A schematic of the test configuration is illustrated in Fig. 3 (a). Based on the test setup, the output characteristic (I_C-$V_{CE,sat}$ under a certain V_G) of the IGBT is obtained and summarized in Fig. 4. Except for the condition of $V_G = 11$ V, relative to the curves at RT, the "knees" of the curves at 77 K shift to higher voltages, while the slopes of the curves at 77 K increase. In other words, the shapes of the curves at 77 K and more "square". Besides, except for the condition of $V_G = 11$ V, $V_{CE,sat}$ at RT is lower than that at RT if I_C is higher than a certain threshold, while $V_{CE,sat}$ at 77 K is lower if I_C is lower than the certain threshold. The threshold voltage increases with V_G. The temperature-related variation is due to two factors: fewer free carriers and higher carrier mobility at cryogenic temperatures. A decrease of the on-state voltage has also been observed thanks to the large increase in the inversion layer carrier mobility that overbalances the increase of the knee forward voltage intrinsic PN junction due to carrier freeze-out [12]. The variation of output characteristics could result in a reduction of conduction losses of 5-20% if I_C is larger than 50 A, as illustrated in Fig. 5.

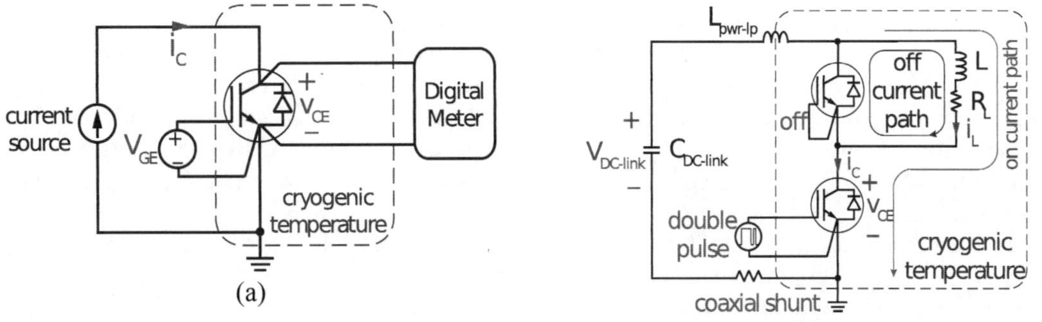

Fig. 3: Test configurations used to for static characteristics (a) and the double pulse test (b).

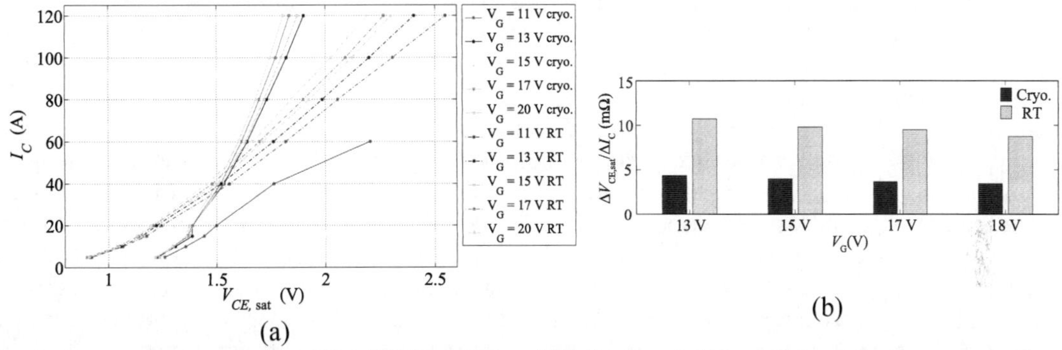

Fig. 4: Output characteristics of the IGBT at RT and 77 K (a) and $\Delta V_{CE,sat}/\Delta I_C$ of the linear part of I_C-$V_{CE,sat}$ curve (b).

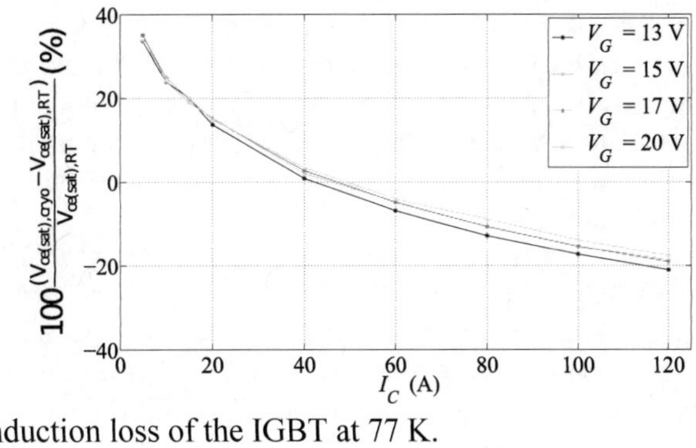

Fig. 5: Variation of conduction loss of the IGBT at 77 K.

In addition, the output characteristic under $V_G = 11$ V presents a different trend from those under higher V_G. The curve for $V_G = 11$ V and 77 K deviates from others at about $I_C = 15$ A, which is significantly lower (about $I_C = 120$ A) under $V_G = 11$ V and RT. The phenomenon means that the threshold voltage $V_{G,th}$) of the IGBT at 77 K is higher than that at RT, which is caused by the lower number of free carriers at 77 K.

The $\Delta V_{CE(sat)}/\Delta I_C$ of linear part of I_C-$V_{CE(sat)}$ curve is provided in Fig. 4 (b), which shows that the relative on-state resistances of the IGBT under cryogenic temperature is much less than those under room temperature.

3.2. Dynamic Characterization

The same stack of IGBTs was characterized using the standardized double-pulse experiment (Fig. 3 (b)). The load inductor is connected across the top IGBT module and consists of 168 turns of 0.823 mm^2 magnet wire resulting in an inductance of 552 μH. The DC resistance is 0.4 Ω at RT and 57 mΩ at 77 K.

The DC link capacitor was kept outside the cryostat and at RT, which results in a total inductance of the main current path of 200 μH. The gate drivers were also operated at RT, which required 50 cm long twisted gate leads, slightly longer than the recommended maximum of 30 cm, yet extra precautions have been taken to avoid any gate resonance. The gate signal was generated in a microcontroller and consisted of the initial on pulse of 150 μs followed by the off pulse of 100 μs, followed by another on pulse of 30 μs followed by another off pulse. The resulting current, voltage, power loss, and gate signal are shown in Fig. 6.

Figure 6 compares the switching performance of the press-pack IGBT at RT and 77 K and shows an improvement of the turn-on dynamic and a reduction of the turn-off time of the IGBT. Figure 6(a) shows the current waveform of the IGBT during the double pulse test. During the turn-on process, the collector current I_C shows a smaller overshoot, which is due to the reverse recovery current of the top anti-parallel diode, at 77 K than at RT. The smaller I_C overshoot at 77 K is due to the anti-parallel diode storing a smaller amount of charge during conduction at lower temperature [24]. The higher on-state value of I_C at 77 K is mainly due to the increased conductivity of the inductive load of the double pulse experiment. A noticeable improvement is observed in the turn-off performance of the IGBT. In this study, the V_G of −10 V was used to turn off the device, and +15 V was used to turn on the device (Figure 6(d)). The negative V_G quickly removes the inversion channel current and eliminates the base drive current. However, the bipolar transistor collector current decays slowly since the excess carriers are removed by recombination. Therefore, the IGBT turn-off current characteristics can be divided into two stages: a rapid decrease stage and a slowly decaying stage [24]. At turn-off, a larger initial drop in I_C is followed by its gradual decay, commonly referred to as the tail current. The two-stage turn-off process was observed both at RT and 77 K (Figure 6 (a)). However, a comparatively faster initial current drop and shorter tail current was observed at 77 K. The faster turn-off is mainly due to lower transistor gain and the reduced lifetime of minority carriers at lower temperature [24]. Due to faster current decay, a higher turn-off voltage overshoot was observed at 77 K as shown in Figure 6 (b). The turn-off voltage overshoot is mainly caused by the induced voltage across the inductive load of the double pulse test setup. Figure 6 (b) also shows that the voltage reaches its steady state after the overshoot faster at 77 K than at RT, which corresponds to the shorter tail current of Figure 6 (a). Furthermore, Figure 6 (c) shows the switching power loss of the IGBT. As expected from the current and voltage waveforms, both the turn-on and turn-off switching energy loss is comparatively smaller at 77 K than at RT. Lower switching energy loss indicates lower thermal stress in the device and thus opens the possibility of operating the IGBT at a higher switching frequency at 77 K.

Based on the test results, the device is comparatively more efficient at 77 K, which was expected since similar findings have been reported in previous work on power semiconductors at cryogenic temperature [25]. The increase in efficiency in the semiconductors is potentially overcompensated by the additional effort to provide cooling at cryogenic temperature. However, the most important finding is that this IGBT stack has been cycled between RT and 77 K more than a dozen times without any degradation. This is new for devices fabricated in a fully industry compatible package.

3.3. Simulation Model of a 200-kW Boost Converter

Based on the encouraging results from the static and double pulse test, a 200-kW boost converter was modeled in PLECS. The input voltage was 1,500 V and the output voltage was 2,000 V. The input inductance was assumed to be 6.3 mH, the output capacitance was 140 μF, and the switching frequency was 2 kHz. The turn-on and turn-off characteristics of the IGBT module was taken from the characterization experiment. At RT, the switching and conduction losses in the IGBT were 373.2 W and 96.1 W respectively (total: 469.3 W). At 77 K, the losses were 147.6 W and 73.4 W respectively (total: 221.0 W). This corresponds to a reduction of losses in the IGBT by 53%. The losses in the inductor could be reduced by a factor of seven if the cross section of the wire is kept identical. Alternatively, the cross section could be reduced to reduce the volume and weight of the inductor.

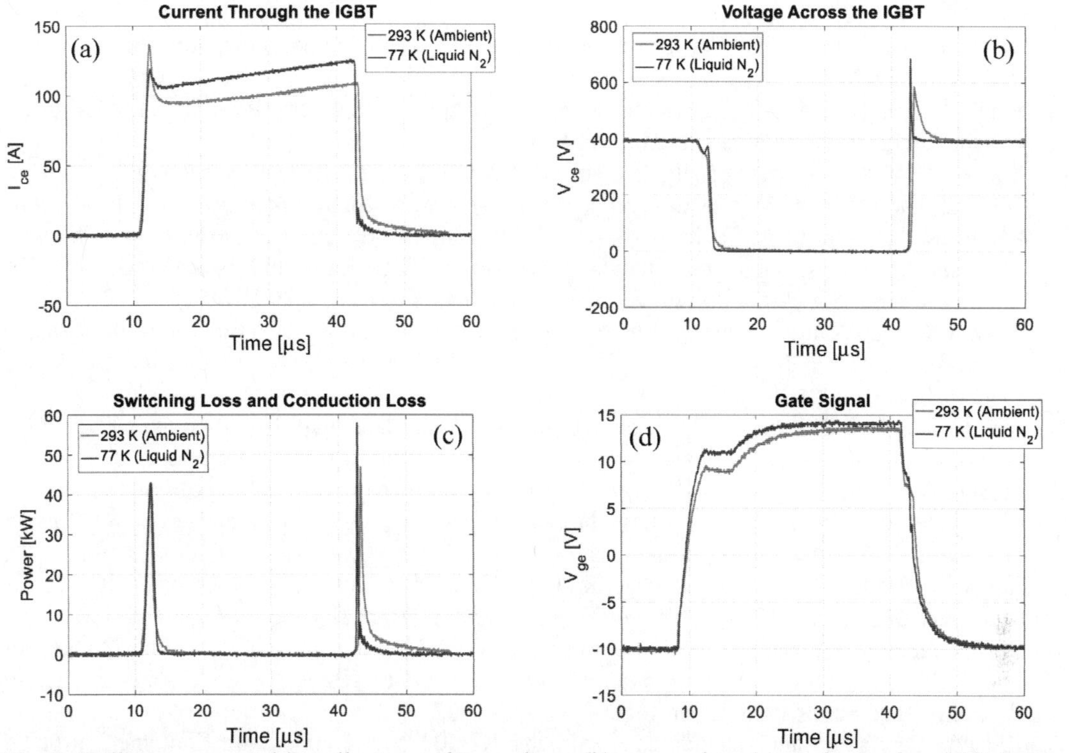

Fig. 6: Collector current (a), collector-emitter voltage (b), power loss (c), and gate signal (d) during the double pulse test.

4. Conclusion

Several challenges exist with the design of cryogenic power electronic devices. This is especially true for applications at medium and high power levels. Semiconductors are the key components of power electronic devices and are the primary focus of this study. To the best of authors' knowledge, this is the first time that a megawatt-class IGBT phase leg has been successfully characterized at cryogenic temperatures. The IGBT phase leg performed well in both static and dynamic tests. In a simulated boost converter configuration, the switching and conduction losses are reduced by 60% and 24%, respectively compared to the values at room temperature. The turn-off time is reduced considerably, potentially enabling higher switching frequencies. Repeated thermal cycles between room temperature and 77 K of the IGBT phase leg did not show any deterioration in performance. The next steps in this project will include characterization of the breakdown voltage at cryogenic temperature and characterizing of the diode. Based on the encouraging results, the authors are planning to design and build a converter based on the tested phase leg to further understand the intricacies of operating power electronics at cryogenic temperatures.

References

[1] Miller J, Santosusso D, Uva M, Woods K, Fitzpatrick B 2013 Naval Superconducting Integrated Power System (SIPS). In: *10th Intelligent Ship Symposium* (Philadelphia: ASNE)

[2] Berg F, Palmer J, Miller P, Husband M, Dodds G 2015 HTS Electrical System for a Distributed Propulsion Aircraft. In: *IEEE Trans. on Appl. Supercond.* 25 (3) 1–5

[3] Rajashekara K, Akin B 2013 A Review of Cryogenic Power Electronics – Status and Applications. In: *Electric Machines Drives Conference (IEMDC)* (Chicago: IEEE) 899–904

[4] Haldar P, Ye H, Efstathiadis H, Raynolds J, Hennessy M J, Mueller O M, Mueller E K 2005 Improving Performance of Cryogenic Power Electronics. In: *IEEE Trans. on Appl. Supercond.*

15 (2) 2370–2075
[5] Singh R, Baliga B J 1996 Cryogenic Operation of Power Bipolar Transistors. In: *Solid-State Electronics* 39 1
[6] Singh R, Baliga B J 1993 Analysis and Optimization of Power MOSFETs for Cryogenic Operation. In: *Solid-State Electronics* 36, 8
[7] Mueller O 1996 Properties of High-Power Cryo-MOSFETs. In: *Conference Record of the 1996 IEEE Industry Applications Conference, Thirty-First IAS Annual Meeting* vol. 3 1443–1448
[8] Leong K K, Bryant A T, Mawby P A 2010 Power MOSFET Operation at Cryogenic Temperatures: Comparison between HEXFET, MDMesh and CoolMOS. In: *22nd International Symposium on Power Semiconductor Devices IC's (ISPSD)* 209–212
[9] Chen S, Cai C, Wang T, Guo Q, Sheng K 2013 Cryogenic and High Temperature Performance of 4H-SiC Power MOSFETs. In: *28th Annual IEEE Applied Power Electronics Conference and Exposition (APEC)* 207–210
[10] Kim H, Lim H J, Cha H 2010 DC Characteristics of Wide-Bandgap Semiconductor Field-Effect Transistors at Cryogenic Temperatures. In: *Journal of the Korean Physical Society* 56 (5)
[11] Forsyth A J, Yang S Y, Mawby P A, Igic P 2006 Measurement and Modelling of Power Electronic Devices at Cryogenic Temperatures. In: *IEE Proceedings – Circuits, Devices and Systems* 153 (5) 407–415
[12] Singh R, Baliga B J 1992 Cryogenic Operation of Asymmetric N-Channel IGBTs. In: *Proceedings of the 4th Intl. Symp. on Power Semiconductor Devices and ICs* 188–193
[13] Rosenbauer F, Lorenzen H W 1996 Behaviour of IGBT Modules in the Temperature Range from 5 to 300 K. In: *Advances in Cryogenic Engineering*, ed. by Kittel P, A Cryogenic Engineering Conference Publication 41. Springer US 1865–1872
[14] Kang X, Caiafa A, Santi E, Hudgins J, Palmer P R 2002 Low Temperature Characterization and Modeling of IGBTs. In: *IEEE 33rd Annual Power Electronics Specialists Conference* 3 1277–1282
[15] Caiafa A, Snezhko A, Hudgins J L, Santi E, Prozorov R 2004 IGBT Operation at Cryogenic Temperatures: Non-Punch-through and Punch-through Comparison. In: *IEEE 35th Annual Power Electronics Specialists Conference* 4 (4) 2960–2966
[16] Wakeman F, Li G, Golland A 2005 New family of 4.5 kV Press-pack IGBTs. In: *PCIM'05*, Nuremberg, Germany
[17] Golland A, Wakeman F 2005 Application of Press-pack IGBTs in Traction Refurbishment. In: *APEC 2005* (Austin: IEEE) 3 2030–2035
[18] Lawless W N, Frederick Clark Jr. C 1986 Capactitive Energy Storage Device for Use at Cryogenic Temperatures. In: *U.S. Patent 4,599,677*
[19] Lawless W N, Frederick Clark Jr. C 1997 Energy Storage at 77 K in Multilayer Ceramic Capacitors. In: *IEEE AES Systems Magazine* 12 (5) 32–35
[20] Pan M J 2005 Performance of capacitors under DC bias at liquid nitrogen temperature. In: *Cryogenics* 45, 463–467
[21] Lawless W N, Clark C F 1988 Electrical and thermal properties of a varistor at cryogenic temperatures. In: *AIP Journal of Applied Physics* 64 (8)
[22] Campbell J E, Eldridge E A, Thompson J K 1974 Handbook on Materials for Superconducting Machinery. In: *ADA002698*
[23] Jensen J E, Tuttle W A, Stewart R B, Brechna H 1980 Brookhaven National Laboratory Selected Cryogenic Data Notebook. In: *BNL 10200-R*, Vol. 1, Sections 1–9
[24] Singh R, Baliga B J 1998 Cryogenic Operation of Silicon Power Devices. In: *Springer*
[25] Zhang Z, Timms C, Tang J, Chen R, Sangid J, Wang F, Tolbert L M, Blablock B J, Costinett D J 2017 Characterization of high-voltage high-speed switching power semiconductors for high frequency cryogenically-cooled application. In: *IEEE Applied Power Electronics Conference and Exposition (APEC)* 1964–1969

IOP Conf. Series: Materials Science and Engineering **279** (2017) 012012 doi:10.1088/1757-899X/279/1/012012

Future cryogenic switchgear technologies for superconducting power systems

C Xu, R Saluja, T Damle, L Graber

Georgia Institute of Technology, Atlanta, GA 30332, United States

Email: chunmengxu@gatech.edu

Abstract. This paper introduces cryogenic switchgear that is needed for protection and control purposes in future multi-terminal superconducting power systems. Implementation of cryogenic switchgear is expected to improve system reliability and minimize overall volume and weight, but such switchgear is not available yet. Design of cryogenic switchgear begins by referring to conventional circuit breakers, a brief review of state-of-the-art switchgear technologies is presented. Then, promising cryogenic interruption media are identified and analysed with respect to physical and dielectric properties. Finally, we propose several cryogenic circuit breaker designs for potential aerospace, marine and terrestrial applications. Actuation mechanism for cryogenic switchgear is also investigated.

1. Introduction

Switching is an indispensable operation in any complex network infrastructure. For power distribution systems, switching ensures the system's flexibility, reliability and efficiency in terms of fault management and optimal configuration. Equipping power systems with appropriate switchgear will not only ensure consistent power delivery, but also prevent spreading damages over the whole system. When it comes to superconducting power systems, only solid-state switchgear is applicable so far. However, solid-state devices have relatively high power loss in on-state conduction, and their isolation capability is very limited. Electromechanical switchgear (or switchgear) has negligible on-state loss and very high isolation capability, but current designs are not suitable for cryogenic switching due to frozen dielectric media, solidified lubricants, cracked polymers, and warped components as a consequence of mismatched coefficients of thermal expansion. Substantial heat load will be added to cryogenic system if ambient-temperature-rated switchgear is directly connected. To tackle these problems, we propose cryogenic electromechanical switchgear as a promising solution to cryogenic switching in this paper.

High temperature superconducting (HTS) power systems for future all-electric aircrafts and ships are also expected to be in need of switchgear: In turboelectric distributed propulsion (TeDP) systems, the lack of lightweight and efficient switchgear makes optimal network architecture unachievable [1, 2]. Solid-state circuit breaker (SSCB) at cryogenic temperature is a potential solution but will add substantial heat load to the cryogenic system due to on-state conduction losses, which will potentially increase total weight meanwhile decrease total efficiency in aircraft and naval applications.

Circuit breaker (CB) is the most advanced and complicate switchgear due to its fault current interruption capability. The cryogenic circuit breaker (CCB) is also the most fascinating device within cryogenic switchgear design. Low weight and high efficiency are expected to make CCBs more competitive in TeDP systems and other applications [3]. On the one hand, CCBs for TeDP systems and all-electric ships can be designed with faster operation speed, because these systems are expected to be fault current limited by using power electronic devices. In terrestrial cryogenic power systems, the capability to interrupt substantial fault current becomes more important. In medium or high-voltage level, CCBs need to work with fault current up to several or tens of kiloamperes, same as the common fault current rating for conventional CBs. Similar to ambient-temperature CBs, CCBs can utilize various cryogenic dielectrics as interruption media, including gas, liquid and vacuum. While the switching chamber of CCBs will always be at cryogenic temperature, the actuator mechanism can either be at room temperature or at cryogenic temperature. We call the corresponding CCBs as "warm-actuated" and "cold-actuated" respectively, which will be further discussed in Section 4.

Section 2 reviews conventional ambient-temperature CBs with respect to interruption medium, chamber design and actuation mechanism, selection criteria for CCBs are generalized. These criteria are applied in potential cryogenic interruption medium identification, which is discussed in Section 3. Section 4 introduces several practical CCB designs, providing demonstrations of switchgear in cryogenic power systems for aerospace, marine and terrestrial applications.

2. Review of Conventional Circuit Breakers

Electromechanical circuit breaker (EMCB) and SSCBs are both conventional CBs working at ambient and elevated temperature. EMCBs have movable contacts and arcs for switching, while SSCBs utilize PN-junction for switching. Our focus is on developing the cryogenic counterpart of conventional EMCBs. According to types of interruption media used, EMCBs can be further divided into three subgroups. Figure 1 is an attempt to show conventional and cryogenic CBs side to side. It includes the interruption media and structure types of conventional CBs, and proposed media and types for cryogenic CBs (further discussed in Sections 3 and 4).

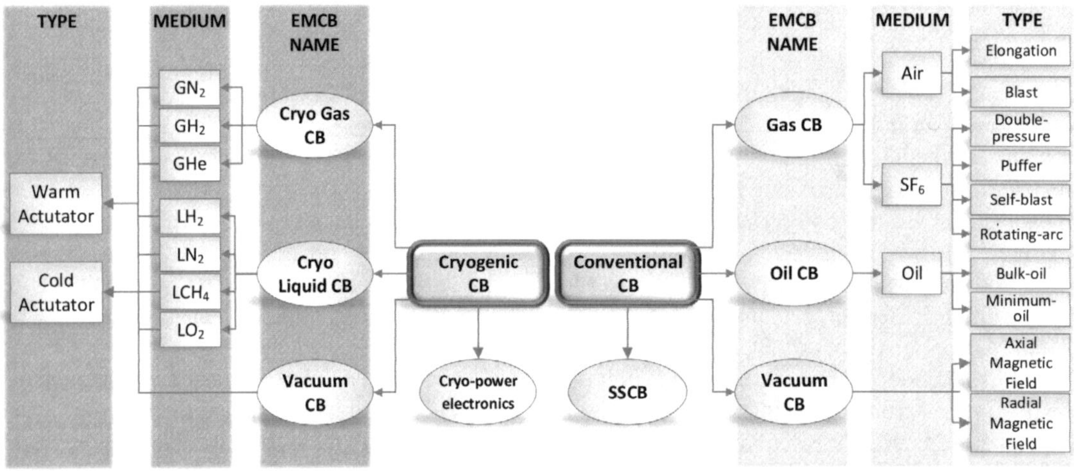

Figure 1. Classification of cryogenic (left) and conventional (right) circuit breakers

2.1. SF_6 Circuit Breakers

The history of gas-insulated CBs started with air-elongation and air-blast CBs. The increased voltage and fault current levels of power system made it necessary to employ strongly electronegative gases for interruption. Today's high-voltage switchgear utilizes sulfur hexafluoride (SF_6) as insulation and interruption medium, because of its high thermal conductivity, high breakdown strength, and fast recovery rate [4].

Based on their arc quenching mechanisms, SF_6 circuit breakers can be divided into four sub-types (Figure 2): double-pressure, puffer, self-blast, and rotating-arc [4]. They all contain a main contact system for lowest contact resistance at closed state, an arcing contact system that provides conducting path for arcs during opening operation, and a nozzle to direct cold interruption medium onto the arc. The main contact system typically surrounds the arcing contact system in a concentric manner. For simplicity, only arcing contacts are shown in Figure 2.

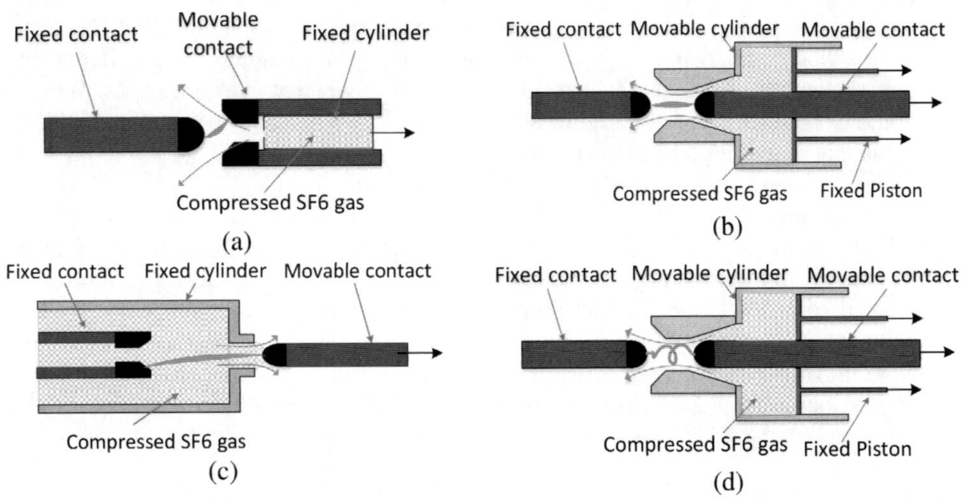

Figure 2. Arcing contact system of SF_6 circuit breakers of (a) double-pressure type (b) puffer type (c) self-blast type and (d) rotating-arc type

Overall, SF_6 is the most common gaseous interruption medium applied in high voltage and ultra-high voltage circuit breakers. SF_6 circuit breaker utilizes various control methods for controlling gas flow in interruption chamber. Correspondingly, in cryogenic gas CBs, gas-control components like puffer and nozzle can also be adopted.

2.2. Oil Circuit Breakers

Mineral oil is a good interruption medium with high dielectric strength and excellent arc quenching properties, so it becomes the best choice for liquid-type CBs. There are two basic structures for oil CBs: bulk-oil and minimum-oil designs (Figure 3) [4]. CBs of bulk-oil design, or dead-tank design, have a grounded tank holding a large volume of oil to immerse conductors and contacts, while minimum-oil designs drive a small amount of oil through contacts solely for extinguishing the arc. Oil circuit breakers have very reliable designs for high voltage operations, and been in service for half of a century. After the 1980s, due to their large volume, large driving force and high flammability, oil breakers were supplanted by vacuum interrupters and SF_6 circuit breakers [4].

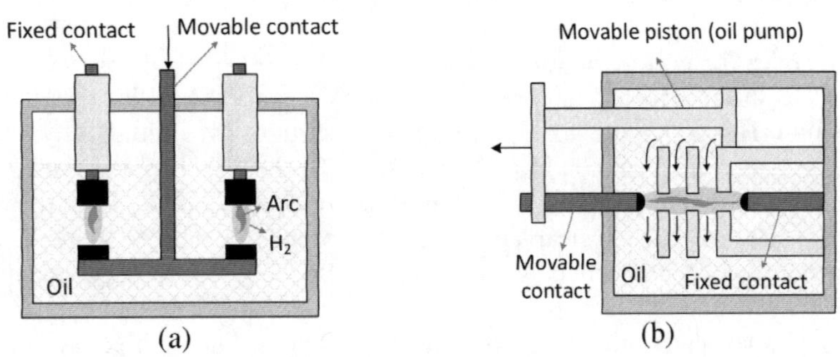

Figure 3. Oil circuit breaker of (a) bulk oil design and (b) minimum oil design

In the oil-type CB, when a fault occurs, movable contacts will draw back from fixed contacts, generating an arc between contacts and triggering nearby oil to evaporate, decompose and ionize. After the flow of cold oil has removed the heat of arcing plasma and the arc cannot be sustained by applied voltage, the hydrogen gas dissociated from mineral oil will quench the arc and re-establish the insulation between contacts. In this way, an opening process for interruption is accomplished. The interruption process is somewhat similar in gas-type CBs and vacuum interrupters, the differences are the interruption media and the actuation mechanism for contact movement in various CBs. With respect to cryogenic liquid CBs, since cryogenic liquids don't degrade or decompose during aging, bulk-oil design is more feasible due to its simpler actuation mechanism.

2.3. Vacuum Interrupters

Vacuum interrupters are best suited for medium voltage applications due to their high reliability, low driving force and small volume. Figure 4 shows the cross section of a typical vacuum interrupter. It consists of a pair of contacts attached to copper conductors that are each connected to a fixed and movable conductor. The movable conductor is connected to an end plate by metal bellows and movable terminal, the terminal is driven by the actuator. The contacts, conductors, and bellows are all enclosed in a ceramic chamber holding high vacuum, the ends of which are metallized and brazed to metal end caps.

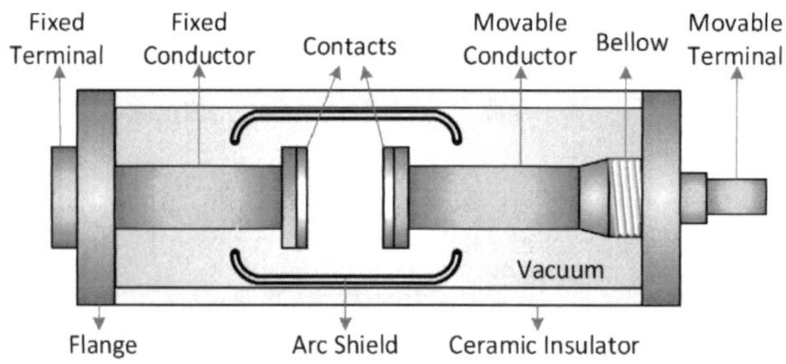

Figure 4. Vacuum interrupter

Vacuum interrupters can operate over a wide temperature range since they need to be able to withstand the vacuum bake-out in the manufacturing process. The materials are expected to tolerate cryogenic temperatures and have matching coefficients of thermal expansion. With an appropriate actuation mechanism, vacuum interrupter will be a promising candidate for cryogenic switching. However, the prevalent problem in cryogenic switchgear, the changed material properties, would be more critical in cryogenic vacuum interrupters. The bellow is expected to become more brittle, which increases the potential of damage by fatigue, so certain parts of interrupter require a re-design to operator in superconducting power systems.

3. Potential dielectrics for cryogenic interruption

Former studies have thoroughly investigated the dielectric properties of cryogenic fluids, such as nitrogen (N_2), helium (He) and hydrogen (H_2) [5, 6]. The permittivity, conductivity, dielectric loss and breakdown strength under different conditions are known [7]. However, there has not been a research focuses specifically on the interruption performances of cryogenic fluids. Therefore, we discuss the "static" properties, that is, the physical and dielectric properties of cryogenic fluids now. Their "dynamic" performances like arc quenching and dielectric recovery require experimental statistics to be fully understood.

One of the most important requirements for an interruption medium is high dielectric strength, also known as breakdown strength. Dielectric strength is the capability of dielectrics to withstand high electric fields, which is crucial for switchgear at open-state. High dielectric strength leads to high survival rate under overvoltage, high reliability in open-state, and low likelihood of dangerous arc

IOP Conf. Series: Materials Science and Engineering **279** (2017) 012012 doi:10.1088/1757-899X/279/1/012012

restrikes and re-ignitions following interruption. Breakdown in gas is characterized by Paschen's Law, which describes the dielectric strength of a gas as a function of the product of gas pressure *p* and gap distance (*d* for uniform field between parallel plate electrodes, or R_1 for non-uniform field between coaxial electrodes). When the gas pressure *p* is replaced by the number of molecules per volume N_v according to ideal gas law, the Paschen's curves are independent from gas temperature [8]. Hence, cryogenic gases perform the same Paschen's curves as ambient-temperature gases. The Paschen's Law of cryogenic gases are shown in Figure 5. Breakdown in liquid cannot be characterized by a pure materials function [5]. Purity, presence of bubbles, time duration of applied voltage can all affect dielectric strength of liquid. Generally, liquids have higher dielectric strength than gases.

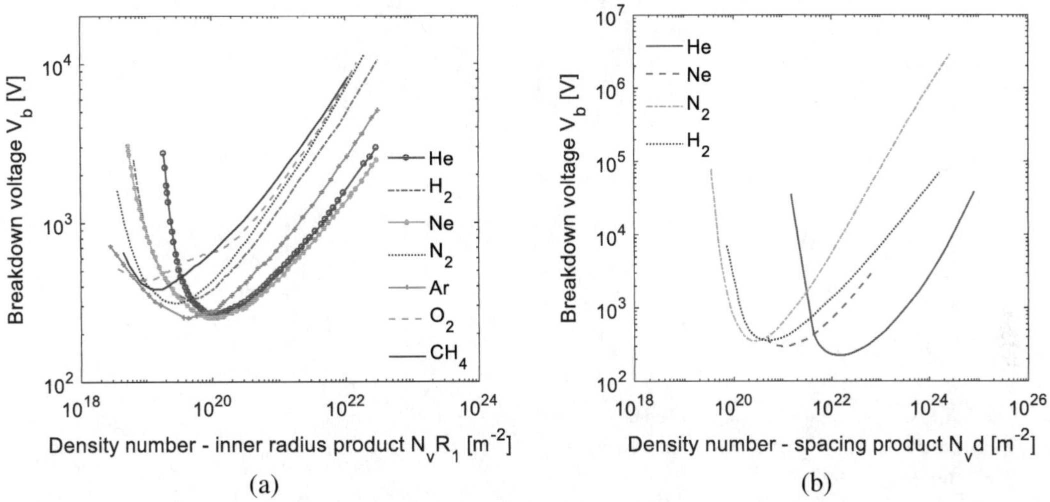

Figure 5. Paschen's curves of cryogenic gases measured at (a) uniform field of parallel-plate electrodes [5, 7] and (b) non-uniform field of coaxial electrodes [8, 9] at room temperature

Besides the dielectric strength, an ideal cryogenic interruption medium should have low boiling temperature, low viscosity, high thermal conductivity and high heat capacity. For cryogenic liquids, the enthalpy of vaporization should be high enough to prevent excessive gas formation. Based on melting and boiling temperature shown in Figure 6, promising cryogenic liquids for interruption are liquid nitrogen (LN_2), liquid fluorine (LF_2), liquid oxygen (LO_2) and liquid methane (LCH_4). The rest of cryogens have a limited liquid temperature range. Furthermore, condensed noble gases (LHe, LNe, and LAr) have low dielectric strength as shown in Figure 5. LF_2 is an outstanding dielectric due to fluorine's strongest electronegativity among all elements. However, the high toxicity, corrosivity, and oxidability exclude LF_2 from industrial application. Similarly, even though LO_2 has strong electronegativity that contributes to dielectric recovery after interruption process, high oxidability makes it hazardous. LCH_4 is considered as a promising candidate, since it can be used both as a cryogen and a fuel for space crafts, aircrafts and ships. Other hydrocarbons such as ethane, propane, ethylene as well as mixtures like liquid natural gas (LNG) are similarly promising. Flammability aspects will need to be considered in designs where evaporated cryogens are vented to the atmosphere.

Cryogenic gases like gaseous hydrogen (GH_2), gaseous helium (GHe), gaseous neon (GNe), gaseous argon (GAr) and gaseous nitrogen (GN_2) are candidates for cryogenic gas CBs. GH_2 is a very appealing gas for arc quenching purposes due to its low viscosity, high thermal conductivity, and relatively high dielectric strength. It is also the functioning component of ambient-temperature transformer oil for arc quenching. GHe shares the high thermal conductivity with GH_2, but suffers from low dielectric strength like all noble gases. For the very low temperature range below 50 K, GHe is the only choice if a non-flammable solution is needed. However, it has been discovered that by adding small amounts of GH_2 to GHe, the dielectric strength of mixture can be substantially improved while maintaining the non-flammable characteristics [10]. There are several more "exotic" cryogenic liquids that would be worth considering: Oxygen difluoride (LOF_2), carbon monoxide (LCO), nitric oxide (LNO). They are not discussed here for toxicity and reactivity reasons.

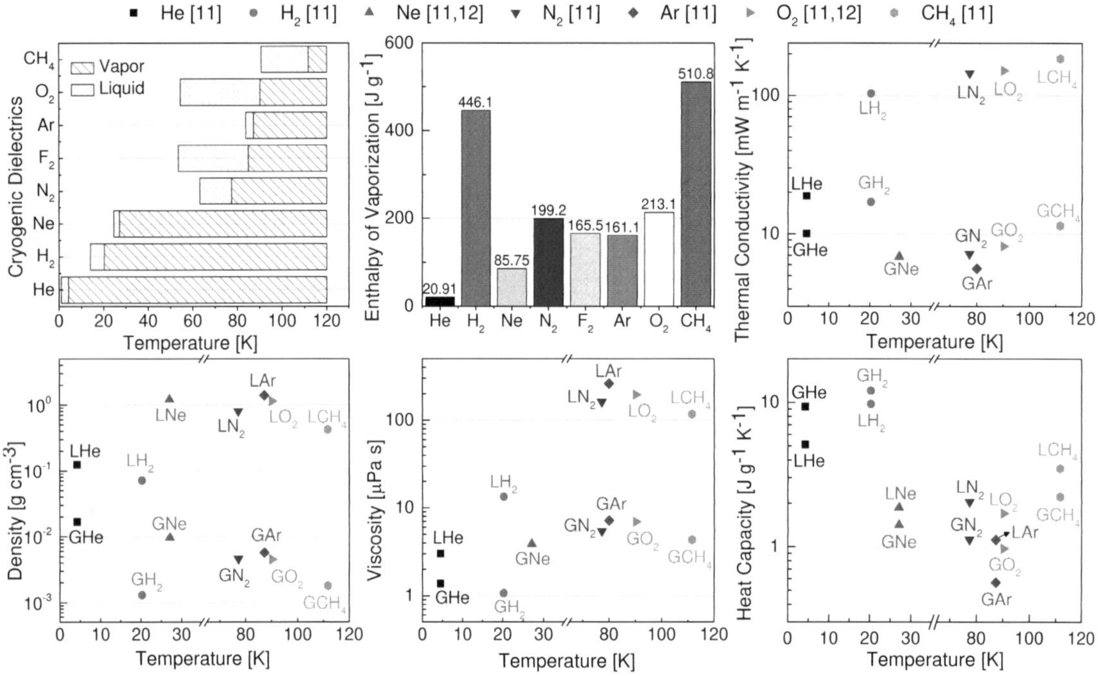

Figure 6. Physical properties of cryogenic fluids at normal boiling point (p = 0.1 MPa) [11, 12]. (a) melting and boiling temperature (b) enthalpy of vaporization (c) thermal conductivity (d) density (e) viscosity and (f) heat capacity at constant pressure. "L" stands for "Liquid", "G" stands for "Gas".

Nitrogen is the most widely-used cryogen in cryogenic engineering, it also exhibits excellent dielectric and interruption performances in superconducting power systems. Researchers have identified GN_2 as potential substitute for SF_6 in gas-insulated switchgear at ambient temperature [13], because N_2 is a strong electron slowing-down gas and can self-heal after breakdown[14]. GN_2 has the highest dielectric strength among cryogenic gases below 80 K, more than 10 times the strength of GH_2 at high pressure ($N_v d = 10^{24}$ m^{-2}). Several studies have tested the interruption performances of LN_2. Morishita et al. [15, 16] invented a puffer-type LN_2 CB which directly blew LN_2 on arcs, showing a higher current interruption capacity and shorter arcing time than CB designs that directly immersed contacts in LN_2. In their experiment, at a gap distance of 10 mm and an applied voltage is 100 V, the maximum interruptible current is 300 A and interruption time is 7 ms. Golde et al. studied applicability [17], contact resistance [18] and short-circuit interruption [19] of commercial vacuum interrupter in LN_2 environment, providing preliminary references for cryogenic vacuum interrupter design.

4. CCB case studies

. The voltage rating and interruption capability of cryogenic switchgear are dependent on interruption medium, actuation mechanism, arc control method, thermal management and so on. Conventional CB structures can be modified in CCBs when they have the same type of interruption medium. Cryogenic liquid CBs can be based on designs similar to conventional oil-type CBs, because the dielectric strength of cryogenic liquids is similar to mineral oil. However, cryogenic gas CBs need numerous modifications from conventional gas CBs. One reason is the large difference in interruption properties between SF_6 and cryogenic gases. The other reason is the pursuit of smaller, lighter and faster switchgear, traditional actuation mechanism of SF_6 breakers would be too heavy for aerospace or naval applications. When it comes to vacuum interrupters, the cryogenic temperature would not influence the dielectric properties of vacuum itself, it may even further drop the vacuum pressure in the interruption chamber. For cryogenic vacuum interrupters, problem concentrates on thermal contraction coordination between different parts in the interrupter, which must be based on the very short list of materials compatible with ultra-high vacuum conditions.

4.1. Cryogenic Liquid Circuit Breakers

Proposed cryogenic liquid CB design is shown in Figure 7. The operation principle is similar to bulk-oil CB as described in section 2.2. The conductors and contacts are immersed in cryogenic liquid to stay at cryogenic temperature, but the actuators can be either at ambient temperature or at cryogenic temperature. The former choice can provide very large driving force and long stroke, while the latter choice has the potential to reduce the overall size. Transition from cryogenic to ambient zone will be isolated by vacuum layer and evaporated gas. This gas cushion contains the vapor generated during arcing process, thus modulates the internal pressure of chamber.

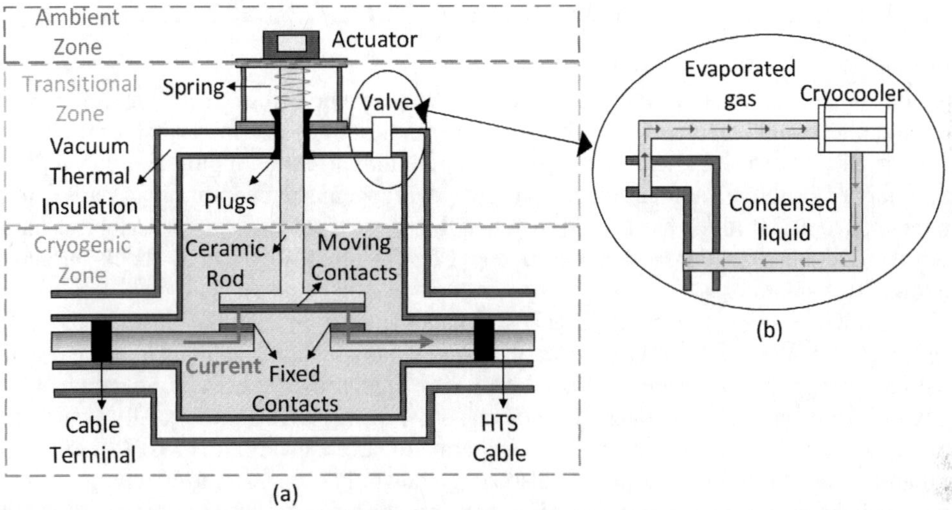

Figure 7. Cryogenic liquid circuit breaker structure: (a) Venting type and (b) condensing type.

However, the evaporated gas is the major threat for cryogenic liquid CB operation. To estimate amount of liquid evaporated during each interruption process, we assume the arcing is restricted in the channel between contacts, so only a small volume of liquid is heated and evaporated, and the evaporation process is isothermal. When the energy generated by arcing is completely absorbed by evaporating the interruption medium, the mass of vapor formed and chamber pressure change can be calculated as

$$m_{evap} = \frac{V_{arc} I_{fault} t_{interruption}}{\Delta H_{vap}} \quad (1)$$

$$\Delta p = \frac{m_{evap}}{M} \frac{RT}{V} \quad (2)$$

Where V_{arc} is the applied voltage to sustain the arc, I_{fault} is the fault current flowing though the arc, t_{interr} is the time used to extinguish the arc, and ΔH_{vap} is the latent heat of vaporization given in Figure 6, M is the molar mass, R is the ideal gas constant, and V is the volume of gas. Table 1 lists the vapor mass for LN_2, LCH_4, LH_2, and LF_2 at arbitrary conditions of 100 V of arcing voltage, 2 kA of fault current, and 20 ms of interruption time. The pressure increase is calculated by the ideal gas law as in equation (2). It can be seen that the pressure increase is fairly moderate for most candidates, between 31.2 kPa for

LCH_4 and 45.9 kPa for LN_2, but a bit more substantial for LH_2 with 285 kPa. While LN_2, LH_2, and LF_2 will recombine after interruption, LCH_4 may dissociate and form some hydrogen gas and carbon sooth along with a few other hydrocarbons. Filtration is needed for long-term use of CH_4 as interruption medium.

For the purpose of withstanding the pressure increase inside the interruption chamber, two different designs are considered. Open loop (venting) and closed loop (condensing) cryogenic liquid CBs are proposed and shown in Figure 7. The venting type CB includes a valve at the top of the tank which releases excess vapor formed during an arc quenching process. The condensing type CB replaces the valve by a cryocooler, which condenses the vapors and directs liquid back into the tank. The size of cryocooler depends on the volume of vapor formed each arcing cycle. This condensing design helps maintain a desired level of fluid in the tank.

Table 1. Evaporated gas and increased pressure of cryogenic liquid CB after arcing

Dielectric fluid	Vapor mass m_{evap} (g)	Pressure increase Δp (kPa)	Dielectric fluid	Vapor mass m_{evap} (g)	Pressure increase Δp (kPa)
LN_2	20.1	45.9	**LH_2**	9.00	285
LCH_4	7.83	31.2	**LF_2**	24.0	40.5

4.2. Cryogenic Vacuum Interrupters

Vacuum interrupters can be used at cryogenic temperature by enclosing the interrupting chamber in a bath of cryogenic liquid, such as LN_2. This can potentially affect several characteristics of vacuum interrupters such as internal and external dielectric strength, current chopping and contact resistance. Also, materials of different thermal expansion rates at cryogenic temperature can cause mechanical stress in the interrupter. While it has been shown that the mechanical stresses due to thermal shock does not affect the vacuum pressure in the interrupter [17], the effect of thermal shock on short-circuit interruption is still unknown.

The breakdown voltage in vacuum largely depends on field emission at small gaps and particle effects at large gaps, along with contact material, geometry and surface quality. The breakdown voltage of LN_2 cooled copper electrodes are shown to be 15-20% higher than ambient temperature at 2 mm gap distance. Golde et al. have demonstrated that the breakdown voltage of a commercially available vacuum interrupter in an LN_2 bath with 11 mm contact separation to have a maximum of 10% higher breakdown voltage than at ambient temperature. It has also been shown that the current chopping of the vacuum interrupter is unchanged by enclosing it in an LN_2 bath, but the arc instability preceding current chopping occurs 10-20% earlier than at ambient temperature due to higher velocity of cathode spots [17].

During arc interruption, the temperature of contacts will increase and generate metal vapor, which make the arc harder to extinguish. A cryogenic vacuum interrupter would result in lower contact erosion, since the contact resistance is lower at cryogenic temperature. The contact resistance of a LN_2 enclosed vacuum interrupter is shown to be about 15% of ambient value under DC current and 30% under AC current [18]. The lower resistance of contacts means lower energy loss caused by heat dissipation. The power loss for a 126 kV vacuum interrupter in a bath of LN_2 calculated to be 15% of the power loss of a 126 kV SF_6 gas insulated switchgear [20].

Vacuum interrupters have desirable characteristics to make it work as a cryogenic switchgear device. However, the effect of thermal shock on short-circuit interruption needs to be further explored, as well the breakdown voltage at cryogenic temperatures and the external dielectric strength of the device when the cooling fluid is at a higher temperature and has bubbles. Furthermore, the material properties will change at cryogenic temperature. This becomes especially important for parts that bend like the metal bellows or slide like conventional actuator. These elements could fail prematurely due to excessive wear and material fatigue in cryogenic environment, thus need extra attention.

4.3. Cold actuated switchgear

Concerning the weight and volume constraints of CCBs used in highly-integrated power systems, the actuation mechanism must be carefully designed to fulfill contact separation within a short period of interruption time. Conventional actuation mechanisms are pneumatic, hydraulic or spring operated. Most of them are assisted with electromagnetic devices, which results in bulky device and slow operation. On the other hand, cryogenic actuation utilizes actuators that can fully function below 100 K, like piezoelectric and magnetostrictive actuators. Cold actuator designs are expected to have weight and volume savings compared to warm actuator designs. However, the driving forces and stroke distance are often reduced at the same time.

Piezoelectric actuators and magnetostrictive actuators are well suitable for cold actuation in cryogenic CBs. Piezoelectric actuators have been historically studied and used in aerospace applications. NASA, for example, designed and tested PMN-PT single crystal piezoelectric stack actuators and flex-tensional actuators for cryogenic applications [21]. Other organizations like ESA [22] and ISRO [23] are also working with piezoelectric actuators in cryogenic atmosphere due to their low power consumption and lubricant-free design. Magnetostrictive actuators are also lubricant-free, which is necessary for cryogenic or vacuum applications. Compared with conventional actuator, piezoelectric and magnetostrictive actuators have limited driving force and short stroke length but light weight and fast response. Considering these features, a piezoelectric-actuator-based high-speed disconnect switch design is being studied for cryogenic applications. The ambient-temperature counterpart is covered by an ongoing research project funded by the National Science Foundation that focuses on fault isolation devices for the FREEDM Systems Center [24] .

5. Conclusion

Apart from a short-term experiment with a vacuum interrupter under laboratory conditions, conventional switchgear does not function at cryogenic temperature. However, switchgear based on liquid and gaseous media as well as vacuum can lend themselves for adaptation to cryogenic operation, major technical issues of this transition are foreseen in this paper. With respect to the interruption medium, a wide variety of cryogenic fluids are expected to performing well as interruption media in cryogenic switchgear. The actuator of cryogenic switchgear can operate at either room or cryogenic temperature, the design with cryogenic actuator could further reduce the heat leakage from ambient into the cryogenic system, resulting in additional weight and space savings for cryogenic switchgear.

Further investigations concentrate on both material and structure perspectives. On the one hand, interruption performances of cryogenic dielectrics will be studied in detail, providing necessary information for chamber design and actuation selection. The optimal contact material and geometry for cryogenic interruption will also be specified. On the other hand, the overall structures for aerospace, marine and terrestrial application will be explored respectively, evaluating their estimated volume, weight and efficiency in real-time operations.

References

[1] Flynn M C, Jones C E, Norman P J and Galloway S J 2016 Fault management strategies and architecture design for turboelectric distributed propulsion. *Int. Conf. on Electrical Systems for Aircraft, Railway, Ship Propulsion and Road Vehicles & Int. Transportation Electrification Conf.*, pp 1-6

[2] Jones C E, Norman P J, Galloway S J, Armstrong M J and Bollman A M 2016 Comparison of candidate architectures for future distributed propulsion aircraft *IEEE Trans. Appl. Supercond.* **26** 3601409

[3] Armstrong M J, Blackwelder M, Bollman A, Ross C, Campbell A, Jones C and Norman P 2015 *Architecture, voltage, and components for a turboelectric distributed propulsion electric grid* (Cleveland) 46-55

[4] van der Sluis L and Kapetanovic M 2014 *Switching in Electrical Transmission and Distribution Systems* (Somerset, GB: Wiley) pp 164-202

[5] Gerhold J 1998 Properties of cryogenic insulants *Cryogenics* **38** 1063-81
[6] Mathes K N 1967 Dielectric Properties of Cryogenic Liquids *IEEE Trans. on Elect. Insul.* **EI-2** 24-32
[7] Gerhold J 1979 Dielectric breakdown of cryogenic gases and liquids *Cryogenics* **19** 571-84
[8] Hackam R 1969 Total secondary ionization coefficients and breakdown potentials of hydrogen, methane, ethylene, carbon monoxide, nitrogen, oxygen and carbon dioxide between mild steel coaxial cylinders *J. of Phys. B: Atomic and Molecular Physics* **2** 216
[9] Hackam R 1969 Total secondary ionization coefficients and breakdown potentials of monatomic gases between mild steel coaxial cylinders *J. of Phys. B: Atomic and Molecular Physics* **2** 201
[10] Park C, Graber L, Kim W, Cheetham P, Kim C, Pamidi S and Rodrigo H 2016 A versatile model for estimating breakdown voltage and its application for cryogenic gas mixtures. *Conf. on Electrical Insulation and Dielectric Phenomena* (Toronto: IEEE) pp 542-5
[11] Haynes W M 2014 *CRC handbook of chemistry and physics* (Boca Raton: CRC press) **6** 21-8, 164
[12] Bich E, Millat J and Vogel E 1990 The viscosity and thermal conductivity of pure monatomic gases from their normal boiling point up to 5000 K in the limit of zero density and at 0.101325 MPa *J. Phys. Chem. Ref. Data* **19** 1289-305
[13] Rokunohe T, Yagihashi Y, Aoyagi K, Oomori T and Endo F 2007 Development of SF_6-Free 72.5 kV GIS *IEEE Trans. Power Del.* **22** 1869-76
[14] Christophorou L G, Olthoff J K and Green D S 1997 Gases for electrical insulation and arc interruption: possible present and future alternatives to pure SF6 *NIST technical note* pp 1-44
[15] Morishita Y, Ishikawa T, Yamaguchi I, Okabe S, Ueta G and Yanabu S 2009 Applications of DC Breakers and Concepts for Superconducting Fault-Current Limiter for a DC Distribution Network *IEEE Trans. Appl. Supercond.* **19** 3658-64
[16] Morishita Y, Koyama T, Yamaguchi I, Okabe S, Ueta G and Yanabu S 2008 Development of DC-current-limiting circuit-breaker with superconducting fault current limiter. *Int. Conf. on Gas Discharges and Their Applications,* (Cardiff: IEEE) pp 97-100
[17] Golde K and Hinrichsen V 2014 Applicability of vacuum interrupters in liquid nitrogen environment. *Int. Symp. on Discharges and Electrical Insulation in Vacuum* (Mumbai: IEEE) pp 397-400
[18] Golde K and Hinrichsen V 2014 Contact resistance of vacuum interrupters for electric power systems in liquid nitrogen at direct and alternating current. *Int. Conf. on Electrical Contacts* (Dresden: IEEE) pp 1-5
[19] Golde K, Hinrichsen V, Gentsch D, Lawall A and Taylor E D 2016 Short-circuit current interruption in liquid nitrogen environment. *Int. Symp. on Discharges and Electrical Insulation in Vacuum* (Suzhou: IEEE) pp 434-8
[20] Wang J, Tan Y, Zhang L, Geng Y, Liu Z and Wang S 2016 Conceptual Design of a Liquid-Nitrogen-Insulated Metal-Enclosed Switchgear *IEEE Trans. on Appl. Supercond.* **26** 5603405
[21] Jiang X, Cook W B and Hackenberger W S 2009 Cryogenic piezoelectric actuator *NASA Technical Reports*
[22] Allegranza C, Gaillard L, Le Letty R, Patti S, Scolamiero L and Toso M 2014 Actuators for space applications: state of the art and new technologies. *Int. Conf. on New Actuators,* (Bremen: Actuator) pp 283-8
[23] Parmar C, Waris M and Pandya V 2016 Design of high voltage full bridge driver for piezoelectric actuator for space applications. *Int. Conf. on Power Syst.* , (New Delhi: IEEE) pp 1-5
[24] Graber L, Smith S, Soto D, Nowak I, Owens J and Steurer M 2015 A new class of high speed disconnect switch based on piezoelectric actuators. *Elect. Ship Technol. Symp.* , (Washington D.C.: IEEE) pp 312-7

Developmental Challenges of SMES Technology for Applications

Charles C. Rong and Paul N. Barnes
U.S. Army Research Laboratory, Adelphi, MD 20783
E-mail: charles.c.rong.civ@mail.mil; paul.n.barnes.civ@mail.mil

Abstract. This paper reviews the current status of high temperature superconductor (HTS) based superconducting magnetic energy storage (SMES) technology as a developmental effort. Discussion centres on the major challenges in magnet optimization, loss reduction, cooling improvement, and new development of quench detection. The cryogenic operation for superconductivity in this technological application requires continued research and development, especially with a greater engineering effort that involves the end user. For the SMES-based technology to more fully mature, some suggestions are given for consideration and discussion.

1. Introduction

Superconducting magnetic energy storage (SMES) technology development has received a renewed focus worldwide, especially with the use of the high-temperature superconductors of BSCCO, MgB_2, and REBCO. The increased developmental investments are directed to applications related to both traditional electric systems and renewable energy conversion. As a short-term energy storage technology, the SMES can provide the benefits of voltage stabilization (momentary voltage spikes and sags), load fluctuation compensation, and improved power quality [1]. In principle, the SMES possesses characteristics of quick response to energy demand and high efficiency [2]. Additional potential niche application areas may include utility in aerospace propulsion [3] and pulsed power source development [4-5].

Although superconducting machines and devices are widely used in laboratories and hospitals at liquid helium temperatures (e.g. NMR and MRI), many applications using superconductors require higher operating temperatures to be more practical. In the case of the SMES, as various other applications, the use of high-temperature superconductors in these systems is not yet adequately developed to take advantage of the higher current carrying capacity under higher magnetic fields at higher operating temperatures, especially above 20 K. This paper reviews some of recent research and development efforts that have brought the SMES incrementally closer to technology maturation, and provides some future considerations for subsequent development.

2. Recent Progress in HTS SMES Development

2.1. Magnet Design Optimization

To improve the magnet performance, efforts focus in areas of coil/magnet configuration optimization, conductor choice, and magnet geometry designs.

2.1.1. Optimization of Solenoid Configuration

For a solenoid magnet, the magnet's energy storage capacity and the upper limit of the attainable magnet field depend partially on the optimization of the solenoid's aspect ratio. A computation effort was carried out [6] based on HTS ribbon conductor with the same specified conditions: a coil volume of 5000 cm^3, a current density of 211 A/mm^2, and a stress of 500 MPa. It was determined that short coils with thick windings will provide a higher stored energy, which in this case was 4237 kJ. If using this conductor with attributes of 500 MPs stress to make a coil weighing 400 kg and operating at 20 K, the specific stored energy will be more than 10 kJ/kg [6]. Development of this magnet will produce a record of storage capacity per unit mass over the most efficient and elegant NbTi coils [7].

This leads to another consideration, the optimization of the coil configuration at the end of the solenoid. Toward this end, models have analysed the transport current performance of a high-temperature superconductor coil and demonstrated that relatively large electric fields appear at the coil's edges [8]. Because of these large magnetic fields, design efforts were undertaken to modify the end coils in order to reduce the component of perpendicular field that the coils are subjected to. One means to accomplish this is to make those coils located at the end of solenoid to be separated with graded heights (i.e. reduced pancake height) [8]. This graded heights of the coils will help minimize the amount of conductor that is subjected to the large fields in these locations at the end of the solenoid magnet. Another separate design approach [9] is to make solenoidal coils with an outward step-shaped configuration near the end, i.e., thinner pancake coil with reduced difference between outside diameter and inner diameter. In essence, the design is to minimize the amount of conductor that is subjected to nearly perpendicular fields in the magnet, with the same goal as the other design [8]. For a fixed amount of tape usage in the entire magnet, this step-shaped coils can increase the critical current to about 1.4 times that with the traditional rectangular-shaped coils [9].

2.1.2. Hybrid HTS Magnet Configuration

The hybrid magnet configuration basically consists of two or more different types of conductors with different superconducting materials. Often some of the coils are wound using the same type of superconducting wire. While the use of a single type of wire is convenient, wires can be specially designed for different sections of the magnet. Consider the previous section where the magnetic field imposes more on the two ends of the solenoid. A magnet may be designed to take into consideration the field's directionality. Most companies are already offering different HTS tapes based on the given application, both in performance and size. Although tailored versions of the same HTS material may be a desired choice, the two examples below make use of different generations of HTS wire, 1G BSCCO and 2G YBCO in the magnet design and fabrication.

Based on the behaviour of the critical current of BSCCO and YBCO tapes in external magnetic fields at 77 K [10], i.e., the I_c of BSCCO decreased more than that of YBCO in the same perpendicular magnetic field, the hybrid magnet was built with YBCO coils on the top and bottom ends and BSCCO in the middle, where the perpendicular field is much smaller than at the ends. This hybrid SMES magnet was shown to have stored energy of 1.6 kJ with operating current of 54 A at 77 K, and 6.0 kJ of 109 A at 69 K, respectively. More recently, a conduction-cooled HTS hybrid magnet was reported [11] for a 150 kJ/100 kW SMES system. In this design the BSCCO coils were placed on the two ends of the magnet with YBCO in the middle, which is a reversal to the design mentioned above [10]. For this design, the critical current of a BSCCO coil falls to about one-third of its original value, and that of a YBCO coil falls to about half of its original value.

Other hybrid magnet designs were reported to have achieved record high levels of magnetic field either with different conductors [12], or with same conductor but multiple geometries [13]. These fruitful hybrid magnet developmental efforts suggest that there is potential to be further explored in the magnet design domain to accomplish specific but challenging goals with different applications.

2.1.3. Solenoidal and Toroidal Geometry

Although the solenoidal design is the most common, it is not the only possible configuration. The practical application of SMES entails not only an issue of efficiency, but also nearby exposure to large magnetic fields, among others. A recent effort focused on the development of a toroidal configuration for a high-temperature superconducting SMES, which offered a power rating of 1 MW for a duration of 5 seconds [14]. The effort provided a comparison of the toroidal design with the solenoidal geometry, taking all major factors into account such as conductor/cable choice, sizing of the magnet bore, corresponding coil design, and ac losses using 2G HTS YBCO tape in the form of a Roebel cable. With the same requirement of stored energy being 5.6 MJ, and overall inductance of 1.25 H for the coil to be able to deliver 5 MJ energy in five seconds, both solenoid and toroid magnet design parameters were presented along with the associated field maps, showing constant stray field lines with the solenoidal geometry, and virtually fully contained field with the toroidal geometry. The total length of HTS tape for the toroid and solenoid was 59.8 km and 68.1 km, respectively.

Numerical calculations and simulations additionally showed that for the same field level, only slightly higher losses per cycle per unit length of conductor occurred for the toroid, despite the much higher transport current of 300 A for the toroid (188 A for solenoid) [14]. This suggests that the losses are dominated by the applied field. The distribution of the length of the HTS tape versus the perpendicular field, to which the HTS tape is subjected when the SMES operates at the maximum current, was calculated for both the toroid and solenoid. By combining the losses per unit length of the conductor corresponding to each level of perpendicular field with the actual length of the conductor that is exposed to that level of field, a quantitative estimate was determined. While this is not a total loss since some other factors are not included in the consideration, it does provide a loss indicator which was 76.7 kJ for the toroidal and 31.0 kJ for the solenoidal design [14]. It is important to note that the toroidal configuration in this example is designed with pancake coils aligned in the toroidal geometry leaving spacings between the coils on the outside of the donut shape. In this case, the perpendicular field on the conductor is over the outside portion of each coil for the toroid, and for the solenoid it is only at the end region of solenoid (albeit the whole coil at the ends), thus, it can be understood that greater ac losses may well be the case for toroid than solenoid. Even though there are relative lower field(s) in the toroidal geometry than in the solenoidal geometry, it is noted that the field(s) is more even across the entire length of the magnet in the toroid.

Further simulation design effort for maximum magnetic energy storage with optimization [15] for both solenoid and toroid showed that the upper limit level of stored energy in solenoidal SMES is estimated to be 2.2 times of that of toroidal SMES with the same length and type of the conductors. However, lower leakage flux density, lower tensile mechanical stresses, and lower magnetization loss--all of these are desirable characteristics—are only provided by the toroidal magnet. It has also been shown that the toroidal magnet for SMES can be further improved by making the coil as mainly racetracks rather than circular shape, and by making additional inner racetracks inside the main racetrack so that the entire coil becomes a constant field unit [16]. This constant field toroidal SMES magnet configuration is estimated to be able to provide 1.6 times more stored energy than the non-constant field counterpart. However, the complexity of the design and the difficulty of manufacturing this device poses a greater challenge at present technology readiness levels [16].

Unlike a hybrid solenoid magnet discussed above, the same toroidal geometry of a 2.5 MJ SMES by either REBCO or BSCCO conductors were also reported [17] to show that YBCO may provide advantages of smaller magnet size, less total conductor length, and smaller ac losses; but the high magnetic flux density in YBCO toroid may lead to very high mechanical stresses.

2.2. *Loss Reduction in Magnet*

When rapidly adding or removing electrical current from the SMES magnet, losses will be generated with these fast current level changes. It is not only due to the hysteresis and coupling losses of the superconductor, but also to the eddy current losses in the metals used to make the high-temperature superconducting wire, for example, substrates and stabilizers. A 'loss indicator' has to be developed for the magnet design effort with the information and knowledge on the loss properties of the superconducting wire/tape as conductor, cable and coil as component, and magnet as device. Some of the reported work is discussed below, although more effort to understand all the factors that contribute to the losses and to understand the interplays/interactions among those factors should come along in the near and far future.

A 28 pancake coil unit made of BSCCO conductor was simulated [18] to reveal the ac loss distribution among the coils under the condition of steady state current I_o = 80 A and a linear dynamic current I_d = 3A in 2 seconds duration. The results showed that higher ac losses occurred with pancake number 1 to 6 and 23 to 28 at the both ends of the solenoid, while 7 to 22 in the middle had smaller ac losses. When

I_d changed from 3 A to 20 A, the increased losses were mostly concentrated in pancake number 1 to 6 and 23 to 28, whereas number 7 to 22 had minimal increase in losses.

Experimental investigation of four YBCO coils [19] that were wound with YBCO with no insulation, YBCO co-wound with copper, YBCO co-wound with stainless steel, and YBCO co-wound with Kapton under varying operating current conditions from 10 A to 80 A at 150 Hz frequency, showed that ac losses for the no-insulation coil (bare wire itself) and copper co-wound coil are much lower than for those using stainless steel and Kapton materials along the conductor, and it was reasoned that lower radial resistance in the formers allows bypassed ac current through the turn-to-turn contact.

Analysis of Eddy current losses on a 2.5 MJ HTS toroidal SMES [20] with YBCO conductor illustrated that engineering effort on the thermal conduction plate and the centre ring was effective to reduce the Eddy current losses. The engineering effort includes making a slit in the bobbin to reduce the induced Eddy current in the metal based thermal conduction plate and in the centre ring, with some trade-off in the mechanical integrity of the SMES structure.

2.3. *Cooling Improvement*

For conduction cooling, it was reported that paraffin impregnation of copper and stainless steel laminated tapes can improve the cooling effects by increasing the contact areas between the copper-stainless steel tape and the cooling copper plate [21].

Mixed cryogen cooling techniques also showed improved thermal stability with the components of solid nitrogen-liquid nitrogen, solid nitrogen-liquid neon, and solid argon-liquid nitrogen [22], with the case of solid cryogen used alone as the baseline for comparison.

Based on cooling requirements needed by different SMES applications, a variety of cryocooling approaches are available for consideration in the design of SMES system, examples include closed re-condensing cryostat, variable operating frequency of the compressor to allow adjustment of refrigeration capacity at 4 K, 20 K, and 77 K, in order to reduce power consumption [23]. High-efficiency conduction cooling technology development was also reported recently [24].

For better cooling control and management, a promising development was reported on built-in cryogenic oscillating heat pipe (OHP) in HTS magnet [25]. The cryogen media employed in the oscillating heat pipe can be hydrogen (17-30 K), neon (26-39 K), or nitrogen (67-91 K). Moreover, a flat-plate cryogenic OHP has been developed, which is suitable for embedding in magnet windings to increase the thermal conductivity and the thermal diffusivity at the same time [26].

2.4. *Quench Detection and Protection*

Traditional quench detection for LTS magnet is based on small voltage signal changes caused by normal transition, but generally this approach is difficult to apply to HTS because of the very slow velocity of normal-zone propagation and relatively large voltage noise from other parts of the system compared to the local normal-zone voltage. However, an accurate and sensitive measurement of the dissipative voltage on an HTS coil was demonstrated by a new detection circuit design with two stage SMES detection scheme [27] that is able to detect a quench regardless of the coil discharge voltage in operation.

A new approach to current detection for HTS coil for quench event was proposed in 2011 [28] and has been in development with simulation and experimentation [29-30]. The concept is to design laminated bundled YBCO conductor that consists of four insulated parallel tapes labelled as 1, 2, 3, and 4, see Figure 1 and 2. This laminated bundled conductor is wound as Pancake Coil 1. Pancake Coils 2, however, is made similarly with four insulated parallel tapes but labelled as 2, 1, 4, and 3. Pancake Coil 1 is connected to Pancake Coil 2 by the four tapes with only the same number as 1 to 1, 2 to 2, 3 to 3, and 4 to 4. In this way, supercurrent carried by tape 1 in Coil 1 will only go to tape 1 in Coil 2, in which tape 1 is no longer in the edge position as in Coil 1, but in the position neighbouring with tape 2 and 4 as indicated by 2, 1, 4, and 3 that are defined for Coil 2. Tapes 2, 3, and 4 in Coil 1 will be also connected to corresponding 2, 3, and 4 in Coil 2 as defined as 2, 1, 4, and 3. Therefore, four tapes 1, 2, 3, 4 in Coil 1 switched their positions in Coil 2 by design as 2, 1, 4, and 3. The design for Coil 3 and 4 with the locations of four tapes is 3, 4, 1, and 2 for Coil 3, and 4, 3, 2, and 1 for Coil 4. Tape 1 will be connected only to tape 1 in all the coils such that supercurrent will travel at four different locations defined in each

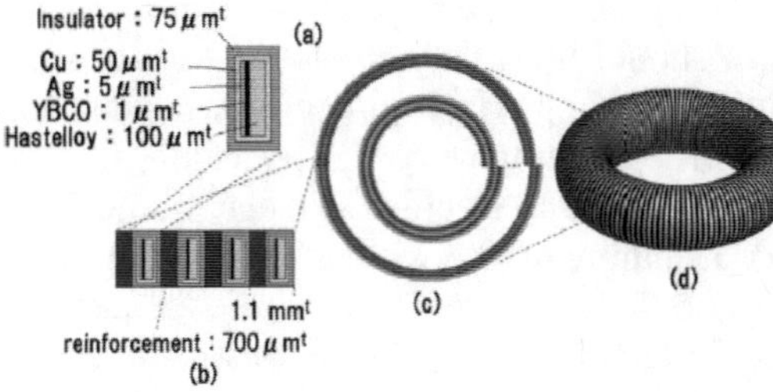

Figure 1. Schematic drawings of (a) YBCO tape, (b) laminated bundle conductor, (c) pancake coil, and (d) toroidal coil. Adapted with permission from Ref. [28], copyright 2011, IEEE.

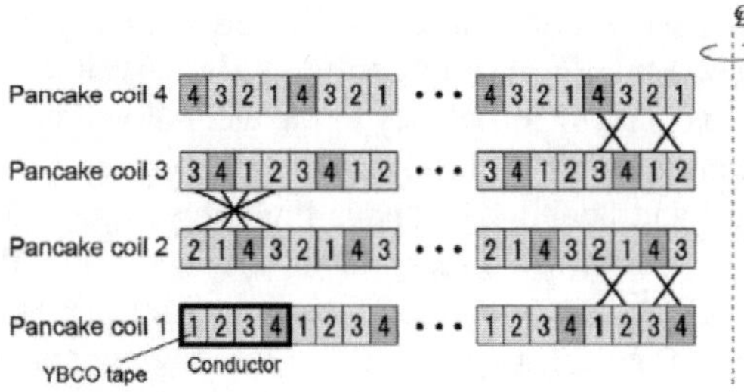

Figure 2. Schematic drawing of transposition in a unit coil. A unit coil consists of four pancake coils, each of which is wound with a laminated bundle conductor composed of four YBCO tapes. Adapted with permission from Ref. [28], copyright 2011, IEEE.

coil from 1, 2, 3, and 4 in Coil 1 and will end up in Coil 4 with 4, 3, 2, and 1. Tape 2, 3, and 4 will follow the similar pattern respectively. This supercurrent transposition will allow observation of any non-uniform current in YBCO laminated bundled conductor, and the non-uniform current occurring can be indicative of a possible quench event. This new coil design with the connection pattern established the base for current based quench detection method [28]. Experimental verification [30] of the concept showed that local normal transition (i.e. the non-uniform current occurring) on one tape could be detected with high accuracy and the associated temperature change was less than 15 K. In principle, once a local normal transition is detected by monitoring the supercurrent level in tape 1, 2, 3, and 4, measures can be taken to protect the tape/coil/magnet.

2.5. Fast Response to Power Demand and High Energy Density

In order to charge-discharge the superconducting coils as fast as possible to meet power demand, a recent paper [31] experimentally examined the effect of improved winding geometry with a quasi-turn-to-turn incomplete insulation that used a narrowed polyimide film of 3.5 mm for a 4.8 mm width YBCO conductor tape. This coil winding additionally created six times more contact area for the coil to contact with cryogen in comparison to conventional fully insulated winding in which only the top and bottom areas of the pancake are actually in contact with the cryogen. Due to this increased cooling contact margin, experimentally the coil could work under over-current conditions and generate relatively higher magnetic field linearly up to ~1.5 times its critical current.

Besides the superconducting magnet, the development of the interface between the dc magnet and the ac power system, as well as the control system made of efficient electronics, are all equally challenging undertakings. As a possible strategy, a fuzzy logic control method was demonstrated [32]. In comparison to metallic oxide semiconductor field effect transistor (MOSFET) based bridge-type chopper controlled SMES system, the SMES with the fuzzy control power electronics will have better robust performance in ideal and disturbance situations (such as in response to power demand in real world). Also communication between the controller and SMES components was shown to be more effective and faster so that shorter response time and safer operation of the SMES system may be ensured [32].

As an energy storage device, the energy density of SMES is always an important characteristic for technology development. Several papers have been published on optimization and design

IOP Conf. Series: Materials Science and Engineering **279** (2017) 012013 doi:10.1088/1757-899X/279/1/012013

considerations of high-energy density SMES [4-5, 33]. To design and develop a high performance SMES with capability to deliver large amounts of energy in a short time, multiple parameter studies and detailed optimizations are needed in order to determine the best suitable topology of SMES [32] to meet those requirements. In terms of optimization, even if only a strongly constrained problem is given, a large space of solutions could still be available [4], which implies that significant effort in simulations and computations are not only highly desirable, but also should be widely employed to enhance the effectiveness and efficiency of the technology development endeavour. In a word, efforts in simulation, computation, and optimization are necessary for HTS SMES development. For example, it is very helpful and beneficial to have the knowledge and understanding that in order to reach the goal of an energy density of 20 kJ/kg (cold mass), we would need, at minimum, the conductor to sustain a hoop stress of 500 MPa with engineering current density (J_e) of 600 A/mm^2, and to fabricate a magnet that has a storage capacity of 1 MJ [5] as the core component for the HTS based SMES, among others.

3. Major Engineering and Technology Challenges

Multi-component/parameter interactions with high sensitivity in undetermined magnitude might contribute to the overall complexity of superconductivity based technology, especially with respect to applications under dynamic conditions that SMES will see. The significant progress reported in the literature contributions cited in this paper (not an exhaustive coverage) suggests more fundamental knowledge be learned and engineering expertise gained for further development of the SMES technology. Although some brief discussion is outlined in the following three sections, many real practical issues, as noted, are more complex than what a few examples can fully represent and illustrate.

3.1. Temperature-related influences

At least for now, effective management of heat flux dynamics in magnet with the best possible heat transfer techniques proves to be of help to the development of SMES technology.

3.2. Multi-component interactions

The superconductors, cables, and magnets are the key components of the SMES, and Ref. [5] provided useful information for building a SMES magnet with a targeted energy density of 20 kJ/kg. Computation and simulation applying well-defined conditions may produce reasonable and meaningful results. In reality, there could be some uncertainties introduced from the process of manufacturing of the conductors/cables, as well as the winding process of the magnet. The consistent performance of the conductors is desired in the first place, and the overall consistency of the magnet as the core component is also expected. The fabrication process to make a magnet may bring in additional variations that should be understood in terms of electric, magnetic, mechanic, and thermal interactions in the configuration of the magnet. And those variables should be well controlled in order to achieve the overall desired performance of the SMES technology. Furthermore, how the dynamic energy input and output conditions presented by real application affects the performance of the SMES magnet should be also learned. After all, a higher degree of optimization effort through extensive testing and design iterations will be essential to ensure the consistency and reliability of a matured technology.

3.3. Specific electromagnetic effects

To better contain the stray magnetic field while also increasing the energy capacity of a racetrack toroidal magnet, the effort described in [16] has shown that the individual coil unit can be made with constant field characters, resulting in 1.6 times more stored energy, see Section 2.1.3. However, due to the concerns of the complexity and difficulty related to the design and manufacturing, it was not recommended as a practical approach by the researchers. This is understandable, given the current knowledge level, engineering ability, and projected economic viability of the constant field racetrack toroidal SMES design. Nevertheless, potential in the current technology landscape still exists and should be further explored in terms of better understanding the electric and magnetic effects in other possible geometries where the magnet components may be arranged in such a way as to achieve any specific effects. This is still a relatively less developed area but with possibility of new discovery and innovation.

4. Summary/Future Considerations

There are some future considerations for the subsequent development of HTS SMES.

4.1. Simulation and computation effort

Extensive simulations and computations should be performed on the SMES system design along with experimental verification and validation. It will be important to map out the three dimensional space of the associated magnetic field for particular designs resulting from the electric current in the SMES coils. Additional developmental issues for modelling and simulation include structural and heat management parameters.

4.2. Testing effort

Conduct comprehensive and thorough testing at component and system levels. A higher degree of optimization through extensive testing and design iterations for all aspects is needed to ensure the consistency and reliability of any superconductor product. This is not simply testing the SMES functionality by itself but especially for its given use. Testing needs to be developed that will determine the utility of SMES prototypes in an electrical grid for achieving its desired purpose. This will entail a comparison to other systems that may achieve similar goals.

4.3. Collaboration

Enhance concerted/coordinated effort to develop collaborations to achieve unrealized synergies that will increase the outcome of the R&D effort, which will accelerate the pace of superconducting technology development. This must include the technology users to ensure proper development of the system.

4.4. Readiness of technology

The ultimate determining factors for a useful technology, including superconductivity, will be the higher maturity levels that can be reached and the system affordability. The superconductivity technology must offer consistent performance in applications with the necessary reliability. Again, this must all be ultimately accomplished with the system being offered at a reasonable cost.

5. References

[1] Buckles W, Hassenzahl W V 2000 Superconducting Magnetic Energy Storage *IEEE Power Engineering Review* **20** 16

[2] Jones C E, Norman P J, Galloway S J, Armstrong M J, Bollman A M 2016 Comparison of Candidate Architectures for Future Distributed Propulsion Aircraft *IEEE Trans. Appl. Supercond.* **26** 3601409

[3] Farhadi M, Mohammed O 2016 Energy Storage Technologies for High-Power Applications *IEEE Trans. Ind. Appl.* **52** 1953

[4] Vincent B, Tixador P, Lecrevisse T, Rey J -M, Chaud X, and Miyoshi Y 2013 HTS Magnets: Opportunities and Issues for SMES *IEEE Trans. Appl. Supercond.* **23** 5700805

[5] Ciceron J, Badel A, Pascal Tixador P, Forest F 2017 Design Considerations for High-Energy Density SMES *IEEE Trans. Appl. Supercond.* **27** 5700705

[6] Rey J M, Bruchon M, Chaud X, Debray F, Lécrevisse T, Mossang E, Tixador P 2011 Geometry Optimization for SMES Solenoids Using HTS Ribbons *IEEE Trans. Appl. Supercond.* **21** 1670

[7] Yamamoto A 2002 A thin superconducting solenoid magnet for particle astrophysics *IEEE Trans. Appl. Supercond.* **12** 438

[8] Ishiguri S, Funamoto T 2011 Performance improvement of a high-temperature superconducting coil *Physica C* **471** 333

[9] Chen X Y, Jin J X 2014 Evaluation of Step-Shaped Solenoidal Coils for Current-Enhanced SMES Applications *IEEE Trans. Appl. Supercond.* **24** 4603404

[10] Zhu J, Qiu M, Wei B, Zhang H, Lai X, Yuan W 2013 Design, dynamic simulation and construction of a hybrid HTS SMES (high-temperature superconducting magnetic energy storage systems) for Chinese power grid *Energy* **51** 184

[11] Ren L, Xu Y, Liu H, Liu Y, Han P, Deng J, Li J, Chen J, Shi J, Tang Y 2015 The Experimental

Research and Analysis of a HTS SMES Hybrid Magnet *IEEE Trans. Appl. Supercond.* **25** 4601605

[12] Markiewicz W D, Larbalestier D C, Weijers H W, Voran A J, Pickard K W, Sheppard W R, Jaroszynski J, Xu A, Walsh R P, Lu J, Gavrilin A V, Noyes P D 2012 Design of a Superconducting 32 T Magnet With REBCO High Field Coils *IEEE Trans. Appl. Supercond.* **22** 4300704

[13] Gupta R, Anerella M, Joshi P, Higgins J, Lalitha S, Sampson W, Schmalzle J, and Wanderer P 2016 Design, Construction, and Testing of a Large-Aperture High-Field HTS SMES Coil *IEEE Trans. Appl. Supercond.* **26** 5700208

[14] Morandi A, Fabbri M, Gholizad B, Grilli F, Sirois F, Zermeño V M R 2016 Design and Comparison of a 1-MW/5-s HTS SMES With Toroidal and Solenoidal Geometry *IEEE Trans. Appl. Supercond.* **26** 5700606

[15] Hekmati A, Hekmati R 2016 Double pancake superconducting coil design for maximum magnetic energy storage in small scale SMES systems *Cryogenics* **80** 74

[16] Radovinsky A L, Bromberg L, Minervini J V, Michael P C, Servais T, Forton E, Pearson E 2016 Constant Field Toroidal SMES Magnet *IEEE Trans. Appl. Supercond.* **26** 5700104

[17] Lee S, Kim W S, Park S H, Lee J K, Park C, Bae J H, Seong K C, Lee H, Lee J H, Choi K, Hahn S 2010 Design of HTS Modular Magnets for a 2.5 MJ Toroidal SMES: ReBCO vs. BSCCO *IEEE Trans. Appl. Supercond.* **20** 1324

[18] Xu Y, Tang Y, Ren L, Jiao F, Song M, Cao K, Wang D, Wnag L, Dong H 2013 Distribution of AC loss in a HTS magnet for SMES with different operating conditions *Physica C* **494** 213

[19] Kim J, Nam S, Jeon H, Kim J, Jang J Y, Ko T K 2016 Experimental Analysis on AC Loss and Fault Current Test of HTS Coils Co-Wound With Various Inserted Materials *IEEE Trans. Appl. Supercond.* **26** 8201605

[20] Lee S, Park S H, Kim W S, Lee J K, Lee S, Park C, Bae J H, Kim S H, Seong K C, Choi K, Hahn S 2011 Analysis of Eddy Current Losses and Magnetization Losses in Toroidal Magnets for a 2.5 MJ HTS SMES *IEEE Trans. Appl. Supercond.* **21** 1354

[21] Kotoyori T, Ohnishi H, Masui Y, Ishiyama A, Tomonori W, Hirano N, Nagaya S, Shikimachi K 2013 Evaluation of Conduction Cooling Effect of Cryocooler-Cooled HTS Coils for SMES Application *IEEE Trans. Appl. Supercond.* **23** 3800604

[22] Song J B, Lee H 2012 Mixed cryogen cooling systems for HTS power applications: A status report of progress in Korea University *Cryogenics* **52** 648

[23] Hirano N, Watanabe T, Nagaya S 2016 Development of cooling technologies for SMES *Cryogenics* **80** 210

[24] Nagaya S, Hirano N, Naruse M, Watanabe T, Tamada T 2013 Development of a High-Efficiency Conduction Cooling Technology for SMES Coils *IEEE Trans. Appl. Supercond.* **23** 5602804

[25] Mito T, Natsume K, Yanagi N, Tamura H, Tamada T, Shikimachi K, Hirano N, Nagaya S 2011 Achievement of High Heat Removal Characteristics of Superconducting Magnets With Imbedded Oscillating Heat Pipes *IEEE Trans. Appl. Supercond.* **21** 2470

[26] Mito T, Natsume K, Yanagi N, Tamura H, Terazaki Y 2013 Enhancement of thermal properties of HTS magnets using built-in cryogenic Oscillating Heat Pipes *IEEE Trans. Appl. Supercond.* **23** 4602905

[27] Badel A, Tixador P, Simiand G, Exchaw O 2010 Quench detection system for twin coils HTS SMES *Cryogenics* **50** 674

[28] Ishiyama A, Ueda H, Aoki Y, Shikimachi K, Hirano N, Nagaya S 2011 Quench Behavior and Protection in Cryocooler-Cooled YBCO Pancake Coil for SMES *IEEE Trans. Appl. Supercond.* **21** 2398

[29] Ueda H, Ishiyama A, Muromachi K, Suzuki T, Shikimachi K, Hirano N, Nagaya S 2012 Quench Detection and Protection of Cryocooler-Cooled YBCO Pancake Coil for SMES *IEEE Trans. Appl. Supercond.* **22** 4702804

[30] Masui Y, Wang X, Ishiyama A, Watanabe T, Hirano N, Nagaya S, Ueda H 2014 Experiment and

Numerical Simulation on Quench Detection in Cryocooler-Cooled YBCO Coil for SMES Application *IEEE Trans. Appl. Supercond.* **22** 4601005

[31] Kim H, Kim J, Lee J, Ko T K 2015 HTS coil with enhanced thermal stability in over-current operation for fast response magnet power application *Supercond. Sci. Technol.* **28** 105006

[32] Wang S, Jin J 2014 Design and Analysis of a Fuzzy Logic Controlled SMES System *IEEE Trans. Appl. Supercond.* **24** 5701205

[33] Tixador P, Nguyen N T, Rey J M, Lecrevisse T, Reinbold V, Trophime C, Chaud X, Debray F, Semperger S, Devaux M, Pes C 2012 SMES Optimization for High Energy Densities *IEEE Trans. Appl. Supercond.* **22** 5700704

Acknowledgements

We are quite appreciative of the insightful comments by two anonymous reviewers of this manuscript. One of them also kindly provided detailed and thorough critique in many aspects, which is particularly acknowledged.

IOP Conf. Series: Materials Science and Engineering **279** (2017) 012014 doi:10.1088/1757-899X/279/1/012014

Experimental investigation into the fault response of superconducting hybrid electric propulsion electrical power system to a DC rail to rail fault

S. Nolan[1,3], C.E. Jones[1], R. Munro[1], P. Norman[1], S. Galloway[1], S. Venturumilli[2], J.Sheng[2], W. Yuan[2].

[1]Department of Electronic and Electrical Engineering, University of Strathclyde, Glasgow, UK.
[2]Department of Electronic and Electrical Engineering, Bath University, Bath, UK.
[3]Steven.nolan@strath.ac.uk

Abstract. Hybrid electric propulsion aircraft are proposed to improve overall aircraft efficiency, enabling future rising demands for air travel to be met. The development of appropriate electrical power systems to provide thrust for the aircraft is a significant challenge due to the much higher required power generation capacity levels and complexity of the aero-electrical power systems (AEPS). The efficiency and weight of the AEPS is critical to ensure that the benefits of hybrid propulsion are not mitigated by the electrical power train. Hence it is proposed that for larger aircraft (~200 passengers) superconducting power systems are used to meet target power densities. Central to the design of the hybrid propulsion AEPS is a robust and reliable electrical protection and fault management system. It is known from previous studies that the choice of protection system may have a significant impact on the overall efficiency of the AEPS. Hence an informed design process which considers the key trades between choice of cable and protection requirements is needed. To date the fault response of a voltage source converter interfaced DC link rail to rail fault in a superconducting power system has only been investigated using simulation models validated by theoretical values from the literature. This paper will present the experimentally obtained fault response for a variety of different types of superconducting tape for a rail to rail DC fault. The paper will then use these as a platform to identify key trades between protection requirements and cable design, providing guidelines to enable future informed decisions to optimise hybrid propulsion electrical power system and protection design.

1. Introduction

It is expected that by 2030, 32,600 new aircraft will be required to replace aging existing fleets and support the anticipated growth in demand for air travel [1]. A number of government agencies have outlined development goals that future aircraft must reach if the industry is to able to grow while meeting regulatory stipulations [2, 3]. These goals encompass CO_2 emissions, noise emissions, efficiency and runway field length. To meet these targets, new and innovative design approaches are required for both the drivetrain and the aircraft body itself. One concept that has the potential to realise significant gains is turbo-electric distributed propulsion (TeDP) [3]. Unlike conventional aircraft, which utilise gas turbine engines to provide thrust, it is proposed that thrust on a TeDP aircraft is produced from electrically driven propulsor motors, with gas engine driven generators utilised to provide this electrical power. A key advantage of TeDP is the flexibility in where the propulsor motors can be placed on the aircraft which can lead to significant aerodynamic and practical advantages [3]. In order for a TeDP architecture to be feasibly placed on an aircraft, the power densities of the components must be extremely high. For this reason, it has been suggested that superconducting materials should be used for the machines as this can improve the power density by a factor of four [4].

While the weight is a significant factor, it is also critical that the design of the electrical power system does not mitigate the overall aircraft performance benefits of TeDP. A key design choice is whether the transmission and distribution system should be AC or DC. Although it has been shown that an AC superconducting system can result in a system with lower mass [5], there are a number of advantages to DC such as allowing for more efficient electrical decoupling of the generation and propulsion sources and offering greater control opportunities. This paper considers the fault response in such a power-electronics interfaced, DC system. The TeDP electrical power system will require appropriate electrical protection methods to be in place so that in the event of an electrical fault, component damage is minimized and the wider system is able to maintain stability.

One key area identified as an issue for protection system design is the discharge of a DC link filter capacitor following a short circuit condition. Due to the lack of impedance offered by superconducting components in the initial moments of a fault, the fault current can be extremely large. Electrically, the capacitor is closer to the fault than the generators. Thus, the peak fault current is shaped in the first instance by the capacitive discharge [6]. This paper explores these initial moments of a DC link capacitor discharge experimentally through superconducting tapes and a cable section in order to assess the response of the superconducting material to a large pulse of current at a range of voltage levels. With respect to these results, this paper will discuss the implications for protection system requirements and the stability of the wider network.

2. Experimental Setup

The aim of the experiment was to capture the response of a capacitive unit discharging through superconducting tapes. To achieve this, the experiment proceeded in two phases. The first phase of the experiment was to charge of a 47 mF, 100V rated capacitor using a controlled voltage source. This was disconnected from the circuit using a mechanical breaker following the charging phase completion. The second phase of the experiment was the discharge of the capacitor. This was realised through the operation of three MOSFETs [7], connected in parallel to ensure that the rated current of the individual MOSFETs was never exceeded. These were controlled via LabVIEW. An inductor was embedded within the discharge circuit to ensure a smooth discharge curve. The discharge circuit was connected to the superconducting tapes through 100 A rated copper cables. All superconducting devices in this experiment used YBCO as the active material. In each experiment the superconducting components were submerged in a bath of liquid nitrogen to achieve an operating temperature of 77 K. Figure 1 shows the circuit used for the experiments. To provide a baseline for comparison, Figure 2 shows the current discharge through a copper cable rated for 100 A continuous current.

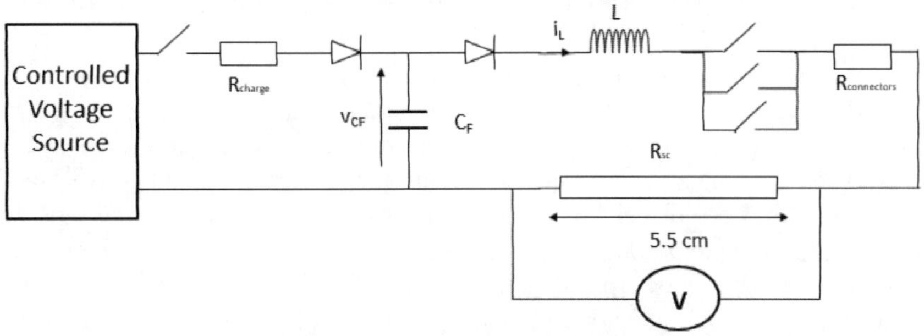

Figure 1. Circuit for Experiment

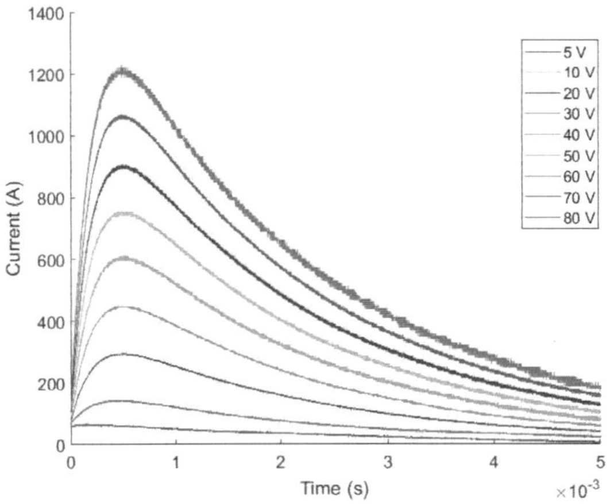

Figure 2. Discharge of capacitor through 100 A Copper Cables

3. Experimental Results

3.1 100 A Copper-Stabilized Tape Discharge

A controlled discharge of the capacitor was carried out through a 100 A Superpower Inc. tape at a range of pre-fault voltage levels from 5V to 80V. The superconducting tape was 4 mm wide, 10 cm long with a 100 μm copper stabilizer. Figure 3 (a) shows the discharge current for each of these tests while figure 3 (b) shows the voltage drop produced across the superconducting tapes.

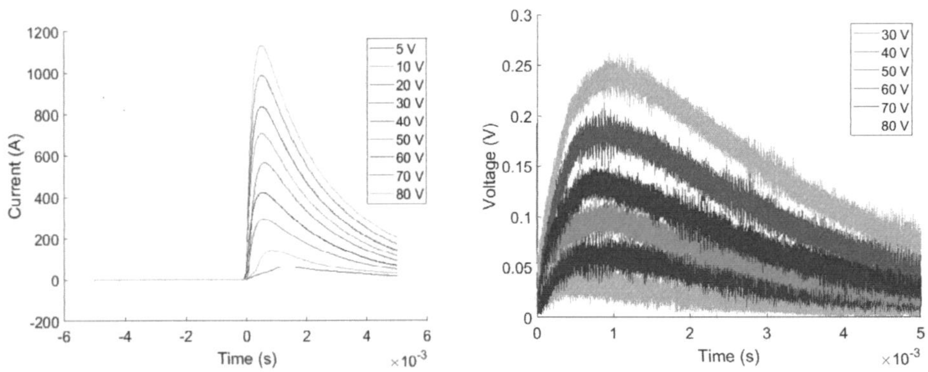

Figure 3. Stabilized Tapes (a) Current Profile and (b) Voltage Developed

It can be seen by comparing the current profile between Figures 2 and 3 (a) that a slight reduction in the peak current occurs due to the electric field generated across the tapes. From Figure 3 (b) it can be seen that above a 30 V charging voltage level, that the superconducting tape starts to produce a noticeable voltage signal between the measurement taps, which are placed across 5.5cm of the tape length. This indicated that quench had taken place. The critical current of the superconducting tape was measured before and after the discharge experiment and the results indicated that no degradation of I_c had taken place. This is because the energy dissipated in the superconductor during the experiment resulted in a temperature rise which was not high enough to cause degradation to take place, normally estimated as 400 K [8]. This demonstrates that these tapes are capable of withstanding short periods of transient overcurrents many multiples of the critical current, I_c. However, they will have a negligible effect on reducing fault current experienced by the system because of the low impedance path offered by the copper stabilizer.

3.2 420 A Cable Discharge

The 420 A superconducting cable was constructed of a 20 mm^2 copper former with 4 copper stabilized superconducting tapes wound around this core, each with a critical current of approximately 100 A at 77 K. The cable measured approximately 30 cm in length. The cable provided negligible impedance to the discharge current as evidenced by the current profile shown on Figure 4.

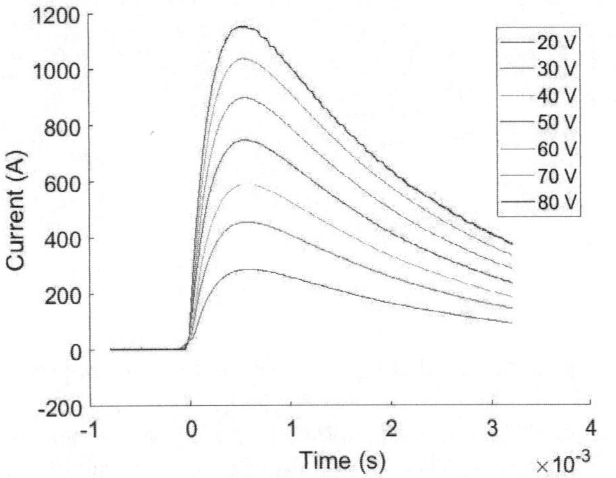

Figure 4. Current Profile of 420 A Cable Discharge.

The voltage profile is almost indiscernible from background noise, and as such is omitted from this paper. This is due to the copper former providing a very low impedance path for the excess fault current to traverse in the case of currents exceeding I_c. By inspection, neither the cable nor the stabilized tape are seen to have a significant impact on the fault current profile when compared to the discharge through the copper cable. Hence, without including additional damping, fault currents would be extremely high due to the initial discharge of DC link capacitors. A consequence of this would be that other components in the system, such as the diodes in the power electronics converters, would have to dissipate high amounts of fault energy. This potentially could result in the damage of components critical to the safe flight of the aircraft. The fast discharge of the capacitor in this very low impedance scenario could cause significant voltage depreciation throughout the system as DC-link capacitors connected across healthy branches begin to discharge into the fault. This would impact on voltage stability. To counter this, extremely fast acting protection systems may present a solution. It can be seen however that the critical current of the superconducting cable and tape does not degrade due to the high fault current, despite it being significantly above the critical threshold, indicating that these components are tolerant of large overcurrents due to the parallel paths provided by the copper materials used in each.

3.3 50 A Superconducting Tape Without Stabilizer Discharge

A controlled discharge was performed through a 50 A YBCO tape without a copper stabilizer, which was replaced by a thin layer (1um) of silver laminated across the superconducting material of the tape. The tape was 10 cm in length and 2 mm wide. Figure 5 (a) shows the current profile resulting from the capacitor discharge, and Figure 5 (b) shows the voltage profile. In contrast to the previous results, it is shown on Figure 5 (a) that there is a reduction in the peak current when compared to the original discharge through standard copper connectors. For the 80 V discharge test this reduction was approximately 700 A. The voltage drop across the superconductor is also orders of magnitude larger in comparison to the discharge through the copper stabilized tape. The voltage signal for the 80 V test (Figure 5 (b)) rapidly drops to zero after 0.0015 s. This is because the temperature reached by the superconducting material exceeded the melting point of the solder on the connections to the voltage measurement equipment, 461 K. Unlike the results presented for the copper stabilised cable, significant critical current degradation was observed following the 70 V discharge test for the tape without a stabilizer. The 80 V discharge completely destroyed the superconducting properties of the YBCO material. Interestingly, no physical damage was apparent upon a visual inspection.

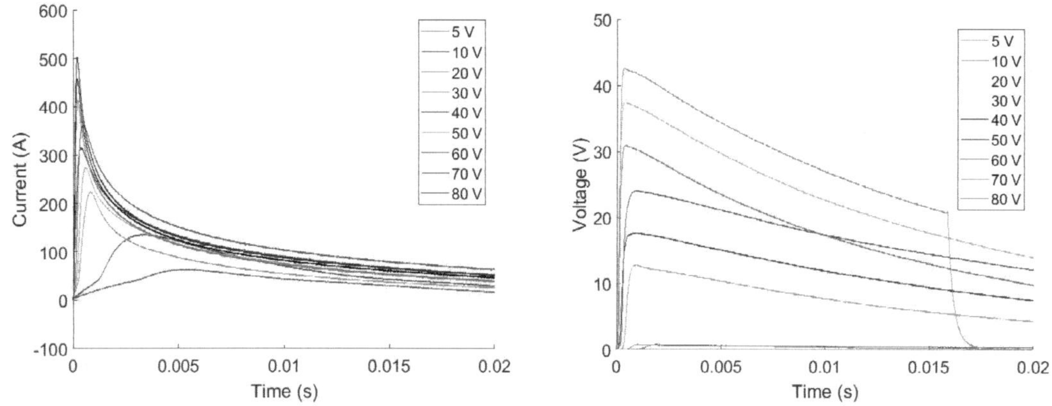

Figure 5. Non-stabilized superconducting tapes (a) Discharge current profile (b) Voltage drop.

Scanning electron microscope (SEM) analysis showed scarring of the superconducting layer at the microscopic level which indicates a possible reason for the destruction to the superconducting properties. Figure 6 shows the critical current of the superconducting tape following successive tests.

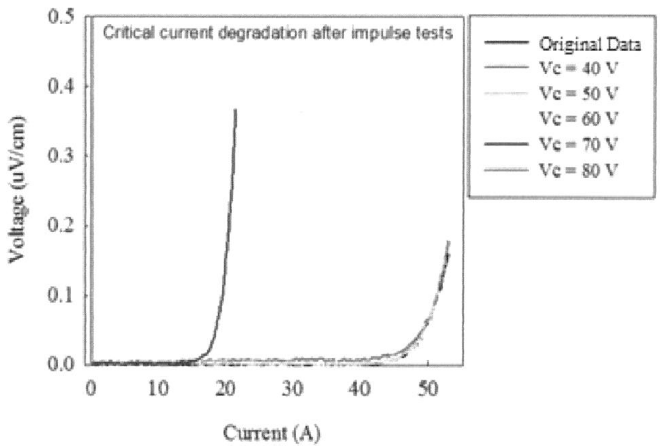

Figure 6. Critical Current Degradation of Unstabilized Tape.

This series of tests showed that the peak discharge current of the filtering capacitor is significantly reduced by using tapes that are highly resistive post quench. By consideration of Ohm's law, this is expected. This demonstrates the potential for using superconducting materials in the design of the transmission network which are highly resistive post quench, as a means to reduce the amount of fault current experienced by components on the network. This may allow for certain components to be de-rated, reducing the weight and volume penalties they incur. Further investigation into the benefits this would bring to the overall performance of the electrical power system is required, but is out with the scope of this paper. The discharge time constant is also significantly larger than the previous experiments that utilised copper to provide shunt paths, due to the higher post quench resistance. This may have benefits for post-fault voltage stability, due to the slower discharge of DC link filters, reducing the speed of fault propagation to healthy branches of the network. However, it must be noted that during this experiment the tape experienced significant thermal and electrical stress, eventually causing the complete breakdown of superconducting properties. In order to take advantage of this potential functionality, components would have to be designed to withstand the maximum current and temperature reached during the fault period. Temperature rise experienced will be directly related to the volume available for dissipating heat. As resistance is inversely proportional to the cable area, a trade-off between fault impedance and temperature rise is created.

3.4 Parallel Tape Arrangement

The parallel tape discharge test utilised 3 SuNAM tapes. Each tape was 6 mm wide, 10 cm in length, possessing a copper stabilizer with a 100 μm thickness, and with a critical current of 175 A. Above 50 V discharge voltage, the quench voltage became noticeable across the tapes. Figure 7 shows the voltage profile for a 50 V discharge test while Figure 8 shows the associated current distribution between the tapes.

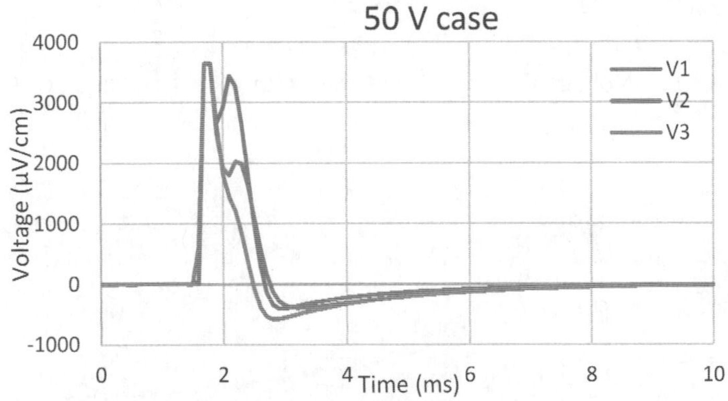

Figure 7. Voltage Across Tapes.

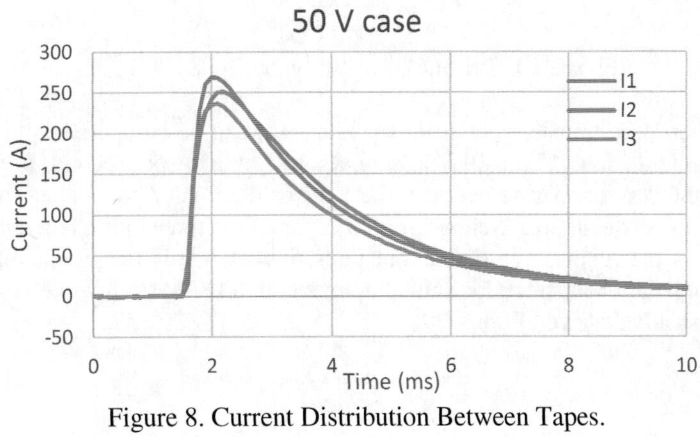

Figure 8. Current Distribution Between Tapes.

The small voltage signal highlights the difficulty of detecting the quench voltage at current levels below 1.3-1.5 I_c. It is also noted that the current is not shared equally between the tapes. As the voltage of the discharge is increased the uniformity of current sharing increases and the current becomes more evenly distributed between the three tapes. This is due to the inductive voltage produced by the tapes, created by the large rate of change of current outweighs the impedance offered by the quenching process and any contact resistances present in the tapes connections. This can make it difficult to detect the quench taking place as the detection mechanism will have to compensate for the inductive voltage produced by large transients. This could be even more prominent in cables with tapes wound helically as the inductance would be significantly larger than in the small tapes considered here. Figures 9 and 10 show the voltage drop across the tapes and the current distribution during the 80 V discharge test.

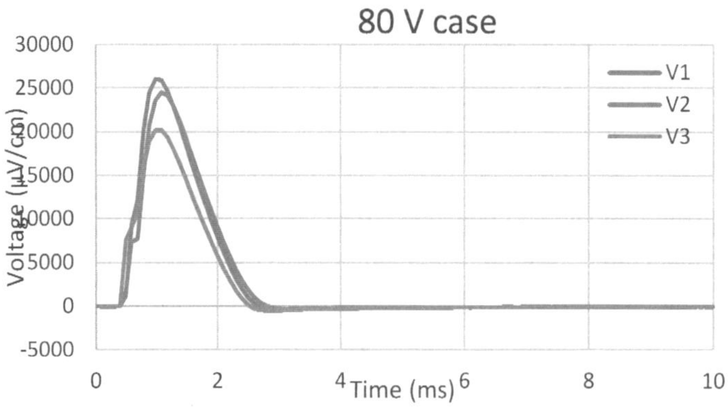

Figure 9. Voltage Drop Across the tapes for 80 V Discharge Test.

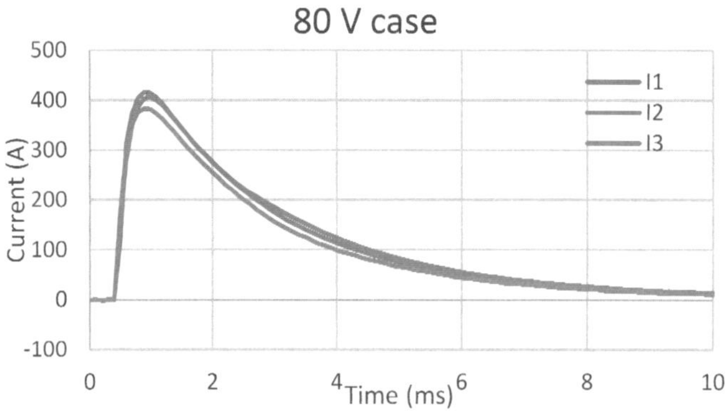

Figure 10. Current Distribution during 80 V Test

Current sharing under large transient conditions (Figure 10) can be compared to Figure 11 where the current is ramped up in a slowly controlled fashion at a rate of 10 A per second. In the load ramp case it can be seen that the current sharing between the tapes is dominated by contact impedance, causing significant differences between the currents carried by each of the three tapes, reaching a peak deviation of 30 %. This shows that under significant transients, inductance is the dominant factor in current distribution between parallel superconducting components in a DC system while contact impedance is more significant in steady state conditions.

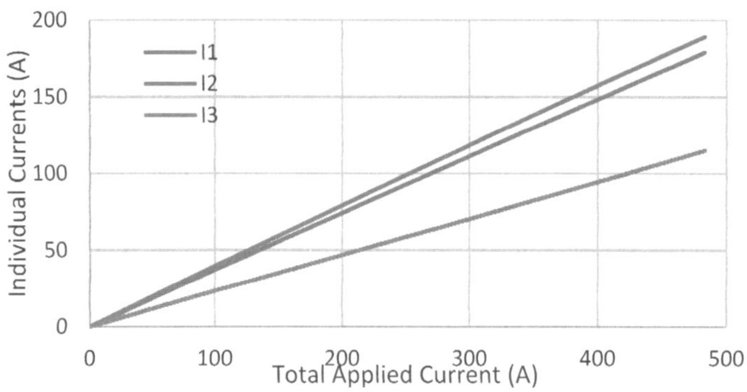

Figure 11. Current Sharing Between Tapes During Ramp Current Test.

4. Conclusion

This paper has shown the resulting current and voltage profiles for a series of capacitive discharge tests conducted in laboratory conditions. It has been shown that tapes and cables that utilise low impedance parallel paths will have negligible impact on fault current, which could result in significant peak currents forming. Although these components are unlikely to suffer degradation as a result of brief transients brought about by a DC link capacitive discharge event, the lack of resistive damping could result in voltage instability throughout all branches connected across a common link. Further work will investigate this potentially hazardous system condition and how solutions such as extremely fast acting circuit breakers can mitigate its effects. Conversely, materials with a high impedance parallel path will significantly reduce the fault current peak. This is at the expense of potential cable degradation if the shunt path is not sized appropriately, creating a protection trade-off between component temperature margins and wider system stability. Current distribution in parallel tapes is also explored, with the results indicating that inductance plays a major role in this process during large transient events while contact impedance dominates steady state situations. Future work will include introducing a DC offset to the fault current such that the fault response of a converter interfaced network can be emulated and protection algorithms can be experimentally evaluated.

5. Acknowledgements

This work was undertaken as part of the Rolls-Royce UTC Programme.

6. References

[1] Airbus, "Flying By Numbers", Airbus, [online] Available: http://www.airbus.com/company/market/forecast/, 2015, accessed on: 12th of May 2016.

[2] European Commission," Flightpath 2050 Europe's Vision for Aviation Report of the High Level Group on Aviation Research", European Union. Available at: http://ec.europa.eu/transport/sites/transport/files/modes/air/doc/flightpath2050.pdf Date Accesed: 03/05/2017

[3] J. L. Felder, H. D. Kim, and G. V. Brown, "Turboelectric Distributed Propulsion Engine Cycle Analysis for Hybrid-Wing-Body Aircraft", 47th AIAA Aerosp. Sci. Meet. Incl. New Horizons Forum Aerosp. Expo., 2009.

[4] Kalsi, Swarn, Singh, "Applications of High Temperature Superconductors to Electric Power Equipment" Chap 4, John Wiley & Sons Ltd, 2011

[5] C. E. Jones, P. J. Norman, S. J. Galloway, M. J. Armstrong and A. M. Bollman, "Comparison of Candidate Architectures for Future Distributed Propulsion Aircraft," in IEEE Transactions on Applied Superconductivity, vol. 26, no. 6, pp. 1-9, Sept. 2016. doi: 10.1109/TASC.2016.2530696

[6] S. Fletcher, "Protection of Physically compact Multiterminal DC systems", Doctoral Thesis, University of Strathclyde, Glasgow, 2013.

[7] Semikron, "SKM 111 AR Datasheet", Semikron, [online] Available at: http://www.gdrectifiers.co.uk/uploads/SEMIKRON_DataSheet_SKM_111_AR_.pdf Date Accessed 26/05/16.

[8] M. J. Kim et al., "Determination of Maximum Permissible Temperature Rise Considering Repetitive Over-Current Characteristics of YBCO Coated Conductors," in IEEE Transactions on Applied Superconductivity, vol. 18, no. 2, pp. 660-663, June 2008. doi: 10.1109/TASC.2008.921397

IOP Conf. Series: Materials Science and Engineering **279** (2017) 012015 doi:10.1088/1757-899X/279/1/012015

A 4 K tactical cryocooler using reverse-Brayton machines

M Zagarola, K Cragin, J McCormick and R Hill

Creare LLC, 16 Great Hollow Road, Hanover, NH 03755, USA

mvz@creare.com

Abstract. Superconducting electronics and spectral-spatial holography have the potential to revolutionize digital communications, but must operate at cryogenic temperatures, near 4 K. Liquid helium is undesirable for military missions due to logistics and scarcity, and commercial low temperature cryocoolers are unable to meet size, weight, power, and environmental requirements for many missions. To address this need, Creare is developing a reverse turbo-Brayton cryocooler that provides refrigeration at 4.2 K and rejects heat at 77 K to an upper-stage cryocooler or through boil-off of liquid nitrogen. The cooling system is predicted to reduce size, weight, and input power by at least an order of magnitude as compared to the current state-of-the-art 4.2 K cryocooler. For systems utilizing nitrogen boil-off, the boil-off rate is reasonable. This paper reviews the design of the cryocooler, the key components, and component test results.

1. Background

Superconducting Electronics (SCEs) and spectral-spatial holography have the potential to revolutionize digital communications for multiple tactical applications. Digital receivers, transmitters, transceivers, and processors have been demonstrated utilizing these technologies with extremely high bandwidth, but must operate at cryogenic temperatures. The most challenging application in terms of size, weight, and power (SWaP) involves small airborne platforms. Passive refrigeration using liquid helium provides the required thermal environment, but is not practical in the field. Commercial mechanical refrigerators, also known as cryocoolers, are large and inefficient and are unable to meet SWaP requirements for small platforms. Space cryocoolers are smaller and lighter, but inefficient at low temperatures and expensive. A new type of tactical refrigeration system is required that has high thermodynamic efficiency, is able to meet SWaP requirements for small platforms, and is optimized for operation with cryogenic electronics.

The technical approach we are developing is a two-stage Reverse Turbo-Brayton (RTB) cryocooler that provides refrigeration at 4.2 K and 25 K, and rejects heat at 77 K through boil-off of liquid nitrogen. The cryocooler concept and projected performance are shown in Figure 1. The cryocooler is predicted to reduce each SWaP parameter by an order of magnitude as compared to the current state-of-the-art 4.2 K cryocooler. In addition, liquid nitrogen is relatively safe, non-toxic, easy to handle, and unlike liquid helium, readily available and transportable. The nitrogen boil-off rate is reasonable. This paper reviews the design of the cryocooler, the key components, and component test results.

IOP Conf. Series: Materials Science and Engineering **279** (2017) 012015 doi:10.1088/1757-899X/279/1/012015

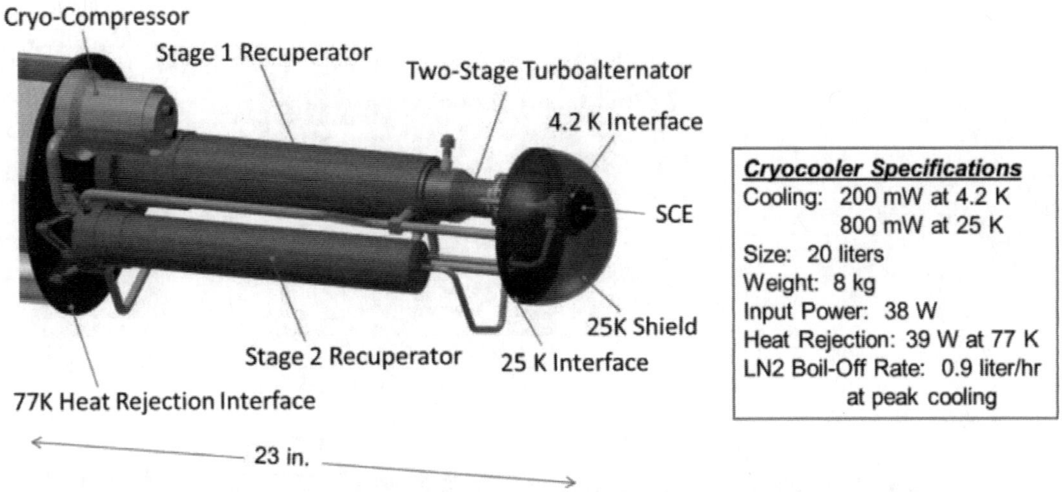

Figure 1. Compact and efficient cryocooler for cooling low-temperature electronics and instruments.

2. Cryocooler Description

A schematic of the RTB cycle is shown in Figure 2. The cycle gas is helium operating between pressures of nominally 0.8 atm and 1.2 atm. The key components within the cycle are the cryo-compressor, two-stage turboalternator, and recuperators. The compressor and turbine utilize high-speed, miniature rotors that are supported on self-acting gas bearings and brushless permanent magnet motors and alternators. The expected operating speeds are 6000 rev/s (360,000 rpm) for the turbine and 7800 rev/s (470,000 rpm) for the compressor, resulting in a compact system with a very high power density and efficiency. The recuperators utilize Creare's silicon slotted plate technology, which has been demonstrated to a Technology Readiness Level of 5 [1, 2]. These recuperators are extremely light and compact. The other components are the aftercooler, thermal interface heat exchangers, control electronics, and tubing, which are all relatively compact and simple components.

Figure 3 shows a notional packaging concept for the cryocooler integrated into a 12 in. pod. The pod is divided into three compartments. The forward compartment houses the ambient electronics; is nominally 12 in. long; and houses the cryocooler control electronics, signal conditioners and processors for the cryo-electronics. The middle compartment houses the cryocooler and is nominally 24 in. long. The aft compartment houses the liquid nitrogen tank and is nominally 36 in. long. The overall length and design of the aft compartment is notional, and will likely be tailored for specific mission durations. The aft compartment shown in Figure 3 provides nominally 32 hrs of operation. The cryocooler will be mounted on the bulkhead between the aft and middle compartments, and will reject heat through the bulkhead directly to the liquid nitrogen. The bulkhead is manufactured from aluminum to ensure that radial temperature gradients are negligible.

Vacuum insulation is required to reduce nitrogen boil-off and thermal parasitics to the cryocooler. Low conductivity penetrations are used to support the cryogen tank. A primary shield that is conductively cooled by liquid nitrogen extends from the aft compartment to the middle compartment encasing the Stage 1 and 2 cryocooler components. The middle compartment is evacuated to reduce thermal parasitics. The overall liquid nitrogen boil-off rate due to parastics is 0.2 liter/hr, which is considerably less than the boil-off rate due to the cryocooler at peak power.

A secondary shield encases the Stage 2 components and operates at nominally 25 K. The surfaces of the shields are coated with a low emissivity coating and wrapped in multi-layer insulation to reduce the radiative heat exchange between the cold and warm surfaces. Penetrations through the shields for fluid, structural, and electrical connections utilize radiation baffles to prevent direct views from warm to hot components. The predicted parasitic heat load to the first cryocooler stage is 30 mW and to the second cryocooler stage about 1 mW.

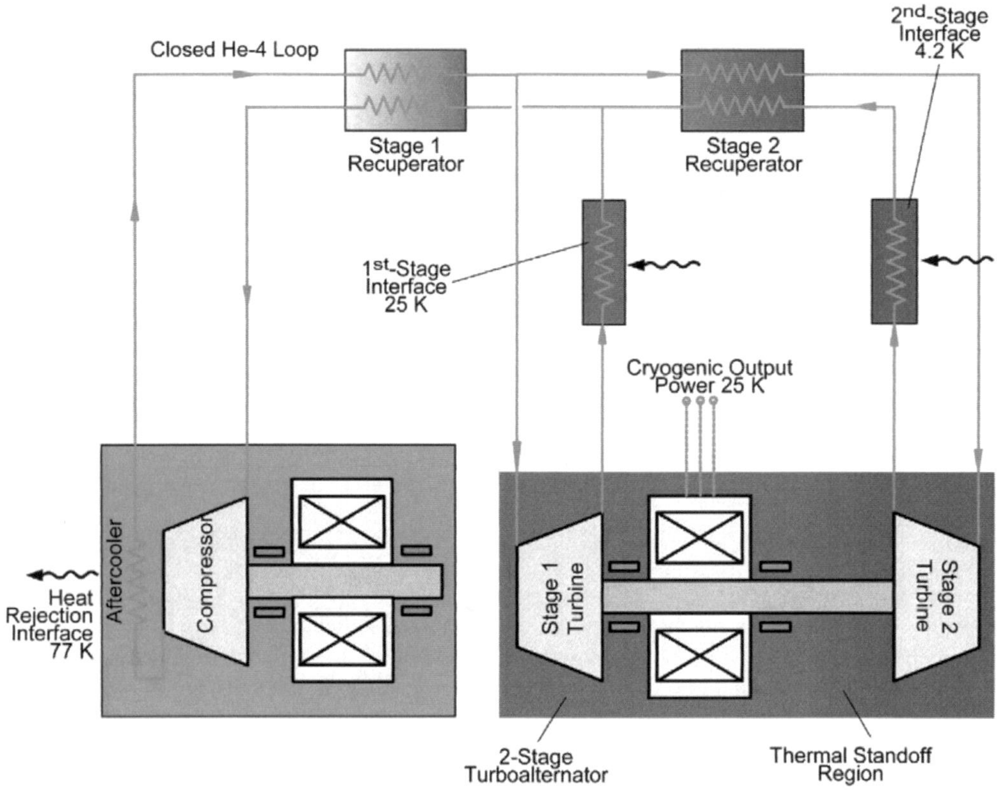

Figure 2. Cycle schematic for RTB cryocooler with cryogenic heat sink.

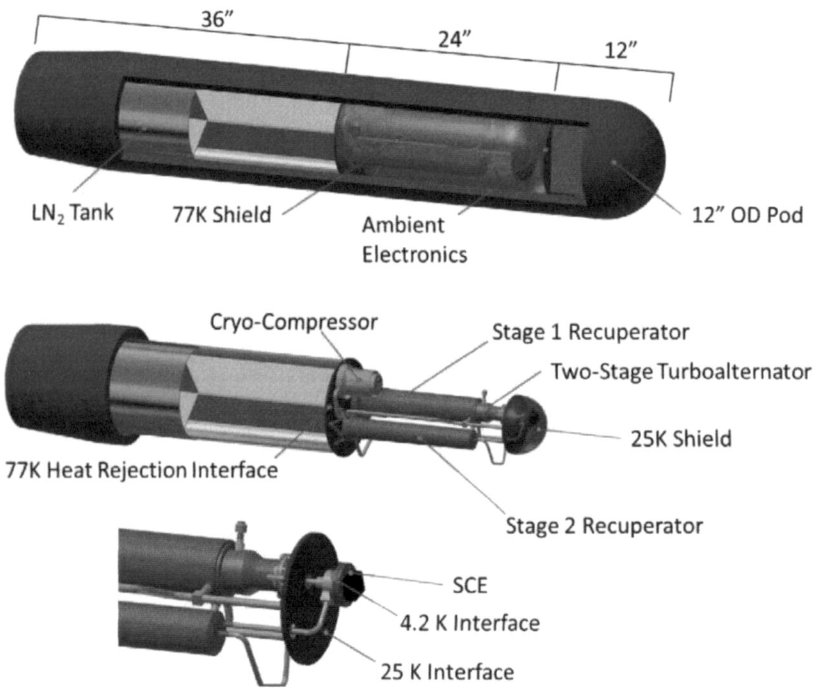

Figure 3. Packaging of cryogenic system into a 12-inch pod.

Based on our packaging and integration concept, we estimated the overall mass of the cryocooler is 8.2 kg. The mass value includes the mass of the shields and cryocooler supports, but does not include the mass of the cryogenic electronics, pod, liquid nitrogen tank, and liquid nitrogen.

In comparison to commercial cryocoolers, the Creare cryocooler requires orders of magnitude less input power and is smaller and lighter. In addition, the underlying cryocooler technology has been developed for the rigors of space flight and is expected to be equally suitable for airborne applications. Table 1 provides a comparison of 4 K cryocoolers. The current state-of-the-art cryocoolers are the SRD/SRDK product line produced by Sumitomo Heavy Industries. These are Gifford-McMahon style cryocoolers that are designed for indoor usage, reject heat at ambient temperature, and have a maintenance interval of 10,000 hrs. The cryocoolers comprise a coldhead and compressor attached using high pressure flexlines. Model SRDK-101D-A11B is the lowest capacity unit and the most compact and lightweight. It requires 1300 W of input power at 60 Hz to simultaneously provide 100 mW of refrigeration at 4.2 K and 5 W of refrigeration at 60 K. The masses and volumes for the cryocooler components are 7.2 kg and 13 liters for the coldhead, and 42 kg and 81 liters for the compressor. Model SRDK-305D-A31C provides 400 mW of refrigeration at 4.2 K, requires 4800 W of input power, weighs 111 kg, and has a volume of 250 liters. These cryocoolers are too big and too heavy and require too much input power for small platforms. In addition, they do not meet the environmental requirements for airborne applications.

Table 1. Comparison of 4 K cryocoolers.

	Creare UAV Cryocooler	Sumitomo SRDK-101D-A11B	Sumitomo SRDK-305D-A31C
Maturity	Design	Commercial	Commercial
Heat rejection temperature	77 K	300 K	300 K
Stage 1 refrigeration	800 mW @ 25 K	5.0 W @ 60 K	20 W @ 40 K
Stage 2 refrigeration	200 mW @ 4.2 K	100 mW @ 4.2 K	400 mW @ 4.2 K
Input power	<50 W	1300 W	4800 W
Mass (refrigeration system)	8 kg	49 kg	111 kg
Volume	20 liters	94 liters	250 liters
LN2 boil-off rate	0.9 liter/hr	N/A	N/A

3. Cryo-Compressor Proof-of-Concept Testing

The key innovation of this cryocooler concept is the operation of the compressor at cryogenic temperatures. This was not previously demonstrated and has several benefits. First, it enables the compressor to achieve a very high compression ratio in a single stage, because the density of helium at 77 K is four times that at room temperature. To achieve this high compression ratio at room temperature would require three centrifugal compressors connected in series with total input power of 250 to 300 W. Second, the high pressure ratio significantly increases the specific cooling power of helium flow, and thus reduces the helium gas flow rate in the system. This in turn reduces the size and mass of the recuperators. Most importantly, the input power to compress the helium is reduced due to the lower enthalpy rise that is required to compress a gas at lower temperatures. These features substantially reduce the size, mass, and input power of the cryocooler, making it ideal for SWaP-sensitive applications.

Proof-of-concept testing was performed using an existing cryo-circulator. The difference between a cryo-circulator and cryo-compressor is the operating speed, and consequently the pressure ratio produced. The circulator operates at lower speeds, pressure ratios, and flow rates than a compressor to serve its intended function. A comparison between the circulator and compressor rotors is shown in Figure 4. The circulator utilizes a 7.6 mm (0.3 in.) impeller and 3.6 mm (0.14 in.) shaft, and operates at up to 2000 rev/s at 70 to 80 K. The cryo-compressor utilizes a 12.7 mm (0.5 in.) impeller and 4.3 mm (0.17 in.) shaft, and

IOP Conf. Series: Materials Science and Engineering **279** (2017) 012015 doi:10.1088/1757-899X/279/1/012015

operates at up to 7800 rev/s at 80 K. The bending critical frequencies are nominally the same so that the shaft dynamics are similar. The shaft surface speed, a critical parameter for bearing performance, is 106 m/s for the cryo-compressor. To obtain the same shaft surface speed for the circulator, we must operate the circulator at a speed of up to 9500 rev/s at 80 K.

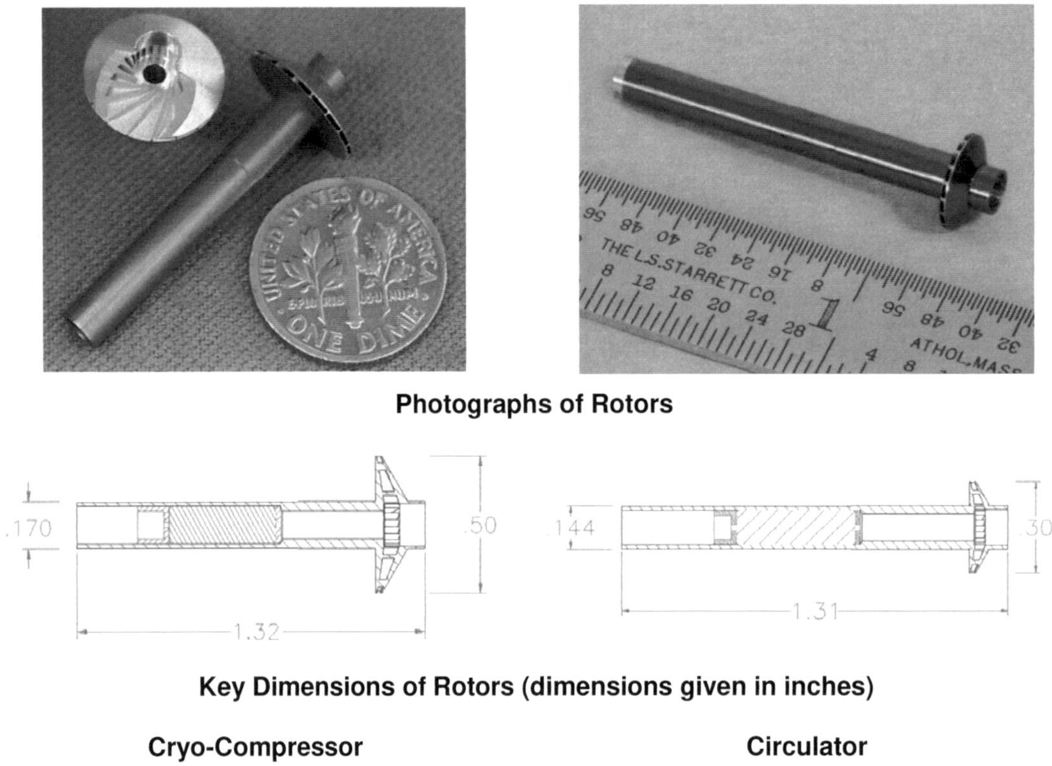

Figure 4. Comparison between cryo-compressor and circulator rotors.

Operation at 77 K is achieved by immersing the circulator assembly in a dewar filled with liquid nitrogen. The tests were performed in an existing open-loop test facility. The test results are shown in Figure 5. The cryo-circulator was cooled from 300 K to 77 K while operating at a speed of 4000 rev/s. Once at 77 K, the speed was gradually increased while monitoring the displacement probes for signs of instability. The speed was increased to a maximum value of 9600 rev/s, at which point we stopped the test. Higher speeds were possible. but we did not want to risk hardware as the performance gains of the cryocooler at higher speeds are not significant. The maximum speed demonstrated is the highest we have yet achieved with a fully functional turbomachine at cryogenic temperatures. Prior demonstrations at higher speeds and at cryogenic temperatures were with simple shafts. Based on these successful results, we initiated the build of the cryo-compressor.

4. Cryo-Compressor Development

The cryo-compressor design is based on an existing permanent magnet motor compressor developed for operation at 300 K [3]. The key changes are selection of compatible materials for operation at cryogenic temperatures and optimization of the aerodynamics for operation at the design conditions. The assembled cryo-compressor is shown in Figure 6.

IOP Conf. Series: Materials Science and Engineering **279** (2017) 012015 doi:10.1088/1757-899X/279/1/012015

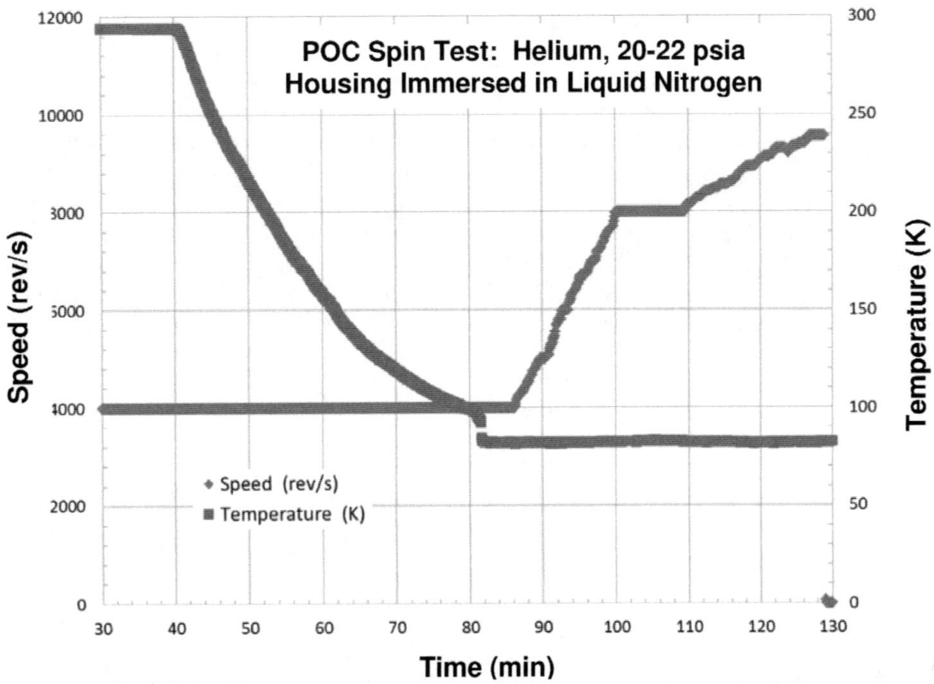

Figure 5. Cooldown curve for cryo-circulator demonstration to high speeds.

Figure 6. Assembled cryo-compressor ready for testing.

Initial testing was performed at ambient temperature. We performed spin tests with the cryo-compressor at ambient temperature in an open-loop helium test loop to assess journal bearing stability up to the design operational speed. We performed several tests where the rotor speed was limited to less than the design speed due to bearing instability. After each test, we incrementally reduced the operating clearance in the journal bearings to enable higher operating speeds. After a few iterations to the journal bearing settings, the rotor attained the design operating speed of 8,000 rev/s.

Following our successful spin test, we performed a test to characterize the compressor performance at room temperature. This test provides an initial indication of the performance and screens for gross performance discrepancies that need to be addressed before investing the time in a cryogenic test. The test data are presented in Figure 7, which compares data taken at room temperature against our performance predictions for the compressor at 80K. The performance of the compressor is characterized by plotting its non-dimensional head rise (head coefficient) versus its non-dimensional flow rate (flow coefficient). The performance of the compressor at cryogenic temperatures is expected to be slightly better due to the higher Reynolds number. From the figure, we note that the performance curve is smooth and regular, and that it is close to our performance predictions for the machine at cryogenic temperature. Proper performance of the compressor at ambient temperature is a necessary but not sufficient condition for proper cryogenic performance. Future testing will be focused on testing at up to 8,000 rev/s at 80 K using a liquid nitrogen heat sink.

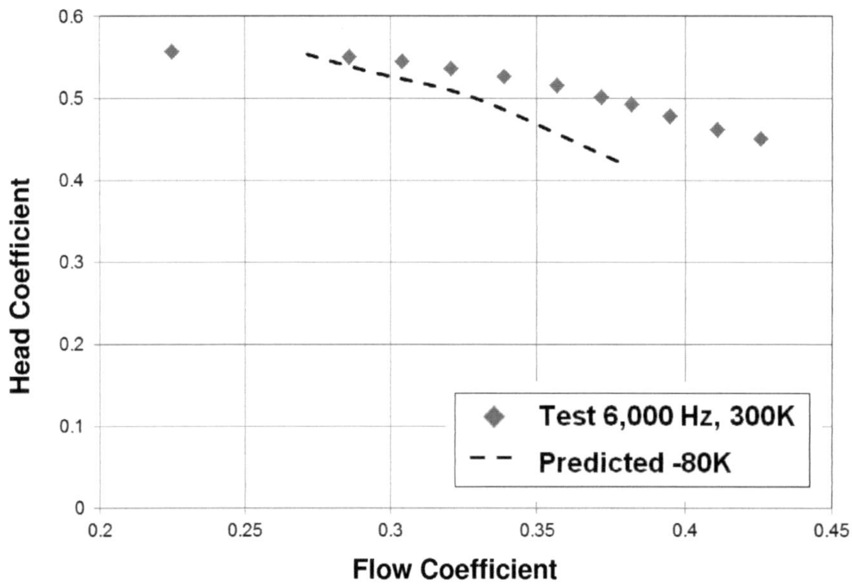

Figure 7. Ambient-temperature test of cryo-compressor compared to cryogenic performance predictions. The test data match performance predictions well, indicating that the compressor is operating close to expectations.

5. Conclusions

Creare is developing a low temperature tactical cryocooler for cooling superconducting electronics and spectral-spatial holography systems. The cryocooler provides refrigeration at 4.2 K, and rejects heat at 77 K through boil-off of liquid nitrogen. The cryocooler is predicted to reduce each SWaP parameter by an order of magnitude as compared to the current state-of-the-art 4.2 K cryocooler. The nitrogen boil-off rate is reasonable, enabling long duration missions using a small dewar of liquid nitrogen.

The key innovation is the use of a cryo-compressor that drastically reduces cryocooler size, weight, power, and cost for tactical missions. Initial proof-of-concept testing indicates that the operational requirements for the cryo-compressor are achievable, and initial performance testing shows that the performance will exceed predictions.

Future work involves testing the cryo-compressor at design operating conditions, completing the development of the remaining cryocooler components, and testing the cryocooler in a laboratory and then operational environment. Successful completion of this program will dramatically improve high-speed digital communications from small platforms.

Acknowledgments
The support of the Office of Naval Research for this work (Contract N00014-14-C-0210) is gratefully acknowledged.

References

[1] Hill R W, Izenson M G, Chen W and Zagarola M 2007 A recuperative heat exchanger for space-borne turbo Brayton cryocoolers *Cryocoolers 14* ed S D Miller and R G Ross Jr (Boulder CO: ICC Press) pp 525–33.
[2] Zagarola M V, Hill R W, Gagne J R and Kaszeta R W 2014 Ultra low power cryo refrigerator for space applications *Cryocoolers 18* ed S D Miller and R G Ross Jr (Boulder CO: ICC Press) p 413.
[3] Zagarola M V, Swift W L, Sixsmith H, McCormick J A and Izenson M G 2002 Development of a turbo Brayton cooler for 6 K space applications *Cryocoolers 12* ed R G Ross Jr (New York NY: Kluwer Academic/Plenum Publishers) pp 571–78.

IOP Conf. Series: Materials Science and Engineering **279** (2017) 012016 doi:10.1088/1757-899X/279/1/012016

Study on the adjustment capability of the excitation system located inside superconducting machine electromagnetic shield

D Xia[1] and Z Xia[2]

[1]Institute of Electrical Engineering, Chinese Academy of Sciences, Beijing, China
[2] Bellevue College, Bellevue, WA, USA

E-mail: xiadong@mail.iee.ac.cn

Abstract. The ability for the excitation system to adjust quickly plays a very important role in maintaining the normal operation of superconducting machines and power systems. However, the eddy currents in the electromagnetic shield of superconducting machines hinder the exciting magnetic field change and weaken the adjustment capability of the excitation system. To analyze this problem, a finite element calculation model for the transient electromagnetic field with moving parts is established. The effects of three different electromagnetic shields on the exciting magnetic field are analyzed using finite element method. The results show that the electromagnetic shield hinders the field changes significantly, the better its conductivity, the greater the effect on the superconducting machine excitation.

1. Introduction

Comparing to conventional electrical machines, the superconducting synchronous machine has smaller volume and weight, higher efficiency and power density, and bigger ultimate capacity. Therefore, it is regarded as one of the most attractive electrical machines in wind power, ship driving and other fields [1]-[3].

Generally, the field winding of superconducting synchronous machines is made of superconducting tapes. Although only DC current flows through the superconducting field winding, the alternating magnetic field generated by the armature winding will reach the field winding. Because the superconducting field winding works in an alternating magnetic field, AC losses are produced in the field winding. The losses reduce the efficiency of the superconducting machines and cause the rise in temperature and the quench of superconducting tapes. In order to shield the superconducting field winding from the alternating magnetic field and reduce its effect on the superconducting field winding, an electromagnetic shield must be used [4]-[8].

The exciting current of the field excitation system must be adjustable, when the synchronous machine is working. As an important part of synchronous machines, the excitation adjustment capability of the field excitation system directly affects the operating characteristics of the machines, and has an important impact on the operation of power systems. Especially for the wind power or the marine power station with a relatively small capacity, a greater fluctuation of grid voltage will be produced when the power system load changes. Therefore, higher requirements for adjustment capability of the field excitation system will be put forward.

Generators are considered as an example below. The main tasks of the field excitation adjustment include [9]: (1) During normal operating conditions, the load of power system always fluctuates, the output power of the synchronous generator must also change accordingly. The exciting current should be adjusted so that the generator output voltage remains within an allowable range. (2) The field excitation system is also responsible for the reactive power rational allocation between parallel operation units, the reactive power output must be changed according to the needs of the power system by adjusting the exciting current. (3) When a short circuit fault occurs during low load operation of the power system, the small short-circuit current may result in that the protection device does not work correctly. The force excitation should be carried out to increase the short-circuit current so that the protection device can act instantly. (4) When the internal fault of the generator or the connected transformer occurs, the fast de-excitation of the generator excitation system is required in order to reduce the damage caused by the fault. Thus, the quick adjustment capability of the excitation system plays a very important role in maintaining the normal operation of generators and power systems.

Since the field winding of the superconducting machine excitation system is located in the electromagnetic shield, its exciting current change will induce eddy currents in the electromagnetic shield tube. The eddy currents will hinder the exciting magnetic field change and weaken the adjustment capability of the excitation system. Currently, the study of the superconducting machine electromagnetic shield focus on the aspect how to prevent the alternating magnetic field produced by the stator armature winding into the rotor [3]-[5], and few studies have looked at this issue of how it affects the fast adjustment capability of excitation system. Therefore, an in-depth research on this issue is very necessary.

2. Finite Element Analysis Model

The electromagnetic shield of superconducting synchronous machines is composed of cylindrical metal material tubes. Since the induced eddy currents inside the metal material tubes must be considered in our analysis, the analysis method for conventional electrical machines is no longer suitable. In addition, the effects of electromagnetic shield on the excitation adjustment capacity belong to the problem of the transient electromagnetic field with moving parts, which further increases the difficulty of the analysis. In order to obtain more accurate results, finite element method is thereby adopted.

The description equation of the analyzed transient electromagnetic field with moving parts can be expressed as follows:

$$\nabla \times \frac{1}{\mu}(\nabla \times A) = J_s - \sigma \frac{dA}{dt} - \sigma \nabla v \qquad (1)$$

where A is magnetic vector potential, J_s conduction current density, μ permeability, σ conductivity, v velocity of moving part.

A 6-pole superconducting synchronous generator was calculated using finite element method. The main parameters and dimensions of the generator are listed in [10]. The generator stator core is made of silicon steel 1010. The stator armature winding is made of conventional copper wires and the number of conductors per slot is 20. The rotor field winding is made of superconducting tapes and the number of turns per pole is 100. The rotor is a coreless structure and the electromagnetic shield tube is located in the rotor surface. The structure of the superconducting synchronous generator is shown in Figure 1.

In order to consider the effect of the structure, material and size of the electromagnetic shield on the quick adjustment capability of the excitation system, we have analyzed three different electromagnetic shield solutions:

The electromagnetic shield consists of a 15 mm thick copper tube (**Scheme 1**); a 5 mm thick stainless steel outer tube and a 10 mm thick copper inner tube (**Scheme 2**); a 10 mm thick stainless steel outer tube and a 5 mm thick copper inner tube (**Scheme 3**).

The electrical conductivity of copper wires is assumed to be 5.8×10^7Siemens/m, the electrical conductivity of stainless steel is assumed to be 1.1×10^6 Siemens/m. The sort from large to small by the conductivity of the three electromagnetic shield solutions is Scheme 1, 2, 3.

IOP Conf. Series: Materials Science and Engineering **279** (2017) 012016 doi:10.1088/1757-899X/279/1/012016

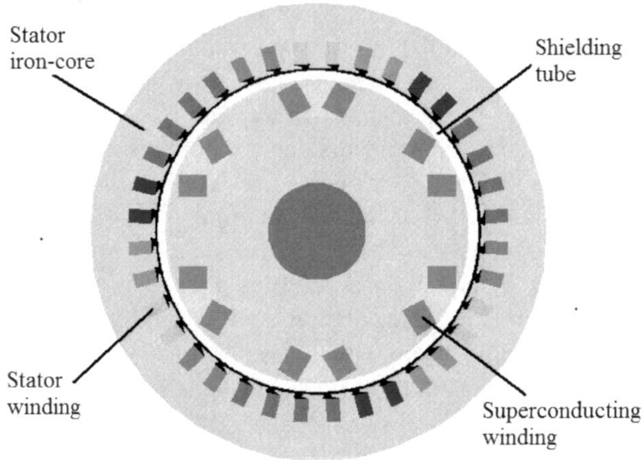

Figure 1. Structure of superconducting synchronous generator

3. Calculation Results and Analysis

Referring to the working state of the field windings in regulating the magnetic field, we have calculated the following three different cases.

3.1 Applying DC voltage to the field winding terminal

In no-load steady state operation, the initial exciting current in the generator field winding is 0. At time t = 0, by applying 500 V of DC voltage to the field winding terminal, the exciting current is produced. Figure 2 and Figure 3 show the magnetic field line and flux density distribution on cross section of the generator at t = 20ms respectively.

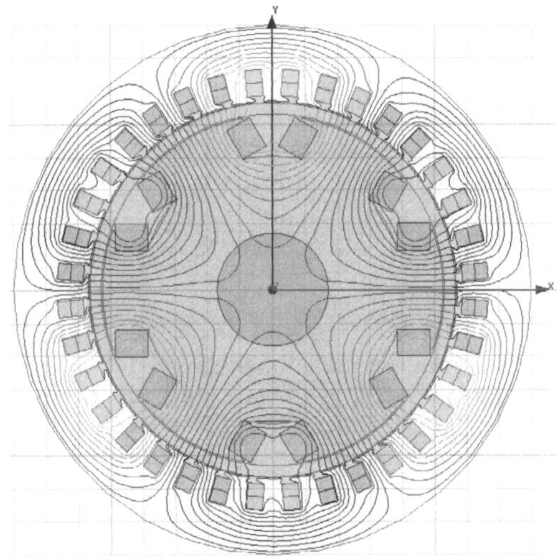

（a）No electromagnetic shielding

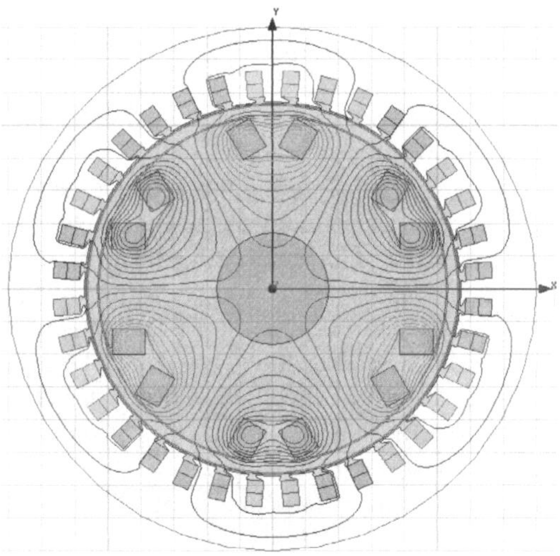

(b) Scheme 1

IOP Conf. Series: Materials Science and Engineering **279** (2017) 012016 doi:10.1088/1757-899X/279/1/012016

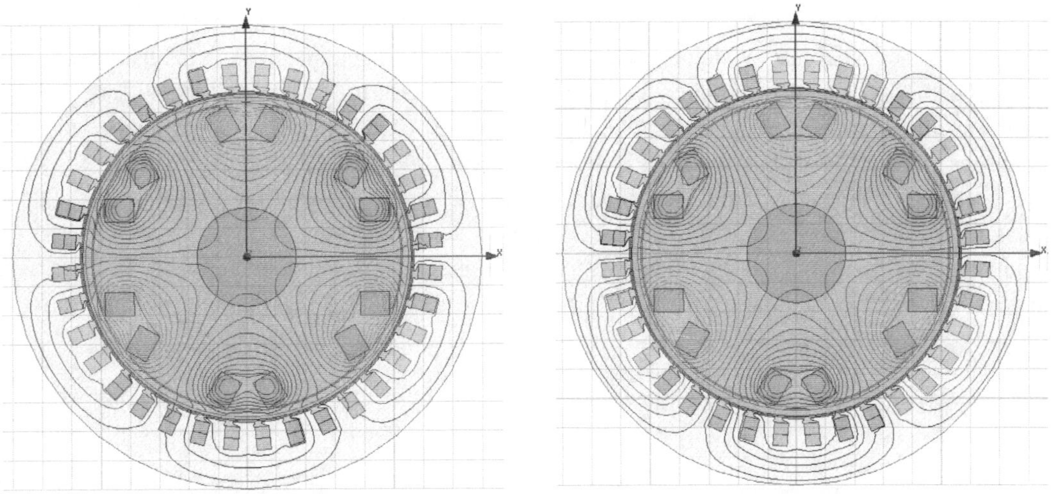

(c) Scheme 2 (d) Scheme 3

Figure 2. Magnetic field line distribution on cross section of generator at t = 20ms

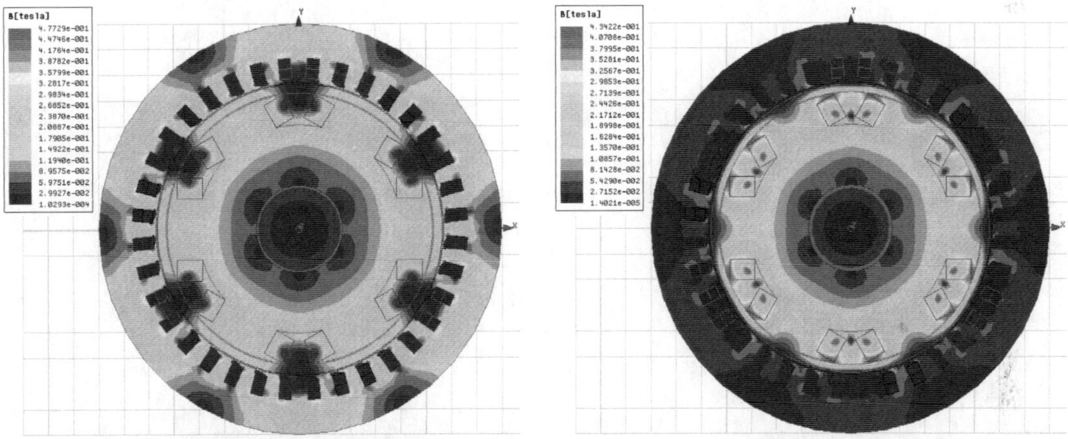

(a) No electromagnetic shielding (b) Scheme 1

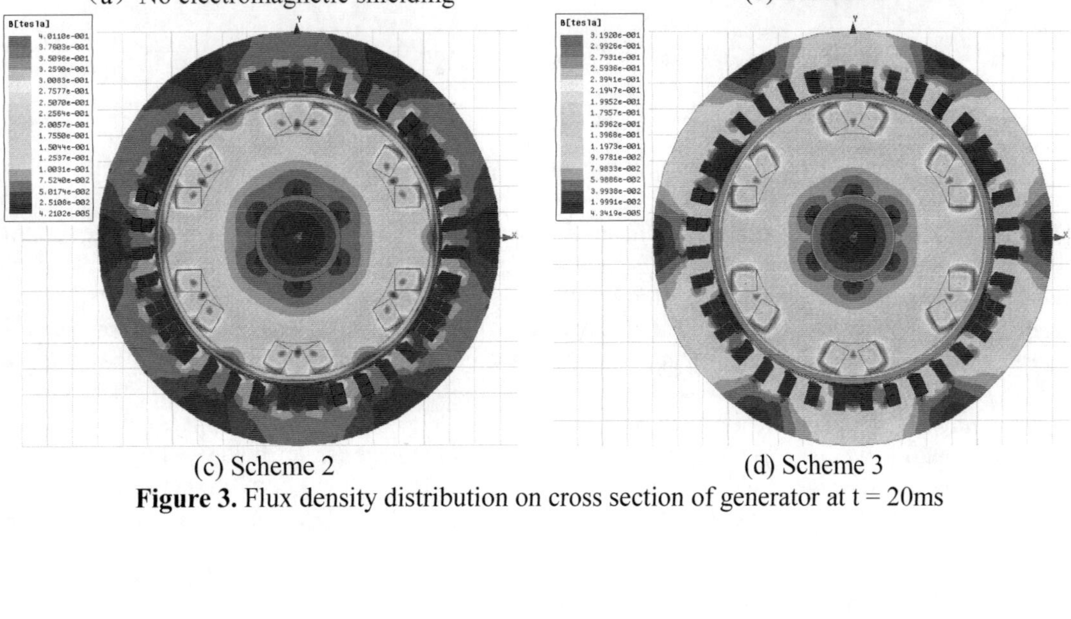

(c) Scheme 2 (d) Scheme 3

Figure 3. Flux density distribution on cross section of generator at t = 20ms

In Figure 2, when the generator has no electromagnetic shield, the magnetic field line distribution in the stator region is the densest, when the generator has an electromagnetic shield, it is relatively sparse, and the magnetic field line distribution in Scheme 1 is the sparsest. In Figure 3, when the generator has no electromagnetic shield, the magnetic flux density in the stator region is the strongest, when the generator has an electromagnetic shield, it is relatively weak, and the magnetic flux density in Scheme 1 is the weakest. This means that the electromagnetic shield hinders the exciting magnetic field change and weakens the adjustment capability of the excitation system.

Figure 4 shows the terminal voltage curves of the A phase stator armature winding in the cases of no electromagnetic shield and the three different electromagnetic shields when the field winding terminal voltage is applied 500 V since t = 0.

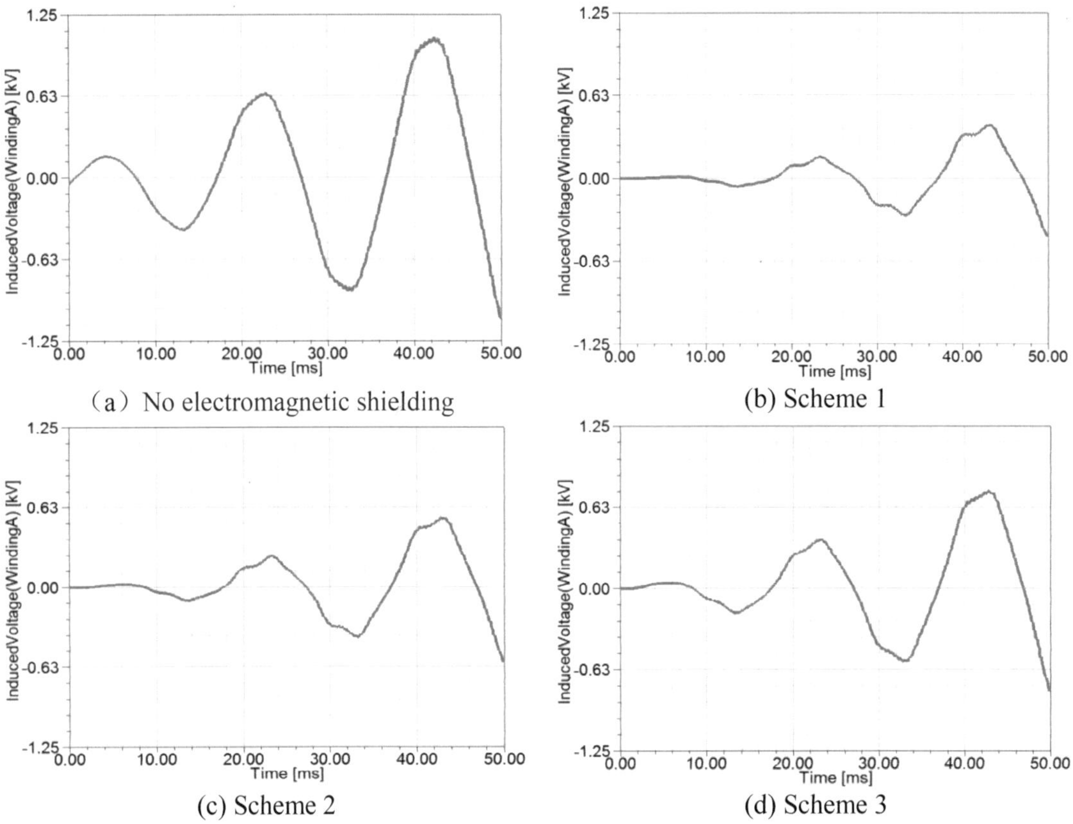

Figure 4. Terminal voltage curves of A phase armature winding

By comparison can it be seen that: When the generator has no electromagnetic shield, the terminal voltage increase of its armature winding is the fastest. When the generator has electromagnetic shield, the terminal voltage increase of the stator armature winding becomes slow, and the terminal voltage increase is the slowest in Scheme 1, but it is relatively the fastest in Scheme 3. It also shows that the better the conductivity of the electromagnetic shield, the greater the effect on the excitation regulation.

3.2 Reversing the field winding terminal voltage

In no-load steady state operation, the initial exciting current in the generator field winding is 1000A. At time t = 0, by reversing the field winding terminal voltage, the exciting current is changed. Figure 5 shows the terminal voltage curves of the A phase stator armature winding in the cases of no electromagnetic shield and the three different electromagnetic shields when the field winding terminal voltage is reversed by the same.

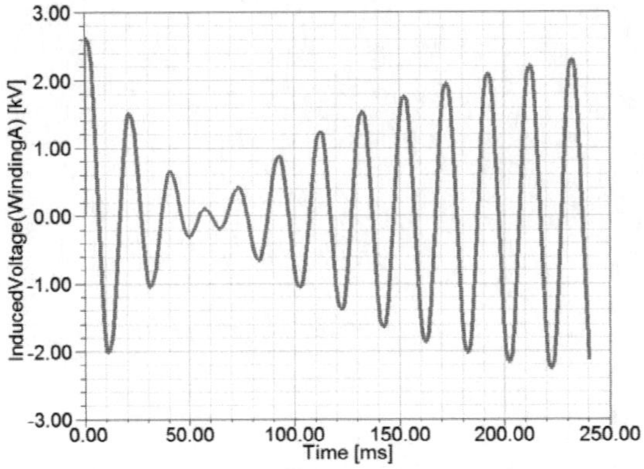

(a) No electromagnetic shielding

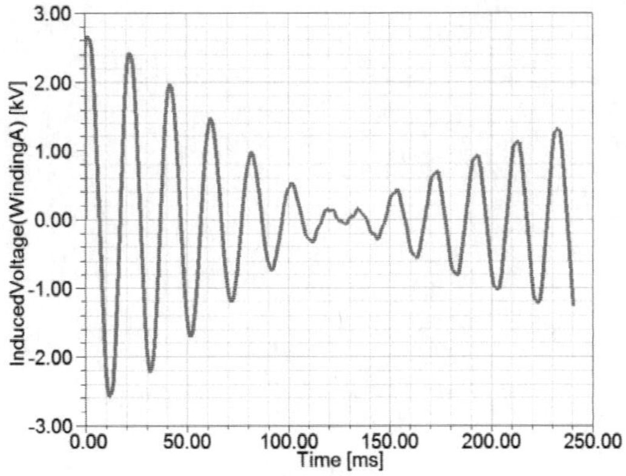

(b) Scheme 1

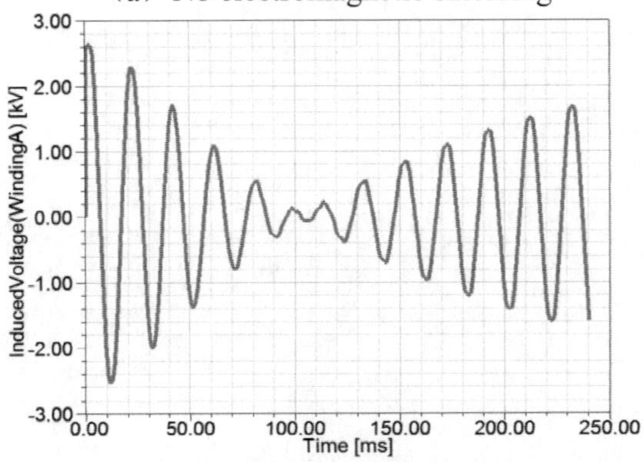

(c) Scheme 2

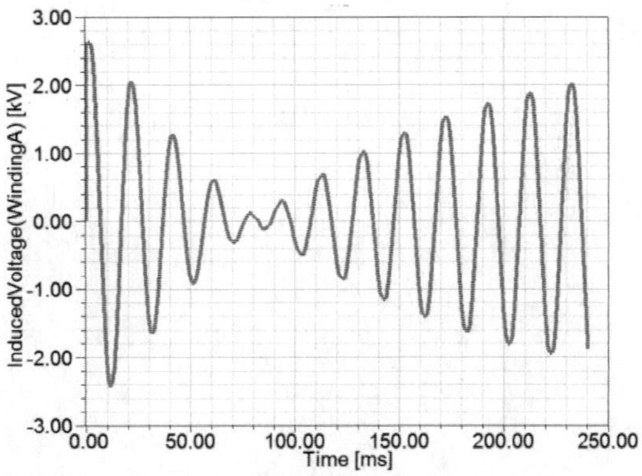

(d) Scheme 3

Figure 5. Terminal voltage curves of A phase armature winding

As can be seen from the figure: When no electromagnetic shielding, the stator armature winding terminal voltage reaches the zero-crossing point at fastest, but for Scheme 1, the most time is required.

3.3 Doubling the field winding terminal voltage

In no-load steady state operation, the initial exciting current in the generator field winding is 1000A. At time t = 0, by doubling the field winding terminal voltage, the exciting current increases.

Figure 6 shows the terminal voltage curves of the A phase stator armature winding in the cases of no electromagnetic shield and the three different electromagnetic shields when the field winding terminal voltage is increased by the same.

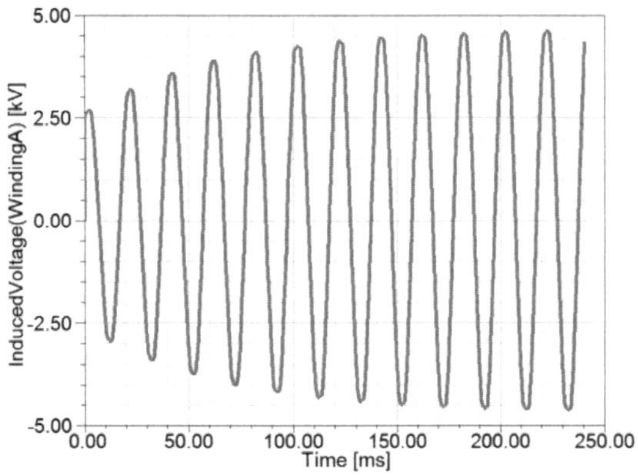

(a) No electromagnetic shielding

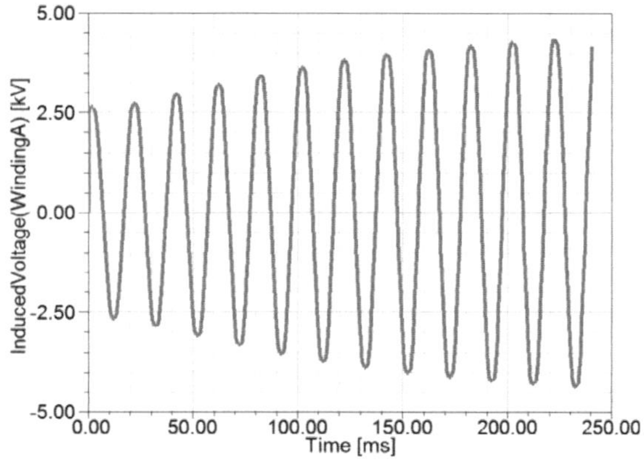

(b) Scheme 1

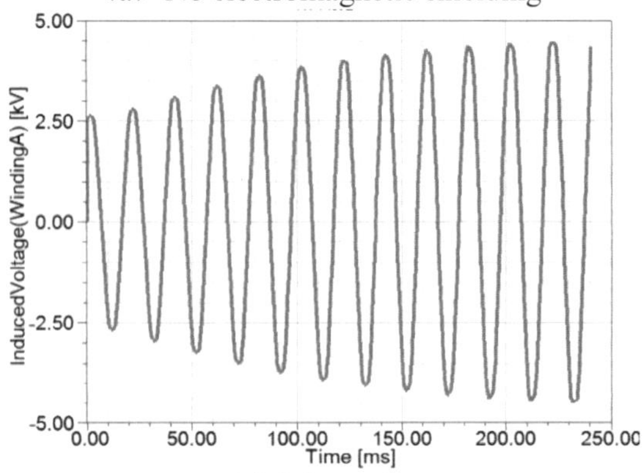

(c) Scheme 2

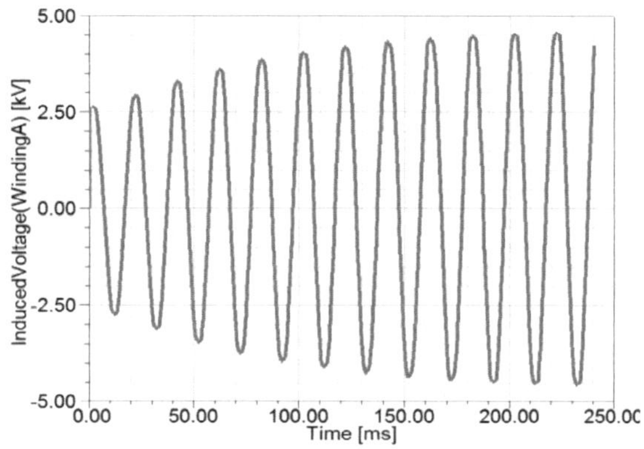

(d) Scheme 3

Figure 6: Terminal voltage curves of A phase armature winding

By comparison can it be seen that: When the generator has no electromagnetic shield, the increase of its armature winding terminal voltage is the fastest. The electromagnetic shield makes the stator armature winding terminal voltage increase slower. For the different solutions of the electromagnetic shield, the terminal voltage increase of the stator armature winding is the slowest in Scheme 1, but it is relatively the fastest in Scheme 3. This means that the better the conductive property of the electromagnetic shield, the bigger the impact on the armature winding terminal voltage increase.

To find out the quantity relationship of the effect of the different electromagnetic shields on the armature winding terminal voltage change, we chose two different time points t = 22.5 and 231.5ms to be compared. Table 1 shows the terminal voltage of the A phase stator armature winding and the terminal voltage ratio against no electromagnetic shield at the two time points.

Table 1. A phase armature winding terminal voltage and ratio

	Time(ms)	Voltage(V)	Ratio	Time(ms)	Voltage(V)	Ratio
No shield	22.5	3202.29	1	231.5	-4599.99	1
Scheme 1	22.5	2743.38	0.857	231.5	-4369.91	0.950
Scheme 2	22.5	2813.12	0.878	231.5	-4484.01	0.975
Scheme 3	22.5	2939.73	0.918	231.5	-4563.05	0.992

It can be seen from the table: The better the conductive property of electromagnetic shield, the smaller the terminal voltage at a same time point. This conclusion is consistent with the previous. The terminal voltage ratio at the time point t = 22.5ms is less than at the time point t = 231.5ms. This indicates that with the time increasing, the influence of the electromagnetic shield becomes smaller and smaller.

4. Conclusion
A 6-pole superconducting synchronous generator without electromagnetic shield and with three different electromagnetic shields was calculated using finite element model for the transient electromagnetic field with moving parts. By means of applying, reversing and doubling the field winding terminal voltage, the exciting current in the field winding of the superconducting generator is altered. The changing process of the magnetic field, the eddy current inside electromagnetic shield and the terminal voltage of the A phase stator armature winding with the exciting current is obtained through the calculation. The study shows, a better conductive property of the electromagnetic shield can improve its shielding capability, while reducing the excitation adjustment capability of superconducting synchronous machines.

Acknowledgement
This work was supported by the National Nature Science Foundation of China (NO. 51377151).

References
[1] Kirtley J L and Furuyama J M 1975 A design concept for large superconducting alternators *IEEE Transactions on Power Apparatus and Systems* vol 94 pp 1264-1269
[2] Snitchler G, Gamble B, King C and Winn P 2011 10 MW class superconductor wind turbine generators *IEEE Transactions on Applied Superconductivity* vol 21 n 3 Part 2 pp 1089-1092
[3] Barnes P N, Sumption M D and Rhoads G L 2005 Review of high power density superconducting generators: Present state and prospects for incorporating YBCO windings *Cryogenics* vol 45 pp 670-686
[4] Miller T J E and Lawrenson P J 1976 Penetration of transient magnetic fields through conducting cylindrical structures with particular reference to superconducting a.c. machines *Proceedings of the institution of Electrical Engineers* vol 123 pp 437-442
[5] Lawrenson P J, Miller T J E and Stephenson J M 1976 Damping and screening in the synchronous superconducting generator *Proceedings of the institution of Electrical Engineers* vol 123 n 8 pp787-794
[6] Leurs L and Stoll R L 1986 Three-dimensional quasi-static magnetic field in superconducting-rotor synchronous generators with a magnetic steel rotor screen *Generation, Transmission and Distribution, IEE Proceedings C* vol 133 pp 69-80
[7] Takuma T, Akita S, Kawamoto T and Yasuda H 1990 Calculation of eddy current in the damper of a superconducting generator *IEEE Transactions on Magnetics* vol 26 pp 921-924
[8] Elhaminia P, Yazdanian M, Zolghadri M R and Fardmanesh M 2011 An analytical approach for optimal design of rotor iron for superconducting synchronous machine *37th Annual Conference on IEEE Industrial Electronics Society* pp 1741-1745
[9] 1996 Electrical engineering handbook *Machinery Industry Press, Beijing* vol. 3
[10] D Xia and Z Xia 2015 Performance analysis of superconducting generator electromagnetic shielding *IOP Conference Series: Materials Science and Engineering* vol. 101 Paper no: 012027

IOP Conf. Series: Materials Science and Engineering **279** (2017) 012017 doi:10.1088/1757-899X/279/1/012017

Terahertz emission from the intrinsic Josephson junctions of high-symmetry thermally-managed $Bi_2Sr_2CaCu_2O_{8+\delta}$ microstrip antennas

Richard A. Klemm[1], Andrew E. Davis[1], Qing X. Wang[1], Takashi Yamamoto[2], Daniel P. Cerkoney[3], Candy Reid[4], Maximiliaan L. Koopman[1], Hidetoshi Minami[5], Takanari Kashiwagi[5], Joseph R. Rain[1], Constance M. Doty[1], Michael A. Sedlack[1], Manuel A. Morales[6], Chiharu Watanabe[1], Manabu Tsujimoto[1], Kaveh Delfanazari[7], and Kazuo Kadowaki[5]

[1]Department of Physics, University of Central Florida, 4111 Libra Drive, Orlando, FL 32816-2385 USA

[2]Institute for Materials Research, Hasselt University, Wetenschapspark 1 B-3590, Diepenbeek, Belguim

[3]Department of Physics and Astronomy, Rutgers University, 136 Frehlinghuysen Rd., Piscataway, NJ, 08854 USA

[4]Lockheed Martin, 5600 Sand Lake Road, Orlando, FL 32819 USA

[5]Graduate School of Pure and Applied Sciences, University of Tsukuba, 1-1-1 Tennodai, Tsukuba, Ibaraki, 305-8571 Japan

[6]Division of Health Science and Technology, Harvard Medical School, 260 Longwood Ave., Boston, MA 02115 USA

[7]Department of Physics, Cavendish Laboratory, University of Cambridge, Cambridge CB3 0HE, UK

E-mail: `richard.klemm@ucf.edu`

Abstract. We show for high-symmetry disk, square, or equilateral triangular thin microstrip antennas of any composition respectively obeying $C_{\infty v}$, C_{4v}, and C_{3v} point group symmetries, that the transverse magnetic electromagnetic cavity mode wave functions are restricted in form to those that are one-dimensional representations of those point groups. Plots of the common nodal points of the ten lowest-energy non-radiating two-dimensional representations of each of these three symmetries are presented. For comparison with symmetry-broken disk intrinsic Josephson junction microstrip antennas constructed from the highly anisotropic layered superconductor $Bi_2Sr_2CaCu_2O_{8+\delta}$ (BSCCO), we present plots of the ten lowest frequency orthonormal wave functions and of their emission power angular distributions. These results are compared with previous results for square and equilateral triangular thin microstrip antennas.

1. Introduction

Until very recently, there has been a region in the electromagnetic (EM) spectrum from about 0.1 to 10 THz over which compact coherent sources have been difficult to produce, due mainly to output power P values below 1 mW, the approximate value desired for many applications. This has been especially true in the more limited region 0.3 to 2.0 THz known as the "terahertz

IOP Conf. Series: Materials Science and Engineering **279** (2017) 012017 doi:10.1088/1757-899X/279/1/012017

gap" [1, 2]. For emission frequencies $f > 2.0$ THz, quantum cascade lasers are operational at the desired power values without cryogenic cooling [3, 4, 5]. Schottky diode frequency mixers, backward-wave oscillators, and frequency multiplier chains are operable for $f \leq 1.2$ THz [6, 7], and intracavity difference frequency generators operate over the f range of 2.6-4.2 THz [8]. Resonant tunneling diodes (RTDs) have been able to operate with sufficient power for $f < 1.4$ THz at room temperature [9, 10], but recently were shown to emit up to 2.0 THz, albeit at P values around 1 μW [11, 12].

A completely different type of source operating in the sub-THz to THz range is due to the ac Josephson effect [13, 14]. The extremely anisotropic, layered, high-transition temperature T_c superconductor $Bi_2Sr_2CaCu_2O_{8+\delta}$ (BSCCO) consists of regularly alternating superconducting and insulating layers [15, 16, 17], each of which acts as an intrinsic Josephson junction (IJJ). The best samples of BSCCO are grown by a modified traveling-solvent floating zone technique [18, 19]. When a dc voltage V is applied across the stack of N active IJJs, it gives rise to an ac current and the emission of a photon at the frequency $f = f_J = 2eV/(Nh)$ due to the ac Josephson effect, where e and h are the electronic charge and Planck's constant, respectively [13]. In addition, a BSCCO mesa structure fabricated from the top of a single crystal behaves as an electromagnetic (EM) cavity, the modes of which depend upon the geometric shape of the mesa [20]. By varying the bias V, the output f changes until it locks onto a standing wave mode of that EM cavity, resulting in coherent emission from the stack of IJJs at $f = f_c(m, n)$, where $f_c(m, n)$ is the frequency of the EM cavity mode indexed by the two integers (m, n) for the particular cavity geometry, enhancing the output power at that f value [20]-[37]. Two reviews of this phenomenon were recently published [38, 39].

However, the introduction of a dc current I into the mesa has also led to severe Joule heating problems, resulting in the formation of inhomogeneous hot spots over which the local temperature $T(\boldsymbol{r}) > T_c$ [40]-[56]. This Joule heating also caused the restrictions $V <$ 1.5 V, $f <$ 1.0 THz, and $P <$ 30μW from a single mesa [57]-[62], although a three-mesa array was reported to have a combined emission P of 0.6 mW [60].

Direct competition with the latest RTD devices [11, 12] has arisen from the construction of "stand-alone mesa sandwich structures". These are fabricated from a BSCCO mesa by cleavage from its BSCCO substrate, depositing Au films on both its top and bottom, and then sandwiching it between two insulating plates [63]-[66]. These sandwich structures allow for much more efficient Joule heat removal in the IJJ emitter devices, allowing bias V values up to 7 V and f values up to 2.4 THz, with strong emissions at particular frequencies above 1 THz [63]-[69]. Strong enhancement of the output power at particular (probably EM cavity) frequencies was also seen in a non-sandwiched rectangular stand-alone mesa and from arrays [59, 60]. Emission from mesas with slightly different heat removal designs while immersed in liquid nitrogen has also been observed [70, 71]. Hence, a more accurate name for a present device is "thermally-managed IJJ microstrip antenna (MSA)".

The important question of the sample homogeneity was raised by comparing the results of thin cylindrical (disk) conventional mesa IJJ emitters and thermally-managed IJJ disk MSAs [49, 55, 57, 65, 72]. In three conventional disk mesas, the strongest emission was observed at that of the lowest frequency transverse magnetic (1,1) disk EM cavity mode, the wave function of which has a line node passing through the disk center at a fixed angle ϕ_0. But the possible azimuthal emission anisotropy was not measured [57].

Moreover, in a more recent thermally-managed IJJ disk MSA using sapphire as the insulating plates, only very weak emission was observed at that TM(1,1) mode frequency [65]. Instead, the strongest emission was observed at about 1.0 THz, which is intermediate between the expected emission frequencies from the (0,1) and (2,1) EM cavity disk modes. As argued in the following, the disk (0,1) wave function exhibits the full rotational invariance of a one-dimensional representation (1DR) of the $C_{\infty v}$ point group appropriate for a perfectly homogeneous disk [73],

Type	Symmetry	n	n'	E	R_2	σ_x	σ_y
A_1	x^2+y^2	even	even	1	+1	+1	+1
A_2	xy	even	odd	1	+1	-1	-1
B_1	xy^2	odd	even	1	-1	+1	-1
B_2	x^2y	odd	odd	1	-1	-1	+1

Table 1. 1DR types, symmetries, allowed wave functions $\psi_{n,n'}(x,y)$ for odd or even $n, n' \geq 0$ of a rectangular microstrip antenna, and the operations of the C_{2v} point group.

and can therefore lead to an EM cavity resonance, whereas the disk TM(1,1) and TM(2,1) wave functions are two-dimensional representations (2DRs) of that point group, which would preclude the build-up of an EM cavity resonance [74, 75]. As noted above, the Au top and bottom layers also appear to amplify the emission power at the EM cavity mode frequencies [59], making the study of EM cavity mode emissions important for the development of useful devices.

2. rectangular microstrip antenna wave functions

In contrast to the three high-symmetry cases considered here, we first consider the simpler rectangular IJJ-MSA of length ℓ and width w, for which the wave (Helmholtz) equation for the magnetic vector potential $\psi \equiv A_z$ is $\nabla^2\psi + (k')^2\psi = 0$. For the physical situation of the emission from the IJJs in BSCCO, the uniform Josephson current is in the z direction, normal to the xy plane of the rectangular MSA. The transverse magnetic (TM) boundary conditions of the electromagnetic wave equation lead to the normal derivative of the wave function vanishing on the boundary. For a rectangular MSA of any material composition with $0 \leq x \leq \ell$, $0 \leq y \leq w$, the normalized wave functions are easily found to be,

$$\psi_{n,n'}(x,y) = (2/\sqrt{\ell w})\cos(n\pi x/\ell)\cos(n'\pi y/w), \qquad (1)$$

and $k'_{n,n'} = \pi\sqrt{(n/\ell)^2 + (n'/w)^2}$, where $n, n' \geq 0$, which clearly satisfy the Neumann TM boundary conditions $\frac{\partial\psi_{n,n'}(x,y)}{\partial x}\Big|_{x=0,\ell} = \frac{\partial\psi_{n,n'}(x,y)}{\partial y}\Big|_{y=0,w} = 0$. The frequencies in the vacuum radiation zone are then given by $f_{n,n'} = c_0 k'_{n,n'}/n_{\rm r}$, where c_0 is the speed of light in vacuum and $n_{\rm r}$ is the index of refraction of the MSA material. For BSSCO, $n_{\rm r} \sim 4.2$.

The wave functions for a perfect rectangular MSA are representations of the C_{2v} point group, with even or odd 180° rotational R_2 symmetry about the centroid at $(\ell/2, w/2)$ and even or odd reflections about the horizontal σ_x and vertical σ_y mirror planes, as detailed in Table 1. [73, 76]. The $n' = n = 0$ case with $f = 0$ cannot radiate. In Fig. 1, contour plots of four simplest examples of rectangular wave functions are presented. All four representations are one-dimensional representations (1DRs), as the trace of the 1×1 identity matrix E is just 1. In Fig. 1(a), $\psi(1,0)(x,y)$ is shown. It has a line node along the y axis, so it is odd (-1) about σ_y, even (+1) about σ_x, and odd (-1) under rotations (R_2) of 180° ($2\pi/2$ radians) about the centroid, so it is representation type B_1. Figure 1(b)is a contour plot of $\psi_{0,1}(x,y)$, which is odd under σ_x and R_2, but even under σ_y, so it is an example of symmetry type B_2. A contour plot of $\psi_{0,2}(xy)$ is shown in Fig. 1(c). This wave function is even under all three group operations, so it is an example of symmetry type A_1. $\psi_{1,1}(x,y)$ is pictured in Fig. 1(d). It is odd under σ_x and σ_y, and even under R_2, so it is an example of symmetry type A_2. Since any even or odd n and even or odd n' fits into this representation table, all of the rectangular MSAs are 1DRs, and can couple to the electromagnetic field, building up a resonance, and emitting with high power.

IOP Conf. Series: Materials Science and Engineering **279** (2017) 012017 doi:10.1088/1757-899X/279/1/012017

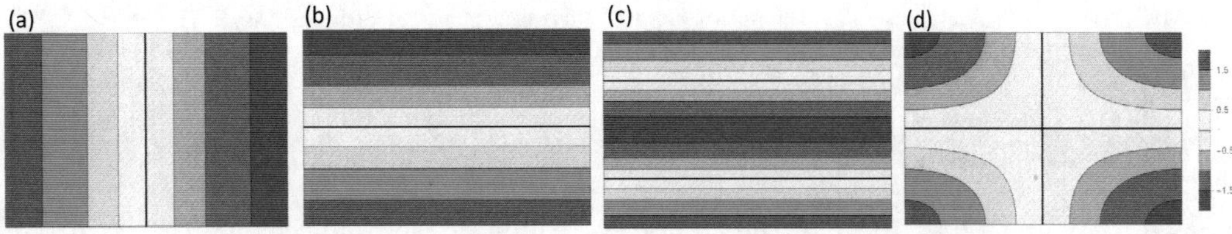

Figure 1. (colour online) Color-coded contour maps of the four simplest rectangular MSA TM wave functions and C_{2v} symmetry types: (a): (1,0), B_1; (b): (0,1), B_2; (c) (0,2), A_1; (d) (1,1), A_2. The color code applies to all four figures. The solid black lines are nodes.

Type	Symmetry	$\psi^{(\pm)}_{n,n'}(x,y)$	odd or even n	E	R_2	R_4	σ_x, σ_y	σ_{d1}, σ_{d2}
A_1	x^2+y^2	$\psi_{n,n}, \psi^{(+)}_{n,n+2p}$	e	+1	+1	+1	+1	+1
A_2	$xy(x^2-y^2)$	$\psi^{(-)}_{n,n+2p}$	o	+1	+1	+1	-1	-1
B_1	x^2-y^2	$\psi^{(-)}_{n,n+2p}$	e	+1	+1	-1	+1	-1
B_2	xy	$\psi_{n,n}, \psi^{(+)}_{n,n+2p}$	o	+1	+1	-1	-1	+1
E	fixed point nodes	$\psi^{(\theta,\pm)}_{n,n+2p+1}$	e,o	2	-1	0	0	0

Table 2. Representation types, symmetries, allowed 1DRs $\psi_{n,n}(x,y)$ and $\psi^{(\pm)}_{n,n+2p}(x,y) = [\psi_{n,n+2p}(x,y) \pm \psi_{n+2p,n}(x,y)]/\sqrt{2}$ for odd or even $n \geq 1$, 2DRs $\psi^{(\theta,+)}_{n,n+2p+1}(x,y$ of the square microstrip antenna, and operations of the C_{4v} point group. See text.

3. Square microstrip antennas

In contrast to the lower-symmetry rectangular MSA, a perfect square MSA of any material composition has much higher symmetry[73, 76, 85]. In Table 2, we list the table of the wave function symmetries under the point group C_{4v}. As noted previously[85], all 1DR wave functions given by Eq. (1) with $\ell = w$ have $|n - n'| = 2p$, where p is an integer. Hence, n and n' are either both even or both odd in 1DRs. These 1DRs have the *same* symmetry under both σ_x and σ_y, and also under both diagonal mirror planes σ_{d1} and σ_{d2}, and although they can be even or odd under rotations R_4 of $2\pi/4$ radians (90° about the centroid, they are even under R_2. These wave functions can excite cavity resonances, and the predicted radiation patterns for the ten lowest-frequency 1DR wave function modes were presented previously[85].

In contrast, wave functions $\psi_{n,n'}(x,y)$ for the square MSA for which $|n-n'| = 2p+1$, (or one of them is even and one of them is odd), are *two-dimensional representations* (2DRs). The simplest examples of these 2DR wave functions are the degenerate lowest frequency $(n,n') = (1,0)$ and (0,1), contour plots of which are pictured in Figs. 2(a) and 2(b), respectively. We note that $\psi)1,0(x,y)$ is odd under σ_y, but even under σ_x and odd under R_2. It has no symmetry under either diagonal mirror plane σ_{d1} or σ_{d2}. Similarly, $\psi_{0,1}(x,y)$ is odd under σ_x and R_2, but even under σ_y, and has no symmetry under either diagonal mirror. However, the particular orthonormal mixtures $\psi^{(\pm)}_{0,1}(x,y) = [\psi_{1,0}(x,y) \pm \psi_{0,1}(x,y)]/\sqrt{2}$ are pictured in Figs. 2(c) and 2(d). These two mixtures now have no symmetry under either σ_x or σ_y, but are still invariant under R_2. In addition, $\psi^{(+)}_{1,0}(x,y)$ pictured in Fig. 2(c) is odd under σ_{d1} (the diagonal from the upper left corner to the lower left corner), but even under the other diagonal mirror plane, σ_{d2}. Correspondingly, $\psi^{(-)}_{1,0}(x,y)$ pictured in Fig. 2(d) is even under σ_{d1}, odd under σ_{d2}, odd under

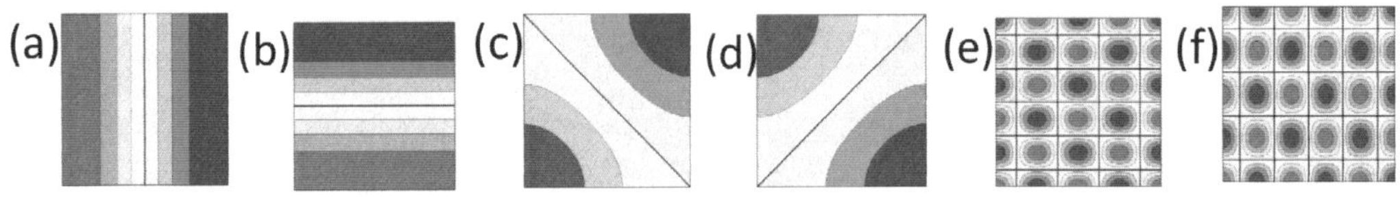

Figure 2. (colour online) Plots of the square wave functions $\psi_{n,n'}(x,y)$ for the (n,n') values (a): (1,0); (b): (0,1); (c): (1,0)+(0,1); (d): (1,0)-(0,1); (e): (4,5); and (f): (5,4).

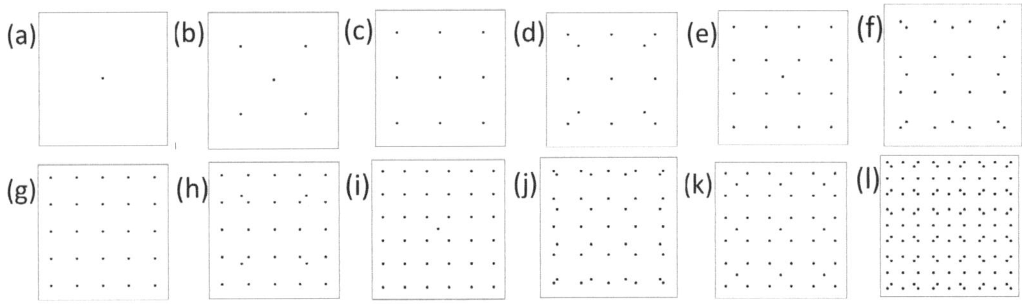

Figure 3. (colour online) Two dimensional representations of the wave functions for a perfect square MSA in order of increasing nominal frequency with (n,n') values: (a)(0,1); (b)(1,2); (c) (0,3); (d) (2,3); (e) (1,4); (f) (3,4); (g) (0,5); (h) (2,5); (i) (1,6); (j) (4,5); (k) (3,6); (l) (6,9).

R_2, and has no symmetry under either σ_x or σ_y. Moreover, we could form a general mix of these two-dimensional degenerate wave functions, forming the two-member orthonormal set

$$\psi^{(\theta,+)}_{n,n+2p+1}(x,y) = \cos\theta\psi_{n,n+2p+1}(x,y) + \sin\theta\psi_{n+2p+1,n}(x,y), \quad (2)$$

$$\psi^{(\theta,-)}_{n,n+2p+1}(x,y) = -\sin\theta\psi_{n,n+2p+1}(x,y) + \cos\theta\psi_{n+2p+1,n}(x,y). \quad (3)$$

The operations of this general pair are given by two-dimensional matrices, and identity matrix in two dimensions has the trace of 2. Thus, the functions pictured in Figs. 2(a)-2(d) could be represented by $\psi^{(\theta,+)}(1,0)$ for $n=p=0$ and respectively $\theta=\pi/2, 0, \pi/4$, and $3\pi/4$. Or, they could be represented by $\psi^{(\theta,-)}(1,0)$ with $\theta = 0, 3\pi/2, 3\pi/4$, and $\pi/4$, respectively. Thus, this is the simplest example of a 2DR wave function. Because θ could be a *random* function of the time, in an IJJ MSA, if $\theta(t)$ changes on the time scale of a single oscillation of the ac Josephson current, the form of the wave function would change, and a cavity resonance could not form. This is the main reason for this study of high-symmetry, thermally-managed IJJ MSAs. But, one might ask, what is a consistent way to represent these 2DR wave functions? One way that we have found is to consider the nodal structure that is invariant under changes in the mixing angle θ. For the (1,0) and (0,1) wave functions pictured in Figs. 2(a)-2(d), for different θ values, there is only one consistent node: the node at the centroid $(\ell/2, \ell/2)$. This nodal structure is pictured in Fig. 3(a). More generally, the ten lowest frequency nodal structures are shown in Figs 3(a)-3(j). Figure 3(j) is obtained from the two wave functions picture in Figs. 2(e) and 2(f), respectively. We note each of these nodal sets satisfies the C_{4v} group operations of the *one-dimensional* representation of type A_1, as the pattern is invariant under σ_x, σ_y, σ_{d1}, σ_{d2}, and R_4. It is further interesting to note that when (n,n') not only satisfy $n' = n+2p+1$, but also are both multiples of an odd integer, then the pattern forms a repeated pattern of the lower number indices. For example, Fig. 3(k) is the pattern for $\psi^{(\theta,\pm)}_{3,6}$, and it consists of a 3×3 array of the same pattern of $\psi^{(\theta,\pm)}_{1,2}$ pictured in Fig. 3(b). Similarly, Fig. 3(l) is the pattern for $\psi^{(\theta,\pm)}_{6,9}$,

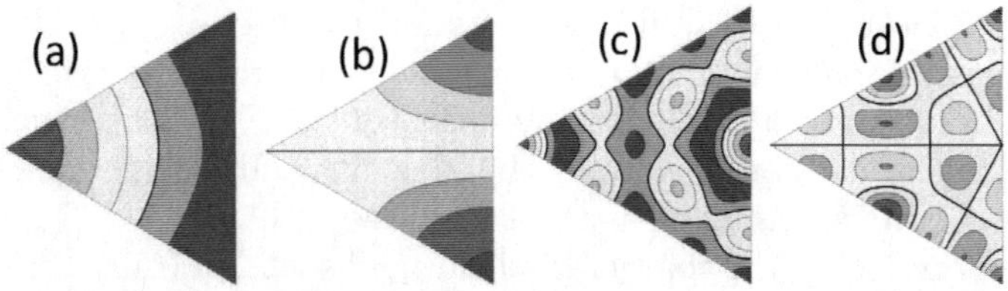

Figure 4. 2DR equilateral triangular mode examples. (a),(b) even and odd (0,1) modes. (c), (d) even and odd (4,2) modes.

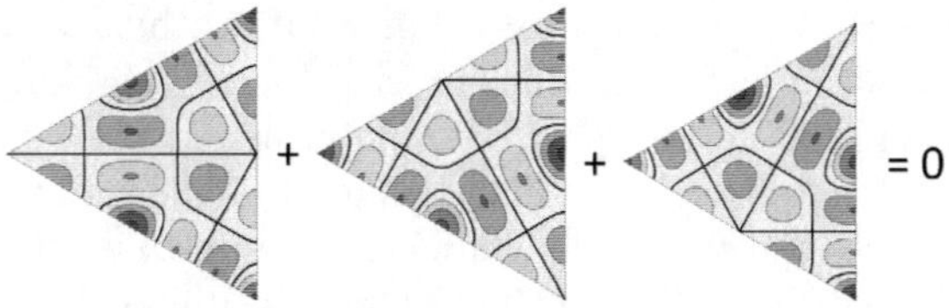

Figure 5. Pictorial summation of the (4,2) odd modes rotated by ±120° about the central axis.

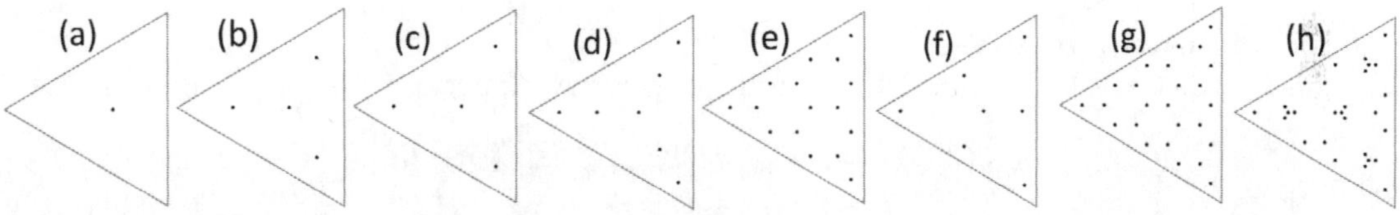

Figure 6. (colour online) Two dimensional representations of the wave functions for a perfect equilateral triangular MSA in order of increasing nominal frequency for the indicated (n, m) pairs: (a)(0,1); (b) (0,2); (c) (1,2); (d) (1,3); (e) (0,4); (f) (2,3); (g) (0,5); (h) (2,4).

which is a 3×3 array of the pattern of $\psi_{2,3}^{(\theta,\pm)}$ shown in Fig. 3(d).

4. Equilateral triangular microstrip antennas

For a perfect equilateral triangular MSA of any material composition, the wave functions are considerably more complicated than for the square[81, 85]. The wave functions for a particular symmetry plane (such as the horizontal axis that bisects an angle and the opposite side), the even wave functions that satisfy the Helmholtz equation with the Neumann (vanishing normal derivatives) boundary conditions are three products of cosine functions of x and y, respectively, and the odd functions are three products of one cosine function of x and a sine function of y[81]. The wavevectors satisfy $k'_{n,m}a = 4\pi\sqrt{n^2 + m^2 + nm}/(3n_{\rm r})$, where a is the length of a side of the triangle. The 1DR wave functions satisfy $|m - n| = 3p$, where p is an integer. For $p = 0$, the 1DR wave functions are non-degenerate, for for $p \neq 0$, they are double degenerate, with functions that are either even or odd about all three mirror planes normal to the planar surface and bisecting one of the angles and its opposite side. But they didn't describe the 2DR wave functions. Here we show how they can be best described.

The lowest frequency wave functions are the (0,1) modes. Those even and odd about the single symmetry plane containing the horizontal axis are pictured in Figs. 4(a) and 4(b), respectively. Similar contour plots of the even and odd (4,2) modes are shown in Figs. 4(c) and 4(d), respectively. Note that these figures only have the C_{3v} group symmetry operation of a single mirror plane, and are thus elements of a 2DR. Previously, it was shown that by forming the sum of the wave function pictured in Fig. 2(a) with its form rotated by $\pm 120°$ about the centroid, one obtains the remarkable value 0 for all points inside the triangle. In Fig. 5, we did this for the odd (4,2) wave function, just as a check on the validity of the C_{3v} group symmetry, proving that this is a 2DR wave function. More generally, the 2DR wave functions are general orthonormal sums of the two degenerate forms, such as those pictured in Figs. 4(a,b) and 4(c,d). As for the consistent method of presenting the characteristic form of such 2DR wave functions, we combine the nodal structure of these two degenerate forms, locating the set of consistent nodal points. In Fig. 8, we have done this for the eight lowest frequency 2DR wave functions. The patterns obtained from the pair pictured in Figs. 4(a,b) is shown in Fig. 6(a). It consists of a single nodal point at the centroid. The pattern obtained analogously from the (02) even and odd wave functions is shown in Fig. 6(b). Similarly, those obained from the even and odd (12), (13), (04), (23), (05), and (24) modes [the latter shown in Figs. 4(c,d)] are shown in Fig. 6(c)-6(h). In each case, the overall pattern has complete R_3 symmetry about the centroid by $\pm 120°$, which is a requirement for the 1DR wave functions. This remarkable finding of the common nodal point array has not been noticed previously to our knowledge. Note that the (24) pattern pictured in Fig. 6(h) is similar to that of the (04) pattern pictured in Fig. 6(3), but four of the points in the figure have split into a triangular pattern of four points. Note that the (04) and (05) nodal patterns shown in Figs. 6(e,g) form honeycomb lattice sections commensurate with the triangle.

5. Disk microstrip antennas

For a perfect disk microstrip antenna of radius a, the wave functions $\psi \equiv A_z$ have the forms

$$\psi_{n,m}(\rho,\phi) = C_{n,m}J_n(k_{n,m}\rho)\cos[n(\phi-\phi_0)], \tag{4}$$

where ϕ_0 is an arbitrary angle, and the $C_{n,m}$ are normalization constants. The wave vectors $k_{n,m}$ are obtained from the Neumann boundary condition $J_n'(k_{n,m}a) = 0$, where $J_n'(z)$ is the first derivative of the regular Bessel function $J_n(z)$. In Fig. 7, contour maps of the 10 lowest frequency disk modes are shown,

We note that in Figs. 7(c) and 7(i) corresponding to the TM(01) and TM(02) cavity modes, the only nodes in the wave functions are circles centered at the origin, so these modes are rotationally invariant about the axis normal to the disk plane that passes through the disk center, and are therefore one-dimensional representations (1DRs) of point group $C_{\infty v}$[73]. The other eight modes all have one or more linear nodes that span the disk and pass through its center, and are thus not invariant under the above rotations. Hence, they are two-dimensional representations (2DRs) of point group $C_{\infty v}$, and will only be realized in imperfect devices.

We first assume that the disk MSA under study has spatial inhomogeneities that break the $C_{\infty v}$ point group symmetry. This could arise from thermal inhomogeneities, such as as hot spot that have been observed in earlier devices that were not thermally managed. In such as case, excitations of any of the cavity modes are possible, and in Fig. 8, we show three-dimensional (3D) plots of the angular distribution of the emission as predicted for BSCCO[74, 75]. In order to distinguish the 2DR wave functions from the 1DR wave functions pictorially, one should imagine that the line nodes bisecting the 2DR wave functions that pass through the origin are free to rotate about the origin, as if the ϕ_0 in Eq. (4) were a random function of the time, which would hinder the wave function from forming a resonance. Hence, the wave function would always vanish at the disk centroid, and possibly at other fixed radii, but would average to zero over a long time. Thus, these line nodes then reduce to a point at the origin (or centroid) of the disk. In Fig. 9, we have shown the reduced 2DRs that represent the 8 wave functions pictured in Fig. 7, plus the TM(13) and TM(23) modes not pictured. Note that Fig. 9(a) corresponds to all TM(n,1) modes with $n \geq 1$. For a disk of unit radius, the radii of the black circles are 0.7187 for the TM(12) mode, 0.7658 for the TM(22) mode, 0.4489 and 0.8218 for the TM(13) mode, and 0.5151 and 0.8443 for the TM(23) mode.

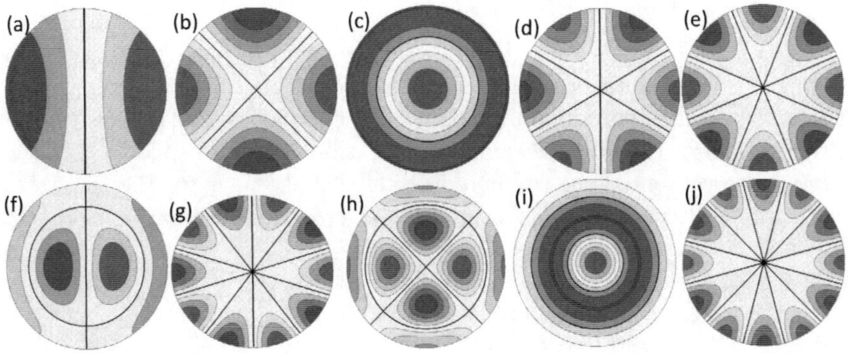

Figure 7. (colour online) Contour plots of the ten lowest frequency TM disk modes. (a) (1,1); (b) (2,1); (c) (0,1); (d) (3,1); (e) (4,1); (f) (1,2); (g) (5,1); (h) (2,2); (i) (0,2); (j) (6,1). The solid black lines and circles are nodes.

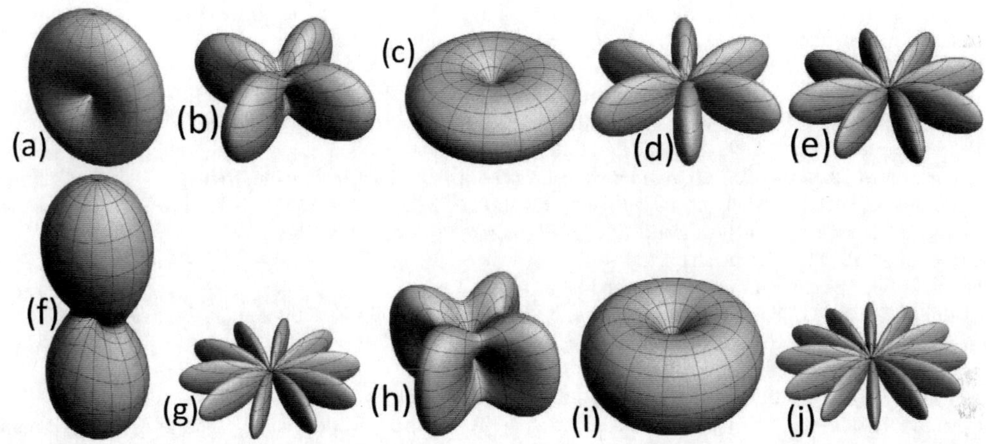

Figure 8. (colour online) Angular distribution of the predicted power distribution from the excitation of each of the ten lowest frequency EM cavity modes of an imperfect disk BSCCO device. (a)(1,1); (b) (2,1); (c) (0,1); (d) (3,1); (e) (4,1); (f) (1,2); (g) (5,1); (h) (2,2); (i) (0,2); (j) (6,1).

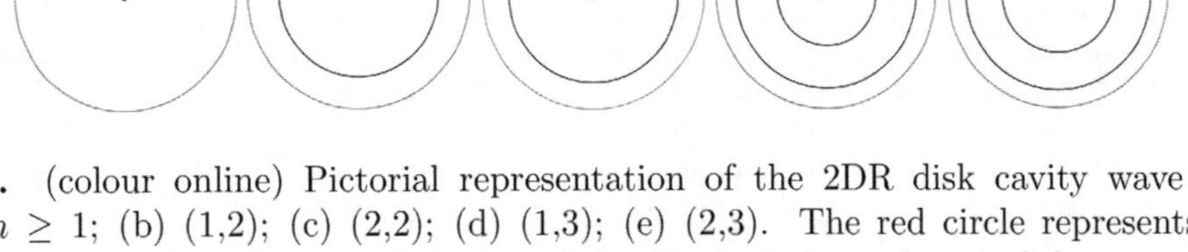

Figure 9. (colour online) Pictorial representation of the 2DR disk cavity wave functions (a)(n,1), $n \geq 1$; (b) (1,2); (c) (2,2); (d) (1,3); (e) (2,3). The red circle represents the disk edge with Neumann boundary conditions, and the black circles and central dots are nodes.

IOP Conf. Series: Materials Science and Engineering **279** (2017) 012017 doi:10.1088/1757-899X/279/1/012017

6. Conclusions

We found a set of nodal points that defines the intersection of the common nodes for the ten lowest frequency degenerate two-dimensional representations of a perfect disk, square, or equilateral triangular microstrip antenna of arbitrary composition. The positions of the nodal points are invariant under the one-dimensional operations of the respective $C_{\infty v}$, C_{4v} and C_{3v} point groups. We also presented the predicted power distributions for the ten lowest frequency cavity modes of a disk intrinsic Josephson junction microstrip antenna.

References

[1] Ferguson B and Zhang X C 2002 *Nat. Mater.* **1** 26–33
[2] Tonouchi M 2007 *Nature Photon.* **1** 97-105
[3] Walther C, Fischer M, Scalari G, Terazzi R, Hoyler N, and Faist J 2007 *Appl. Phys. Lett.* **91** 131122
[4] Fathololoumi S, Dupont E, Chan C W I, Wasilewski Z R, Laframboise F R, Ban D, Mátyás A, Jirauschek C, Hu Q, and Liu H C 2012 *Opt. Express* **20** 3866–76
[5] Kiessling J, Breunig I, Schunemann P G, Buse K, and Vodopyanov K 2013 *New J. Phys.* **15** 105014
[6] Maestrini A *et al.* 2010 *IEEE Trans. Microw. Theory Tech.* **58** (2) 1925–32
[7] Maestrini A *et al.* 2012 *IEEE Trans. THz Sci. Tech.* **2** 177–85
[8] Lu Q Y, Slivken S, Bandyopadhyay N, Bai Y, and Razeghi M 2014 *Appl. Phys. Lett.* **105** 201102
[9] Koyama Y, Sekiguchi R, and Ouchi T 2013 *Appl. Phys. Express* **6** 064102
[10] Kanaya H, Sogabe R, Maekawa T, Suzuki S, and Asada M 2014 *J. Infrared Milli. Terahz. Waves* **35** 425–31
[11] Maekawa T, Kanaya H, Asada M, and Suzuki A 2016 *Appl. Phys. Express* **9** 024101
[12] Asada M and Suzuki S 2015 *Jap. J. Appl. Phys.* **54** 020309
[13] Josephson B D 1962 *Phys. Lett.* **1** 251–53
[14] Barbara P, Cawthorne A B, Shitov S V, and Lobb C J 1999 *Phys. Rev. Lett.* **82** 1963–66
[15] Kleiner R, Steinmeyer R, Kunkel G, and Müller P 1992 *Phys. Rev. Lett.* **68** 2394–97
[16] Kleiner R and Müller P 1994 *Phys. Rev. B* **49** 1327–41
[17] Klemm R A 2012 *Layered Superconductors* vol 1 (Oxford: Oxford University Press)
[18] Mochiku T and Kadowaki K 1994 *Physica C* **235-240** 523–24
[19] Mochiku T, Hirata K, and Kadowaki K 1997*Physica C* **282-287** 475–76
[20] Ozyuzer L *et al.* 2007 *Science* **318** 1291–93
[21] Kadowaki K *et al.* 2008 *Physica C* **468** 634–39
[22] Minami H, Kakeya I, Yamaguchi H, Yamamoto T, and Kadowaki K 2009 *Appl. Phys. Lett.* **95** 232511
[23] Ozyuzer L *et al.* 2009 *Supercond. Sci. Technol.* **22** 114009
[24] Kashiwagi T *et al.* 2011 *J. Phys. Soc. Jpn.* **80** 094709
[25] Kashiwagi T *et al.* 2012 *Jap. J. Appl. Phys.* **51** 010113
[26] Benseman T M, Koshelev A E, Gray K E, Kwok W K, Welp U, Kadowaki K, Tachiki M, and Yamamoto T 2011 *Phys. Rev. B* **84** 064523
[27] Koseoglu H, Turkoglu F, Simsek Y, and Ozyuzer L 2011 *J. Supercond. Nov. Magn.* **24** 1083–86
[28] Kurter C, Ozyuzer L, Proslier T, Zasadzinski J F, Hinks D G, and Gray K E 2010 *Phys. Rev. B* **82** 224518
[29] Li M Y *et al.* 2012 *Phys. Rev. B* **86** 060505R
[30] Minami H, Tsujimoto M, Kashiwagi T, Yamamoto T, and Kadowaki K 2012 *IEICE Trans. Electron.* **E95-C** 347–54
[31] Kakeya I, Omukai Y, Yamamoto T, Kadowaki K, and Suzuki M 2012 *Appl. Phys. Lett.* **100** 242603
[32] Turkoglu F, Koseoglu H, Demirhan Y, Ozyuzer L, Preu S, Malzer S, Simsek Y, Müller T, Yamamoto T, and Kadowaki K 2012 *Supercond. Sci. Technol.* **25** 125004
[33] Yuan J *et al.* 2012 *Supercond. Sci. Technol.* **25** 075015
[34] Tsujimoto M, Minami H, Delfanazari K, Sawamura M, Nakayama R, Kitamura T, Yamamoto T, Kashiwagi T, Hattori T, and Kadowaki K 2012 *J. Appl. Phys.* **11** 123111
[35] An D Y *et al.* 2013 *Appl. Phys. Lett.* **102** 092601
[36] Kadowaki K *et al.* 2013 *Physica C* **491** 2–6
[37] Benseman T M, Koshelev A E, Kwok W K, Welp U, Vlasko-Vlasov V K, Kadowaki K, Minami H, and Watanabe C 2013 *J. Appl. Phys.* **113** 133902
[38] Welp U, Kadowaki K, and Kleiner R 2013 *Nat. Photon.* **7** 702–10
[39] Kakeya I and Wang H B 2016 *Supercond. Sci. Tech.* **29** 073001
[40] Niratisairak S, Haugen O, Johansen T H, and Ishibashi T 2008 *Physica C* **468** 442–46
[41] Wang H B, Guénon, Yuan J, Iishi A, Arisawa S, Hatano T, Yamashita T, Koelle D, and Kleiner R 2009 *Phys. Rev. Lett.* **102** 017006
[42] Wang H B *et al.* 2010 *Phys. Rev. Lett.* **105** 057002

[43] Guénon S *et al.* 2010 *Phys. Rev. B* **42** 214506
[44] Gross B *et al.* 2012 *Phys. Rev. B* **86** 094524
[45] Klemm R A, LaBerge E R, Morley D R, Kashiwagi T, Tsujimoto M, and Kadowaki K 2011 *J. Phys.: Condens. Matter* **23** 025701
[46] Minami H, Watanabe C, Sato K, Sekimoto S, Yamamoto T, Kashiwagi T, Klemm R A, and Kadowaki K 2014 *Phys. Rev. B* **89** 054503
[47] Watanabe C, Minami H, Sekimoto S, Yamamoto T, Kashiwagi T, Klemm R A, and Kadowaki K 2014 *J. Phys.: Condens. Matter*, vol. 26, 2014, Art no. 172201.
[48] Watanabe C *et al.* 2015 *Appl. Phys. Lett.* **106** 042603
[49] Watanabe C *et al.* 2016 *Supercond. Sci. Tech.* **29** 065022
[50] Krasnov V M 2011 *Phys. Rev. B* **83** 174517
[51] Yurgens A A 2011 *Phys. Rev. B* **83** 184501
[52] Tsujimoto M, Kambara H, Maeda Y, Yoshioka Y, Nakagawa Y, and Kakeya I 2014 *Phys. Rev. Applied* **2** 044016
[53] Kashiwagi T *et al.* 2014 *Appl. Phys. Lett.* **104** 022601
[54] Kashiwagi T *et al.* 2014 *Appl. Phys. Lett.* **104** 082603
[55] Demirhan Y *et al.* 2015 *Vacuum* **120** (SI) 89-94
[56] Zhou X J *et al.* 2015 *Phys. Rev. Applied* **3** 044012
[57] Tsujimoto M, Yamaki K, Deguchi K, Yamamoto T, Kashiwagi T, Minami H, Tachiki M, Kadowaki K, and Klemm R A 2010 *Phys. Rev. Lett.* **105** 037005
[58] Yamaki K, Tsujimoto M, Yamamoto T, Furukawa A, Kashiwagi T, Minami H, and Kadowaki I 2011 *Opt. Express* **19** 3193–201
[59] Tsujimoto M *et al.* 2012 *Phys. Rev. Lett.* **108** 107006
[60] Benseman T M, Gray K E, Koshelev A E, Kwok W K, Welp U, Minami H, Kadowaki K, and Yamamoto T 2013 *Appl. Phys. Lett.* **103** 022602
[61] Kitamura T *et al.* 2014 *Appl. Phys. Lett.* **105** 202603
[62] Sekimoto S, Watanabe C, Minami H, Yamamoto T, Kashiwagi T, Klemm R A, and Kadowaki K 2013 *Appl. Phys. Lett.* **103** 182601
[63] Ji M *et al.* 2014 *Appl. Phys. Lett.* **105** 122602
[64] Kashiwagi T *et al.* 2015 *Appl. Phys. Lett.* **106** 092601
[65] Kashiwagi T *et al.* 2015 *Appl. Phys. Lett.* **107** 082601
[66] Kashiwagi T *et al.* 2015 *Phys. Rev. Applied* **4** 054018
[67] Nakade K, Kashiwagi T, Saiwai Y, Minami H, Yamamoto T, Klemm R A, and Kadowaki K 2016 *Sci. Rep. (Nature)* **6** 23178
[68] Asai H and Kawabata S 2014 *Appl. Phys. Lett.* **104** 112601
[69] Rudau F *et al.* 2015 *Phys. Rev. B* **91** 104513
[70] Hao L Y *et al.* 2015 *Phys. Rev. Applied* **3** 024006
[71] Minami H, Watanabe C, Kashiwagi T, Yamamoto T, Kadowaki K, and Klemm R A 2016 *J. Phys.: Condens. Matter* **28** 025701
[72] Kadowaki K *et al.* 2010 *J. Phys. Soc. Jpn.* **79** 023703
[73] Tinkham M 1964 *Group Theory and Quantum Mechanics* (New York: McGraw-Hill)
[74] Klemm R A and Kadowaki K 2010 *J. Supercond. Novel Magn.* **23** 613–16
[75] Klemm R A and Kadowaki K 2010 *J. Phys.: Condens. Matter* **22** 375701
[76] Boas M L 2006 *Mathematical Methods in the Physical Sciences* 3rd ed. (Hoboken, NJ: Wiley)
[77] McCartin B J 2002 *Math. Probl. Eng.* **8**517–39
[78] McCartin B J 2011 *Laplacian Eigenstructure of the Equilateral Triangle*(Ruse, Bulgaria: Hikari, Ltd.)
[79] Stambaugh N and Semon M 2013 *Can. J. Phys.* **99** 1–19
[80] Klemm R A, Delfanazari K, Tsujimoto M, Kashiwagi T, Kitamura T, Yamamoto T, Sawamura M, Ishida K, Hattori T, and Kadowaki K 2013 *Physica C* **49** 30-34
[81] Cerkoney D P *et al.* 2017 *J. Phys.: Condens. Matter* **29** 015601
[82] Cerkoney D P 2015 *Terahertz radiation form high-temperature superconducting BSCCO mesas of various geometries* B. S. Honors in the Major Thesis, University of Central Florida (unpublished).
[83] Balanis C A 2005 *Antenna Theory, Analysis and Design* 3rd ed. (Hoboken: Wiley)
[84] Morales M A 2015 *Angular dependence of the emission from the intrinsic Josephson junction in pie-shaped wedge triangular BSCCO mesas* B. S. Honors in the Major Thesis, University of Central Florida (unpublished).
[85] Klemm R A, Davis A E, and Wang Q X 2017 *IEEE J. Sel. Topics Quant. Electron.* **23** 8501208
[86] Kashiwagi T *et al.* 2017 *Supercond. Sci. Tech.* **30** 074008

Measurement of a Conduction Cooled Nb_3Sn Racetrack Coil

HS Kim[1], C Kovacs[1], J Rochester[1, 3], MD Sumption[1], M Tomsic[2], X Peng[2], D Doll[2]

[1] Department of Materials Science and Engineering, The Ohio State University, Columbus, Ohio 43210, USA

[2] Hyper Tech Research Incorporated, 539 Industrial Mile Road, Columbus, Ohio 43228, USA

[3] Author to whom any correspondence should be addressed. Electronic mail: rochester.11@osu.edu.

Abstract. Use of superconducting coils for wind turbines and electric aircraft is of interest because of the potential for high power density and weight reduction. Here we test a racetrack coil developed as a proof-of-concept for cryogen-free superconducting motors and generators. The coil was wound with 1209 m of 0.7-mm-diameter insulated tube-type Nb_3Sn wire. The coil was epoxy-impregnated, instrumented, covered with numerous layers of aluminized mylar insulation, and inserted vertically into a dewar. The system was cooled to 4.2 K, and a few inches of liquid helium was allowed to collect at the bottom of the dewar but below the coil. The coil was cooled by conduction via copper cooling bars were attached to the coil but also were immersed in the liquid helium at their lower ends. Several current tests were performed on the coil, initially in voltage mode, and one run in current mode. The maximum coil I_c at 4.2 K was 480 A, generating 3.06 T at the surface of the coil. The coil met the design targets with a noticeable margin.

1. Introduction

Superconducting motors and generators can have a much higher power density than their conventional counterparts. Such a high power density is an advantage for offshore wind turbines, where the size of the nacelle and tower are important cost factors [1-4]. For all-electric aircraft, higher power density and higher power per unit weight are the driving factors for the consideration of superconducting motors and generators [5-14]. This is enabled by superconducting machines because the superconducting windings can achieve magnetic fields above those of conventional windings, and energy density goes as the square of the field. For superconducting generators, some designs have superconducting rotors (inside), but conventional stators (surrounding the rotor). However, for fully cryogenic designs both the rotor and stator are superconducting. In these designs, the rotors generate field experienced by the stators and the stators are exposed to relatively high dB/dt, which leads to potentially large losses in the superconductor. A new design by Haran [15] uses field coils which surround a rotating induction coil. The induction coil is Cu, but the field coils are made from either Nb_3Sn or YBCO, and generate high fields, about 6-8 T in present designs. This large value leads to very high power densities, while keeping the losses relatively small, since the superconductors only experience a ripple field. Presently, Nb_3Sn is the more affordable option, and leads to very high power density designs.

The design for this machine calls for a set of racetrack-like windings placed to form a cylindrical field winding in side of which the induction coils rotate. In a previous paper [15], a design for one of these Nb_3Sn racetrack coils was given, along with field and strain modelling, and the coil construction

Table 1. Wire and coil specifications

Coil Material	
Former	Copper
Outer cover	316 stainless steel
Coil Parameters	
Winding pack Length	618.1 mm
Winding Pack Height	70.7 mm
Winding Pack Width	9.5mm
No. Turns	956.75
Total Conductor Length	1209 meters
Turns/layer	11
No. Layers	87
Wire	
Strand	Nb_3Sn, tracer number 3635
Insulation	s-glass
Number of Filaments	180
OD bare wire	0.7 mm
OD with insulation	0.8 mm
Non-Cu	46.5%
Heat Treatment	625C/120 h, vacuum

and winding was described. In this paper, we describe the test results on that coil, including current and field measurements at the target temperature as well as at higher temperatures.

In the final machine as designed, the system would use cryocoolers with the coil conduction cooled using thermal straps. In this work, we use conduction cooling, but for simplicity use a liquid helium reservoir as the cooling basin connected to the cooling straps, rather than a cryocooler. Nevertheless,

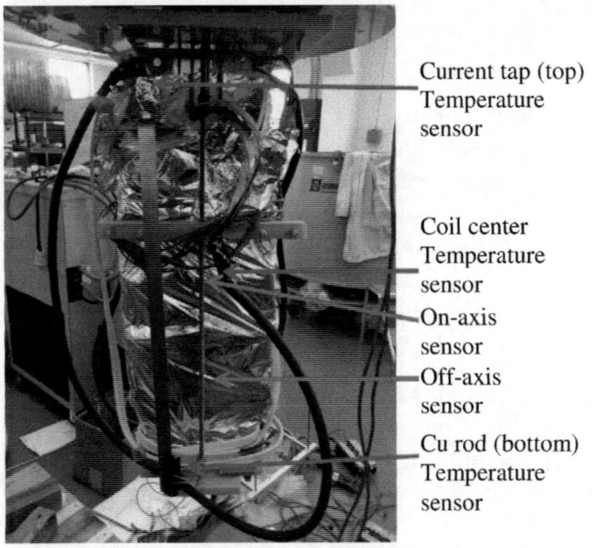

Figure 1. View of the coil instrumentation. Shown here are the three temperature sensors: one each at the current tap, coil center, and lower end of the Cu rod; as well as the two Hall sensors: one on-axis and one off-axis.

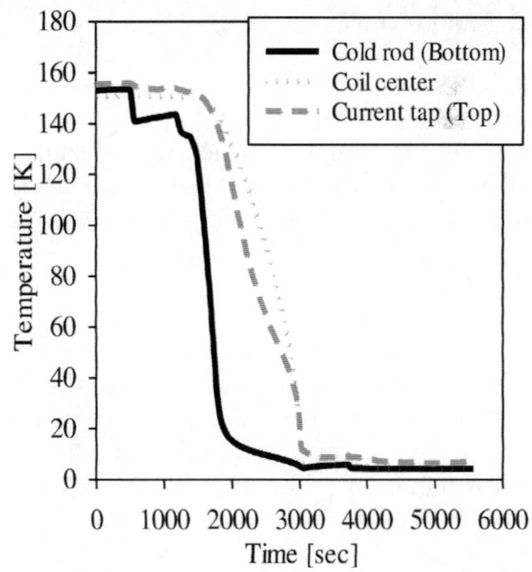

Figure 2. Temperature at each sensor during cooldown

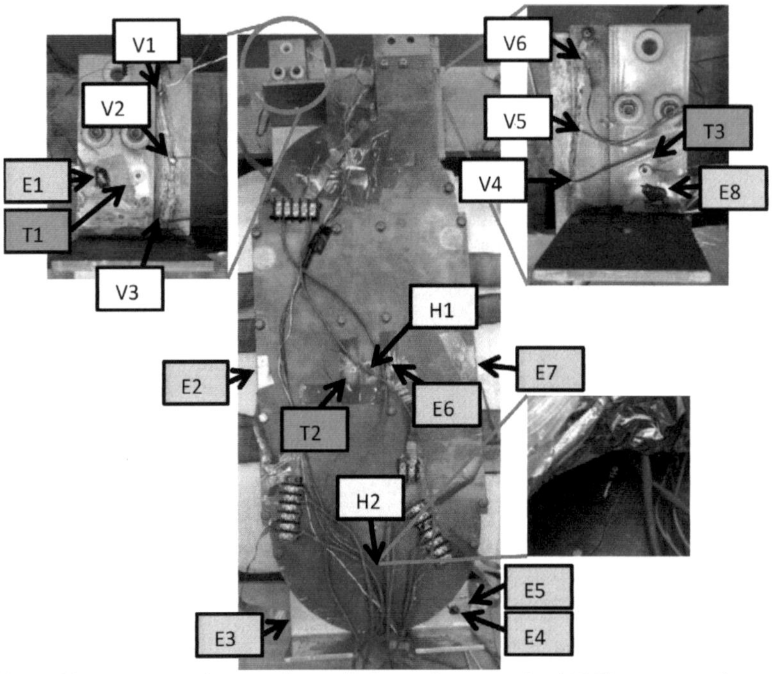

Figure 3. Location of instrumentation on the coil. Locations marked "V" notate voltage taps; "H", Hall sensors; "T", Cernox temperature sensors; and "E", Type-E thermocouples.

we are able to see the performance of the coil in conduction cooled mode, as well as the temperature gradient of the coil.

The coil measured here is the one which was modelled and described previously in [15]. Herein we report the results from the successful manufacture and testing of a prototype coil.

2. Manufacture

2.1. Conductor

The conductor (strand 3635), a tube-type strand manufactured by Hyper Tech Research, Inc. has 0.7 mm OD before insulation and contains 180 superconducting filaments. The filament size was 35 μm and the non-Cu area fraction was 46.5%. Pure Cu was used as the interfilamentary matrix. The wire was insulated with s-glass before winding.

2.2. Coil Winding

The coil was wound counter-clockwise with 1209 m of wire, with 11 turns per layer and 956.75 turns in total. The winding pack had a length of 618.1 mm, a height of 70.7 mm, and width of 9.5 mm. Actual winding payoff tension was maintained at 4.95 lbs. (nominal 5 lbs.).

After winding, the coil was heat treated in vacuum at 625°C for 120 hrs plus ramp-up and cool-down. Current leads were soldered with 40/60 Pb-Sn solder. The coil was then impregnated by vacuum infusion with epoxy at a pressure of 710 Torr, with flow duration of 7 hours and curing time of 24 hours. The epoxy was cured by heating tape with blanket at 80-90 °C.

2.3. Instrumentation

Measurements were enabled by the attachment of sensors to the coil (refer to Figures 1 and 3). Three Cernox temperature sensors were used; one at the current leads at the top, one at the center point of the coil, and one at the bottom end of the Cu support rods. Two Hall sensors were attached for magnetic flux measurement: one on-axis, and one off-axis at the point of maximum magnetic field. Several

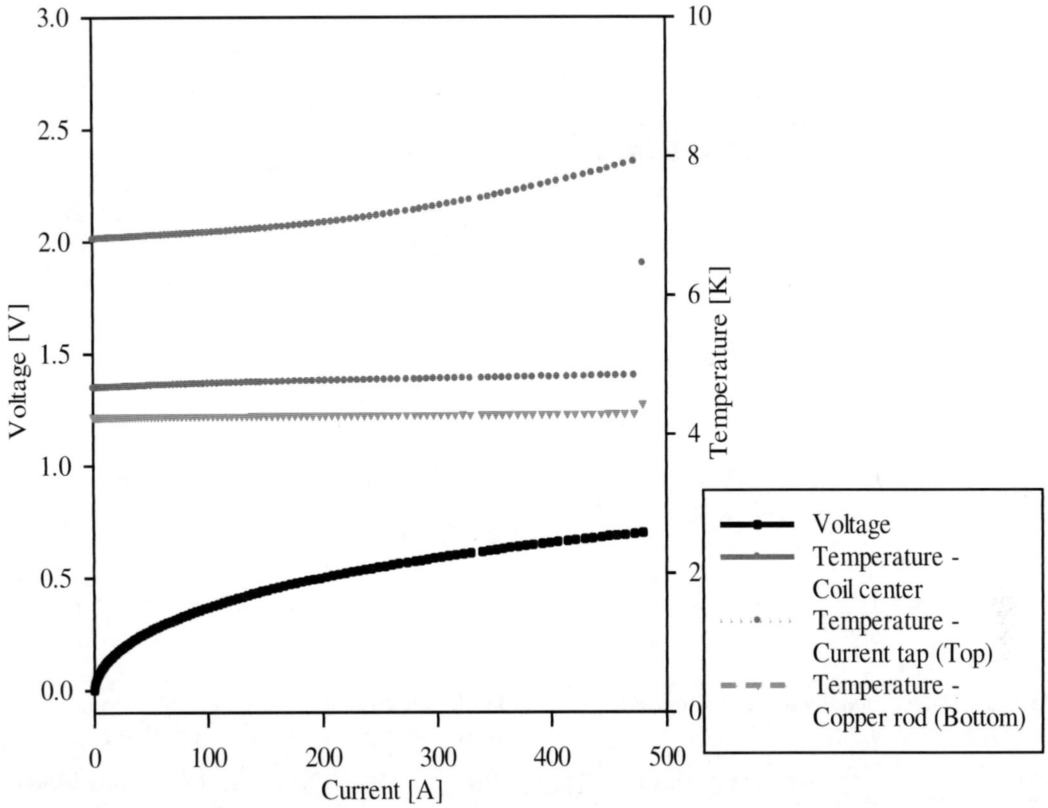

Figure 4. Current vs voltage and temperature in voltage mode (Run 4)

Type-E thermocouples were attached using Stycast 2850FT and insulated from the coil using thin cigarette paper. Voltage taps were also attached to the coil at intervals.

3. Coil Setup and Cooldown

The coil was mounted vertically in a dewar, sitting on a G10 plate which hangs on threaded rods from the top flange of the dewar. Insulated copper current leads ran from the top of the dewar, down into the liquid helium (just a few inches at the bottom of the dewar, below the coil itself) and then back up to the coil. The coil was covered with multilayer insulation to reduce radiative and convective heat transfer into the coil. The coil itself has a Cu former, with Cu cooling flanges which extend out the bottom of the coil; these flanges also serve in this instance as the feet of the coil, which rest on the G10 plate. An additional pair of Cu busbars runs to the top of the dewar (once can be seen). In addition, a dedicated set of flexible copper cables preferentially cool the leads.

At the start of the measurement, the coil was placed in the dewar and liquid nitrogen was injected slowly into the bottom of the dewar, cooling the coil overnight. The nitrogen was allowed to boil off before beginning the liquid helium fill.

After cooling, a few inches of liquid helium were allowed to accumulate at the bottom of the dewar, below the coil. The helium cooled the "feet" of the dewar, as well as the busbar, and these cooled the coil by conduction. The lowest temperatures achieved were 4.21 K at the coil bottom, 4.49 K at the coil center, and 6.46 K at the coil top (top of current lead). This gives a temperature difference across the coil of 2.25 K.

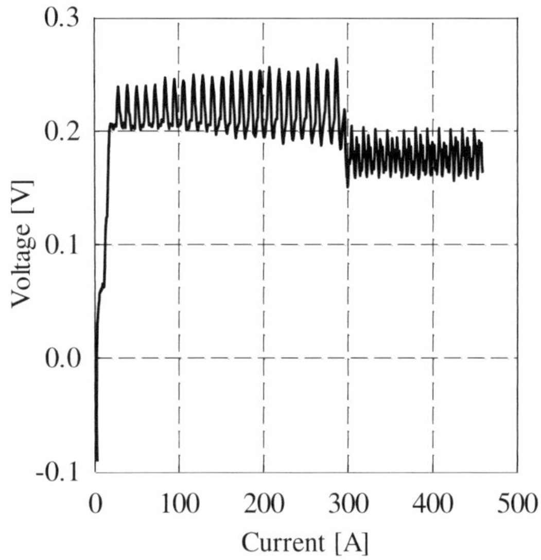

Figure 5. *I-V* data at 4.2 K in current mode

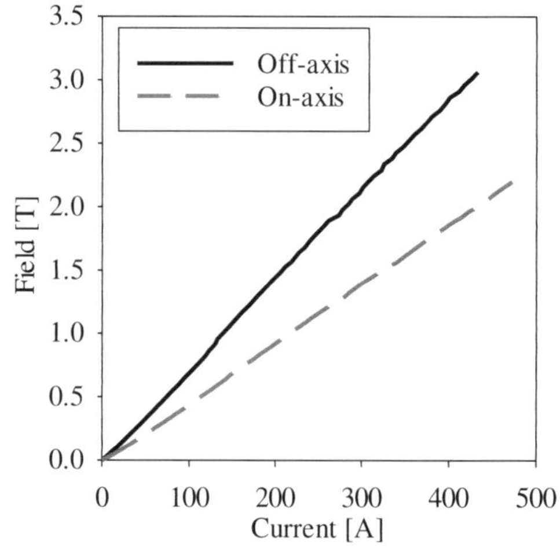

Figure 6. Field (*B*) vs current (*I*) at 4.2 K

Table 2. Critical/quench current of coil, and maximum field at 4.2 K

Run (#)	I_c, I_q [A]	Max Field [T] On-axis	Max Field [T] Off-axis
1	437	2.00	-
2	440	-	3.06
3	459	-	-
4	480	2.20	-

Table 3. Critical/quench current of coil and maximum field at various temperatures

Run (#)	Coil Temp. [K]	I_c, I_q [A]	Max Field [T] On-axis
5	5.56	457	2.09
6	8.02	384	1.75
7	10.83	265	1.20

4. Experiment

After the initial cool down, several current runs were performed. Initially the current was applied using voltage mode. It is a very safe way to run the magnet and is done to prevent the possibility of overvoltage during the run. The *I* vs *V* and *T* curve for the final run at 4.2 K (after some positive training of the coil) is shown in Figure 4. The voltage is not zero because it has an inductive component (we are measuring a magnet). The curves are not flat with an offset because the ramp rate changes during measurement as an artifact of the fact that we use voltage mode. The voltage ramping rate was 0.01 V/s, and the average current ramping rate above 400 A was approximately 1.1 A/s. The maximum take-off voltage trigger was set at 1 V. We can see that the temperature of the coil rises slightly for the coil end near the current taps; this is to be expected for a cryocooled coil unless very high additional cooling is present. All runs were fairly similar, although some positive training was seen. The transition was by quench, again not unexpected for an LTS coil running at high currents to I_c in conduction cooled mode.

The next step was a measurement of the coil in current mode. The benefit of this is that it provides a flat *I-V* curve, although that comes with an inductive voltage offset. The *I-V* curve for this run is shown in Figure 5. Current was applied with an initial ramping rate of 1 A/s, which was reduced to 0.8 A/s near 300 A. This reduction in current ramp rate explains the change in noise and inductive offset at 300 A. The maximum take-off voltage trigger was set at 300 mV.

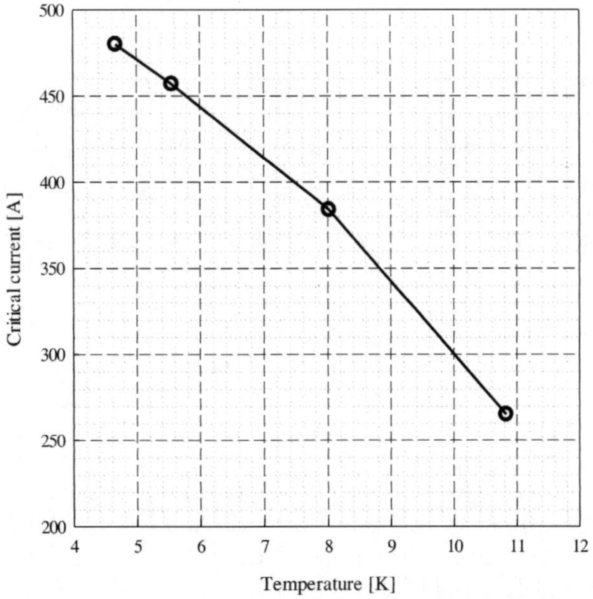

Figure 7. Critical current vs temperature of coil

The coil showed some positive training, but the best values obtained are shown in Table 2. The highest I_c was 480 A. This is above the target current of 435 A, with some margin. There is also a temperature margin, since the top portion of the coil was at 7.9 K at transition (see below). The target B (at the highest spot on the coil exterior) was expected to be 3 T at 435 A, and that is consistent with our measurement of 3.06 T at 440 A.

We note that the on-axis (H1, Figure 3) and maximum (H2, Figure 3) magnetic fields could not be measured simultaneously. They were measured on Run #4 and Run #2, respectively. The maximum on-axis field was measured at 480 A, while the maximum off-axis field was measured at 440 A (see Figure 6).

After the run in current mode, further voltage mode runs were performed to measure I_c as a function of temperature. Figure 7 shows the coil critical current as a function of temperature at the coil bottom. It should be kept in mind that the temperature at the top of the coil was about 2.5-3 K higher than at the bottom. This would suggest that the actual coil I_c at 4.2 K is closer to 550 A.

5. Summary
A Nb_3Sn coil was fabricated, cooled, and tested. A small amount of liquid helium below the coil was used as the cold reservoir with the coil cooled by conduction through the cooling legs. The operational current target of $I = 435$ A was achieved, with both current (45 A) and temperature (3.8 K) margin. The field values reached 3 T at the outer can of the coil, as expected by modelling. The coil met the needed targets for the winding design with a noticeable margin.

References
[1] A. B. Abrahamsen, N. Mijatovic, E. Seiler, T. Zirngibl, C. Træholt, P.B. Nørgard, N.F . Pedersen, N.H. Andersen, and J. Østergard, "Superconducting wind turbine generators", Supercond. Sci. Technol. 23 (2010) 034019.

[2] B. B. Jensen, N. Mijatovic, , and A. B. Abrahamsen, "Development of superconducting wind turbine generators", Journal of Renewable and Sustainable Energy 5, (2013) 023137; doi: 10.1063/1.4801449

[3] H. Ohsaki, L. Quéval, and Y. Terao, "Design and Characteristic Analysis of 10 MW Class

Superconducting Wind Turbine Generators with Different Types of Stator and Rotor Configurations", IEEE Trans Appl. Supercond. (2013) 398.

[4] X. Song and B. Bech, "Design Study of Fully Superconducting Wind Turbine Generators", IEEE Trans. Appl. Supercond. 25 (2015) 5203605

[5] R.H. Jansen, G.V. Brown, JL. Felder, K.P. Duffy, "Turboelectric Aircraft Drive Key Performance Parameters and Functional Requirements", Proc. Of Propulsion and Energy 2015, At Orlando, Florida

[6] P. Gemin, T. Kupiszewski, A. Radun, Y Pan, R. Lai, D. Zhang, R. Wang, X. Wu, Y. Jiang, S. Galioto, K. Haran, W. Premerlani, J. Bray, and A. Caiafa, "Architecture, Voltage and Components for a Turboelectric Distributed Propulsion Electric Grid (AVC-TeDP), NASA Technical report, NASA/CR—2015-218713.

[7] C. A. Luongo, P. J. Masson, T. Nam, D. Mavris, H. D. Kim, G. V. Brown, M. Waters, and D. Hall, "Next Generation More-Electric Aircraft: A Potential Application for HTS Superconductors", IEEE Trans. Appl. Supercond. 19 (2009) 1055.

[8] J. L. Felder, G. V. Brown, H.D. Kim, J. Chu, "Turboelectric Distributed Propulsion in a Hybrid Wing Body Aircraft", Proceedings of the XX International Symposium on Air Breathing Engines 2011 : (ISABE 2011) : Gothenburg, Sweden, 12-16 September, 2011, pp. 1340

[9] P. J. Masson, G. V. Brown, D. S. Soban and C. A. Luongo, "HTS machines as enabling technology for all-electric airborne vehicles", Supercond. Sci. Technol. 20 (2007) 748–756 doi:10.1088/0953-2048/20/8/005

[10] F. Berg, J. Palmer, P. Miller, M. Husband, and G. Dodds, "HTS Electrical System for a Distributed Propulsion Aircraft", IEEE Trans. Appl. Supercond. 25 (2015) 5202705

[11] F. Berg, J. Palmer, P. Miller, and G. Dodds, "HTS System and Component Targets for a Distributed Aircraft Propulsion System", IEEE Trans. Appl. Supercon. 27 (2017) 3600307

[12] T. Haugan, "Development of Superconducting and Cryogenic Power Systems and Impact for Aircraft Propulsion", Presentation, Energy Materials and Applications, Orlando FL, 25 Jan 2013.

[13] P.N. Barnes , M.D. Sumption, G.L. Rhoads, "Review of high power density superconducting generators: Present state and prospects for incorporating YBCO windings", Cryogenics 45 (2005) 670–686

[14] P.N. Barnes, G.L. Rhoads, J.C. Tolliver, M.D. Sumption, and K.W. Schmaeman, "Compact, Lightweight, Superconducting Power Generators", IEEE Transactions on Magnetics 41 268-273 (2005).

[15] D. Loder, R. Sanchez, M. Feddersen, K. Haran, M. Sumption, M. Tomsic, Y. Jinji and D. Doll, "A Conduction Cooled Nb_3Sn Racetrack Coil: Design, Construction, and Testing," 2016 IEEE Power and Energy Conference at Illinois (PECI), 2016.

IOP Conf. Series: Materials Science and Engineering **279** (2017) 012019 doi:10.1088/1757-899X/279/1/012019

Improvement of small to large grain A15 ratio in Nb_3Sn PIT wires by inverted multistage heat treatments

Christopher Segal, Chiara Tarantini, Peter J Lee and David C Larbalestier

Applied Superconductivity Center, NHMFL, Florida State University, Tallahassee, FL 32310, USA

E-mail: segal@asc.magnet.fsu.edu

Abstract. The next generation of superconducting accelerator magnets for the Large Hadron Collider at CERN will require large amounts of Nb_3Sn superconducting wires and the Powder-In-Tube (PIT) process, which utilizes a $NbSn_2$-rich powder core within tubes of Nb(7.5wt%Ta) contained in a stabilizing Cu matrix, is a potential candidate. However, the critical current density, J_c, is limited by the formation of a large grain (LG) A15 layer which does not contribute to transport current, but occupies 25-30% of the total A15 area. Thus it is important to understand how this layer forms, and if it can be minimized in favor of the beneficial small grain (SG) A15 morphology which carries the supercurrent. The ratio of SG/LG A15 is our metric here, where an increase signals improvement in the wires A15 morphology distribution. We have made a critical new observation that the initiation of the LG A15 formation can be controlled at a wide range of temperatures relative to the formation of the small grain (SG) A15. The LG A15 can be uniquely identified as a decomposition product of the $Nb_6Sn_5(Cu_x)$, surrounded by a layer of rejected Cu, thus the LG A15 is not only of low pin density, but is not continuous grain to grain. We have found that in single stage reactions limited to 630 °C - 690 °C, the maximum SG A15 layer thickness prior to LG A15 formation is very sensitive to temperature, with a maximum around 670 °C. This result led to the design of four novel heat treatments which all included a short, high temperature stage early in the reaction, followed by a slow cooling to a more typical reaction temperature of 630 °C. We found that this heat treatment (HT) modification increased the SG A15 layer thickness while simultaneously suppressing LG A15 morphology, with no additional consumption of the diffusion barrier. In the best heat treatment the SG/LG A15 ratio improved by 30%. Unfortunately, J_c values suffered slightly, however further exploration of this high temperature reaction region is required to understand the limits to A15 formation in Nb_3Sn PIT conductors.

1. Introduction

The Powder-In-Tube (PIT) manufacturing process has been used to produce multifilament Nb_3Sn wires for over 30 years [1]. Starting with the earliest variants at ECN, followed by Shape Metal Innovations (SMI), and today Bruker EAS, three morphologies of Nb_3Sn A15 grains are always formed during the reaction; a small grain (SG), a large grain (LG), and fully disconnected core grains (figure 1). The small grain ($\sim$ 150 nm) region forms from Sn contained in Nb_6Sn_5 diffusing into the Nb(Ta) tube, while the the large (>1 μm) and core grains were observed to have formed from decomposition of the $Nb_6Sn_5(Cu_x)$ layer which precedes the A15 reaction [2, 3]. It should be noted that although LG A15 does form in the internal Sn method [4], it is typically < 5 vol% of the A15 area. Given this state of the art conductor, the recommendations to improve the

PIT method are to consume more diffusion barrier without loss of RRR [5] and to suppress the formation of LG A15 as it does not contribute to transport current but occupies 25-30% of the total A15 area [6, 7]. A metric we will use here to determine if the microstructure has improved is the SG/LG ratio of the A15 morphologies, where a higher number is better. In order to suppress the LG A15, we first needed to develop a better understanding of its formation, as well as the formation of the A15 precursor phases Nausite [8], $NbSn_2$ and Nb_6Sn_5 by microstructural and chemical analysis.

2. Experimental Details

The wire studied is 0.78 mm in diameter from billet B31284 manufactured by Bruker EAS, and contains 192 filaments with diameter of ~39 μm set in a Cu stabilizer (Figure 1). Each filament consists of a thick Nb-7.5wt%Ta tube clad in Cu and filled with a Sn-rich powder mixture typically containing $NbSn_2$, Sn and Cu [9]. A thin Cu sleeve separates the powder from the inner tube wall. Such a monofilament is drawn, cut and stacked multiple times until the desired architecture is achieved; here a 192 filament wire.

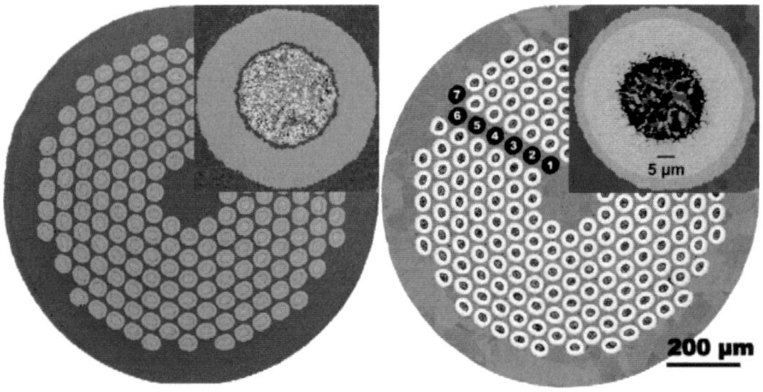

Figure 1: FESEM-BSE images of polished, transverse cross-sections of an unreacted (left) and reacted (right) Bruker PIT wire 1 mm in diameter with 192 superconducting filaments within a matrix of stabilizing Cu. Inset image is of a typical round filament

To study the phase and morphology evolution in these Nb_3Sn wires, a series of quench experiments were performed in which short samples of wire were removed from the reaction furnace at key times to better understand the microstructural evolution. Wires were cut into 10 cm lengths, the ends were crimped shut, and sealed in an evacuated quartz tube in which a small amount of argon (~ 300 torr) was introduced to promote heat transfer. These samples were then stacked in the furnace, and individually pulled out into an ice bath to "freeze" the microstructure. The wires were then mounted in a conductive puck and polished for imaging in a Scanning Electron Micropscope (SEM).

Digital Imaging was performed using a Zeiss 1540 Crossbeam Field Emission Scanning Electron Microscope (FESEM). The back scattered electron (BSE) intensity in the FESEM is sensitive to atomic number (Z) and phases with a higher effective Z are brighter in the BSE image, as seen in Figure 1. These data combined with knowledge of the phase diagrams for Cu-Sn [10] and Nb-Sn [11, 12] allow quick identification of the Cu, Nb-7.5wt%Ta, and the $Nb(Ta)_3Sn$ phases. We can also measure the area fractions of each phase with the widely used open-source software package, FIJI [13] (based on Image J [14]), allowing us to make high precision comparisons between samples.

Critical current (I_c) measurements were made in a 15 T solenoid using four-point measurements on 4.5 cm long straight samples with voltage taps in the center 1 cm apart. A voltage criterion of 1 μV/cm was used to determine the superconducting transition. To compare wires with different architectures we can normalize the critical current over the non-Cu area to determine a critical current density (J_c). Furthermore, we can measure the quality of the superconducting A15 phase by normalizing I_c over only the current carrying small-grain A15 layer to determine a layer J_c ($J_{c-layer}$).

3. Results

3.1. Phase Evolution under 630 °C, prior to A15 formation

The phase diagrams for Cu-Sn and Nb-Sn allow us to anticipate the phases that will be produced as the reaction proceeds. For the following discussion the values of interest are similar, and either phase diagram can be used. First we consider the starting wire in an unreacted state as seen in the left side of Figure 1. As the heat treatment begins, Sn is rapidly absorbed into the α-Cu sleeve and transforms to the most Sn rich Cu-Sn compound, η (Figure 2). The η phase metlts at 408 °C, and as the reaction proceeds, the η begins converting into ϵ which remains as the dominant Cu-Sn phase until sufficiently depleted of Sn towards the end of the reaction. In addition to the described Cu-Sn phase formation, there is a ternary Sn-Nb-Cu phase called Nausite [8] that forms as early as 200 °C [15]. The structure and composition of Nausite has recently been reported to be $(Nb_{0.75}Cu_{0.25})Sn_2$ [16].

Above 408 °C, both the Nausite and ϵ phase grow in thickness until about 560 °C, at which point the Nausite decomposes into $NbSn_2$, rejecting its Cu. The $NbSn_2$ is rather short lived, and at 630 °C transforms to Nb_6Sn_5, containing up to 5at%Cu, in less than an hour (Figure 2).

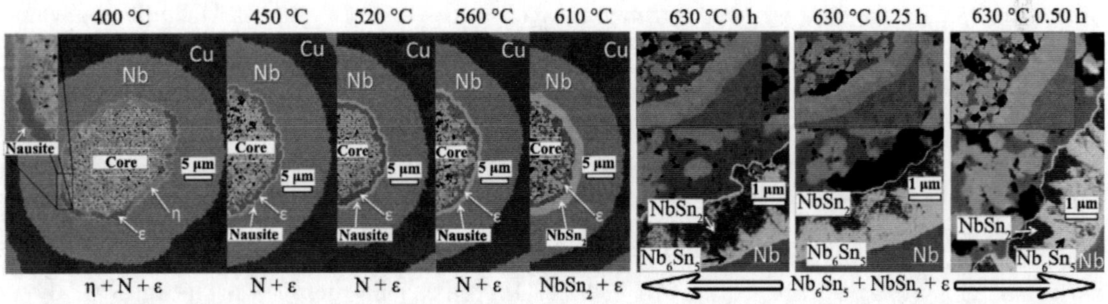

Figure 2: FESEM-BSE images of polished, transverse cross-sections showing phase evolution up to 610 °C. In the 630 °C panels the Nb-Sn phases have bee contrast enhanced at the DB interface to better distinguish phases, original images are inset. A thin light ring of Nausite (N) has formed on heating to 400 °C and continues to grow until about 610 °C when it decomposes into $NbSn_2$. The Cu-Sn phase has a transformation from η to ϵ + liquid at 408 °C which remains as the dominant Cu-Sn phase until sufficiently depleted of Sn towards the end of the reaction.

3.2. A15 growth at 630 °C

At 630 °C, the dominant phase kinetics are for the $Nb_6Sn_5(Cu_x)$ to feed Sn to the Nb-Ta tube and form SG A15. Already at only five hours, we see a thin layer of SG A15 form between the DB and the $Nb_6Sn_5(Cu_x)$ layer at the edge of the core (Figure 3). At 10 h the SG A15 layer thickness more than doubled to 0.85 μm. By 12 h, the $Nb_6Sn_5(Cu_x)$ begins to decompose into large grains of A15 at the interface with the SG A15 layer, ejecting Cu into mesoscopic precipitates at the large grain boundaries. The appearance of the Cu is not surprising as it is well known that Nb_3Sn does not dissolve any Cu [17]. By 30 h much LG A15 has formed with the surrounding Cu ejected during the transformation. Meanwhile, SG A15 continues to grow. At 58 h there is a SG A15 layer ~3.4 μm thick, the LG A15 layer growing appreciably, with a now fully continuous Cu-Sn phase (ϵ) separating the A15 phases from the remaining $Nb_6Sn_5(Cu_x)$. The $Nb_6Sn_5(Cu_x)$ is fully consumed around 100 h. However, there is significant Sn left stranded in the core that could form more A15 by continued reaction. Once the reaction is completed around 240 h, 58.1% of the filament area is A15. This 58.1% volume of A15 is composed of

3 morphologies; 40.7% SG A15 (5-5.5 μm thick), 14.4% LG A15 (2.5 μm thick), and 3.0% of porous, disconnected core grains of A15. Important to note is that only the SG A15 carries the transport current [7], meaning that 30% of the A15 is useless in the current conductor design. From the critical viewpoint of making PIT more attractive as a high J_c conductor, we need to fully understand why the LG A15 forms, and if it can be suppressed in favor of the desirable SG A15.

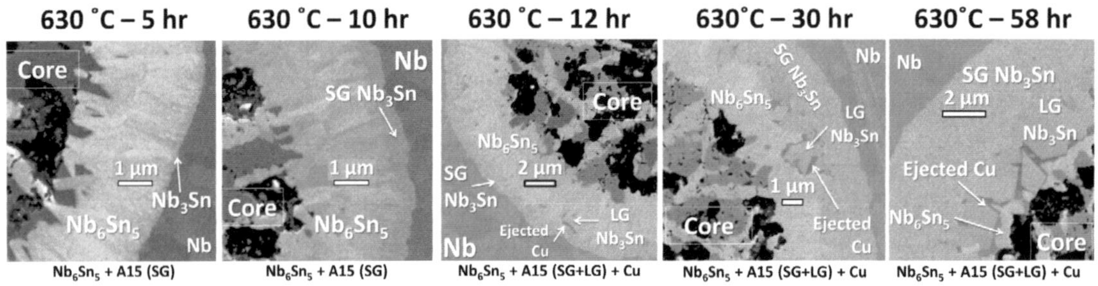

Figure 3: FESEM-BSE images of polished, transverse cross-sections showing A15 phase growth at 630 °C. The SG A15 layer steadily grows, with LG A15 nucleating around 12 h. By 58 h most of the Nb_6Sn_5 is gone, and we have continuous layers of SG A15, LG A15, and Cu phases which separates the Nb_6Sn_5 from being in direct contact with the SG A15 layer.

3.3. Heat treatment effects on A15 morphology early in the reaction

The temperature of the A15 reaction has a substantial influence on the formation of LG A15 early in the reaction [3]. At 675 °C reference [3] found a much thicker layer of SG A15, than what was described here at 630 °C when only one large A15 grain has formed (Figure 4). This is a surprising result because at 630 °C the thickness of the SG A15 layer is less than a micron when LG A15 starts nucleate, but at 675 °C the SG A15 layer has doubled in thickness to about 2 μm.

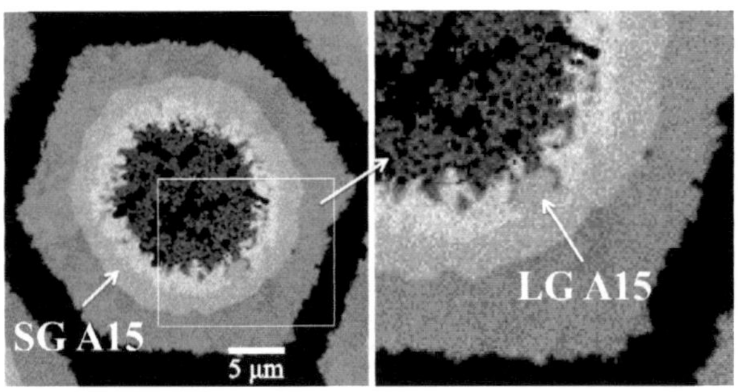

Figure 4: One filament from a wire produced by SMI heat treated at 675 °C for 4 h. There is already a thick layer of SG A15, and one large A15 grain has formed in the Nb_6Sn_5 layer with Cu ejected around it [3].

Previously, we have analyzed the SG/LG ratio of 10 PIT wires differing by either HT or billet design, and found that the total A15 volume fraction varies only slightly from 55-58% in fully reacted samples [5], with the SG/LG ratio varying from 2.4-3.0. However, the vintage SMI PIT sample in Figure 4 shows that the ratio of SG/LG A15 does vary substantially early in the reaction as a function of HT temperature, when LG A15 is just forming. Consequently we performed a study on the phase morphology variation as a function of temperature early in the HT, performing single stage HT's which all ramped up at 100 °C/h to a dwell temperature of either 630, 650, 670, or 690 °C for between 50-250 h. For this experiment we obtained about 50 m of conductor at 0.78 mm wire diameter through the LARP collaboration (details in section 5).

The focus of these heat treatments was to observe the early stages of the reaction by quenching samples to compare the rates of SG and LG A15 growth as a function of dwell temperature.

In all cases, the SG A15 formed first, and the LG A15 formed later in the A15 reaction, and more rapidly at higher temperatures. From our quench experiment at 630 °C we found that the LG A15 started forming around 12 h, with a SG A15 layer thickness of 0.85 μm. At 650 °C the SG A15 layer was 1.5 μm thick and the LG A15 first precipitates out after 8.3 h. At 670 °C the LG A15 begins forming after only 5 h, and the SG A15 layer formed is 2 μm thick. At 690 °C the SG layer thickness is only 1.6 μm with LG A15 forming in less than two hours (Figure 5). If we take this data and create an isothermal transformation diagram for A15 morphology, we see a clear boundary for LG formation (Figure 6). This isothermal transformation diagram incited us to develop a new type of multi-temperature A15 reaction in which we start with a brief high temperature excursion to grow a thick SG A15 layer, then drop the temperature to minimize LG A15 formation.

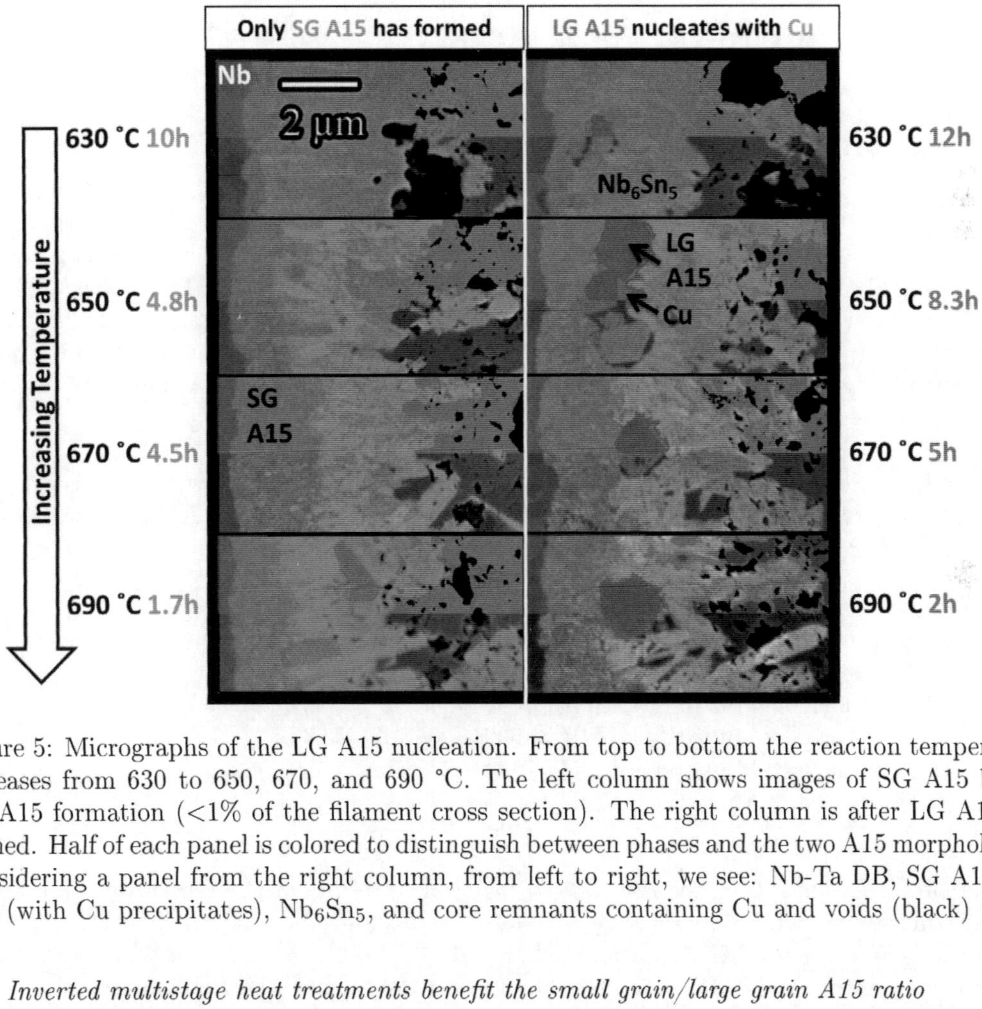

Figure 5: Micrographs of the LG A15 nucleation. From top to bottom the reaction temperature increases from 630 to 650, 670, and 690 °C. The left column shows images of SG A15 before LG A15 formation (<1% of the filament cross section). The right column is after LG A15 has formed. Half of each panel is colored to distinguish between phases and the two A15 morphologies. Considering a panel from the right column, from left to right, we see: Nb-Ta DB, SG A15, LG A15 (with Cu precipitates), Nb_6Sn_5, and core remnants containing Cu and voids (black)

3.4. Inverted multistage heat treatments benefit the small grain/large grain A15 ratio

In an attempt to avoid or delay the LG nucleation as suggested by the previous experiment in section 3.3, we designed four "inverted multistage" HT's. A typical HT from the manufacturer may contain multiple stages, but always increasing in temperature, a standard BEAS HT for example is 620 °C 100 h + 640 °C 12 h. Here we invert this by first going to a high temperature, and then cooling to the final A15 reaction temperature. The components of this new HT design can be summarized as:

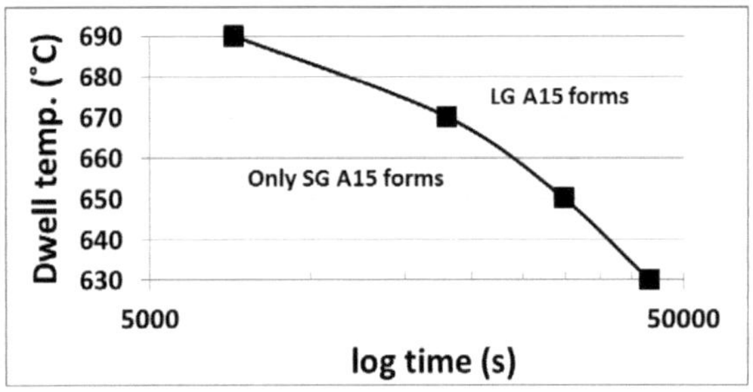

Figure 6: Isothermal transformation curve showing how temperature and time effect the onset of large grain A15 formation in single dwell HT's. The line is generated from the right column of Figure 5 and is an approximate boundary for when LG A15 will form.

- Ramp-up to the highest reaction temperature (HRT) (670 °C - 690 °C).
- Short dwell at HRT (including no dwell in some cases) (0 - 3.5 h)
- Ramp-down to the low-temperature reaction (LTR) dwell. (5 - 20 °C/h)
- Long LTR dwell (190 - 220 h)

For the purpose of this experiment we wanted to optimize the HRT, HRT dwell time, and ramp-down to the LTR dwell such that large grains of A15 have not yet formed, and we have maximized the SG A15 layer thickness.

3.4.1. Microstructures early in the the inverse multistage heat treatment Early in the inverse multistage HT's, The SG/LG ratio is quite variable since LG A15 can be substantially delayed while growing SG A15. The results are summarized in Table 1. Our first HT (A) nucleated large grains about 13 h into the LTR dwell at 630 °C with a SG layer thickness of 2.2 μm, a 10 % improvement over the best single step SG A15 layer thickness observed at 670 °C with no LG A15 (Figure 7). HT C was the best at delaying LG A15 while producing the most SG A15 early in the reaction as seen in Figure 7.

Table 1: Early microstructure of the four wires heat treated using inverse multistage heat treatments, and two single stage HT wires included for reference. The SG A15 layer thickness is measured when there is 1vol% LG A15.

HT (°C)	HRT	SG layer thickness (μm)
630	630	0.85
670	670	2.0
Inverse Multistage HT A	690	2.2
Inverse Multistage HT B	690	2.7
Inverse Multistage HT C	690	3.0
Inverse Multistage HT D	670	2.2

3.4.2. Phase and microstructure after full reaction The results of the four inverse multistage HT's after a full reaction are compiled in Table 2. The most notable were the SG A15 improving from 40 to 44% while the LG A15 decreased from 13.3 to ~11.8%, the core A15 being halved. This led to an improved SG/LG A15 ratio in all HT's, the highest being 3.9 compared to 2.8 in a standard HT. Heat treatments B and C had very similar SG/LG A15 ratios, however HT C had an additional 1.5% of total A15 (Figure 8).

3.4.3. Electrical properties after full reaction To compare I_c values of the inverse multistage HT's to the standard HT of 620/100 + 640/90. I_c values of HT's B, C, and D varied by less than 1%, from 489-493 A; about 2% less than the standard wire (Table 2). However, I_c is

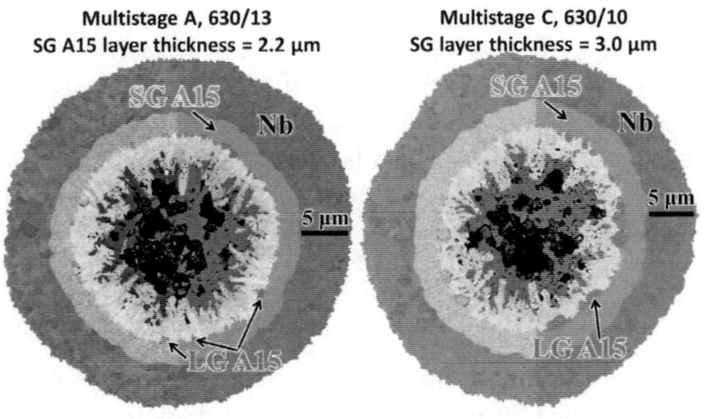

Figure 7: HT A and C are compared as the LG A15 is forming. Filaments are half colored to distinguish between phases. The SG layer thickness of HT C after 10 h in the LTR dwell is 3 μm.

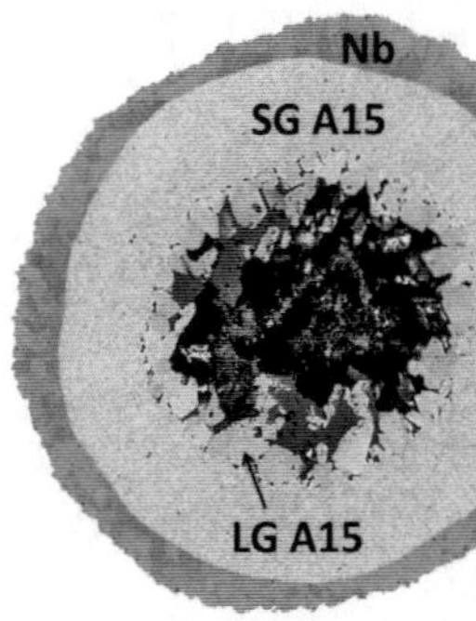

Figure 8: A filament from HT C reacted at a LTR dwell at 630 °C for 210 h. The large A15 grains were reduced, and are very loosely connected to the SG A15 layer.

strongly dependent on the amount of SG A15, and its grain size. We can better understand the differences between HT's by normalizing critical current to the SG A15 layer for each sample since only that layer carries current. A wire reacted with a standard HT has a $J_{c-layer}$(12 T, 4.2 K) of 5564 A/mm^2, while the best HT (C) had a $J_{c-layer}$(12 T, 4.2 K) of only 4926 A/mm^2, a substantial drop. Heat treatments C and D had very similar layer J_c(12 T, 4.2 K) which is 10% lower than a wire with standard HT. By staying below 670 °C we were able to recover some of the layer J_c (12 T, 4.2 K), increasing from about 4950 A/mm^2 in HT B & C to 5150 A/mm^2 in HT D, however, less A15 is produced overall.

Table 2: Fully reacted microstructure and J_c(12 T, 4.2 K) of wires heat treated using inverse multistage heat treatments, and the standard HT BEAS wire included for reference.

			Electrical Properties			Phase or morphology volume %						
HT (°C)	HRT (°C)	Time in LTR dwell (h)	I_c(A)	J_c non-Cu	J_c layer	DB	A15 total (S+L+C)	Remnant core	A15 SG	A15 LG	A15 core	SG/LG ratio
620+640	-	100+90	501	2237	5564	23.4%	56.0%	20.6%	40.2%	13.3%	2.5%	3.0
Inverse Multistage HT A	690	220	-	-	-	25.4%	56.3%	18.4%	43.5%	12.0%	0.8%	3.6
Inverse Multistage HT B	690	190	489	2183	4958	24.4%	56.0%	19.4%	44.0%	11.2%	0.8%	3.9
Inverse Multistage HT C	690	210	491	2192	4926	24.0%	57.3%	18.7%	44.5%	11.6%	1.2%	3.8
Inverse Multistage HT D	670	210	493	2201	5137	24.5%	56.6%	18.9%	42.8%	12.4%	1.3%	3.5

4. Discussion

For all heat treatments the SG A15 formed before LG A15. We found that the SG/LG A15 ratio can be increased, but only by going about 60 °C above the standard HT temperatures. The final A15 volume of all inverse multistage HT wires were similar, ~43-44.5%, but produced different $J_{c-layer}$ values, attributed to both the HRT of the HT as well as the temperature at which A15 is formed. Heat treatments with HRT of 690 °C showed about a 12% decrease in 12 T $J_{c-layer}$, while an HRT of 670 °C showed only an 8% decrease over the standard HT. We believe this difference is due to the average grain size being higher for the SG A15 which forms $>$ 670 °C, and was the motivation for HT D having a maximum of 670 °C, in an effort to keep the grain size small. This was successful as the $J_{c-layer}$ improved from HT C, but not as high as the standard BEAS HT. While all HT's improved the SG/LG A15 ratio, the lowest was HT

D which had an HRT of only 670 °C. Our observation is that higher temperatures (>670 °C) tend to either 'push' the core material to the tube wall, or agglomerates voids in the center of each filament. In either case, the result is that nearly all Sn in the core has a clearly connected diffusive path to react with the Nb(Ta) tube. At lower temperatures there tends to be many small voids in the core, leading to Sn being trapped and eventually transforming into A15 as seen in Figure 1. With the SG/LG ratio now improved, we need to find an optimized HRT which still allows the most Sn to react with the tube wall, but without causing the small grains of A15 to grow too large.

5. Conclusion

One possible method to improve Nb_3Sn PIT wires is to suppress LG A15 formation in favor of SG A15. There is a complex succession of phases which form and decompose until finally the A15 reaction temperature is reached (> 630 °C), where the dominant kinetics are for $Nb_6Sn_5(Cu_x)$ to feed Sn into the NbTa tube to form SG A15. A higher reaction temperature allows greater delivery of Sn to the Nb-Ta tube for more SG A15, but has yet to prevent the LG A15 from forming, which seems to be the exhaustion point of the original $NbSn_2$ core. A series of isothermal HT studies encouraged us to design new inverted multistage HT's with a short high temperature spike in the beginning to rapidly grow the SG A15 before LG A15 forms. These HT's achieved a 50% increase to SG A15 layer before LG A15 nucleated out of the $Nb_6Sn_5(Cu_x)$ in the early hours of the reaction. After a full reaction, the SG A15 layer increased by 10%, while the LG A15 decreased 10-15%, and the core A15 was halved, improving the SG/LG ratio from 3.0 to 3.9. Unfortunately, J_c(12 T,4.2 K) suffered, dropping by as much as 10%. We believe that large variations in the SG/LG A15 ratio early in the A15 reaction window can be improved by further optimization of inverted multistage HT's. While the J_c(12 T) is sensitive to grain size, typically restricting reactions <670 °C, the higher temperature HT's may be more favorable for 15-16 T applications once fully optimized.

Acknowledgments

Acknowledgments All experiments were carried out on Billet 31284, supplied to us by LBNL through the US LHC Accelerator Research Program (LARP), which is a BNL, FNAL, LBNL, and SLAC collaboration with CERN for the High Luminosity LHC program. The wire was received at 0.79 mm diameter in the unreacted condition. This work was supported by the US Department of Energy (DOE) Office of High Energy Physics under award DE-SC0012083, by CERN, and by the National High Magnetic Field Laboratory (which is supported by the National Science Foundation under NSF/DMR-1157490), the State of Florida, and CERN under grants KE1920/TE and RF02226

References

[1] Beijnen C van and Elen J. "Multifilament Nb_3Sn superconductors produced by the E.C.N. technique". Version 2. In: *IEEE Transactions on Magnetics* 15.1 (1979), pp. 87–90.

[2] Veringa H, Hoogendam P and Wees A. "Growth kinetics and characterization of superconducting properties of multifilament materials made by the ECN powder method". Version 2. In: *IEEE Transactions on Magnetics* 19.3 (May 1983), pp. 773–776.

[3] Hawes CD. "Investigations of the Inhomogeneity of a Powder-In-Tube Nb_3Sn Conductor". PhD thesis. University of Wisconsin - Madison, 2000.

[4] Michiel MD and Scheuerlein C. "Phase transformations during the reaction heat treatment of powder-in-tube Nb_3Sn superconductors". Version 2. In: *Superconductor Science and Technology* 20.10 (Oct. 1, 2007), p. L55.

[5] Segal C et al. "Evaluation of critical current density and residual resistance ratio limits in powder in tube Nb_3Sn conductors". Version 2. In: *Superconductor Science and Technology* 29.8 (2016), p. 085003.

[6] Godeke A et al. "State of the art powder-in-tube niobium–tin superconductors". Version 2. In: *Cryogenics* 48.7 (July 2008), pp. 308–316.

[7] Tarantini C et al. "Composition and connectivity variability of the A15 phase in PIT Nb_3Sn wires". Version 2. In: *Superconductor Science and Technology* 28.9 (Sept. 1, 2015), p. 095001.

[8] Naus M, Lee P and Larbalestier D. "The influence of the starting Cu-Sn phase on the superconducting properties of subsequently reacted internal-Sn Nb_3Sn conductors". Version 2. In: *IEEE Transactions on Applied Superconductivity* 11.1 (Mar. 2001), pp. 3569–3572.

[9] Godeke A. "Performance Boundaries in Nb_3Sn Superconductors". PhD thesis. University of Twente, 2005.

[10] Fürtauer S et al. "The Cu–Sn phase diagram, Part I: New experimental results". Version 2. In: *Intermetallics* 34 (Mar. 2013), pp. 142–147.

[11] Charlesworth JP, Macphail I and Madsen PE. "Experimental work on the niobium-tin constitution diagram and related studies". Version 2. In: *Journal of Materials Science* 5.7 (July 1970), pp. 580–603.

[12] Toffolon C et al. "Reassessment of the Nb-Sn system". Version 2. In: *Journal of Phase Equilibria* 23.2 (Mar. 1, 2002), pp. 134–139.

[13] Schindelin J et al. "Fiji: an open-source platform for biological-image analysis". Version 2. In: *Nature Methods* 9.7 (July 2012), pp. 676–682.

[14] Schneider CA, Rasband WS and Eliceiri KW. "NIH Image to ImageJ: 25 years of Image Analysis". In: *Nature methods* 9.7 (July 2012), pp. 671–675.

[15] Pong I, Oberli LR and Bottura L. "Cu diffusion in Nb_3Sn internal tin superconductors during heat treatment". Version 2. In: *Superconductor Science and Technology* 26.10 (Oct. 1, 2013), p. 105002.

[16] Martin S et al. "The crystal structure of $(Nb_{0.75}Cu_{0.25})Sn_2$ in the Cu-Nb-Sn system". Version 2. In: *Intermetallics* 80 (2017), pp. 16–21.

[17] Yamashina T and Kajihara M. "Quantitative Explanation for Uphill Diffusion of Sn during Reactive Diffusion between Cu-Sn Alloys and Nb". Version 2. In: *MATERIALS TRANSACTIONS* 47.3 (2006), pp. 829–837.

Study of Thermomechanical Properties of The Epoxy-Impregnated Cable Composite for a 15 T Nb_3Sn Dipole Demonstrator

Pei Li, Steve Krave, Alexander Zlobin

Fermi National Accelerator Laboratory, Batavia, IL, 60510

peili@fnal.gov

Abstract. The knowledge of the thermomechanical properties of the composite of cable/insulation/epoxy impregnation are important for the design, fabrication and operation of superconducting accelerator magnets. As a part of the 15 T dipole magnet development at Fermi National Accelerator Laboratory (FNAL), we studied the thermomechanical properties of cable stack that represents the cable composites in the 15 T dipole. The measurements include thermal contraction and strain-stress characterization under compressive load along the principal directions. The cable stack samples show hysteresis behaviour in loading-unloading cycles, which is found to be most dramatic along the azimuthal direction. Also, the choice of insulation material/procedure is found to strongly impact the bonding between cables and epoxy/cable layers. The cable stacks measured in this study use E-glass tape wrapping insulation and show weaker bonding to cables than similar cable stacks using S-2 glass sleeves insulation previously studied.

1. Introduction

Superconducting magnets that generate 15 T field is crucial to the realization of hadron collider with energy level beyond that of LHC. Fermi National Accelerator Laboratory (FNAL) started the development of a 15 T Nb_3Sn dipole demonstrator magnet based on the "cosine theta" design [1-3].

In the development of the 15 T dipole, the knowledge of the thermomechanical properties of all the components are very important. The supporting structures of the magnet are fabricated from metals and their thermomechanical properties are well known. On the on the hand, the cable composite (Nb_3Sn cable/insulation/epoxy impregnation) is more complicated because it contains several components of drastically different thermomechanical properties and its fabrication is a multiple-variable procedure which can result in significant variation of properties. To provide reliable and relevant thermomechanical properties data to the design and fabrication of magnets, it is important to perform measurements on samples that best represent the cable composite used in the magnets.

2. Sample Description

The thermomechanical properties measurements were performed on cable stacks that were fabricated using the same technology and materials as in the 15 T dipole. Specifically, the cable studied here is used to wind the two outer layers of the dipole [1]. The detailed information of the cable and insulation is shown in table 1.

Table 1. Specification of the cable stacks

Specification	
Nb_3Sn strand	0.7 mm 108/127 RRP strand
cable	40 strands, with core
insulation	E-glass tape wrapping, 0.075 mm thick, 12.7 mm wide
Binder	CTD1202X, cured for 120 °C/30 min in air
Reaction procedure	210 °C/48 h, 400 °C/48 h, 640 °C/48 h, in flowing Ar
Epoxy impregnation	CTD101K

To prepare the cable stack sample, the cable was cut into 6.5 inch long pieces. After cutting, the ends of the cable pieces were sealed by tungsten inertia gas (TIG) welding to prevent the leakage of tin during heat treatment. A 0.075 mm thick, 12.7 mm wide E-glass insulation tape was wrapped around the cable pieces, with a 40-50% turn-to-turn overlap. 10 wrapped cable pieces were stacked in a reaction fixture, with alternating the wide and the narrow edges. A ceramic binder CTD1202X was uniformly brushed over the stacked cables. The reaction fixture was first heated in air at 120 °C for 30 minutes to cure the binder and then reacted in a tube furnace (210 °C 48 hrs/400 °C 48 hrs/640 °C 48 hrs) in flowing Ar gas. The reacted cable stack was impregnated using CTD101K epoxy. Each impregnated cable stack has a usable length of 6 inch and a cross section of 15.6 mm by 17.2 mm. A full cable stack was further cut into shorter (0.6-2 inches) segments for measurements using a low-speed diamond saw. A cable stack sample has three principal directions, *i.e.* azimuthal, radial and axial, as illustrated in figure 1.

The deformation of samples was measured with strain gages (type WK350-13-125AD-W by MicroMeasurement) and the gage installations were performed following the standard procedure using M-bond (type AE10).

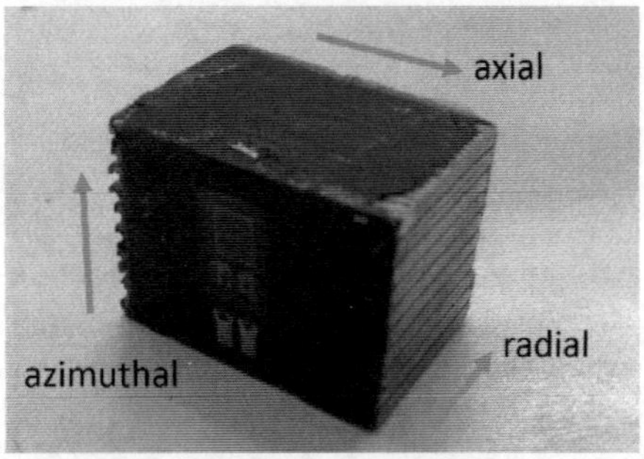

Figure 1. A cable stack sample with the three principal orthogonal directions labeled. On this sample, a gage is installed to measure the azimuthal strain.

3. Thermal Contraction Measurement

To measure the thermal contraction of cable stacks, gages were mounted along the principal directions. The samples were gradually cooled from room temperature (~293 K) to 77 K over 30 minutes and the change of gage resistance, ΔR, was measured.

The measured ΔR includes: (i) the resistance change of gage material with temperature; (ii) resistance change proportional to the thermal contraction difference between the gage grid alloy and the tested

IOP Conf. Series: Materials Science and Engineering **279** (2017) 012020 doi:10.1088/1757-899X/279/1/012020

material. To extract the thermal contraction of the sample, a material with known thermal contraction should be measured with the same type of strain gage as reference [4]. We used high-purity quartz as the reference material, of which the thermal contraction was taken to be zero. As verification, we also measured a piece of Al-6061 and the thermal contraction from 293 K to 77 K is 0.37%, comparing with the theoretical value of 0.389%. The thermal contraction results of the cable stacks are listed in table 2.

Table 2. Thermal contraction results

Sample	Thermal contraction (%), 293 K to 77 K
Cable stack, radial	0.436
Cable stack, azimuthal	0.19
Cable stack, axial	0.215
Al-6061	0.37
Al-6061 (calculated)	0.389

4. Mechanical Property Measurements

The strain-stress curves of cable stacks were tested under compression along its principal orthogonal directions in a calibrated hydraulic loader at room temperature. The samples were compressed in stainless steel fixtures. The strain of the samples was measured by strain gages. Each test included three loading-unloading cycles between zero and the pre-set peak load. A typical strain-stress curve from a piece of Al-6061is shown in figure 2. Above 5 MPa, the strain-stress curve is linear and the fitted Young's modulus is 71.2 GPa, comparing with the typically reported value of about 70 GPa.

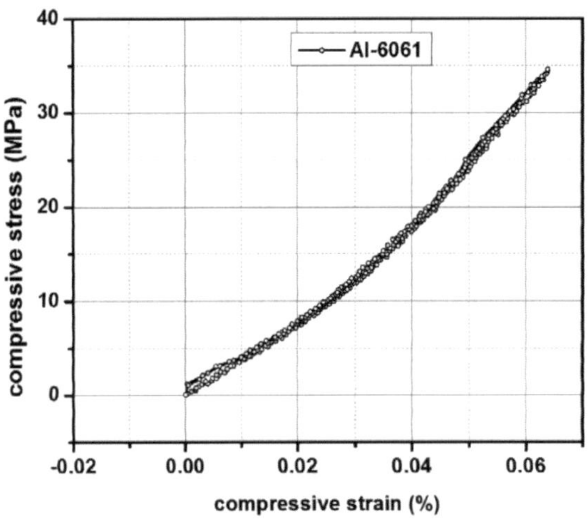

Figure 2. The strain-stress curve of a piece of Al-6061sample. Three loading-unloading cycles are shown.

4.1. Strain-stress characterization of cable stacks under azimuthal load

The strain-stress curves of cable stacks under azimuthal load are shown in figure 3. Figure 3(A) shows the results of the initial, low-level azimuthal loading of a sample. As described above, three cycles between zero and the peak load (~11 MPa) were performed. At the end of the first loading-unloading cycle, with zero stress, the sample is under ~0.005% compressive strain. This shows that an azimuthal load of ~11 MPa already induces plastic deformation in the sample. Also, we notice that in the subsequent two cycles, the sample shows hysteresis behaviour. The loading curves (of the second and the third cycle) and the unloading curves (of all the three cycles) follow two different traces. On the other hand, these cycles do not further increase the plastic deformation in the sample. Fitting of these curves shows the Young's modulus along the azimuthal direction to be ~ 17 GPa. Figure 3(B) shows the results of gradually increasing the peak azimuthal load. The residual plastic deformation increases with the peak load. With a peak load of 140 MPa, the plastic deformation is about 0.75%. These cycles also show similar hysteresis behaviour observed in the low-level loading tests.

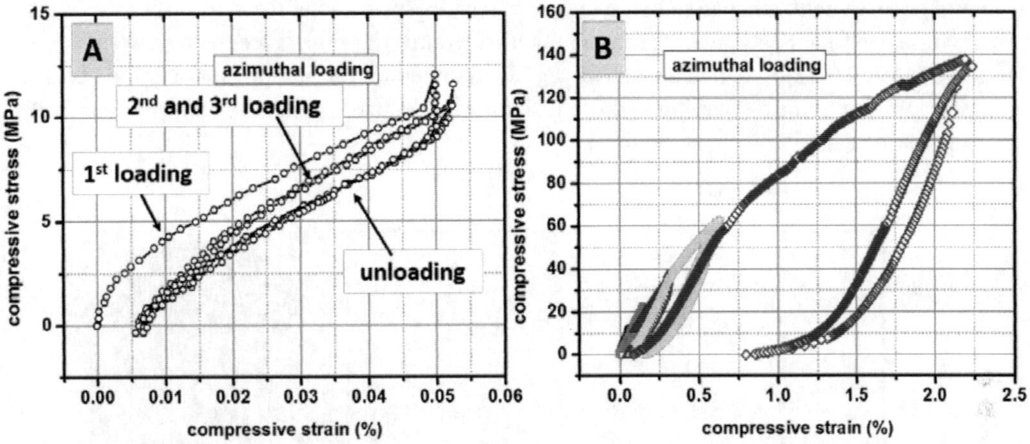

Figure 3. Strain-Stress curves of cable stack under azimuthal load. (A) The initial three loading-unloading cycles up to 11 MPa. (B) Cyclical loading with increasing the peak load to 11, 20, 40, 60 and 140 MPa respectively.

4.2. Strain-stress characterization of cable stacks under radial load

The strain-stress curves of cable stacks under radial load are shown in figure 4. Figure 4(A) shows the results of the initial radial loading of a sample to a peak load of ~ 11 MPa. After three cycles, the sample shows no plastic deformation and the hysteresis between loading and unloading curves are less significant comparing with azimuthal loading. The fitted Young's modulus is about 49.2 GPa. Figure shows the results of increasing the peak load. With increasing the peak load, the sample begins to show plastic deformation and the hysteresis behaviour becomes more obvious. To date, we have completed radial loading-unloading cycles up to 35 MPa and the residual plastic deformation is about 0.01%.

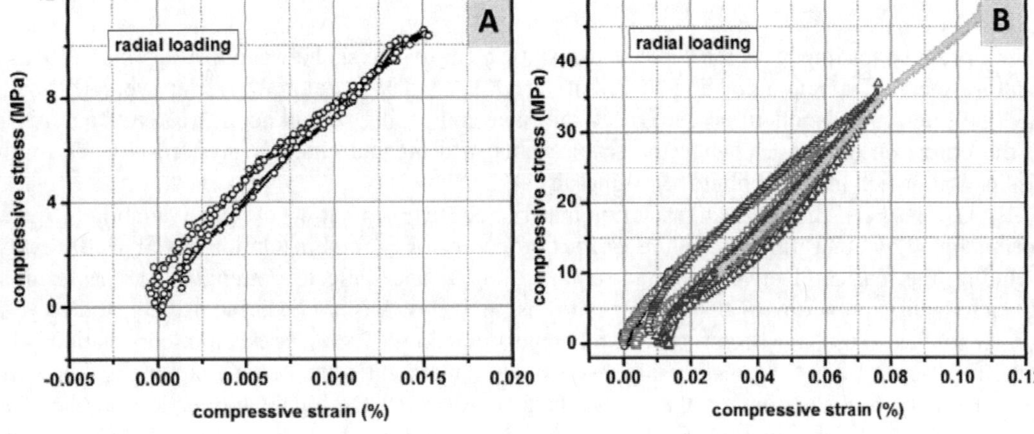

Figure 4. Strain-stress curves of cable stack under radial load. (A) the initial three loading-unloading cycles of a sample up to 11 MPa. (B) Cyclical loading of the sample with increasing peak load.

With further increasing the peak radial load, we observed sample failure. An example is shown in figure 5. As shown by the strain-stress curve, when load reaches about 50 MPa, the compressive strain measured by the gage suddenly begins to drop and continues to decrease until the peak load of ~ 62 MPa. During unloading, the measured strain remains constant around -0.03%. Since compressive strain is considered positive here, a negative value indicates tensile strain. After the removal of the sample from the test fixture, we found bulging of the sample, with the outmost of the layers bent outward and partially delaminated from the stack. Since the strain gage is mounted on the sample surface, the observed drop of compressive strain during loading and the residual tensile strain is because of the local deformation of the outmost layers.

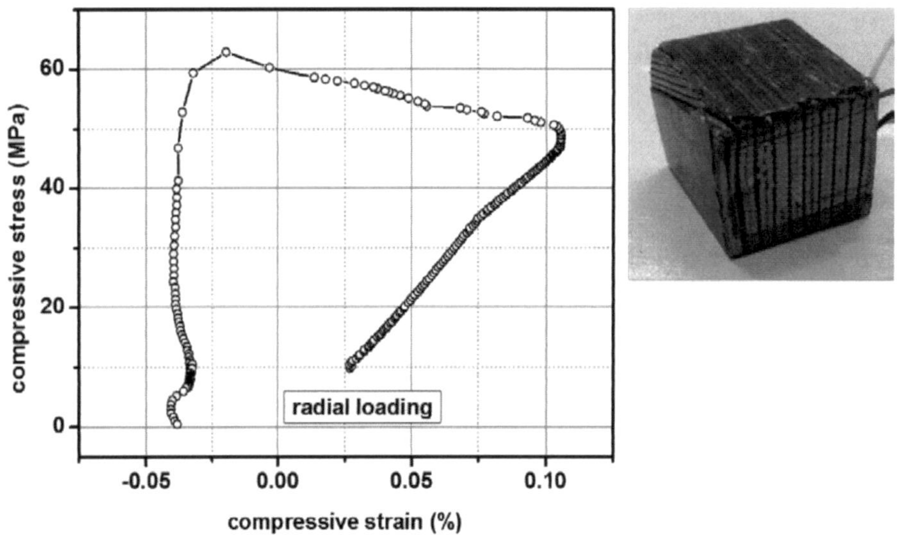

Figure 5. The failure of a sample during radial loading. Left, the strain-stress curve. The sample failure occurred at ~ 50 MPa and further loading caused the measured strain to drop to ~0.03%, which remained constant during unloading. Right, photo of the sample after loading test. Bulging of the sample and the delamination of the outmost layers were found. See the text for more details.

5. Discussion

In compressive loading tests, the cable stacks show clear hysteresis between loading and unloading. In contrast, the loading and unloading curves of Al-6061, an elastic material, overlap well. This suggests that in the design and modelling of magnets, the cable composites should not be treated as purely elastic. In the operation of magnet, when the strain status of a magnet changes, hysteresis leads to energy dissipation, which may contribute to magnet instability.

E-glass has been chosen as an insulation material because of its low cost and availability [5] and has been adopted by other magnets. While various properties of E-glass insulation have been investigated, including thermal conductivity, thermal contraction [6] and dielectric strength [5], the strain-stress characterization properties of cable composites using E-glass has not been intensively studied yet. The similar studies so far have been focused on composites using S-2 glass and ceramic insulation [4]. This is the first time Nb_3Sncable stacks using this insulation method (E-glass tape wrapping and binder) was systematically measured for their thermomechanical properties. During the fabrication and the handling of the cable-stack samples with E-glass insulation, we noticed that the bonding between cable and epoxy/insulation layer is weaker than similar samples with S-2 glass insulation. While the load tests of control samples with S-2 glass are in progress, in a recent study on similar cable stacks with S-2 glass sleeve insulation [7], where the test condition is similar to this work, radial load was successfully applied beyond 100 MPa without sample failure. In contrast, we found that 50 MPa radial load can cause delamination in our cable stacks using E-glass wrapping insulation. On the other hand, the test method used does not fully represent the stress condition of the cable composite in real magnets. The loading fixture does not provide constraints to sample surfaces that are parallel to load whereas the cable composite is constrained by the mechanical structure in magnets. To extend the tests into high stress (> 200 MPa) regime, which is of more interest to the development of high field magnets, loading fixtures that provide mechanical constraints should be used.

6. Conclusion

Thermal contraction and strain-stress characterization were performed on cable stacks that represents the cable composite in the 15 T dipole magnet being developed at FNAL. The cable stacks show hysteresis behaviour in loading-unloading cycles. The bonding to cables of E-glass wrapping insulation is found to be weaker than S-2 glass sleeve insulation previously studied, which leads to early delamination in samples compressed without mechanical constraints at 50 MPa radial load. We will extend tests in high stress regime (> 200 MPa) regime with improved test fixture and samples fabricated from the inner layer cables of the 15 T dipole (28 1.0 mm strands) as well as control samples with S-2 glass will be included.

References

[1] Andreev N, Barzi E, Carmichael J, Kashikhin V, Novitski I, Turrinoni D and Zlobin A, *Fermilab technical report*, TD-16-004.

[2] Kashikhin V, Andreev N, Barzi B, Novitski I and Zlobin A, *IOP Conf. Series: Materials Science and Engineering* 101 (2015) 012055 .

[3] Novitski I, Andreev N, Barzi E, Carmichael J, Kashikhin V, Turrioni D, Yu M and Zlobin A, *IEEE Trans. Appl. Supercond.* 26(4):1 2016

[4] Chichili D, Arkan T, Ozelis J and Terechkine I, *IEEE Trans. Appl. Supercond.* 10(1):1317 2000

[5] Bossert R, Ambrosio G, Andreev N, Whitson G and Zlobin A, *AIP Conf. Proc.* 986:161 2008

[6] Imbasciati L, Volpini G, Ambrosio G, Chichili D, Pedrini D, Previtali V, Rossi L and Zlobin A, *IEEE Trans. Appl. Supercond.* 13(2):1788 2003

[7] Bossert R, Krave S, Ambrosio G, Andreev N, Chlachidze G, Nobrega A, Novitski I, Yu M and Zlobin A, *AIP Conf. Proc.* 1574:132 2014

IOP Conf. Series: Materials Science and Engineering **279** (2017) 012021 doi:10.1088/1757-899X/279/1/012021

Effect of sheath material and reaction overpressure on Ag protrusions into the TiO_2 insulation coating of Bi-2212 round wire

I Hossain[1,2], J Jiang[1], M Matras[1,2], U P Trociewitz[1], J Lu[1], F Kametani[1,3], D Larbalestier[1,2,3] and E Hellstrom[1,2,3]

[1] National High Magnetic Field Laboratory, Florida State University, Tallahassee, Florida, FL 32310, USA
[2] Materials Science and Engineering, Florida State University, Tallahassee, Florida, FL 32310, USA
[3] Department of Mechanical Engineering, FAMU-FSU College of Engineering, Florida State University, Tallahassee, Florida, FL 32310, USA

Email: sh15c@my.fsu.edu

Abstract. In order to develop a high current density in coils, Bi-2212 wires must be electrically discrete in tight winding packs. It is vital to use an insulating layer that is thin, fulfils the dielectric requirements, and can survive the heat treatment whose maximum temperature reaches 890 °C in oxygen. A thin (20-30 μm) ceramic coating could be better as the insulating layer compared to alumino-silicate braided fiber insulation, which is about 150 μm thick and reacts with the Ag sheathed Bi-2212 wire during heat treatment. At present, TiO_2 seems to be the most viable ceramic material for such a thin insulation because it is chemically compatible with Ag and Bi-2212 and its sintering temperature is lower than the maximum temperature used for the Bi-2212 heat treatment. However, recent tests of a large Bi-2212 coil insulated only with TiO_2 showed severe electrical shorting between the wires after over pressure heat treatment (OPHT). The origin of the shorting was frequent silver protrusions into the porous TiO_2 layer that electrically connected adjacent Bi-2212 wires. To understand the mechanism of this unexpected behaviour, we investigated the effect of sheath material and hydrostatic pressure on Ag protrusions. We found that Ag protrusions occur only when TiO_2-insulated Ag-0.2%Mg sheathed wire (Ag(Mg) wire) undergoes OPHT at 50 bar. No Ag protrusions were observed when the TiO_2-insulated Ag(Mg) wire was processed at 1 bar. The TiO_2-insulated wires sheathed with pure Ag that underwent 50 bar OPHT were also free from Ag protrusions. A key finding is that the Ag protrusions from the Ag(Mg) sheath actually contain no MgO, suggesting that local depletion of MgO facilitates local, heterogeneous deformation of the sheath under hydrostatic overpressure. Our study also suggests that predensifying the Ag(Mg) wire before insulating it with TiO_2 and doing the final OPHT can potentially limit Ag protrusions.

1. Introduction

$Bi_2Sr_2CaCu_2O_x$ (Bi-2212) is the only high temperature superconductor available in a round wire form that has a very high critical current density (J_c is around 10^5 A cm^{-2} up to 45 T) and irreversibility field (>100 T) at 4.2 K [1,2]. This round wire geometry is preferred by magnet designers and builders as it can be easily twisted and cabled into desired shapes. Thus Bi-2212 round wire is an attractive candidate for very high field applications that are beyond the range achievable by using Nb_3Sn and Nb-Ti technology [3-6]. In order to attain high J_c, the Bi-2212 round wire undergoes an overpressure heat treatment (OPHT) at 50 bar with a 1 bar oxygen partial pressure (PO_2) that prevents the formation of large gas bubbles that significantly degrade J_c in 1 bar processing[4, 7]. High superconducting intergrain connectivity is developed by the highly-aligned grains resulting from the strong quasi-biaxial texture that forms in a narrow filament cavity in the round wire architecture [8]. Bi-2212 wires are made with the Bi-2212 powder in contact with pure Ag because of the good chemical compatibility between them. But, pure Ag is weak (yield strength 45 MPa at 4.2 K after OPHT), so the wire is strengthened by using a stronger outer sheath that is usually a dispersion strengthened Ag-0.2 wt% Mg (hereafter Ag(Mg)) [9].

One of the essential components of any superconducting magnet is electrical insulation, which prevents electrical shorting while charging the magnet to the operating current or discharging during a quench. For Bi-2212, the insulation is applied to the as-drawn wire and it must survive being wound into a coil, as well as the heat treatment, which has a maximum temperature around 890 ^{0}C in an oxidizing environment. Also, it has to be chemically compatible with the Ag(Mg) sheath and with Bi-2212, and it has to have sufficient dielectric strength so that the thin layer of insulation provides adequate electrical standoff within the coil winding pack while maintaining a high conductor packing factor. The alumino-silicate braid Bruker-Oxford Superconducting Technology has used to insulate Bi-2212 wires is quite thick (about 150 μm) and reduces the winding current density in coils. This insulation material also absorbs Ag from the sheath and Cu from the Bi-2212 core during the heat treatment, which reduces J_c by 15-20% compared to bare wire [10]. Considerable efforts have been made over the years to develop an effective insulation layer for Bi-2212 that is thin, chemically compatible, has good adhesive properties, and is viable for long length coating [11-15]. TiO_2 insulation developed by nGimat LLC and by Kandel *et al.* appeared to be a successful solution for insulating Bi-2212 [16, 17]. The thin (10-30 μm) TiO_2 coating has sufficient scrape resistance in both as-deposited and heat treated condition, is chemically compatible with Bi-2212 and Ag(Mg), and has high breakdown voltage and dielectric strength [17-19].

However, a test coil at the National High Magnetic Field Laboratory (NHMFL) that used Ag(Mg) Bi-2212 wire insulated with TiO_2 suffered from severe electrical shorting after OPHT. Post-mortem microstructural studies revealed that severe Ag protrusion through the insulation layer caused the shorting (figure 1). Here we report on a detailed investigation to determine what causes this Ag protrusion. We used Bi-2212 wires with both Ag(Mg) and pure Ag sheaths and carried out heat treatments at 50 bar (OPHT) and at 1 bar total pressure. Our results suggest that the combination of a Ag(Mg) sheath that is coated with TiO_2, and the overpressure that is applied during the OPHT causes the Ag protrusions. We also found that we can potentially limit Ag protrusions into TiO_2 by predensifying the Ag(Mg) wire before coating it with TiO_2 and then winding the coil and doing the final OPHT [20].

2. Experimental procedure

All the wires used in this study were fabricated by Bruker-Oxford Superconducting Technology (Bruker-OST) using the powder in tube (PIT) method with Bi-2212 powder in a composition of $Bi_{2.17}Sr_{1.74}Ca_{0.89}Cu_{2.00}O_x$ produced by Nexans. The wire architecture was a double restack with the Bi-2212 powder in contact with pure Ag. The outer sheath was either Ag-0.2wt%Mg (referred to as Ag(Mg) wire) or pure Ag to study the effect of different sheath materials. The diameter of the as-drawn Ag(Mg) wire and pure Ag sheathed wire without insulation were 1 mm and 1.3 mm respectively. The as-drawn wires were dip coated with the TiO_2 slurry at the NHMFL, followed by a burn out heat treatment at 450 ^{0}C in 1 bar flowing O_2 to remove the organics from insulation layer [18]. A systematic variation in coating thickness from the dip coating process was observed along the wire circumference. The thinnest region was around 13 μm thick, which continuously increased to 23 μm thickness at the opposite side of the wire. Careful observation of transverse cross-sections revealed that the coating thickness was the same (between 13 and 23 μm) after 1 bar and 50 bar heat treatment.

The wires were then heat treated in a flowing gas system following the heat treatment procedure shown in figure 2 [21]. Heat treatments were carried out at both 1 bar and 50 bar (OPHT) total pressure.

The 1 bar heat treatments were done in pure O_2 and the OPHT were done in a Ar-2%O_2 gas mixture at 50 bar total pressure, both of which maintained PO_2 = 1 bar during the heat treatment. 8 cm long samples were used because earlier observations showed that Ag protrusions also occurred in these short samples. For the 1 bar heat treatment, the wire ends were kept open to allow gas (typically CO_2 and H_2O) that was present in the as-drawn wire to escape out the ends, thus avoiding swelling and degradation of the wire. In contrast, both ends of the wires that underwent OPHT were hermetically sealed with a Ag-bead using a torch.

Cross-sections of the heat treated wires were dry polished using SiC papers to 800 grit, followed by final polishing in an automatic vibratory polisher (Buehler Vibromet 2) using a suspension of 50 nm alumina powder in methanol. A Zeiss 1540EsB scanning electron microscope (SEM) was used to observe the microstructure. We prepared thin lamellae for STEM observations using a focused ion beam (FIB) in the Zeiss 1540 EsB. A JEOL ARM200cF was used to perform high-angle annular dark field scanning transmission electron microscopy (HAADF-STEM) imaging.

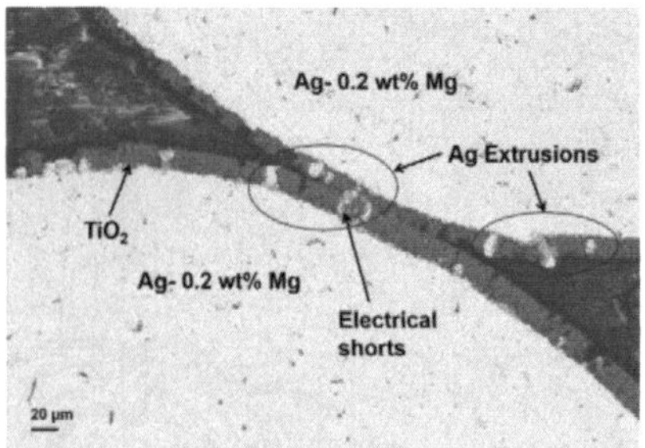

Figure 1. Transverse cross section from the shorted coil. The circled regions show Ag protrusions (white) in the TiO_2 insulation. Some of the protrusions shown in the middle of the figure connect adjacent Bi-2212 wires, causing the shorting.

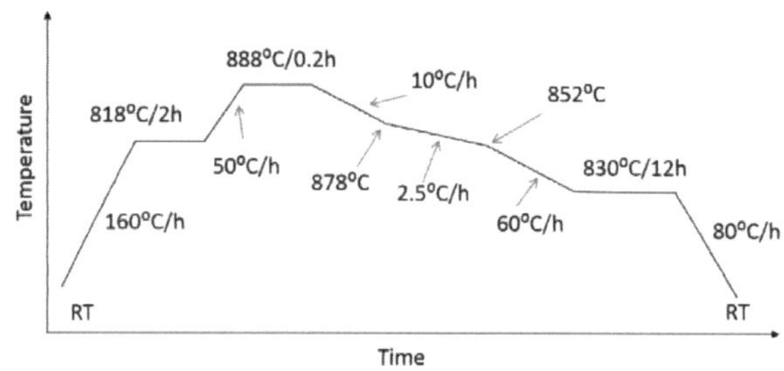

Figure 2. Heat treatment schedule for Bi-2212. This heating schedule was used for the 50 bar OPHT and for the 1 bar heat treatment.

3. Results

3.1. Effect of external pressure

To investigate the effect that external pressure during the heat treatment has on Ag protrusion into TiO_2, samples of Ag(Mg) wire were heat treated at 1 and 50 bar total pressure. As seen in the longitudinal cross-section in figure 3, the Ag(Mg) wire that underwent OPHT has frequent Ag protrusions into the TiO_2, but there are fewer protrusions than in the shorted coil shown in figure 1. We observed a longitudinal section of the interface between the sheath and the insulation that was about 4.5 cm long, and found that the Ag protrusions were randomly distributed along the entire length. The Ag protrusions were irregular in shape and size, and most of them did not penetrate the entire thickness of the TiO_2 layer. Previous studies in our lab showed that cracks and porosity formed in the insulation layer during the sintering of TiO_2 particles [17, 18]. Thus, the Ag protrusions did not crack the insulation, rather they grew into the pre-existing gaps and pores in the TiO_2 insulation.

No Ag protrusions were observed in the Ag(Mg) wire heat treated at 1 bar (figure 4), even though cracks and porosity in the TiO_2 layer were present to the same extent as observed in the OPHT sample (figure 3). The interface between the sheath and insulation was well defined.

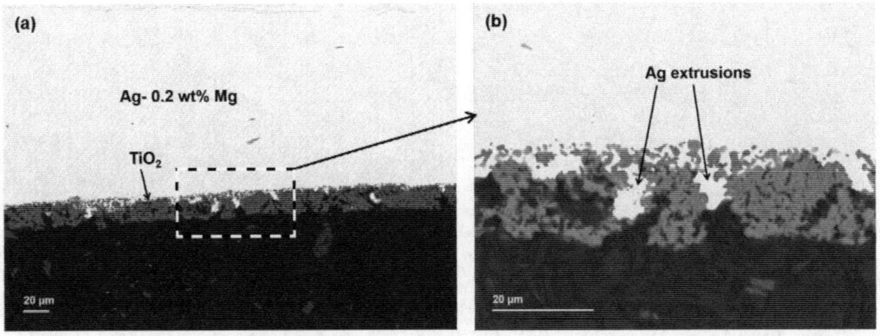

Figure 3. (a) SEM image of the interface between the Ag(Mg) sheath and the TiO_2 insulation in a Ag(Mg) wire after 50 bar OPHT (longitudinal cross section). (b) Higher magnification image of a region in (a).

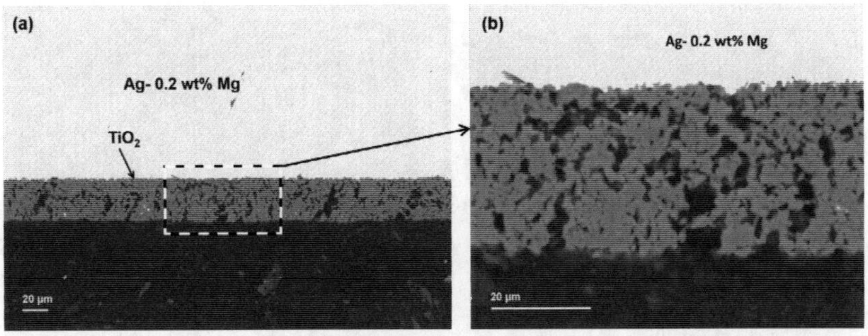

Figure 4. (a) SEM image of the interface between the Ag(Mg) sheath and the TiO_2 insulation in a Ag(Mg) wire after 1 bar processing (longitudinal cross section). (b) Higher magnification image of a region in (a). Note: The thickness of the TiO_2 layer is greater than 23 µm in this longitudinal cross section because the section is not at the exact middle of the wire.

3.2. *Effect of sheath material*
We also applied OPHT on wire with a pure Ag sheath to understand what role the composition of the sheath plays in Ag protrusion. Figure 5 shows there were no Ag protrusions and that there was a well-defined interface between the pure Ag sheath and the TiO_2 insulation layer. This was identical to the behaviour of Ag(Mg) wire heat treated at 1 bar (figure 4).

3.3. *HAADF-STEM images*
The HAADF-STEM image of figure 6(a) shows the distribution of MgO in two adjacent grains in the Ag-Mg sheath at a depth of ~8 µm from the surface of the bare wire (this wire was not coated with TiO_2 insulation) in a OPHTed Ag(Mg) wire. It is apparent from the figure that the density and distribution of MgO varies significantly between the two grains, one of them being devoid of MgO while the other had a uniform distribution of spherical MgO precipitates. In general, we found that the density of MgO particles is significantly lower near the wire surface than in the middle of the Ag(Mg) sheath. Figure 6(b) shows the HAADF-STEM image of a Ag protrusion in the TiO_2. Interestingly, the Ag protrusion appeared to be completely devoid of MgO precipitates. A sharp interface was observed between the Ag and TiO_2, which indicates no chemical reaction occurred between Ag and TiO_2, confirming that they are chemically compatible.

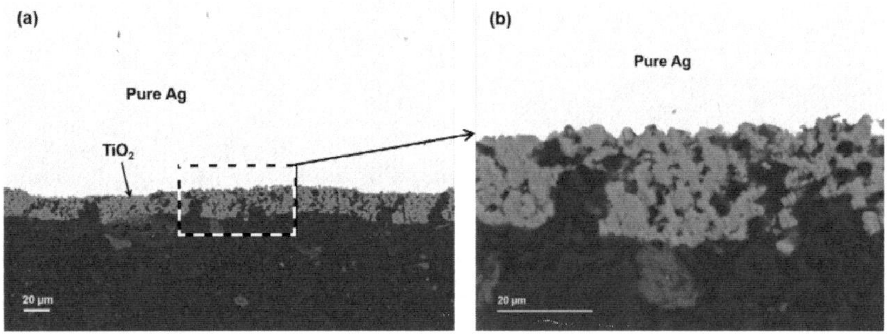

Figure 5. (a) SEM image of the interface between the Ag sheath and the TiO_2 insulation in a wire with a pure Ag sheath after OPHT (longitudinal cross section). (b) Higher magnification image of a region in (a).

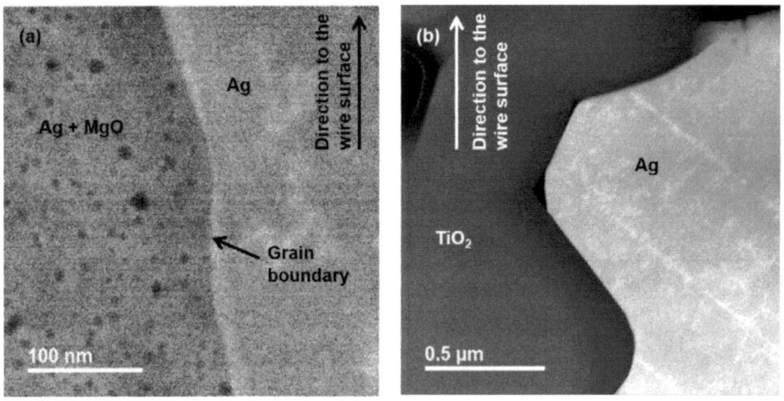

Figure 6. HAADF-STEM images of (a) grains within the Ag(Mg) alloy sheath. The darker round shaped particles are MgO. This sample was OPHTed without a TiO_2 coating. (b) A Ag protrusion in the TiO_2 insulation layer. The brighter contrast inside the Ag is due to defects introduced by the Ga ion beam during sample preparation by using focused ion beam. This samples was OPHTed with a TiO_2 coating.

3.4. Effect of predensification

To understand if densifying the wire before the application of insulation can prevent Ag protrusion into TiO_2, a small coil with Ag(Mg) wire was heated to 820 ^{0}C for 2 h at 50 bar (1 bar PO_2) with a heating and cooling rate of 160 ^{0}C/h following the procedure in [19]. The coil was then unwound, the TiO_2 insulation coating was applied, followed by rewinding, organic burn out, and OPHT at 50 bar. As figures 7 (a, b) show, no Ag protrusions were observed in this predensified wire. However, small, smooth, regularly-spaced, undulations on the surface of the Ag(Mg) sheath were observed. We have also checked the microstructure of Ag(Mg) wires without insulation that had been predensified separately and did not undergo OPHT (figure 7 (c, d)). They showed undulations of similar shape but smaller size and shorter periodicity.

4. Discussion

The Ag(Mg) sheath is used for Bi-2212 wires, because, when the alloy is oxidized, MgO precipitates form in the Ag that strengthen the Ag sheath. It was expected that the MgO particles in the heat-treated wires would be The Ag(Mg) sheath is used for Bi-2212 wires, because, when the alloy is oxidized, MgO precipitates form in the Ag that strengthen the Ag sheath. It was expected that the MgO particles in the heat-treated distributed uniformly; however, we found a non-uniform MgO particle distribution in the Ag(Mg) after the OPHT. In particular, the MgO density is significantly reduced and nonuniform near the wire surface. This nonuniform distribution of MgO likely causes a local variation in mechanical properties between different regions in the sheath. We expect the regions with higher MgO content to be stronger and harder than regions with lower MgO content. Our hypothesis is that, while the external pressure provided by the isostatic gas pressure during OPHT is isotropic, the regions with different MgO content respond differently to this applied pressure, causing MgO-free Ag to be squeezed out of the matrix and into the insulation. We expect that the MgO-poor, weaker regions in the Ag(Mg) form the protrusions, which is based on the lack of MgO precipitates in the Ag protrusions shown in figure 6(b). At this point, we do not know what causes this variation in MgO distribution. The wire with the pure Ag sheath that was OPHTed had no MgO, so we expect the mechanical properties of the sheath to be uniform, so there was no Ag protrusion.

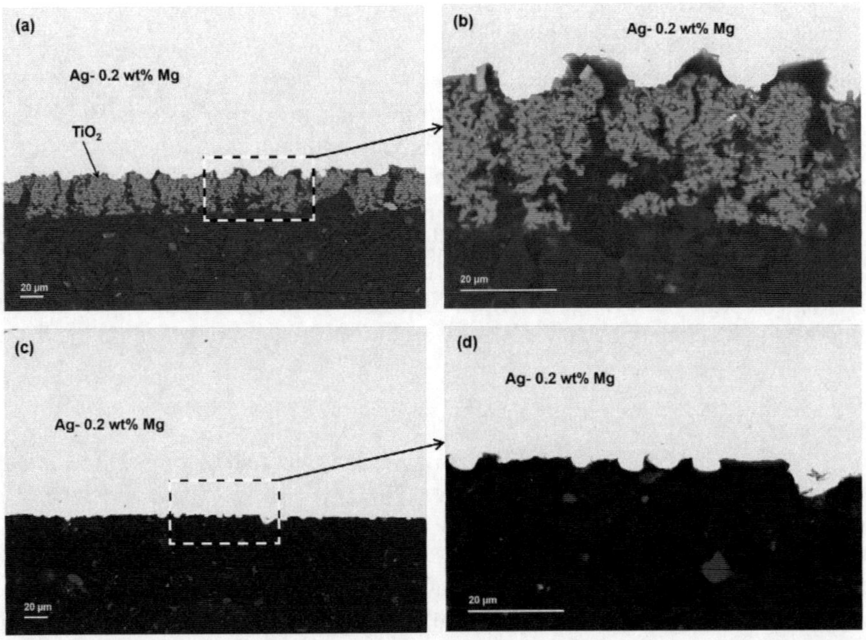

Figure 7. (a, b) SEM image of the interface between the Ag(Mg) sheath and the TiO_2 insulation in a sample that was predensified, coated with TiO_2, and then OPHTed. (c, d) The surface of the Ag(Mg) sheath in a sample that was only predensified and was not coated with TiO_2 (all images are showing longitudinal cross section).

In 1 bar processed Ag(Mg) wire, there is no external pressure gradient to cause Ag to protrude. We confirmed the need for a pressure gradient to form Ag protrusions by doing OPHT on a Ag(Mg) wire with open ends at 50 bar. With open ends, a pressure gradient could not form between the outside and inside of the wire. As with the 1 bar processed Ag(Mg) wire, no Ag protrusions formed even with 50 bar total pressure when the wire ends were open..

Matras *et al.* showed that 50 bar OPHT reduces the diameter of Bi-2212 round wires by around 4% [22]. They found that about 80% of this diameter reduction occurs while holding for 2 h at 820 ^{0}C, which is below the melting point of Bi-2212, during the standard heat treatment. Figures 7 (a, b) showed no Ag protrusions in the sample that was predensified at 820 °C before being insulated with TiO_2, and then given a full OPHT. The quasiperiodic undulations on the surface of this initially predensified wire (figures 7 (a, b)) appeared underneath the insulation layer. In addition, the predensified Ag(Mg) wire without insulation shows similarly shaped (although smaller size) undulations (figure 7 (c, d)). These observations suggest that the surface undulations were formed during the predensification step where the majority of densification takes place. HAADF-STEM analysis of these undulations revealed that they too contain almost no MgO precipitates.

Based on our observations, we propose that the following conditions are needed to form Ag protrusions: (1) a sheath made from Ag(Mg) (2) the wire has to be coated with TiO_2 insulation before the majority of the densification due to overpressure takes place. Our results suggest that it may be possible to limit Ag protrusions by predensifying the Ag(Mg) wire before applying the TiO_2 insulation, followed by full OPHT.

During the review process of this paper, one of the reviewers suggested that as the wire diameter decreased in diameter by 4%, the TiO_2 layer had to move in the circumferential direction to adhere to the shrinking wire, and this TiO_2 motion scraped up nonuniformities on the surface of the Ag that became the Ag protrusions in the TiO_2. However, the question is whether there is enough movement of the TiO_2 layer to cause Ag protrusions the size that we see in our wire. We have designed an experiment to check this hypothesis.

5. Conclusions

We studied the formation of Ag protrusions through the TiO_2 insulation layer in Bi-2212 round wires using two different sheath materials and under various heat treatment conditions. Our results suggest Ag protrusions occur only in Ag-0.2 wt% Mg sheathed wires during heat treatment under a substantial overpressure. HAADF-STEM observations showed non-uniform distribution of MgO in the Ag(Mg) sheath near the surface of the wire, which potentially leads to local variations in mechanical properties in different regions of the sheath causing Ag to protrude during OPHT. The results of our study suggests that predensifying Ag(Mg) wire below the Bi-2212 melting temperature before applying the insulation and then doing a full OPHT may be an effective way to limit Ag protrusions into the TiO_2 insulation layer.

Acknowledgement:
This work was performed at the National High Magnetic Field Laboratory, which is supported by National Science Foundation Cooperative Agreement No. DMR-1157490 and the State of Florida and also supported by the US Department of Energy Office of High Energy Physics under grant DE-SC0010421. National Institute of General Medical Sciences of the National Institutes of Health also supported the research reported here under Award Number R21GM111302.The authors are grateful to the members of the Bi-2212 group in the Applied Superconductivity Center for their valuable input.

References:

[1] Miao H, Marken KR, Meinesz M, Czabaj B and Hong S 2005 Development of round multifilament Bi-2212/Ag wires for high field magnet applications. *IEEE Transactions on applied superconductivity*. **15**(2):2554-7.

[2] Chen B, Halperin WP, Guptasarma P, Hinks DG, Mitrović VF, Reyes AP and Kuhns PL 2007 Two-dimensional vortices in superconductors. *Nature physics*. **3**(4):239-42.

[3] Godeke A, Cheng D, Dietderich DR, Ferracin P, Prestemon SO, Sabbi G and Scanlan RM 2007 Limits of NbTi and Nb_3Sn, and Development of W&R Bi–2212 High Field Accelerator Magnets. *IEEE Transactions on Applied Superconductivity*. **17**(2):1149-52.

[4] Larbalestier DC, Jiang J, Trociewitz UA, Kametani F, Scheuerlein C, Dalban-Canassy M, Matras M, Chen P, Craig NC, Lee PJ and Hellstrom EE 2014 Isotropic round-wire multifilament cuprate superconductor for generation of magnetic fields above 30 T. *Nature materials*. **13**(4):375-81.

[5] Weijers HW, Trociewitz UP, Marken K, Meinesz M, Miao H and Schwartz J 2004 The generation of 25.05 T using a 5.11 T $Bi_2Sr_2CaCu_2O_x$ superconducting insert magnet. *Superconductor Science and Technology*. **17**(4):636.

[6] Godeke A, Cheng DW, Dietderich DR, Hannaford CR, Prestemon SO, Sabbi G, Wang XR, Hikichi Y, Nishioka J and Hasegawa T 2009 Progress in wind-and-react Bi-2212 accelerator magnet technology. *IEEE Transactions on Applied Superconductivity*. **19**(3):2228-31.

[7] Kametani F, Shen T, Jiang J, Scheuerlein C, Malagoli A, Di Michiel M, Huang Y, Miao H, Parrell JA, Hellstrom EE and Larbalestier DC 2011 Bubble formation within filaments of melt-processed Bi2212 wires and its strongly negative effect on the critical current density. *Superconductor Science and Technology*. **24**(7):075009.

[8] Kametani F, Jiang J, Matras M, Abraimov D, Hellstrom EE and Larbalestier DC 2015 Comparison of growth texture in round Bi2212 and flat Bi2223 wires and its relation to high critical current density development. *Scientific reports*. **5**.

[9] Smith DR and Fickett FR. Low-temperature properties of silver 1995 *Journal of Research-National Institute of Standards and Technology*.**100**:119.

[10] LoSchiavo MP 2010 Processing issues of $Bi_2Sr_2CaCu_2O_8$ round wire involving leakage and alumino silicate insulation. *MS thesis*. Florida State University.

[11] Celik E, Avci E and Hascicek YS 2000 High temperature sol–gel insulation coatings for HTS magnets and their adhesion properties. *Physica C: Superconductivity*. **340**(2):193-202.

[12] Celik E, Akin Y, Mutlu IH, Sigmund W and Hascicek YS 2002 $BaZrO_3$ insulation coatings for HTS coils. *Physica C: Superconductivity*. **382**(4):355-60.

[13] Mutlu IH, Celik E and Hascicek YS 2002 High temperature insulation coatings and their electrical properties for HTS/LTS conductors. *Physica C: Superconductivity*. **370**(2):113-24.

[14] Celik E, Avci E and Hascicek YS 2004 Growth characteristics of ZrO_2 insulation coatings on Ag/AgMg sheathed Bi-2212 superconducting tapes. *Materials Science and Engineering: B*. **110**(2):213-20.

[15] Xue Y, Mark S, Shoup S, Marken KR, Miao H, Maarten M, Gourlay SA and Scanlan R 2003 Development of CCVD ceramic insulation for Bi-2212 superconducting wires and Rutherford cables. *IEEE transactions on applied superconductivity*. **13**(2):1796-9.

[16] Ishmael S, Luo H, White M, Hunte F, Liu XT, Mandzy N, Muth JF, Naderi G, Ye L, Hunt AT and Schwartz J 2013 Enhanced Quench Propagation in $Bi_2Sr_2CaCu_2O_x$ and $YBa_2Cu_3O_{7-x}$ Coils via a Nanoscale Doped-Titania-Based Thermally Conducting Electrical Insulator. *IEEE Transactions on Applied Superconductivity*. **23**(5):7201311.

[17] Kandel H, Lu J, Jiang J, Chen P, Matras M, Craig N, Trociewitz UP, Hellstrom EE and Larbalestier DC 2015 Development of TiO_2 electrical insulation coating on Ag-alloy sheathed $Bi_2Sr_2CaCu_2O_{8-x}$ round-wire. *Superconductor Science and Technology*. **28**(3):035010.

[18] Lu J, McGuire D, Kandel H, Xin Y, Chen P, Jiang J, Trociewitz U, Hellstrom E and Larbalestier DC 2016 Ceramic insulation of $Bi_2Sr_2CaCu_2O_{8-x}$ round wire for high-field magnet applications. *IEEE Transactions on Applied Superconductivity*. **26**(4):1-5.

[19] Chen P, Trociewitz UP, Dalban-Canassy M, Jiang J, Hellstrom EE and Larbalestier DC 2013 Performance of titanium oxide–polymer insulation in superconducting coils made of Bi-2212/Ag-alloy round wire. *Superconductor Science and Technology*. **26**(7):075009.

[20] Matras MR 2016 Investigation of Ag-sheathed multi-filamentary $Bi_2Sr_2CaCu_2O_{8-x}$ superconducting round wires processed with overpressure, for high field magnets. *PhD thesis*. Florida State University.

[21] Jiang J, Starch WL, Hannion M, Kametani F, Trociewitz UP, Hellstrom EE and Larbalestier DC 2011 Doubled critical current density in Bi-2212 round wires by reduction of the residual bubble density. *Superconductor science and technology*. **24**(8):082001.

[22] Matras MR, Jiang J, Larbalestier DC and Hellstrom EE 2016 Understanding the densification process of $Bi_2Sr_2CaCu_2O_x$ round wires with overpressure processing and its effect on critical current density. *Superconductor Science and Technology*. **29**(10):105005.

IOP Conf. Series: Materials Science and Engineering **279** (2017) 012022 doi:10.1088/1757-899X/279/1/012022

Tensile properties and critical current strain limits of reinforced Bi-2212 conductors for high field magnets

M. Brown[1], E. Bosque[1], D. McRae[2], R.P. Walsh[2], J. Jiang[1], E. E. Hellstrom[1], Y. Kim[1], U. Trociewitz[1], A. Otto[3], D. C. Larbalestier[1]

[1]Applied Superconductivity Center, National High Magnetic Field Laboratory, Florida State University, Tallahassee, FL 32310 USA

[2]Magnet Science & Technology Division, National High Magnetic Field Laboratory, Florida State University, Tallahassee, FL 32310 USA

[3]Solid Material Solutions, Chelmsford, MA 01863 USA

Email: mdb06h@my.fsu.edu

Abstract. We study here the effect of axial strain on the degradation of the critical current I_c for bare and reinforced, overpressure processed Bi-2212 conductors. We show that reinforcement markedly improves the conductor's stress limit, doubling it from ~150 MPa in the bare conductor to ~300 MPa when reinforced. We find also that certain processes used to reinforce the conductor slightly reduce the I_c degradation strain limit from ~0.6% to ~0.4%. Stress vs strain data taken from the samples studied here has been used to create a finite element model to explore the feasibility of using a reinforced Bi-2212 strand (produced by Solid Material Solutions) in a small test coil. The model predicts an IC limited coil with a maximum hoop strain of 0.31%, well below the experimentally verified strain limit, and is designed to lead to Bi-2212 coils that are not strain limited, but I_c limited.

1. Introduction

Bi-2212 is a high temperature superconductor (HTS) currently being developed in round wire form for use in high-field magnet applications [1] such as general purpose high-field research, nuclear magnetic resonance (NMR) [2], and the next generation of particle accelerators [3], [4]. HTS conductors are very important for operation in high fields above those attainable by the ubiquitous low temperature superconductors (LTS), Nb_3Sn and Nb-Ti. Bi-2212 is unique in that it is the only HTS conductor capable of high critical current density J_c available in a macroscopically isotropic, round-wire form. This allows for its use in well understood cabling technologies commonly used for round wire LTS like Rutherford cabling [5].

From the perspective of using Bi-2212 for high-field magnet applications, the overpressure heat treatment (OPHT) process can reliably achieve high overall wire critical current densities, J_E, in present state-of-the-art Bi-2212 strands. However, high J_E values and the desire to use Bi-2212 in high fields then focuses attention on the low elastic modulus (E = 70 GPa) and low yield strength ($\sigma_{0.02}$ < 100 MPa) of the Ag/Ag alloy matrix that sheathes the Bi-2212 filaments. This leads to strain- rather than I_c-limited performance in many high field solenoid designs. J_E degrades as a function of strain due to progressive fracture of the Bi-2212 filaments, making mechanical characterization of reinforced strands highly desirable in the ongoing effort to increase the stress limits of the conductor at higher fields [6]–[8]. Reinforcement methods for high field coils vary: for example, the coil as a whole, the conductor, or both can be reinforced. Here we report on reinforcement at the strand level, which involves bonding a high-strength metallic alloy strip to the broad flat faces of rectangularly rolled Bi-2212 (Figure 1) [9] in a form referred to as strengthened rectangular wire (SRW).

Presently, Bi-2212 conductors are made using the Powder-In-Tube (PIT) method [10], requiring the drawing of billets of Bi-2212 powder and Ag matrix composites into thin (0.8 – 1.3 mm diameter) wires in which the Bi-2212 powder precursor is only ~70% dense [11]. The critical current (I_c) of Bi-2212 wires can be significantly enhanced by implementing an over-pressure heat treatment (OPHT) at 10-100 bar of inert gas containing 1 bar pO_2 [12]. The OP densifies the Bi-2212 precursor in the Ag creep regime[13], and stifles the formation of gas bubbles which form from the agglomerated porosity in the filaments, obstructing the flow of current and greatly limiting I_c of the wire, especially in long length wires from which hot residual gas trapped in the pores cannot escape [14].

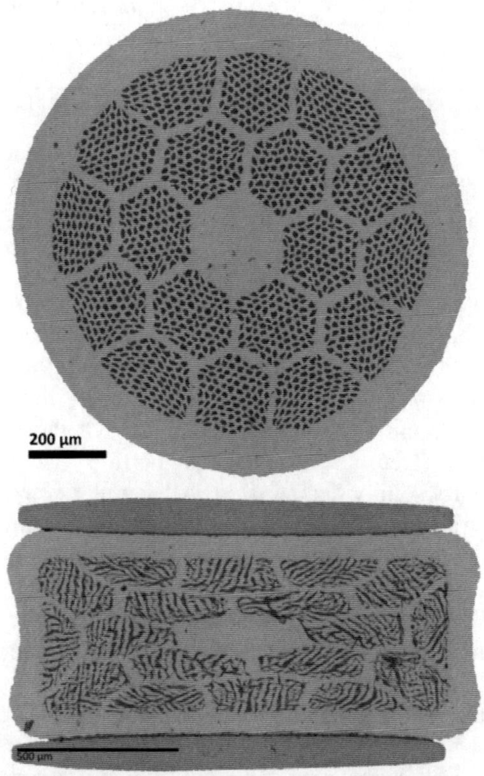

Figure 1: Transverse cross-section of a densified pmm160719-1 round wire (top) and fully heat treated strengthened rectangular wire (SRW) strand cross section; pmm160719-1 wire, rolled to a rectangular cross section and bonded to reinforcement strips on the long flat edges of the conductor (bottom).

2. Experimental Details

2.1. *Bi-2212 and Ag-Mg Alloy Composite Samples*

The Bi-2212 wire studied here (PMM160719-1, OST-1) was produced as a round wire by Oxford Superconducting Technologies and then made rectangular and reinforced by Solid Material Solutions (SMS). The as-delivered 1.2 mm diameter wire has a double stack, 85 x 18 filament geometry with a pure Ag matrix around each filament and an outer sheath of Ag-0.2wt%Mg alloy. During OPHT, the sheath becomes dispersion strengthened during its elevated-temperature exposure to oxygen leading to the formation of MgO_2, which provides some reinforcement for the conductor.

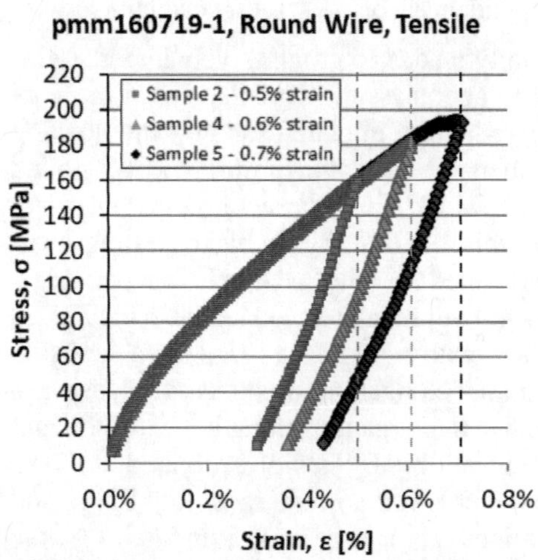

Figure 2: Stress vs. strain curves for multiple heat treated pmm160719-1 round wire samples loaded in tension each to a progressively larger strain value. One sample was left for testing I_C directly after heat treatment without any mechanical deformation.

The wires were tested in four conditions. Some were left in the as-drawn round state, while others

Table 1: Geometries and mechanical properties

Sample Name	Transverse Cross Section	Reinforcement bond HT'd	Reinforced	Filament Geometry	Transverse Area [mm^2]	I_c (5T, 4.2K) [A]	J_E (5T, 4.2K) [A·mm^{-2}]	Modulus, E [GPa]
OST-1	round	no	no	85 x 18	1.083	881	814	83.6 ± 2.6
SMS-1	rectangular	no	no	85 x 18	0.643	429	667	81.7 ± 1.6
SMS-2	rectangular	yes	no	85 x 18	0.653	449	687	78.5 ± 4.3
SMS-SRW	rectangular	yes	yes	85 x 18	0.923	466	505	126 ± 3.8
SMS Laminate	flat	yes	-	-	0.120	-	-	254 ± 6.4

were rolled to a rectangular cross section. Some rectangular wires were reinforced with a strip of high strength alloy on both wide surfaces using a bonding heat treatment at 822 °C for 24 hours. Of those that were reinforced, a portion were purposefully delaminated. All samples were strained first before I_c testing. Our evaluations provide a basis to measure the effects of reshaping, reinforcing with a high strength alloy, and heat treating to secure the reinforcement to the conductor. The reinforcing material was also mechanically tested to compare to the SRWs.

All wires were over-pressure processed in argon gas containing 1 bar pO_2 at 50 bar total pressure at the NHMFL. Important sample properties are presented in Table 1.

Figure 3: MTS servo-hydraulic test machine at the NHMFL, and tensile test set-up.

2.2. *Stress vs. Strain*

In order to measure the tensile strain dependence of I_c, the specimens were axially loaded to progressively larger strains that encompass the expected strain range of degradation. Figure 2 shows the loading sequences used for pmm160719-1 round wire, which are representative for all wires in this study. Mechanical testing was performed at the NHMFL on an MTS servo-hydraulic test machine (Figure 3) fitted with a 1kN load cell in a cryostat. Sample elongation was measured using a 50.8 mm gauge length extensometer (designed by J. Shepic) calibrated at 77 K. Samples were submerged in LN_2, and strained at a rate of 0.5 mm/min to the desired strain, at which point the sample was carefully removed for later testing and microscopy. Stress vs. strain curves were also taken separately for the reinforcement

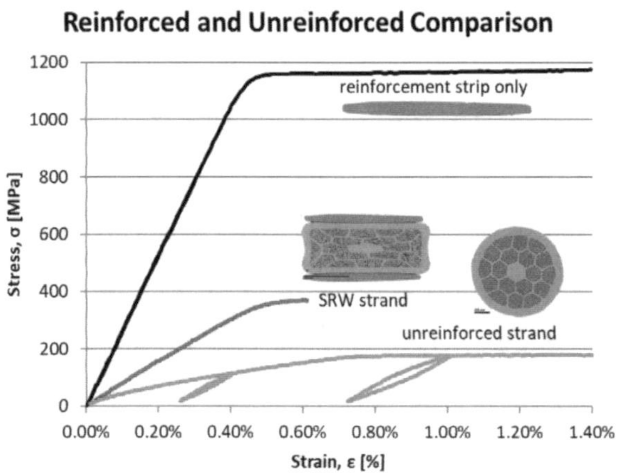

Figure 4: Stress- strain curves for reinforcement strip, reinforced OPHT SRW strand, and unreinforced OPHT pmm160719-1 round wire. All tests were performed in LN_2 at

material itself (Figure 4). The moduli reported in Table 1 are taken from unloading portions of the stress-strain curves.

2.3. Measurement of I_c

After samples had been axially loaded, they were I_c tested using the four-probe transport method in a magnetic field of 5 T at 4.2 K with a 1 μV·cm^{-1} voltage criterion. For the round wires, the magnetic field was applied perpendicular to the wire axis, and for the rectangular wires the magnetic field was applied perpendicular to the wire axis and the long flat faces. The overall critical current density for each sample, J_E, was calculated using the whole wire cross sectional area after OPHT, including the high strength laminate for strengthened wires.

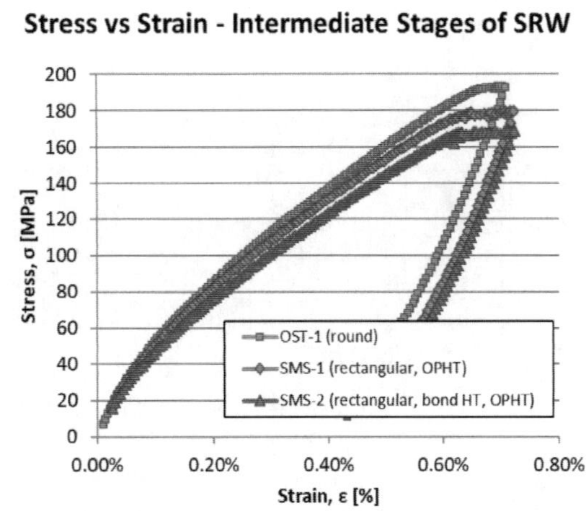

Figure 5: Stress vs. strain curves for intermediate wires in the SRW process: pmm160719-1 round wire (OPHT), the 17R1 wire rolled to rectangular cross section (OPHT), and the C67aR1 wire also rolled to rectangular cross section (bond HT & OPHT).

Cross sectional areas of the round wire were calculated using sample diameters measured over a 2cm length by an INEXiv automated digital measurement microscope. Only those wires determined to have been fully densified were used for this study. Cross sectional areas and filament area for the rectangular samples were calculated using digital images taken with an Olympus digital microscope. Samples were also imaged in a Zeiss 1540EsB Crossbeam field emission scanning electron microscope (FESEM) using a backscatter electron (BSE) detector.

3. **Results**

3.1. Stress vs Strain

The primary stress-strain data are shown in Figure 4. The unreinforced wire transitions into a constant stress deformation mode at about ~0.6% strain and 180 MPa during which the filaments break and some work-hardening of the Ag compensates the loss of strength caused by broken Bi-2212 filaments. The laminated and bonded SRW starts to yield at about 300 MPa and ~0.4% strain, while the reinforcement itself has a yield of about 1150 MPa at a strain of ~0.4%.

Figure 5 shows tensile data of wires in intermediate stages of SRW production. First we show the round wire from which all SRW wires are produced, pmm160719-1, next the 17R1 wire, which has been rolled to a rectangular cross section by SMS, and last, the C67aR1 wire after being both rolled and subjected to bonding the high strength laminate to the long edges of the conductor. The laminate was then peeled away to test the effect of the bonding HT on the properties of the conductor itself. All wires in this study, including these, were heat treated together by OPHT at 50 bar.

The data in Figure 5 show that each processing step does slightly reduce both the elastic modulus and the onset of yield. Modulus values taken from the initial slope at the point of unloading the samples can be found in Table 1. These values have ranges that may overlap, but from Figure 5 we can see that although the unloading slopes closely resemble one another, the effective moduli before first yield are clearly different. The stress at which each sample goes plastic also decreases with each additional processing step.

3.2. I_C & Limits of Strain-Induced I_C Degradation

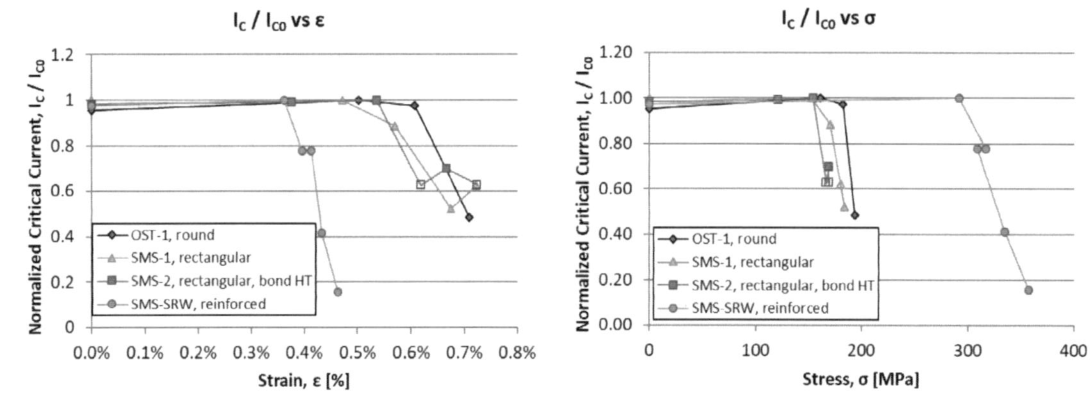

Figure 6: I_C / I_{C0} vs. strain (left) and stress (right) for each sample showing the point at which I_c begins to degrade significantly. Each data point represents one sample, strained to a defined maximum value at 77 K and then released. I_c was then tested at 4.2 K in a 5T background. It is clear the damage begins at about twice the stress in the reinforced wire but also that the critical strain is reduced from ~0.6% to ~0.4%.

Coils made from unreinforced Bi-2212 conductors will often be strain limited by fracture of the Bi-2212 filaments induced by Lorentz stresses since it has been shown that I_C degrades as strain reaches ~0.6 % in round, unreinforced strand [7]. Figure 6 shows I_C / I_{C0} vs strain (left) and stress (right) data taken for each sample type. The OST-1 round wire shows the highest strain limit slightly above 0.6 %, whereas I_C for SMS-1 and SMS-2 drops off at less than 0.6 % strain. This indicates two things: first, that rolling the conductor to rectangular shape can slightly degrade the conductor's strain limit. Second, SMS-2, having gone through a bonding HT at 820°C before OPHT, does not show a reduced strain limit, implying that, although a bond HT mildly affects the mechanical properties of the wire (Figure 5), it does not affect the strain properties of the filaments themselves. More significantly, we see a decrease in the strain limit of reinforced SMS-SRW wire from ~0.6% to ~0.4%, but Figure 6 (right) also shows that the reinforcement effectively doubles the stress at which degradation starts to occur from 150-160 MPa to ~300 MPa. This is for a wire with 21.5% reinforcement in the cross section.

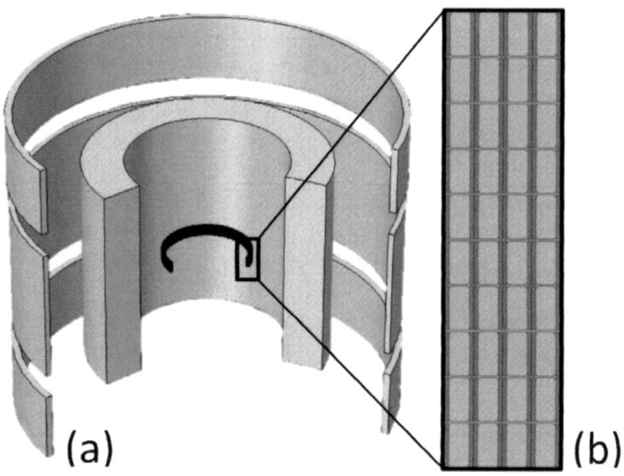

Figure 7: Cross sectional view of SRW test coil modelled using Comsol Multiphysics. (a) Large blue coils represent cryocooled LTS magnet which produces an 8 T background field. (b) The light grey regions are the conductor itself, the dark grey regions are strips of reinforcement bonded to both sides of the conductor, and the blue indicates interstitial regions filled with epoxy.

Table 2: SRW Comsol Model Parameters

Layers:	4	**Build:**	3 mm
Turns/Layer:	10	**Temp:**	4.2 K
Height:	13.5 mm	**Current:**	450 A
ID:	124 mm	J_E:	686 A/mm^2

3.3. Finite Element Coil Modelling

A finite element model (FEM) of a potential Bi-2212 SRW test coil using these stress-strain data was built using Comsol Multiphysics. Figure 7 shows an axisymmetric longitudinal coil cross section built of SRWs, and depicts the proposed winding pack operating in an ~8 T

IOP Conf. Series: Materials Science and Engineering **279** (2017) 012022 doi:10.1088/1757-899X/279/1/012022

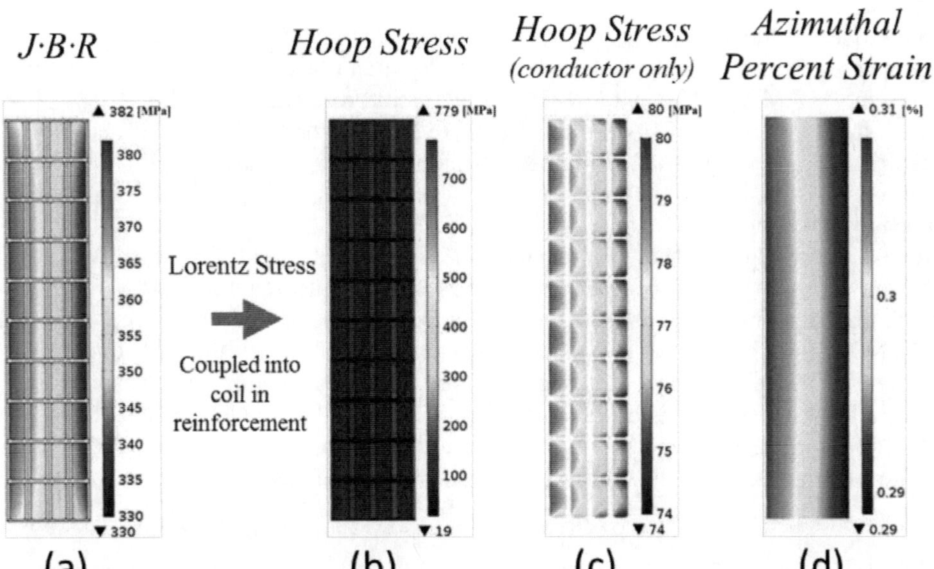

Figure 8: Coil model using an SRW-SMS simulated in Comsol Multiphysics. The figure shows (a) the calculated stress using JBR, (b) the distribution of stress within the cross section implying stress concentration in the high strength laminate, (c) the simulated final values of stress within the conductor only, and (d) the azimuthal percent strain throughout the coil cross section. Table 2 shows the coil model parameters. Note that from Tables 1 & 2, the coil is simulated very near its highest I_C value (5 T, 4.2 K), but the peak azimuthal strain (hoop strain) seen by the conductor is 0.31%, well below the 0.4% limit required to degrade I_C for SRW short samples.

background field at 4.2 K. A detailed set of test coil parameters are included in Table 2. Figure 8 (a) shows the unsupported turn *J·B·R* Lorentz source averaged over the Ag/Bi-2212 conductor area (without the reinforcement) in the 8 T background field at a Ag/Bi-2212 conductor current density, J_E, of 687 A/mm^2. In this unsupported turn approximation, a peak stress of ~360 MPa occurs at the inside radius and it exceeds the maximum strength of the unreinforced wire, ~180 MPa, by about two times. However, by bonding the rectangular wire with a high strength laminate and epoxy impregnating the whole coil pack, the stress distribution is greatly changed. Figure 8 (b) reports the structural mechanical result of distributing this load between conductor and high modulus laminate, where the peak stress now reaches ~700 MPa, still well below the measured yield stress of ~1150 MPa in Figure 4. It is clear the large differences in elastic moduli concentrate stress in the reinforcing laminates. In Figure 8 (c), focus is narrowed to the Ag/Bi-2212 component alone, where the stress is everywhere less than 100 MPa, well below filament fracture stress. Finally, the resultant azimuthal strains are shown in Figure 8 (d) and they are everywhere less than 0.31%, well below the onset strain of fracture of 0.4% shown in Figure 4. In summary, when ramped to 450 A (just above the expected I_C for SRWs at 8 T) the coil is now I_C, rather than strain limited.

4. Discussion

The results presented here modelling the use of the conductor in a comparatively large bore (124 mm) test bed shows the huge value of laminating weak Ag-sheathed Bi-2212 conductors with a laminate with three times the modulus of Ag (250 vs. 70 GPa). Lamination with just 20% of the cross-section doubles the fracture strength of the conductor before filament fracture to more than 300 MPa by reducing the stress on the Ag/Bi-2212 components of the conductor. The process for lamination is flexible, allowing greater or lesser amounts to be applied, thus making this a versatile technique for controlling damage in high field Bi-2212 solenoid magnets.

5. Conclusion

This study used mechanical testing and measurement of I_C to compare the coupled mechanical and electrical properties of reinforced Bi-2212 conductor. We found that the conductor stress limits could be doubled from ~150 MPa in bare wires to 300 MPa in reinforced wires with only ~20% of laminate. Simulation of the value of the high modulus laminated conductor for a test coil showed its strong value for reducing strain on the Bi-2212 filaments in an impregnated winding.

6. Acknowledgements

We gratefully acknowledge the help of technical staff, A. Francis and G. Bradford. Additionally, we are thankful to D. Abraimov for his constant willingness for enlightening discussion.

This work was supported by a grant from the US Department of Energy, Office of High Energy Physics (DE-SC0010421) and from the NHMFL, which is supported by the NSF under NSF/DMR-1157490 and by the State of Florida. Research reported in this publication was also supported by the National Institute of General Medical Sciences of the National Institutes of Health under Award Number R21GM111302. The content is solely the responsibility of the authors and does not necessarily represent the official views of the National Institutes of Health.

References

[1] Y. Huang, H. Miao, S. Hong, and J. A. Parrell, "Bi-2212 Round Wire Development for High Field Applications," *IEEE Trans. Appl. Supercond.*, vol. 24, no. 3, pp. 1–5, Jun. 2014.

[2] H. W. Weijers *et al.*, "High Field Magnets With HTS Conductors," *IEEE Trans. Appl. Supercond.*, vol. 20, no. 3, pp. 576–582, Jun. 2010.

[3] A. Godeke *et al.*, "Wind-and-react Bi-2212 coil development for accelerator magnets," *Supercond. Sci. Technol.*, vol. 23, no. 3, p. 34022, 2010.

[4] A. Ballarino, "Prospects for the use of HTS in high field magnets for future accelerator facilities," *CERN Document Server*, 11-Nov-2014. [Online]. Available: http://cds.cern.ch/record/1968557. [Accessed: 15-Jun-2017].

[5] E. Barzi, V. Lombardo, D. Turrioni, F. J. Baca, and T. G. Holesinger, "BSCCO-2212 Wire and Cable Studies," *IEEE Trans. Appl. Supercond.*, vol. 21, no. 3, pp. 2335–2339, Jun. 2011.

[6] A. Godeke *et al.*, "Critical current of dense Bi-2212 round wires as a function of axial strain," *Supercond. Sci. Technol.*, vol. 28, no. 3, p. 32001, Mar. 2015.

[7] R. Bjoerstad *et al.*, "Strain induced irreversible critical current degradation in highly dense Bi-2212 round wire," *Supercond. Sci. Technol.*, vol. 28, no. 6, p. 62002, 2015.

[8] U. P. Trociewitz, "Making strain effects inside Bi-2212/Ag conductors visible," *Supercond. Sci. Technol.*, vol. 28, no. 7, p. 70501, 2015.

[9] A. Otto, "Reinforced 2212 multifilament superconducting wire with low aspect shape, cables thereof, and 2223 silver tape-based cables," WO2016130975 A2, 18-Aug-2016.

[10] H. Miao, K. R. Marken, M. Meinesz, B. Czabaj, and S. Hong, "Development of round multifilament Bi-2212/Ag wires for high field magnet applications," *IEEE Trans. Appl. Supercond.*, vol. 15, no. 2, pp. 2554–2557, Jun. 2005.

[11] J. Jiang *et al.*, "Doubled critical current density in Bi-2212 round wires by reduction of the residual bubble density," *Supercond. Sci. Technol.*, vol. 24, no. 8, p. 82001, Jun. 2011.

[12] D. C. Larbalestier *et al.*, "Isotropic round-wire multifilament cuprate superconductor for generation of magnetic fields above 30 T," *Nat. Mater.*, vol. 13, no. 4, pp. 375–381, Apr. 2014.

[13] T. Shen, A. Ghosh, L. Cooley, and J. Jiang, "Role of internal gases and creep of Ag in controlling the critical current density of Ag-sheathed $Bi_2Sr_2CaCu_2O_x$ wires," *J. Appl. Phys.*, vol. 113, no. 21, p. 213901, Jun. 2013.

[14] F. Kametani *et al.*, "Bubble formation within filaments of melt-processed Bi2212 wires and its strongly negative effect on the critical current density," *Supercond. Sci. Technol.*, vol. 24, no. 7, p. 75009, 2011.

[15] A. Kajbafvala, W. Nachtrab, R. Kumar, F. Hunte, T. Wong, and J. Schwartz, "High strength oxide dispersion strengthened silver aluminum alloys optimized for $Bi_2Sr_2CaCu_2O_x$ round wire," *Supercond. Sci. Technol.*, vol. 26, no. 12, p. 125012, 2013.
[16] A. Kajbafvala, W. Nachtrab, T. Wong, and J. Schwartz, "$Bi_2Sr_2CaCu_2O_x$ round wires with Ag/Al oxide dispersion strengthened sheaths: microstructure–properties relationships, enhanced mechanical behavior and reduced Cu depletion," *Supercond. Sci. Technol.*, vol. 27, no. 9, p. 95001, 2014.
[17] M. Matras, "Investigation of Ag-sheathed multi-filamentary $Bi_2Sr_2CaCu_2O_x$ superconducting round wires processed with overpressure, for high field magnets," PhD thesis, Materials Science Program, The Florida State University, 2016.

IOP Conf. Series: Materials Science and Engineering **279** (2017) 012023 doi:10.1088/1757-899X/279/1/012023

Structure and superconducting characteristics of magnesium diboride, substitution of boron atoms by oxygen and carbon

Tetiana Prikhna[1], Vitaliy Romaka[2], Michael Eisterer[3], Andrii Shapovalov[1], Artem Kozyrev[1], Gennadiy Grechnev[4], Viktor Boutko[5], Wilfried Goldacker[6], Tobias Habisreuther[7], Oleksii Vakaliuk[8] and Bernd Halbedel[8]

[1]V. Bakul Institute for Superhard Materials of the National Academy of Sciences of Ukraine (NASU), 2, Avtozavodskaya Str., Kiev 07074, Ukraine
[2]Lviv Polytechnic National University, 12 Bandera Str., Lviv, 79013, Ukraine
[3]Atominstitut, TU Wien, Stadionallee 2, 1020 Vienna, Austria
[4]B. Verkin Institute for Low Temperature Physics of the National Academy of Sciences of Ukraine, Kharkov 61103, Ukraine
[5]Donetsk Institute for Physics and Engineering named after O. O. Galkin of the National Academy of Sciences of Ukraine, R. Luxemburg Str. 72, Donetsk-114, 83114, Ukraine
[6]Karlsruhe Institute of Technology (KIT), 76344 Eggenstein-Leopoldshafen, Germany
[7]Leibniz Institute of Photonic Technology e. V., Albert-Einstein-Straße 9, 07745 Jena, Germany
[8]Technische Universität Ilmenau, Institute of Materials Engineering, Gustav-Kirchhoff-Strasse 6, D-98693 Ilmenau, Germany

e-mail of the corresponding author: prikhna@ukr.net

Abstract. An x-ray analysis of MgB_2-based materials shows that they contain MgB_2 and MgO phases. According to a quantitative Auger analysis (taken after removing the oxidized surface layer by Ar ion etching in the microscope chamber) the MgB_2 phase contains some amount of oxygen that approximately corresponds to the composition $MgB_{2.2-1.7}O_{0.4-0.6}$. Rietveld refinement of the MgB_2 phase, based on EDX data with varying B/O content, leads to the composition $MgB_{1.68-1.8}O_{0.2-0.32}$. Ab-initio modelling of boron substitution by oxygen in MgB_2 (ΔH_f =-150.6 meV/atom) shows that this is energetically favourable up to the composition $MgB_{1.75}O_{0.25}$ (ΔH_f =-191.4 meV/atom). In contrast to carbon substitution, where very small levels of doping can dramatically affect the superconducting characteristics of the material with concomitant changes in the electron density, oxygen substitution results in very little change in the superconducting properties of MgB_2. The formation of vacancies at the Mg site of both MgB_2 and substituted $MgB_{1.75}O_{0.25}$ was modelled as well, but has shown that such processes are energetically disadvantageous (ΔH_f of $Mg_{0.875}B_2$ and $Mg_{0.75}B_{1.75}O_{0.25}$ are equal to -45.5 and -93.5 meV/atom, respectively).

1. Introduction

Magnesium diboride has a simple hexagonal structure in which layers of boron atoms alternate with magnesium atom layers [1]. Differently from Y- or Bi-based high temperature superconductors the larger coherence length of MgB_2 makes the superconducting current flow in the material insensitive to the presence of grain boundaries. In fact, grain boundaries and inclusions of secondary phases having sizes commensurable with the coherence length serve as pinning centers for Abrikosov vortices and thus their higher density should correlate with higher critical current densities.

SEM and Auger spectroscopy of the structures of MgB_2-based materials [2-6] prepared by us under different conditions (in flowing Ar, by hot pressing at 30 MPa, by spark plasma sintering at 16-96 MPa, under high quasihydrostatic pressure (2 GPa) – high temperature (600-1100 °C), by high pressure-high temperature treatment of wires, by magnetron sputtering of MgB_2 targets and deposition of thin films) showed that admixed oxygen is present in all materials to a rather high extent [7] due to the high affinity of Mg to oxygen. In bulk superconducting MgB_2 with high critical current densities (see, for example, Fig. 1a) and high critical magnetic fields (Fig. 1d) usually more than two phases are present according to the SEM and Auger data (Fig. 1c): (1) the matrix phase with near MgB_2 stoichiometry which contains a small amount of admixed oxygen; (2) inclusions or grains of higher magnesium borides MgB_x, x>>2 (MgB_4, MgB_7, MgB_{12}, MgB_{17}, MgB_{20}, Mg_2B_{25}, etc.); (3) nanolayers (if the synthesis temperature was low) or separate oxygen enriched inclusions (if it was higher) with near MgBO stoichiometry [7]. The reflections of the higher magnesium borides cannot be detected by x-ray powder diffraction because the inclusions are small, dispersed in the matrix and have big and complicated unit cells. Thus, a lot of small reflections result from the grains of higher borides, which are not seen in the background of the intensive MgB_2 reflections with simple hexagonal structure. The analysis using the FullProf Suite program package for the x-ray patterns revealed the presence of phases with MgB_2 (main phase) and MgO (to a much smaller extent) structures (see, for example Fig. 1a). According to the results of crystal structure refinement, the 2*d* crystallographic site, which should be solely occupied by boron atoms in pure MgB_2, is occupied by a statistical mixture of boron and oxygen in our samples [7]. In all cases the B/O ratio in the $MgB_{2-x}O_x$ solid solution remains practically the same and has nearly the same lattice parameters. No additional reflections that would correspond to the superstructure of $MgB_{2-x}O_x$ were observed. Thus, the analysis using Rietveld refinement allowed us to conclude that the composition of the phase with MgB_2 structure was $MgB_{1.68\text{-}1.74}O_{0.26\text{-}0.32}$ [7].

A positive effect in enhancing J_c and the upper critical field in MgB_2 materials can be obtained by dopants, in particular by carbon, carbon nanotubes, nanodiamonds or by carbon based additives, e.g. SiC, ethyltoluene, carbohydrates, malic acid, and many others [8-13]. Some carbon goes into the MgB_2 structure and slightly reduces the transition temperature, but the upper critical magnetic field and the irreversibility field increase. If the amount of carbon introduced into MgB_2 is too high, the reduction of the transition temperature annihilates the advantages of its addition.

The present paper also describes the results of *ab initio* calculations of the electronic structure and stability of the MgB_2 compounds with partial oxygen and carbon substitution for boron. The results presented here extend our fist results which were described in [7, 14].

2. Experimental

The microstructure of the materials was characterized by a JAMP–9500F (combined SEM and Auger analyzers) allowing to investigate non-oxidized MgB_2 surfaces (by etching 1×1 mm^2 square surfaces by Ar ions), to perform a quantitative Auger analysis of approximately 10 nm-sized inclusions and to construct depth profiles of the element distribution. The critical current densities, J_c, of the bulks and thin films were estimated from data obtained by an Oxford Instruments vibrating sample magnetometer (VSM) using the Bean model [15]. The values of the upper critical magnetic field, B_{c2}, and the irreversibility field, B_{irr}, were estimated from resistivity measurements (90% and 10% of the resistivity just above the beginning of the transition) using the four probe technique at a current of 10-100 mA and in fields from 0 to 15 T.

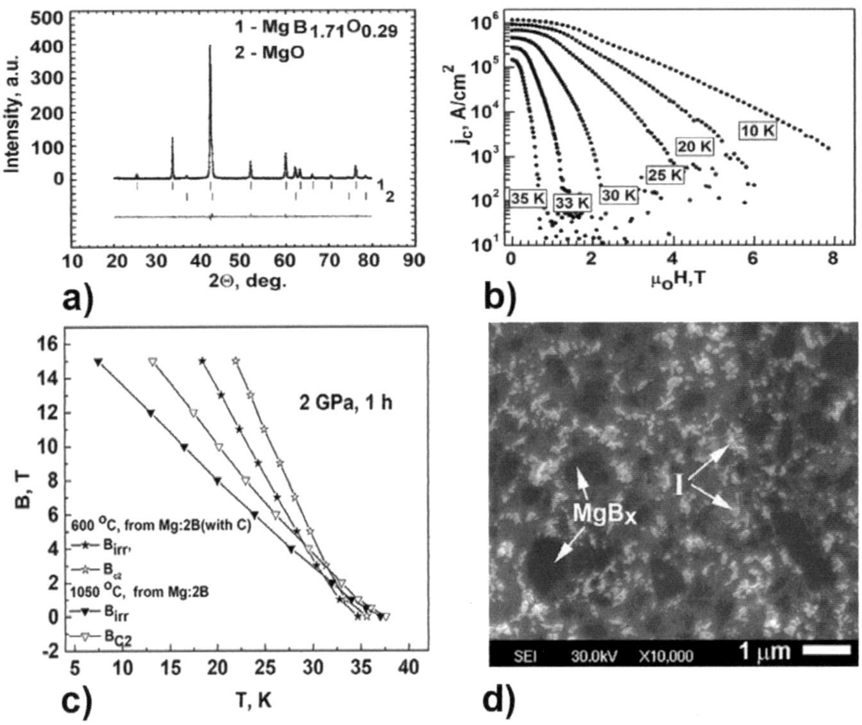

Figure 1. (**a**) - X-ray diffraction pattern, (**b**) – dependence of critical current densities, J_c, on magnetic field, $\mu_o H$, at 10, 20, 25, 30, 33, 35 K and (**d**) - SEM (SEI mode) structure of the material synthesized at 2 GPa, 1050 °C for 1 h from Mg:2B mixture (approximate composition according to the quantitative Auger analysis: matrix – $MgB_{2.2\text{-}1.7}O_{0.4\text{-}0.6}$, "I" – oxygen enriched inclusions appear brightest - $MgB_{0.6\text{-}0.8}O_{0.8\text{-}0.9}$; and the boron enriched inclusions "MgB_x" appear darkest – $MgB_{7\text{-}13}O_{0.02\text{-}0.06}$); (**c**) - dependence of the upper critical magnetic field (B_{c2}) and of the irreversibility field (B_{irr}) of MgB_2-based materials synthesized at 2 GPa for 1h: at 600 °C from Mg:2B (amorphous, contained 3.5 wt% of C) and 1050 °C from Mg:2 B (amorphous, contained 0.47 wt% of C).

The electronic structure of $MgB_{2\text{-}x}O_x$ and $MgB_{2\text{-}x}C_x$ was obtained using *ab initio* calculations for 2×2×2 and 2×2×1 supercells within the density-functional theory (DFT) [16] by employing the full potential codes: the linearized augmented plane wave method (FP-LAPW with generalized gradient approximation (GGA) [17] implemented in Elk [18] and WIEN2k [19]) and the full-potential LMTO method (FP-LMTO, RSPt implementation [20, 21] with local density approximation (LDA) [22]). All constructed supercells were based on the experimental parameters of the hexagonal MgB_2 cell, a=0.30844 nm and c=0.35215 nm. To analyze the adjusted (refined) positions of atoms and the related binding energies E_b in $MgB_{2\text{-}x}O_x$ and $MgB_{2\text{-}x}C_x$, WIEN2k [19] was applied.

For the thermodynamic calculation of the enthalpy of formation of binary and ternary compounds, a crystal structure optimization was performed including pure B (116 *k*-points), Mg (84 *k*-points), and the O_2 molecule [18]. The *k*-point mesh size for binary and ternary solids was set to 20×20×10 giving in total 264 (MgB_2) and 462 ($MgB_{1.75}O_{0.25}$) *k*-points in the irreducible part of the unit cell.

X-ray phase analysis was carried out with PowderCell v2.3 [21] using the powder patterns obtained on an X'Pert Pro/PANanalytical diffractometer (Cu K_α radiation, λ = 0.15418 nm) with Bragg-Brentano geometry. Rietveld refinement was performed using the FullProf Suite program package [24].

IOP Conf. Series: Materials Science and Engineering **279** (2017) 012023 doi:10.1088/1757-899X/279/1/012023

3. Results and discussion

The discussion in the paper refers to the established by us processing of bulks and wires. Figures 2a, b show isosurfaces of the electron localization function (elf) of MgB_2 (Fig. 2a) and $MgB_{1.75}O_{0.25}$ (Figs. 2b) at elf = 0.7. The strong electron localization in MgB_2 between boron atoms corresponds to strong covalent bonding within the boron network. Substitution of B atom by O leads to electron localization around O atoms and thus bonding polarization (Figs. 2b). Oxygen atoms affect not only the B-O bonds but also nearby B-B bonds by a deformation of the elf. Figure 2c shows the binding energies, E_b, of $MgB_{2-x}O_x/C_x$, i.e. of MgB_2 with oxygen or carbon substitutions at boron positions calculated using WIEN2k. Figure 2d demonstrates the Z - contrast image of coherent oxygen-containing inclusions in MgB_2 [010] bulk material, experimentally obtained by the authors of [26], which confirms that oxygen (when its amount is small) prefers to reside at each second boron plane thus leaving each second boron plane pristine. Figure 3 shows the maps of the electronic density distribution (calculated using WIEN2k) for the pristine MgB_2 structure (Fig. 3a), $MgB_{2-x}O_x$ (Figs. 3b-d) and $MgB_{1.5}C_{0.5}$ (Figs. 3e, f). Figs. 3b-d demonstrate the cases for $MgB_{2-x}O_x$ for x=0.25, (Fig. 3b shows the boron plane with the embedded oxygen atoms in nearby positions, while oxygen atoms are absent in the second boron plane (not shown) of the same unit cell.) and x=0.5. (Fig. 3c shows the boron plane without embedded oxygen atoms, while the second (alternate) boron plane contains these atoms.) Fig.3d displays a selected cut of the unit cell inclined to the basal boron planes to show the boron plane with as well without oxygen and the Mg atoms of the same unit cell. Figure 3e presents the boron plane of the $MgB_{1.5}C_{0.5}$ compound with the position of the embedded carbon atoms resulting in the lowest binding energy. Fig. 2c and Fig. 3f demonstrate selected cuts of the $MgB_{1.5}C_{0.5}$ unit cell to show the boron plane with carbon and the Mg atoms of the same unit cell.

It should be mentioned that for all values of x used for the calculations of the density of states (DOS) near the Fermi level of $MgB_{2-x}O/C_x$ (up to x=1) the densities of state were above zero, i.e. all the compounds show metal-like behaviour. In the case of $MgB_{2-x}O_x$ the lowest DOS (about 0.46 states/eV/f.u.) was found for $MgB_{1.75}O_{0.25}$ if the oxygen atoms are in neighbouring positions [7]. If the concentration of oxygen is higher than in $MgB_{1.5}O_{0.5}$ the MgB_2 structure is destroyed. In the case of $MgB_{2-x}C_x$ the lowest DOS (about 0.3 states/eV/f.u.) was found for $MgB_{1.5}C_{0.5}$, when carbon distributed homogeneously (i.e. is introduced in both basal boron planes of the same 2×2×2 MgB_2 supercell [25]). The maps of the electron density distribution were calculated under the assumption that the oxygen or carbon atoms substituting boron can be shifted from the positions occupied by boron because of the size difference of the atomic radii and the amount of electrons in their atoms.

The binding energy was calculated under the assumption of a homogeneous (curves 1 and 3) distribution of oxygen or boron in the MgB_2 structure and, in parallel, for the case of a so-called ordered distribution of the substituted atoms. This means that we calculated maps for substitutions where oxygen atoms occupy nearby or alternative boron positions and when they move into one or two boron planes of the unit cell.

The curves 2 and 4 (Fig. 2c) correspond to the lowest E_b for each concentration of oxygen atoms (distributed in definite order or randomly). The lowest E_b is observed when oxygen moves into nearby positions or form zigzag chains (curve 2, Fig. 2c), while in the case of carbon there was no difference (curves 3 and 4, Fig. 2c). Therefore, the lower E_b in the case when oxygen moves into nearby positions can explain the tendency of its segregation in the MgB_2 structure and the formation of oxygen-enriched inclusions as well as why a rather high amount of oxygen can be present in superconducting MgB_2 with a high transition temperature.

The imbedding of carbon into the MgB_2 structure is preferable in any case. It does not make any difference whether it substitutes boron in a special order or homogeneously, i.e. introducing carbon into both boron planes of the unit cell. Thus, if the concentration of carbon atoms is high, the transition temperature and superconductivity of MgB_2 are essentially suppressed. Fig. 1c shows the upper critical magnetic fields and the irreversibility fields of MgB_2-based material prepared from boron to which carbon was specially added (3.5 wt%). Usually 0.31-0.47 wt% of carbon are present in amorphous boron. Despite the very high critical magnetic fields, the T_c of the material is reduced to 34 K and not

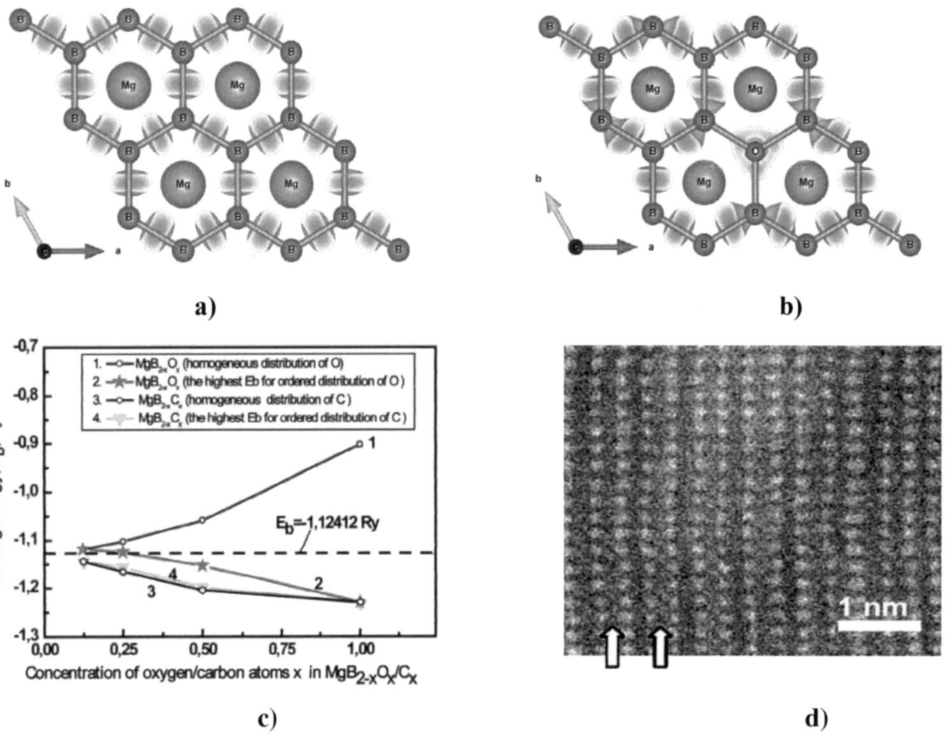

Figure 2. (a, b) - Isosurfaces of the electron localization function (elf) of MgB_2 and $MgB_{1.75}O_{0.25}$ at elf = 0.7; (c) - Dependence of the binding energy, E_b, on the oxygen concentration, x, in $MgB_{2-x}O_x/C_x$: 1, 3- homogeneous oxygen and carbon substitutions for boron atoms, respectively; 2, 4 - the lowest binding energy for each x in the case of ordered oxygen and carbon substitutions (for example, in the nearby positions or in pairs), respectively; (d) - Z - contrast image of coherent oxygen-containing inclusions in MgB_2 [010] bulk [26]. Bright atoms - Mg. The contrast increases in each second row and is due to the presence of oxygen in each second boron plane. The white arrows show the columns of atoms in which oxygen is present.

Table 1. Calculated enthalpies of formation of MgB_2, MgO, $Mg_{1-x}B_2$, and $MgB_{2-x}O_x$

No	Compound	Enthalpy of formation, ΔH_f (meV/atom)
1	MgB_2	-150.6
2	$Mg_{0.875}B_2$	-45.5
3	$Mg_{0.75}B_2$	+74.6
4	$Mg_{0.75}B_{1.75}O_{0.25}$	-93.5
5	$MgB_{1.75}O_{0.25}$	-191.4
6	$MgB_{1.5}O_{0.5}$	-162.6
7	MgO	-2719.7
8	$MgO_{0.5}B_{0.5}$	-82.0

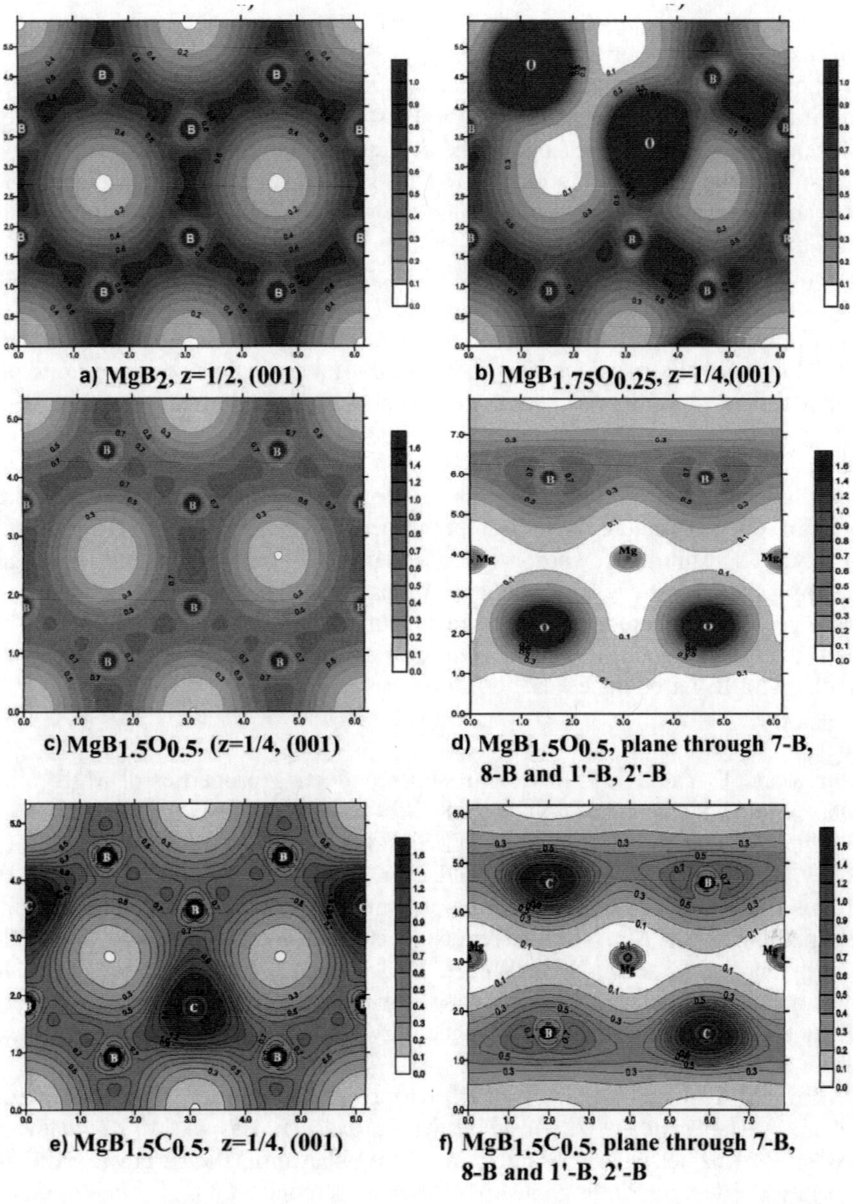

Figure 3. Maps of electron density distribution for: (a) - MgB_2 (z=1/2, (001)), (b) - $MgB_{1.75}O_{0.25}$ (z=1/4,(001) [7]), (c) - $MgB_{1.5}O_{0.5}$ (z=1/4, (001)); z- coordinates of the plane of a 2×2×2 supercell, where z is given in units of the *c* parameter of a 2×2×2 MgB_2 supercell [25]; (d) – $MgB_{1.5}O_{0.5}$ in the transversal plane under an angle to the basal boron planes of the hexagonal unit cell, in order to show the boron plane without imbedded oxygen atoms together with the Mg plane (i.e., the plane which goes through 7-B, 8- B and 1'-B, 2'-B positions of a 2×2×2 supercell [25]); (e) - $MgB_{1.5}C_{0.5}$ (z=1/4, (001)); (f) - $MgB_{1.5}C_{0.5}$ in the transversal plane under an angle to the basal boron planes (the plane which goes through 7-B, 8- B and 1'-B, 2'-B positions of a 2×2×2 supercell).

all carbon added to boron is imbedded into the MgB_2 structure, only about 0.050 at%, while according to our estimations the amount of imbedded carbon was around 0.0028 at% in materials with T_c near

38 K. However, these values are obtained from the lattice parameter which is likely influenced by the simultaneous oxygen substitution as well.

Table 1 shows the results of our calculations of the formation enthalpy, ΔH_f, for MgB_2, MgO, $Mg_{1-x}B_2$, and $MgB_{2-x}O_x$. The calculations of MgB_2 structures with a deficit of Mg were performed because this could as well be the reason for the deviation of composition from stoichiometric MgB_2. But, as the results presented in the Table 1 show, this is energetically disadvantageous. Moreover, the formation of vacancies in the Mg crystallographic site with simultaneous substitution of B by O is still less favorable than the formation of pure MgB_2 or even $MgB_{2-x}O_x$.

The formation enthalpy of MgO is very high (exothermal process) and the introduction of boron into the MgO structure dramatically decreases $-\Delta H_f$. The formation of $MgB_{0.6-0.8}O_{0.8-0.9}$ inclusions found in our experiments remains so far unclear. Further calculations are in progress for a clarification. However, phases which are unexpected from enthalpy considerations of the isolated compound may form in a multicomponent system, such as TiH_2 in titanium doped MgB_2 [27].

4. Conclusions

Ab initio calculations of the DOS, of maps of the electron density distribution and of the formation enthalpy of $MgB_{2-x}O_x/C_x$ structures have been performed. The results show that it is most favourable in the case of oxygen substitution for boron to form the $MgB_{1.75}O_{0.25}$ compound (if oxygen substitutes boron in nearby positions). Very similar compositions were observed experimentally in MgB_2-based materials with very good superconducting characteristics. Further modelling is necessary to clarify the formation of nanolayers and inclusions with near MgBO stoichiometry, which were observed experimentally in MgB_2-based materials.

5. References

[1] C. Buzea and T. Yamashita, Review of superconducting properties of MgB_2, *Supercond. Sci. Technol.* 14(11), R115 (2001) doi:10.1088/0953-2048/14/11/201.

[2] T. Prikhna, M. Eisterer, H.W. Weber, W. Gawalek, V. Kovylaev, M. Karpets, V. Moshchil, A. Kozyrev, T. Basyuk, X. Chaud, W. Goldacker, V. Sokolovsky, J. Noudem, A. Borimskiy, V. Sverdun, E. Prisyazhnaya, Temperature-pressure induced nano-structural inhomogenities for vortex pinning in bulk MgB_2 of different connectivity, *Physica C* 503, 109-112 (2014).

[3] T.A. Prikhna, M. Eisterer, H.W. Weber, W. Gawalek, V.V. Kovylaev, M.V. Karpets, T.V. Basyuk and V.E. Moshchil, Nanostructural inhomogeneities acting as pinning centers in bulk MgB_2 with low and enhanced grain connectivity, *Supercond. Sci. Technol.* 27(4), 044013 (9pp) (2014).

[4] T. Prikhna, M. Eisterer, W. Gawalek, A. Kozyrev, H.W. Weber, V. Sokolovsky, X. Chaud, J. Noudem, T. Habisreuther, V. Moshchil, M. Karpets, T. Basyuk, V. Kovylaev, J. Dellith, V. Sverdun, R. Kuznietsov, C. Shmidt, T. Vitovetskaya, L. Polikarpova, Synthesis Pressure–Temperature Effect on Pinning in MgB_2-Based Superconductors, *J. Supercond. Nov. Magn.* 26, 1569–1576 (2013).

[5] T. Prikhna, W. Gawalek, M. Eisterer, H. Weber, M. Monastyrov,V. Sokolovsky, J. Noudem, V. Moshchil, M. Karpets, V. Kovylaev, A. Borimskiy, V. Tkach, A. Kozyrev, R. Kuznietsov, J. Dellith, C. Shmidt, D. Litzkendorf, F. Karau, U. Dittrich, M. Tomsic, The effect of high-pressure synthesis on flux pinning in MgB_2-based superconductors, *Physica C: Superconductivity.* 479, 111-114 (2012).

[6] T. Prikhna in: *Structure and properties of bulk MgB_2 // Superconducting Wires: Basics and Applications,* edited by R. Fluekiger, World Scientific Series in Applications of Superconductivity and Related Phenomena., p. 131-157 (668 pp.) (2016).

[7] T.A. Prikhna, V.V. Romaka, A.P. Shapovalov, M. Eisterer, V. Sokolovsky, H.W. Weber, G.E. Grechnev, V.G. Boutko, A.A. Gusev, A.V. Kozyrev, W. Goldacker, V.E. Moshchil, V.B. Sverdun, T. Habisreuther, C. Schmidt, V.V. Kovylaev, V.E. Shaternik, M.V. Karpets, A.V. Shaternik, Structure and Properties of MgB_2 Bulks, Thin Films, and Wires, *IEEE Transactions*

on Applied Superconductivity, 27(4), 1 (2017) DOI: 10.1109/TASC.2016.2638201

[8] E.W. Collings, M.D. Sumption, M. Bhatia, M.A. Susner and S.D. Bohnenstiehl, Prospects for improving the intrinsic and extrinsic properties of magnesium diboride superconducting strands, *Supercond. Sci. Technol.* 21(10) 103001 (2008).

[9] S.X. Dou, O. Shcherbakova, W.K. Yeoh, J.H. Kim, S. Soltanian, X.L. Wang, C. Senatore, R. Flukiger, M. Dhalle, O. Husnjak and E. Babic, Mechanism of Enhancement in Electromagnetic Properties of MgB_2 by nano SiC doping, *Phys. Rev. Lett.,* 98, 097002 (2007) doi: 10.1103/PhysRevLett.98.097002.

[10] J.H. Kim, S. Oh, Y-Uk Heo, S. Hata, H. Kumakura, A. Matsumoto, M. Mitsuhara, S. Choi, Y. Shimada, M. Maeda, J.L. MacManus-Driscoll, S.X. Dou, Microscopic role of carbon on MgB_2 wire for critical current density comparable to NbTi, *NPG Asia Materials* 4, e3 (2012); doi:10.1038/am.2012.3.

[11] X.L. Wang, S. Soltanian, M. James, M.J. Qin, J. Horvat, Q.W. Yao, H.K. Liu and S.X. Dou, Significant enhancement of critical current density and flux pinning in MgB_2 with nano-SiC, Si, and C doping, *Physica C* 408-410, 63-67 (2004).

[12] Y.B. Zhang, X.J. Shan, X.W. Bai, T.Y. Liu, H.M. Zhu and Ch.B. Cai, In situ synthesis and current-carrying characteristics of superconducting MgB_2–B_4C composites with MgB_2 fractions ranging from 18% to 85%, *Supercond. Sci.Technol.,* 25, 095003 (2012).

[13] H. Yamada, N. Uchiyama, A. Matsumoto, H. Kitaguchi and H. Kumakura, The excellent superconducting properties of *in situ* powder-in-tube processed MgB_2 tapes with both ethyltoluene and SiC powder added, *Supercond. Sci. Technol.* 20, L30 (2007).

[14] T. Prikhna, A. Shapovalov, M. Eisterer, V. Shaternik, W. Goldacker, H.W. Weber, V. Moshchil, A. Kozyrev, V. Sverdun, V. Boutko, G. Grechnev, A. Gusev, V. Kovylaev, A. Shaternik, Pinning in high performance MgB_2 thin films and bulks: Role of Mg-B-O nano-scale inhomogeneities, *Physica C: Superconductivity and its applications* 533, 36–39 (2017).

[15] Bean, C. B., Magnetization of high field superconductors, *Rev. Mod. Phys.* 36, 31-36 (1964).

[16] R.G. Parr and W. Yang, Density-Functional Theory of Atoms and Molecules, Oxford Univ. Press, (1989), e-book.

[17] J.P. Perdew, S. Burke, and M. Ernzerhof, Phys. Rev. Lett. 77, 3865 (1996).

[18] ELK, Program package; http://elk.sourceforge.net/.

[19] P. Blaha, K. Schwarz, G.K.H. Madsen *et al.,* WIEN2K, An augmented plane wave + local orbitals program for calculating crystal properties, Techn. Univ., Wien (2001).

[20] J.M. Wills, M. Alouani, P. Andersson, A. Delin, O. Eriksson, and A. Grechnev, Full-Potential Electronic Structure Method. Energy and Force Calculations with Density Functional and Dynamical Mean Field Theory, Berlin: Springer, (2010) e-book.

[21] http://fplmto-rspt.org/

[22] U. von Barth, L. Hedin, A local exchange-correlation potential for the spin polarized case I, *J. Phys. C: Solid State Phys*. 5, 1629, (1972).

[23] W. Kraus, G. Nolze, Powder Cell for Windows. – Berlin (1999).

[24] Rodríguez-Carvajal, J. Recent Developments of the Program FULLPROF, in Commission on Powder Diffraction (IUCr). Newsletter, 26, 12-19 (2001)

[25] T.A. Prikhna, A.P. Shapovalov, G.E. Grechnev, V.G. Boutko, A.A. Gusev, A.V. Kozyrev, M.A. Belogolovskiy, V.E. Moshchil and V.B. Sverdun, Formation of nanostructure of magnesium diboride based materials with high superconducting characteristics, *Low Temperature Physics* 42(5), 486–505 (2016).

[26] R.F. Klie, J.C. Idrobo, N.D. Browning, A. Serquis,Y.T. Zhu, X.Z. Liao, and F.M. Mueller, Observation of coherent oxide precipitates in polycrystalline MgB_2, *Appl. Phys. Lett.* 80, 3970-3972 (2002).

[27] S. Haigh, P. Kovac, T. A. Prikhna, Ya.M. Savchuk, M. Kilburn, C. Salter, J. Hutchison, C. Grovenor, Chemical interactions in Ti doped MgB_2 superconducting bulk samples and wires *Supercond. Sci. Technol.* 18, 1–7 (2005) doi:10.1088/0953-2048/18/0/000

IOP Conf. Series: Materials Science and Engineering **279** (2017) 012024 doi:10.1088/1757-899X/279/1/012024

Pressure-induced property improvement of magnesium diboride wire

F Wan[1], M D Sumption[1], M A Rindfleisch[2] and E W Collings[1]

[1]Center for Superconductor and Magnetic Materials, Department of Materials Science and Engineering, The Ohio State University, Columbus, OH 43210, USA.
[2]Hyper Tech Research Inc, Columbus, OH 43228, USA

Email Address: wan.108@osu.edu

Abstract. The 4.2 K non-barrier transport J_c values of the powder-in-tube (PIT) *in-situ* MgB_2 wires have been enhanced by cold high pressure densification (CHPD). With respect to the control wire, the 1.5 GPa pressure induced the J_c improvement of the wire at 4.2 K and 10 T by 25 %. The J_c enhancement induced by the CHPD may result from two aspects. First, the CHPD resulted in the reduction of the transverse MgB_2 core area. Second, the grain connectivity of the PIT *in-situ* MgB_2 wire was improved by the CHPD, which is reflected by the reduced porosity.

1. Introduction

Compacted-powder pellets and the cores of powder-in-tube (PIT) wires are inherently porous. Thus the packing densities of sintered *ex-situ* processed wires, based geometrically on the packing density of a uniform-sized starting powder fill, would be around 65% (although powders of mixed sizes will fill space more efficiently). The expected density of *in-situ*-processed material is complicated by shrinkage associated with the $Mg+2B \rightarrow MgB_2$ reaction. As a result it can be shown that a mixture of equi-sized Mg and B particles will after reaction lead to an MgB_2 density of 49%, although the use of coarse Mg particles could raise the final density to 62% [1]. Porosity in the starting powder fill (and staring porosity plus reaction in the case of *in-situ* material) impedes current flow and prevents the J_c of the bulk or wire from attaining its intrinsic intragranular value. Current transport between grains may be reduced by the presence of insulating partial coatings [2]. These effects can combine to yield very low connectivities [3], for example 20~40% [1] which would offer the possibility of transport J_c increases by factors of 5~2.5. The problems of porosity and intergranular coatings call for separate solutions. Many researchers have addressed the former and, as reviewed below, have attempted to squeeze out porosity by applying high pressures to bulk and wire samples both during the reaction heat treatment as well as prior to it.

Shields *et al.* [4] reported that hot isostatic pressing (HIP) induced the magnetic critical current densities as high as 1.3×10^6 A/cm^2 at 0 T and 9.3×10^5 A/cm^2 at 1 T of the MgB_2 bulks at 100 MPa and 950 °C, 20 K. Serquis *et al.* [5] presented that the J_c values of HIP – processed PIT MgB_2 wires reached 10^6 A/cm^2 at 5 K and 0 T and the order of 10^4 A/cm^2 at 1.5 T and 26.5 K. Susner *et al.* [6] reported that a MgB_2 monofilamentary strand hot-pressed by 458 MPa achieved the J_c value of 1.81×10^5 A/cm^2 at 4.2 K and 5 T, with respect to 0.86×10^5 A/cm^2 of the unpressed strand.

Cold high pressure densification (CHPD) has been considered as a method for improving the grain connectivity of the PIT *in-situ* MgB_2 wire as evidenced by the enhancements of relative mass density, electrical resistivity and transport J_c [7, 8]. Flükiger and Hossain [7] indicated that the CHPD induced the J_c improvement by 53 % at 4.2 K and 10 T for a binary Fe/MgB_2 round wire. Flükiger *et al.* [9] also presented that the J_c value of monofilamentary *in situ* MgB_2 wire doped with $C_4H_6O_5$, heat-treated with 4.0 h/600 °C, and cold-densified by the pressure of 1.5 GPa reached about 4.3 × 10^4 A/cm^2 at 4.2 K and 10 T, with respect to 2.2 × 10^4 A/cm^2 of the control wire. In this paper, the influence of the CHPD on the current-carrying capacity of the 2 % - C doped PIT *in-situ* MgB_2 wires was investigated. The MgB_2 wire heat-treated by 1.0 h/675 °C and cold-densified by the pressure of 1.5 GPa was successfully prepared with 4.2 K non-barrier transport J_c value of 3.6 × 10^4 A/cm^2 at 10 T and the order of 10^5 A/cm^2 at 7 T.

2. Experimental

2.1 Sample Preparation

Several coils of 2 % carbon-doped unreacted PIT *in-situ* MgB_2 wires were manufactured and provided by the Hyper Tech Research, Inc. The outer sheath and chemical barrier of the MgB_2 wire are copper and niobium, respectively. The outer diameter of the MgB_2 wire is 0.834 mm; the fill factor of MgB_2 core is 25.7 % and the ratio of Mg:B is 1:2.

The wires to be pressed were held between the jaws of a six-inch Kurt vice. The jaws were faced with blocks of heat treated M2 steel and plates of tungsten carbide as shown in Figure 1. Held between this assembly was the wire to be pressed and upper and lower blades of tool steel 0.75 mm thick. Also located between the jaws was a 10 ton hydraulic short-body ram. Vertical pressure was applied to the tool-steel blades, and hence the subject wire, by a 20-ton hydraulic ram. Vertical pressures of 1.0, 1.5, and 2.0 GPa were applied to the wires which at the same time were exposed to 2.0 GPa horizontal pressure, Table 1.

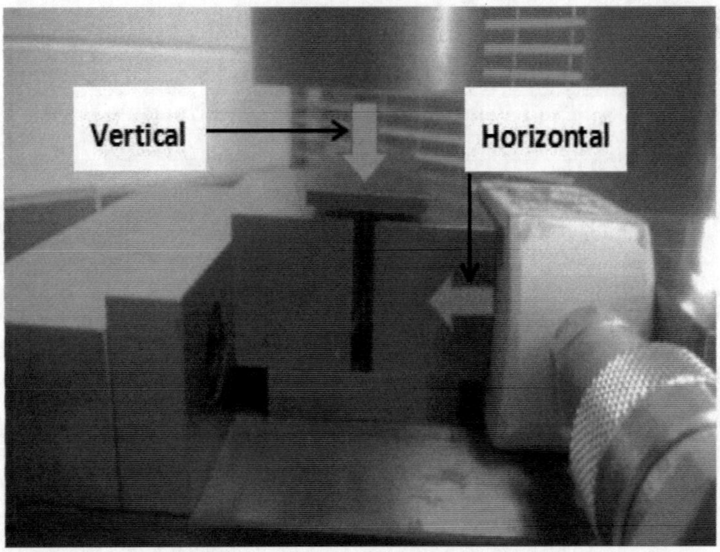

Figure 1. The assembly used for applying pressures on the MgB_2 wires. The directions of vertical pressure and horizontal pressure were defined.

During vertical pressing there is a tendency for the wires to expand horizontally and to extrude past the blades as shown in Figure 2.

A fixed 2.0 GPa horizontal pressure continually applied prevented this from happening and enabled clean rectangular cross sections to be maintained, Figure 3.

All the wires were encapsulated in quartz tubes under 200 torr Ar and heat treated for 1h/675°C.

Figure 2. The extrusion of the wire vertically cold-pressed by 2.0 GPa.

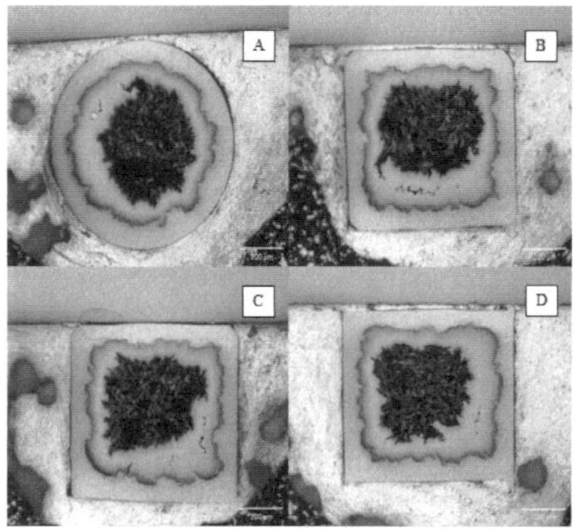

Figure 3. Transverse cross sectional areas of the MgB_2 wires (A) W00 (B) W10 (C) W15 (D) W20.

Table 1: Pressure schedule for MgB_2 wire

Sample No.	Vertical Pressure (GPa)	Horizontal Pressure (GPa)
W00	0.0	0.0
W10	1.0	2.0
W15	1.5	2.0
W20	2.0	2.0

2.2 Measurement

The transport critical current densities of the wires were measured as a function of applied magnetic fields up to 12 T at 4.2 K in a pool of boiling liquid helium by the four-probe technique. The wires were 5 cm long, the gauge length between voltage taps was 5 mm, and the voltage criterion was 1.0 μV/cm.

3. Result and Discussion

3.1. Transport Result

Figure 4 shows the 4.2 K non-barrier transport critical current densities of the cold-pressed and control wires as a function of applied magnetic field up to 12 T. All cold-densified wires have J_c values above 3.0×10^4 A/cm^2 at 10 T. Compared with control wire, the 4.2 K non-barrier transport J_c values of W10

IOP Conf. Series: Materials Science and Engineering **279** (2017) 012024 doi:10.1088/1757-899X/279/1/012024

(1.0 GPa cold-pressed wire) and W15 (1.5 GPa cold-pressed wire) have been enhanced at all applied magnetic fields; for example, the 4.2 K, 10 T non-barrier transport J_c values of the W10 and W15 are 15 % and 25 % higher than that of W00. The best J_c values were obtained by the wire W15 at all applied magnetic fields and are found to be 3.6×10^4 A/cm^2 at 10 T and 1.1×10^5 A/cm^2 at 7 T.

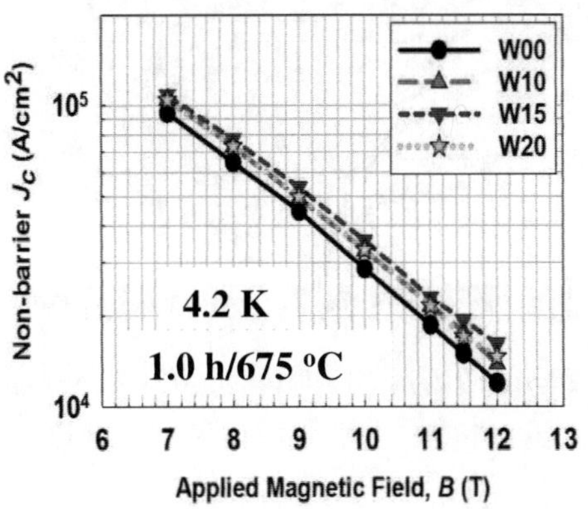

Figure 4. Non-barrier transport critical current density J_c of the MgB_2 wires as a function of applied magnetic field at 4.2 K.

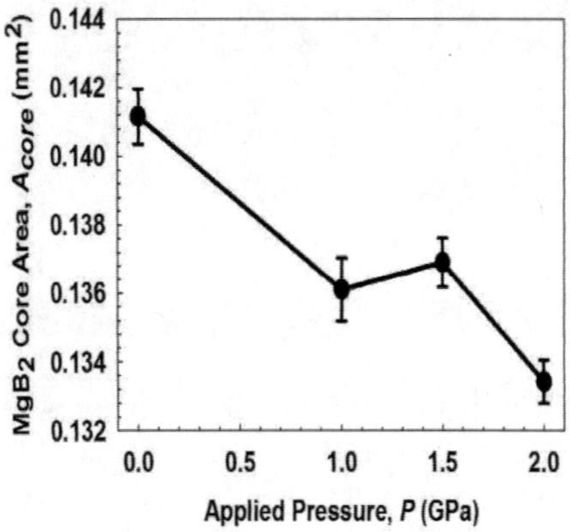

Figure 5. Average transverse MgB_2 core area of the PIT *in-situ* MgB_2 wire with error bars versus applied pressure.

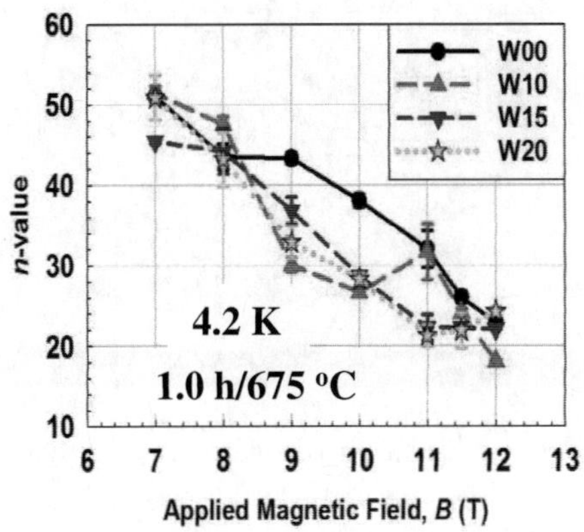

Figure 6. Average *n*-values of the MgB_2 wires with error bars as a function of applied magnetic field at 4.2 K.

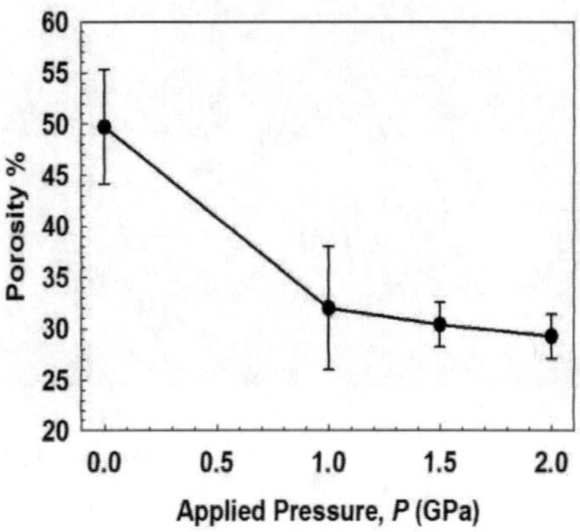

Figure 7. Average porosity of the PIT *in-situ* MgB_2 wire with error bar versus applied pressure.

Figure 8. The SEM image in SE mode of the transverse MgB_2 core area of the W00. Two large voids were marked in this figure.

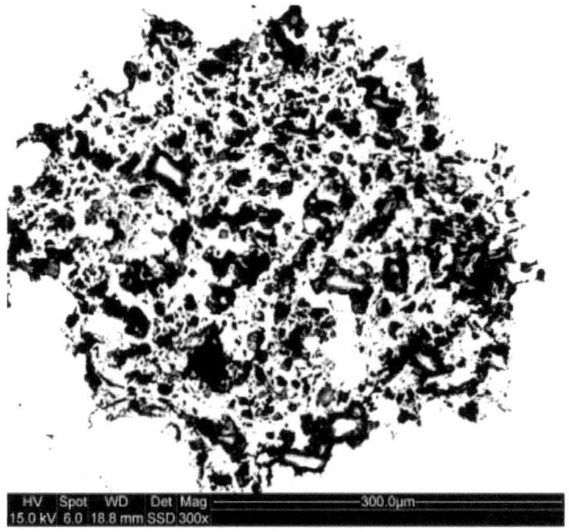

Figure 9. The SEM image in BSE mode of the transverse MgB_2 core area of the W00.

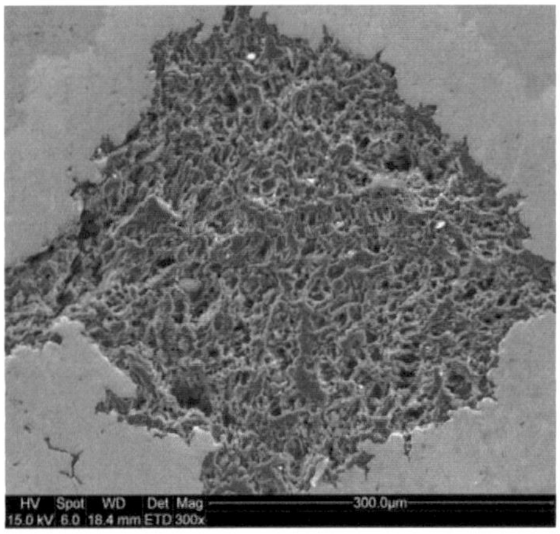

Figure 10. The SEM image in SE mode of the transverse MgB_2 core area of the W15.

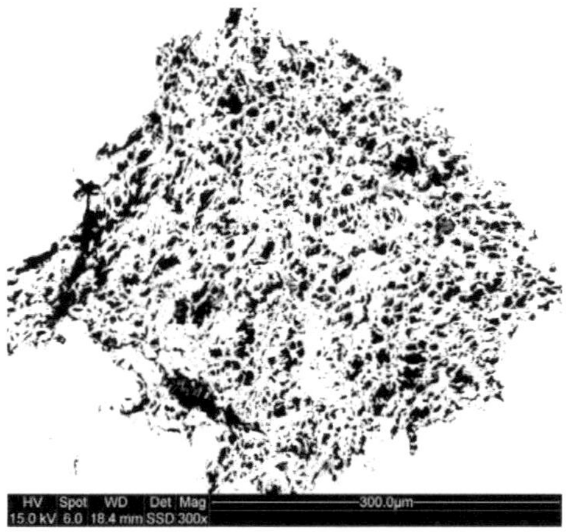

Figure 11. The SEM image in BSE mode of the transverse MgB_2 core area of the W15.

The J_c enhancements of the cold-pressed MgB_2 wires are partially due to the reduction of the transverse MgB_2 core area induced by the CHPD as shown in Figure 5. The pressure of 2.0 GPa led to the largest reduction of the transverse MgB_2 and was found to be 5.5 %. If the reduced transverse MgB_2 core area is the unique factor to inducing the J_c enhancements of the MgB_2 wires, the J_c of W20 will be improved by a factor no higher than 5.8 %. However, the J_c enhancement of W20 was at least 10.4 % at the range of 7 to 12 T indicating that the CHPD has the potential of improving other properties of the PIT *in-situ* MgB_2 wire as well as reducing transverse MgB_2 core area leading to the enhancements of its J_c values. It can be seen that the 4.2 K non-barrier transport J_c values of W20 were lower than those of the W15 at all applied magnetic fields suggesting the too large pressure might introduce defects in the MgB_2 wire and therefore induced the J_c degradation of the MgB_2 wire. W20 had the 4.2 K non-barrier transport J_c values of 3.4×10^4 A/cm^2 and 1.1×10^5 A/cm^2 at 10 T and 7 T, respectively, which were still higher than those of W00. Figure 6 shows the 4.2 K n-values of the cold-densified and control wires as a function of applied magnetic field up to 12 T. The largest error in the measurement of n-value is 11%. At 10 T the cold-densified wire has a 34% decrease in n-value, but the n-value of W00 is close to those of the cold-densified wires at 8 T. The field dependence of n-value suggests that the CHPD might not induce higher level of extrinsic defects in PIT *in-situ* MgB_2 wire.

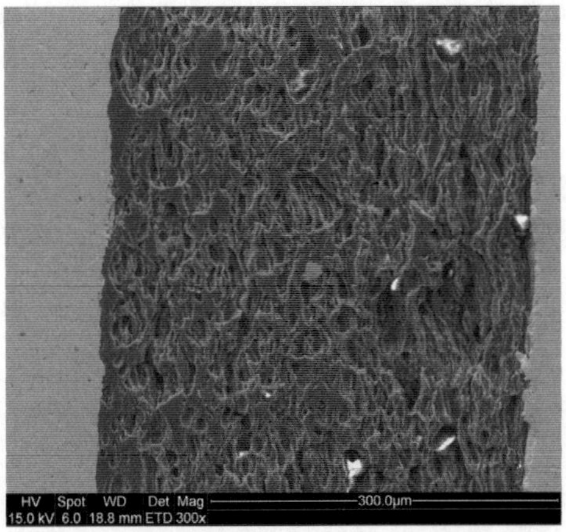

Figure 12. The longitudinal cross section area of W00.

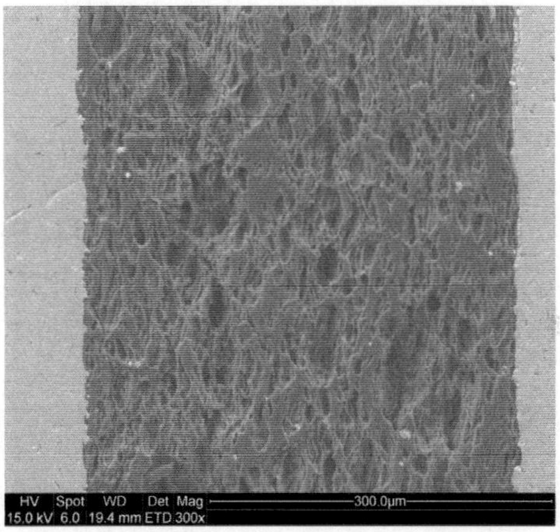

Figure 13. The longitudinal cross section area of W15.

3.2. Microstructures

Figure 7 shows the average porosity of the wire versus the applied pressure. It can be seen that the wire processed by the CHPD have lower porosity than the control wire. Figure 8 - Figure 11 demonstrate the SEM images with secondary - electrons (SE) and back – scattered electrons (BSE) modes of the transverse MgB_2 core area of W00 and W15. In the BSE SEM image as in Figure 9 and Figure 11, the dark parts represent the voids existing in the wires and light parts represent the MgB_2 stringers in the wires. The SEM images in BSE mode were taken under same brightness and contrast. Consequently, the porosity of the wire can be obtained by calculating the area fraction of the dark parts in the BSE mode. The SEM images in SE mode as shown in Figure 8 and Figure 10, there exists some large voids in W00, but no large void was observed in W15. Based on these results, it can be concluded the CHPD has the ability to squeeze out the porosity of the PIT *in-situ* MgB_2 wire suggesting the enhanced grain connectivity of the MgB_2 wire. On the other hand, the CHPD has the limited ability to further reduce the porosity of the wire when the applied pressure was larger than 1.0 GPa. For example, the reduction of the average porosity of the MgB_2 wire was only 8.5 % when the applied pressure was increased from 1.0 GPa to 2.0 GPa; however, it was reduced by 35.6 % when the applied pressure was increased from 0.0 GPa to 1.0 GPa. The longitudinal cross section areas of W00 and W15 are shown in Figure 12 and Figure 13. Susner *et al.* [10] described that continuous MgB_2 stringers in PIT *in-situ* wire were aligned along drawing direction and elongated voids which were left behind by melted Mg powder. It can be seen that the CHPD does not alter the morphology of the continuous MgB_2 stringers in the PIT *in-situ* MgB_2 wire.

4. Conclusions

Cold high pressure densification can improve the 4.2 K non-barrier transport J_c of the PIT *in-situ* MgB_2 wire in present paper. The best 4.2 K non-barrier transport J_c performance was obtained by W15 at all applied magnetic fields. The *n*-values of all cold-densified wires are above 30 when the applied magnetic field is below 9 T. The pressure dependence of the porosity of the PIT *in-situ* MgB_2 wire was discussed. The reduced porosity of the MgB_2 wire processed by the CHPD reflects that the J_c enhancement induced by the CHPD might result from the improved grain connectivity of the PIT *in-situ* MgB_2 wires.

5. References

[1] Collings E W Sumption M D Bhatia M Susner M A and Bohnenstiehl S D 2008 *Supercond. Sci. Technol.* **21** 103001
[2] Jiang J Senkowicz B J Larbalestier D C and Hellstrom E E 2006 *Supercond. Sci. Technol.* **19** 33-6.
[3] Yamamoto A Shimoyama J Kishi K and Matsushita T 2007 *Supercond. Sci. Technol.* **20** 658-66.
[4] Shields T C Kawano K Holdom D L and Abell J S 2002 *Supercond. Sci. Technol.* **15** 202-5.
[5] Serquis A Civale L Hammon D L Liao X Z Coulter J Y Zhu Y T Jaime M Peterson D E Mueller F M Nesterenko V F and Gu Y 2003 *Appl. Phys. Lett.* **82** 2847-9.
[6] Flükiger R Hossain M S A and Senatore C 2009 *Supercond. Sci. Technol.* **2** 085002.
[7] Susner M A Sumption M D Bhatia M Tomsic M J Rindfleisch M A Collings E W 2008 *AIP Conference Proceedings* **986** 375.
[8] Hossain M S A Senatore C Flükiger R Rindfleisch M A Tomsic M J Kim J H and Dou S X 2009 *Supercond. Sci. Technol.* **22** 095004
[9] Flükiger R Hossain M S A Kulich M and Senatore C 2011 *Physica C.* **471** 1119-23.
[10] Susner M A Daniels T W Sumption M D Rindfleisch M A Thong C J and Collings E W 2012 *Supercond. Sci. Technol.* **25** 065002.

IOP Conf. Series: Materials Science and Engineering **279** (2017) 012025 doi:10.1088/1757-899X/279/1/012025

Influence of oxygen doping on critical fields in MgB_2 bulk superconductors

D Zhang[1], X Li[1], F Wan[1], C J Thong[2], M A Rindfleisch[2], M J Tomsic[2], M D Sumption[1], E W Collings[1]

[1] Center for Superconducting and Magnetic Materials, Department of Materials Science and Engineering, The Ohio State University, Columbus, OH 43210, USA

[2] Hyper Tech Research Incorporated, Columbus, OH 43212, USA

E-mail: zhang.5952@osu.edu

Abstract. In this work we studied the influence of SnO_2 doping on the critical fields and temperatures of MgB_2 bulk samples. Bulk samples were made by mixing *ex-situ* MgB_2 powder with 5 wt% SnO_2 powder and then pressing the mixed powders into pellets using a pressure of 2000 psi. The bulk pellets were sintered at 900°C in a furnace under flowing Ar. The samples were quenched to room temperature after dwelling at 900°C for 5h. XRD, SEM, and magnetic measurements were made on doped and control samples. XRD showed a decomposition of the SnO_2 and very slight reductions in the a-axis and c-axis lattice parameters of the MgB_2 phase. *M-T* (Magnetization-Temperature) curves showed a decrease in B_{c2} of approximately 1 T in the temperature range of 24 K - 39 K with SnO_2 additions as compared to the control samples.

1. Introduction

MgB_2 superconducting materials are of significant interest because of their relatively high critical temperature (~ 39 K), low cost, and simple two-gap nature. MgB_2 has strong interest for low to moderate field (0-3 T) applications in the 4-30 K temperature range, including MRI [1,2], superconducting motors and generators [3,4], and fault current limiters [5]. However, high field applications are limited by its relatively low upper critical field (B_{C2}) and low irreversible field (B_{irr}) which leads to low critical current densities (J_c) at high magnetic fields as compared to Nb_3Sn or YBCO. Much research has focused on the chemical doping of MgB_2 aimed at increasing either critical fields, critical currents, or both [6-13]. The most successful dopant has been C, added directly or as a compound, which releases C during reaction, which increases B_{C2} [12-13]. A few other dopants have been investigated which may add small amounts of flux pinning [14-15], but the increases in J_c were marginal since most dopants (except C) tend to segregate at grain boundaries. Some promising work on Dy_2O_3 has shown increases in J_c, although the mechanism is not fully clear [16-19]. However, a certain level of nanoscale intragrain defects were observed [17] in Dy_2O_3 doped MgB_2. In any case, the critical field increases seen in bulk and wire samples, while significant, have not reached the levels seen in earlier doped thin film work where critical fields of around 60 T and higher were seen at 4.2 K [7,8]. Oxygen doping appears to be part of the mechanism for the critical field increases seen in thin films, but so far, such increases have not been substantially translated to wires or bulks. Dou et al [6] used Sb_2O_3 powders as additions to in-situ route MgB_2 in an attempt to increase the critical fields, and in fact modest increases (about 1.3 T at 23.75 K) were seen in B_{C2}. In this work, we attempt to dope bulk samples with oxygen by including a SnO_2 powder additive to MgB_2 *ex-situ* powders. SnO_2 is an oxide which becomes unstable above 500 °C, and thus we will use this as a dopant in an attempt to provide an internal oxidation source for our MgB_2 *ex-situ* powder samples. We chose SnO_2 not only because of its above-noted decomposition, but also because Sn additions have been shown to be relatively benign additions to MgB_2 [20-21]. Below we carry out an initial study which attempts to increase the critical fields via oxygen doping using SnO_2 as the carrier.

2. Experimental details

Two MgB_2 samples were prepared using ex-situ powders which were pressed into pellets and then heat treated. One was made using pure pre-reacted MgB_2 powders (99% pure). A second, doped sample used pre-reacted MgB_2 powders (99% pure) mixed with 5wt% SnO_2 powders (99% pure) in order to investigate the influence of SnO_2 additions on the critical fields of MgB_2 bulks. First, the powders were mixed. Then the mixed powders were pressed into pellets (8.8 mm OD and 5 mm in height) under a load of 2000 psi at room temperature. The pellets were then sintered. The temperature was ramped up at 5°C/h, then held at 900°C under flowing Ar, and then the sample holder was pulled directly out of the furnace.

Microstructures of the MgB_2 pellets (after powderization) were characterized by scanning electron microscopy (SEM) using an FEI Sirion FEG in secondary mode. Powder X-ray diffraction (XRD) was performed using a Rigaku MiniFlex 600. XRD patterns were collected from 20 degrees to 80 degrees with a rate of 0.02 degree/step. The phases in the materials were identified, and lattice parameters were obtained for the MgB_2 phase.

Magnetization-Temperature (*M-T*) curves were measured on the samples from 5 K to 45 K using the vibrating sample magnetometer (VSM) mode of a Quantum Design Physical Property Measurement System (PPMS). In order to perform the measurements, the sample was first taken above T_c, a field of 0 T was applied, and the sample was cooled in zero field to 5 K. A field of 0.1 to 5 T was then applied, and the temperature slowly increased (5 K/h) until 45 K was reached; this constituted the zero field cooled measurement (ZFC). Once 45 K was reached, the field was left on while the temperature was again decreased through the transition and finally to 5 K; this was the field cooled (FC) measurement. The critical temperature at a given field was defined to be the point at which the deviation from the normal state magnetization (taken as a fraction of the total magnetization change during the ZFC measurement) was 5%. The irreversibility temperature was defined as the point at which the ZFC and FC measurements at a given field diverged [22]. Since T_c (T_{irr}) at a given field can be interpreted as B_{c2} (B_{irr}) at a given temperature, these results are presented as B_{c2} and B_{irr} vs T below.

3. Results and Discussions

3.1. Microstructure of the MgB_2 pellets

Figure 1 shows the SEM (secondary electron, SE mode) of the undoped sample (left) and the SE-SEM of the 5wt% SnO_2 doped sample (right). Both samples have been ground into fine powder, after which the powder was coated with a layer of gold for the SEM since there was charging effect due to bad connectivity in the samples. The grains are roughly equiaxed and SnO_2 additions did not make an appreciable change in the sample morphology. Furthermore, no noticeable secondary phases were seen from SEM images.

Figure 1 SEM images of undoped MgB_2 bulk sample (a) and the SnO_2 doped MgB_2 bulk sample (b). Pellet samples were powderized and gold coated before SEM.

3.2. Structure and phase analysis

Figure 2 shows the powder x-ray diffraction patters for the SnO_2 doped MgB_2 sample and the control sample. It can be seen that in both samples MgB_2 is a major phase, and that some MgO was also formed during the sintering process. From the doped sample pattern, we do not see SnO_2, but we do see free Sn, indicating the expected decomposition of SnO_2 to Sn and oxygen. Further, the active Mg reacted with Sn to form the intermetallic compound Mg_2Sn.

Based on the XRD data, lattice parameters for these samples were extracted. The lattice parameters of the undoped sample are a= 3.086 Å and c=3.525 Å, while those of the doped sample are: a=3.085 Å, and c=3.524 Å. There is no discernable lattice shift in a-axis and c-axis lattice parameters in MgB2 phases for the doped sample, which is different from Dou's work [3].

Our XRD results, while not speaking directly to oxygen doping per se, do allow us to look for MgB_2 lattice parameter shifts and the presence of gross second phases. Also, it allows us to look for presence of SnO_2 as well as free Sn – allowing us to see if SnO_2 has decomposed.

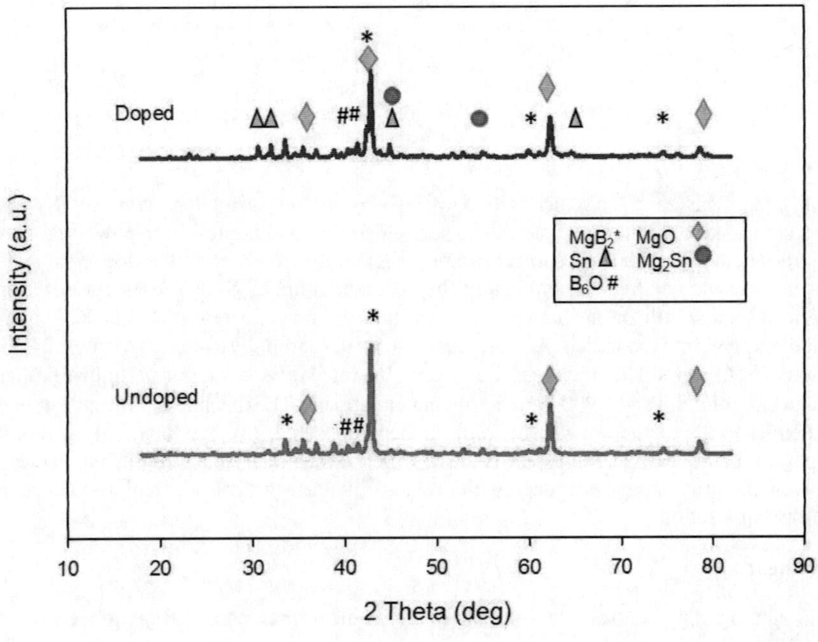

Figure 2 XRD patterns for 5wt% SnO_2 doped MgB_2 and control MgB_2 samples.

3.3. Magnetic response of the MgB_2 pellets

B_{irr} and B_{c2} as a function of temperature for the 5 wt% SnO_2 doped MgB_2 and the control MgB_2 bulk samples are shown in Figure 3. From the graph, we can see that in the temperature window between 24 K and 39 K, the doped sample has lower upper critical fields and irreversible fields than the control. In the 24-39 K window, there is approximately 0.5 T reduction in B_{irr} and B_{c2} for the doped sample as compared to the undoped sample. This result is different from Dou's result [6], where Sb_2O_3 additions increased the B_{C2} from 4.1 T (undoped sample) to 5 T at 29.2 K. Thus, under our specific preparation conditions we observed a decrease in the critical fields (within the temperature window we explored), rather than an increase with SnO_2 doping. This work only explored one doping level and one reaction heat treatment (HT), so we may not have found the optimum conditions. Also, we used an ex-situ route, rather than an in-situ one, and this may be an important factor.

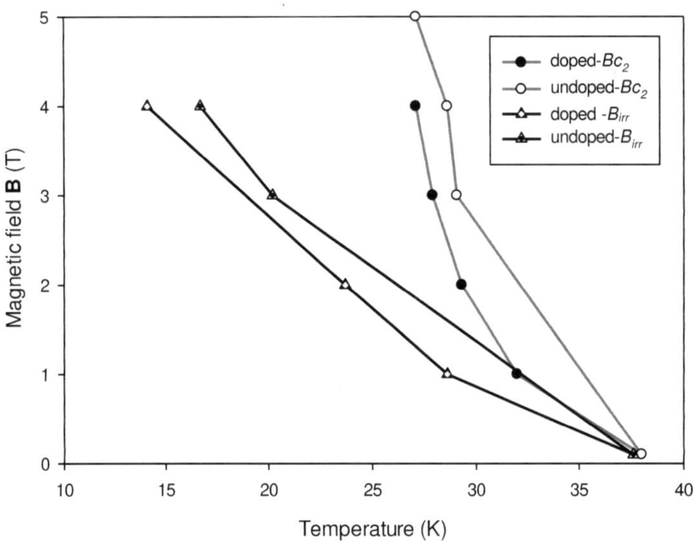

Figure 3 Temperature dependence of B_{c2} and B_{irr} of pure MgB_2 sample and SnO_2 doped sample.

4. Conclusion

An MgB_2 bulk sample has been fabricated with 5 wt% SnO_2 doping using an *ex-situ* processing route, and it has been compared to a control sample. SEM performed on the samples after powderizing shows similar grain structure in the doped and control sample. XRD measurements on the doped sample show no SnO_2, but the presence of free Sn, indicating the decomposition of SnO_2 during the reaction HT. XRD measurements show little or no change in c-axis lattice parameter, suggesting a lack of significant incorporation of oxygen into the lattice. A slight decrease in B_{c2} and B_{irr} was observed in the 24 K-39 K range. Addition of SnO_2 has slightly decreased B_{c2} and B_{irr} of MgB_2 given our particular preparation conditions of an ex-situ route and a 900 ^{o}C reaction heat treatment. The mechanism is unclear, and may be connected either to the oxygen or the Sn from the decomposed SnO_2. We have only explored one reaction temperature and one reaction type (ex-situ), which leaves a large amount of parameter space to explore. Given the promising results seen with previous in-situ work [6], we think oxygen additions are worth further exploration.

Acknowledgement

This work was supported by the National Institute of Biomedical Imaging and Bioengineering, under grant R01EB018363.

References

[1] Tomsic M, Rindfleisch M, Yue J, McFadden K, Phillips J, Sumption MD, Bhatia M, Bohnenstiehl S, Collings E W 2007 *International journal of applied ceramic technology* **4** 250-259

[2] Sumption M D, Bhatia M, Buta F, Bohnenstiehl S, Tomsic M, Rindfleisch M, Yue J, Phillips J, Kawabata S, Collings E W 2005 *Supercond. Sci. Technol.* **18 (7)** 961

[3] Nakamura T, Yamada Y, Nishio H, Kajikawa K, Sugano M, Amemiya N, Wakuda T, Takahashi M, Okada M 2012 *Supercond. Sci. Technol.* **25 (1)** 014004

[4] Marino I, Pujana A, Sarmiento G, Sanz S, Merino JM, Tropeano M, Sun J, Canosa T 2015 *Supercond. Sci. Technol.* **29 (2)** 024005

[5] Ye L, Majoros M, Campbell A M, Coombs T et al 2007 *Supercond. Sci. Technol.* **20 (7)** 621

[6] Zhang Y, Dou S X 2011 *J. Mater. Res.* **26** **(21)** 2701-2706
[7] Gurevich A, Patnaik S, Braccini V, Kim KH, Mielke C, Song X, Cooley LD, Bu SD, Kim DM, Choi J H, Belenky L J 2003 *Supercond. Sci. Technol.* **17** **(2)** 278
[8] Eom C B, Lee M K, Choi J H, Belenky L J 2001 *Nature* **411** **(6837)** 558
[9] Sumption M D, Bhatia M, Rindfleisch M, Tomsic M, Soltanian S, Dou S X, Collings E W 2005 *Appl. Phys. Lett.* **86** **(9)** 092507
[10] Dou S X, Soltanian S, Horvat J, Wang X L, Zhou S H, Ionescu M, Liu H K, Munroe P, Tomsic M 2002 *Appl. Phys. Lett.* **81** **(18)** 3419-3421
[11] Yeoh W K, Kim J H, Horvat J, Xu X, Qin M J, Dou S X, Jiang C H, Nakane T, Kumakura H, Munroe P 2006 *Supercond. Sci. Technol.* **19** **(6)** 596
[12] Li G Z, Sumption M D, Zwayer J B, Susner M A et al 2013 *Supercond. Sci. Technol.* **26** **(9)** 095007
[13] Susner M A, Yang Y, Sumption M D et al 2010 *Supercond. Sci. Technol.* **24** **(1)** 012001
[14] Prokhorov V G, Svetchnikov V L, Park J S et al 2009 *Supercond. Sci. Technol.* **22** **(4)** 045027
[15] Shcherbakova O V, Pan AV, Wang J L et al 2007 *Supercond. Sci. Technol.* **21** **(1)** 015005
[16] Mikheenko P, Chen S K, MacManus-Driscoll J L 2007 *Appl. Phys. Lett.* **91** 202508
[17] Li G Z, Sumption M D, Rindfleisch M A, Thong C J, Tomsic M J, Collings E W 2014 *Appl. Phys. Lett.* **105** **(11)** 112603
[18] Yang Y, Sumption M D, Collings E W 2016 *Scientific reports* **6** 29306
[19] Brunner B, Rosová A, Kováč P, Reissner M, Dobročka E 2016 *Supercond. Sci. Technol.* **30** **(2)** 025004.
[20] Akamaru S, Ishikawa F, Nishimura K, Abe T, Matsuyama M 2013 *Materials Transactions* **54** **(12)** 2258-2264
[21] Zongqing M et al 2010 *Supercond. Sci. Technol.* **23** 025005
[22] Gurevich A 2003 *Phys. Rev. B* **67** **(18)** 184515

IOP Conf. Series: Materials Science and Engineering **279** (2017) 012026 doi:10.1088/1757-899X/279/1/012026

AC Loss Analysis of MgB2-Based Fully Superconducting Machines

M Feddersen[1], K S Haran[2], and F Berg[3]

[1] Zunum Aero, 811 Kirkland Ave Ste 201, Kirkland, WA 98033
[2] University of Illinois, Urbana-Champaign, 306 N Wright St, Urbana, IL 61801
[3] Airbus Central R&T, Willy-Messerschmitt-Straße 1, 82024, Taufkirchen, Munich, Germany

mattf@zunum.aero, kharan@illinois.edu

Abstract.
Superconducting electric machines have shown potential for significant increase in power density, making them attractive for size and weight sensitive applications such as offshore wind generation, marine propulsion, and hybrid-electric aircraft propulsion. Superconductors exhibit no loss under dc conditions, though ac current and field produce considerable losses due to hysteresis, eddy currents, and coupling mechanisms. For this reason, many present machines are designed to be partially superconducting, meaning that the dc field components are superconducting while the ac armature coils are conventional conductors. Fully superconducting designs can provide increases in power density with significantly higher armature current; however, a good estimate of ac losses is required to determine the feasibility under the machines intended operating conditions. This paper aims to characterize the expected losses in a fully superconducting machine targeted towards aircraft, based on an actively-shielded, partially superconducting machine from prior work. Various factors are examined such as magnet strength, operating frequency, and machine load to produce a model for the loss in the superconducting components of the machine. This model is then used to optimize the design of the machine for minimal ac loss while maximizing power density. Important observations from the study are discussed.

1. Introduction

Superconducting (SC) machines show promise for significant increases in specific power over conventional designs. SC machines have been proposed and demonstrated for offshore wind and ship propulsion [1–3]; the dramatic increase in specific power also enables weight-sensitive transportation applications, such as electric aircraft, to be considered [4].

Most SC machines developed to date are partially superconducting where some components, typically armature coils, remain conventional (copper or aluminum). Fully SC machines, which have both the field and armature components built from superconductors, can further improve specific power if the increased losses at cryogenic temperatures can be accommodated. The AC losses introduced in fully SC machines are particularly difficult to manage, and these losses are compounded in the high frequency cases present in many electric machines. As a result, fully SC machines are known to be well suited for low frequency

IOP Conf. Series: Materials Science and Engineering **279** (2017) 012026 doi:10.1088/1757-899X/279/1/012026

applications such as offshore wind, however in higher speed uses the decision is not as clear. Direct drive turbo-electric aircraft or distributed propulsion aircraft will require machine speeds upwards of 1000rpm, which may or may not be feasible with fully SC machines given the current state of superconductor technology.

The aim of this paper is to determine the frequency and speed range at which fully SC machines offer better performance than similarly sized partial SC machines based on available conductors. This is targeted towards the eventual adoption of superconducting machines in aircraft, which require higher operating speeds and specific power. This analysis is performed through a combination of electromagnetic finite element analysis (FEA) and AC loss calculations.

2. Aircraft application background

There has been significant recent interest in the application of SC technology to future electric aircraft, with a number of publications by major aerospace organisations signaling engagement with the subject [5, 6]. This is motivated by a new interest in the aerospace industry in the electrification of flight. Radical, energy-saving designs can be enabled by electric propulsion, provided very lightweight and efficient electrical components are available.

Early studies on the system-level application of SC machines and propulsive power distribution by Brown [7] and Berg et al. [8] impress the need for ambitious power densities of over 20 kW/kg, which may require fully SC machines to meet. However, the final choice of using fully SC machines will be subject of an aircraft-level optimum taking, among others, the following points into consideration:

A) The mass, volume and power demand of the on-board cryocooling system is as crucial to the aircraft feasibility as the machine performance. Thus, the use of fully SC machines in aircraft will be subject to the losses caused in the machine, the operating temperature required and the cryo-cooling technology available.

B) Fully superconducting machines may help provide high power density at lower speeds, allowing direct drive of propellers without the need for gearboxes, which have additional weight, cooling, efficiency and cost implications.

C) For aircraft system integration, an SC armature can have a major benefit of allowing connection directly to SC cables without the use of heavy current leads and further heat ingress.

3. Machine design background

A general equation that loosely describes the power produced by an electric machine is

$$P = knABD^2L \tag{1}$$

where k is a design constant, A is the electric loading, B is the magnetic loading, D is the machine radial diameter, L is the machine active length or depth, and n is the rotating speed.

For a given machine volume, improvements in power density can be achieved either electromagnetically (increase A and/or B in Eq. 1 or with machine speed (n).

Superconducting machines achieve higher power for a given size by increasing electromagnetic loading with the high current capabilities of SC wire. Magnetic loading can be improved by replacing permanent magnets or copper field coils with SC coils, which can produce magnetic flux density on the order of 6-10 T [9], and electric loading can potentially be increased by replacing traditional copper armature coils with SC wire.

In this paper, SC machines will be categorized in two groups. The first group, partial SC machines, has a superconducting field component to increase magnetic loading, but the three-phase armature in the machine is still made with copper or other normally conducting materials. In the second group are fully SC machines, which have superconducting wire in both the field and the armature components to increase both the magnetic and electric loading.

Partial SC machines have demonstrated increases in specific power with newer designs proposing upwards of 40 kW/kg. Despite this, there are still significant challenges in SC wire manufacturing, machine cooling, and torque tube design [10].

4. AC Loss

In the fully superconducting machine topology described in this paper, the SC armature coils will carry time-varying current while being exposed to a rotating magnetic field with some phase delay. These conditions will generate heat within the armature coils, a phenomenon known as superconductor AC loss.

4.1. Main types of loss

Loss due to alternating transport current (without changing external field) has been analyzed by Norris and is used extensively to predict SC behavior [11]. Similarly, loss due to a changing external magnetic field (without alternating current) is covered by Carr and Wilson for both bulk superconductors and twisted wire [12, 13]. Both of these losses are characterized as hysteresis losses.

In a practical superconducting wire, SC filaments are embedded in a normally conductive matrix. The matrix will create eddy currents when exposed to alternating fields, which will create loss proportional to the resistivity of the matrix and the inductance of the twisted wire. This will be referred to as a collective coupling loss.

All of these losses can be calculated analytically for SC wires under time-varying magnetic fields using Equations 2 and 3 [13]. Each loss component Q is calculated in Joules per cycle per meter of wire; Q_h is the loss due to hysteresis and Q_c is the coupling loss.

$$Q_h = \frac{B_m^2}{2\mu_0}\frac{\lambda}{\omega^2\tau_\alpha^2+1}\Gamma\Big(\frac{\beta}{(\omega^2\tau_\alpha^2+1)^{\frac{1}{2}}}\Big) \tag{2}$$

$$Q_c = \frac{B_m^2}{2\mu_0}\frac{\alpha\pi\omega\tau_\alpha}{(\omega^2\tau_\alpha^2+1)} \tag{3}$$

where B_m is the peak-to-peak amplitude of the magnetic flux swing experienced by the wire, λ is the fraction of the wire that is superconducting, ω is the rad/s frequency of the alternating field, τ is the LR time constant of the wire cross section, α is the wire's internal eddy current

IOP Conf. Series: Materials Science and Engineering **279** (2017) 012026 doi:10.1088/1757-899X/279/1/012026

shielding factor, β is the ratio of the applied magnetic field to the penetration field strength of the superconducting filaments, and Γ is a normalized loss function for hysteresis based on β.

4.2. Simultaneous field and current

Equation 2 does not account for the increased hysteresis loss due to the addition of transport current under time varying fields, which is what the fully SC machine armature would experience. WJ Carr showed that the resulting increase in AC loss can be approximated with a small, nearly constant multiplication factor for the case of large applied field (large β) and small transport current [14]. Since a fully superconducting machine would experience strong magnetic fields across the armature and would likely operate well below critical current levels to maintain a factor of safety, this approximation should suffice for this study.

While not extensively tested for circular wires, HTS tapes under simultaneous time varying fields and currents have been analyzed and shown to match Carr's results for large applied field ($\beta > 1$) [15].

4.3. Rotating field and phase shift

Rotating fields have been shown analytically to reduce the penetration field B_p in SC wire by Prigozhin. This will increase the associated losses by a small scalar multiple near $\frac{\pi}{2}$ [16]. Phase shift effects on simultaneous applied field and transport current losses have been analyzed for SC tapes in various conditions [17, 18]. In general, a phase shift of 90 degrees between the field and current results in a loss minimum while a phase shift of 0 or close to 0 results in a maximum. Since this phenomenon is difficult to analytically model, it is left out of the calculations. A controlled synchronous machine will often operate close to a 0 degree phase shift (i.e. $I_a = I_q$) to maximize torque, so the absence of a phase shift factor is a good worst approximation of the loss.

5. Design Comparison

A partial SC machine design from [4] consists of SC coils carrying direct current to produce magnetic field of 8-9 T maximum. In the air gap this field is up to 3.3 T. The armature carries three-phase current through copper wires. A notable feature of this design is the swap between stator and rotor roles, to ease the cryogenic cooling of the superconducting coils. The stator in this machine consists of the SC field coils and provides the dc field, and the rotor consists of the copper armature coils to create a rotating field. To eliminate ferromagnetic shielding, a secondary set of SC coils is placed on the perimeter of the machine to counteract the magnetic field outside of the machine. This secondary set of coils act as an active shielding mechanism. The basic concept is illustrated with the schematic in Figure 1 below.

The partial SC machine from [4] was designed for aerospace applications, with specifications of 12MW output at 8000 rpm. Figure 2 shows a cutaway section of a CAD model of the motor. From this design, a fully SC machine is proposed that utilizes superconducting wire instead of copper for the three-phase armature. For reasonable cooling

IOP Conf. Series: Materials Science and Engineering **279** (2017) 012026 doi:10.1088/1757-899X/279/1/012026

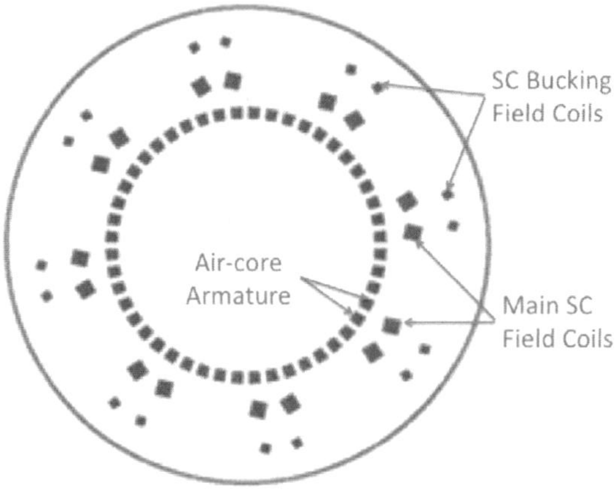

Figure 1: Partial SC motor cross section, showing the positioning of the armature, field coils, and shielding coils

requirements, MgB2 is chosen as the superconducting material operating at 20 K. A summary of initial changes is presented in Table 1.

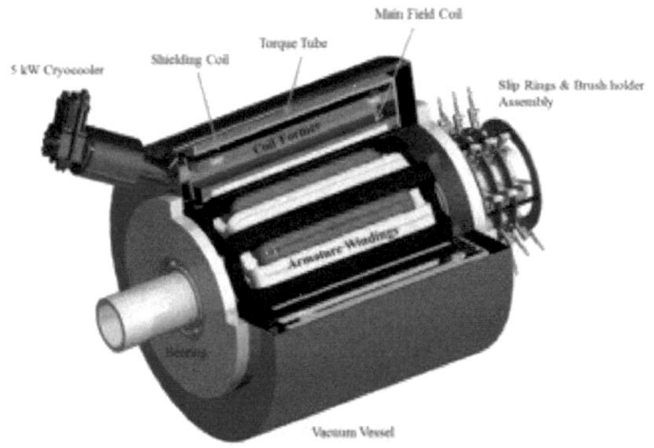

Figure 2: 3D render of the partial SC motor design

MgB2 was chosen since it has demonstrated high critical current density at medium temperatures with reasonable AC loss. While the current density is lower than Nb3Sn at comparable temperatures, it is still high enough to significantly increase the electric loading in the machine. In Table 1, the field coil current density acts as an indicator of the magnetic loading since the coil and magnetic circuit are fixed, and the armature coil current density acts as a measure of the electric loading.

MgB2 wire, discussed in [19], was chosen for this analysis for its low ac loss and strong critical current characteristics.

5.1. AC Loss Computation

FEA simulations were performed on the machine model using Flux2D to determine the torque, air-gap flux density, and magnetic field distribution on the armature and stator coils.

IOP Conf. Series: Materials Science and Engineering **279** (2017) 012026 doi:10.1088/1757-899X/279/1/012026

Table 1: Changes from partial SC to full SC design

Parameter	Partial	Full
SC material	Nb3Sn	MgB2
Operating temperature	4 K	20 K
Wire diameter	700 μm	600 μm
Field coil current density	400 A/mm2	200 A/mm2
Armature current density	7 Arms/mm2	70 Arms/mm2

An example of the B field distribuion over an armature bar is shown in Fig. 3. This illustrates the change in the B field on a slice of the armature cross section over an electrical half-cycle. In the figure, the vertical axis is the B field magnitude, the horizontal axis is the angular position of the rotor, and the axis going into the page is the radial position on the armature. The B field was initially set to 3 T in the armature to match the partially SC machine design, and was reduced further in subsequent analysis to keep AC losses at a manageable level.

The B field FEA results shown in Fig. 3 were fed into Equations 2 and 3 to determine the hysteresis and coupling losses on a wire-by-wire basis within the armature pole. These losses were then multiplied by the geometric parameters of the machine (active length, number of poles) to determine the total ac loss in the armature.

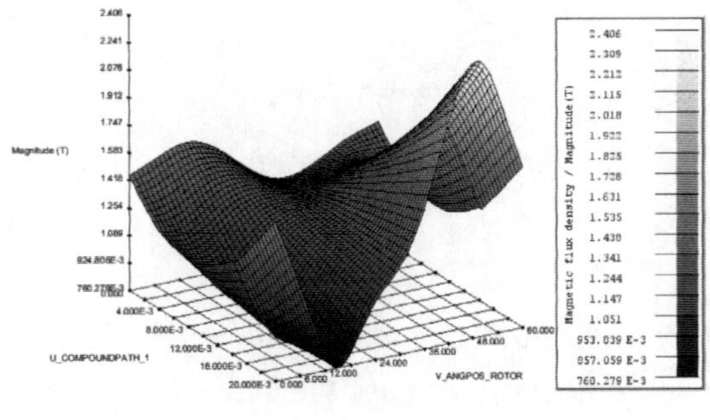

Figure 3: Example B field distribution across a section of the armature bar over an electric half-cycle

A cryogenic cooling factor based on available cryo-cooler data [10] was applied to the loss calculations to account for the additional power required to maintain the operating temperature of 20 K. 100 W of input cooling power was estimated to extract 1 W of heat at 20 K.

This process of geometry design, FEA, and AC loss calculation was performed for four different machine geometries derived from the actively-shielded design described in [20] each representing a potential machine application.

The first is an 8 pole machine, intended for higher speed aerospace propulsion applications. The second and third are 16- and 20-pole machines with larger diameters intended for ship propulsion or hydro-electric generators. The fourth design has 46 poles, and is intended for a low speed, high power application such as offshore wind turbines. For each of these four designs, length was varied with speed to maintain the same output power, and the AC loss was calculated across these variations.

5.2. Results

The resulting cooling power required to compensate for armature AC losses is shown in Fig. 4 with a reference to rotor speed, which is proportional to electrical frequency. In general, higher speeds lead to increased eddy/coupling losses, which is to be expected. At speeds commonly encountered in industrial or aerospace applications (>1800 rpm), the cooling power is of the same order as the whole rating of the electrical machine. This is of course, undesirable.

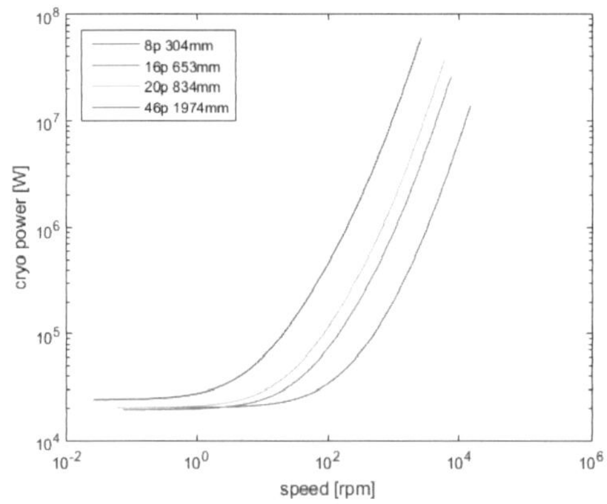

Figure 4: Log-log plot of the required cryogenic power vs rotor speed for the four machine designs at 10 MW

Since the machines were all evaluated at a constant power rating, it should be noted that increasing the speed of the machine will decrease the active length, and thus size and weight, of the machine to maintain the same power. This in turn increases the specific power, or the ratio of power to weight. This suggests that an attractive option is to increase the speed to decrease the specific power. However, there is a tradeoff between specific power and AC loss, as the loss will increase exponentially with speed, indicated in Fig. 4. To compensate for this, the magnetic loading (B) and electrical frequency (or pole count) need to be reduced for the high speed applications, negating power density advantages. To clarify this tradeoff, a metric dividing the power density by the required cryogenic power is calculated and presented in Fig. 5. This metric is called the power-density-to-cryo (PDC) ratio. Since weight data was unavailable for this study, a volumetric power density (MW/m3) was utilized.

From this metric, optimal operating speeds are observed for each of the machine designs. As expected, the larger 46-pole machine has this optimum at lower speeds while the 8-pole machine has a higher peak at higher speeds.

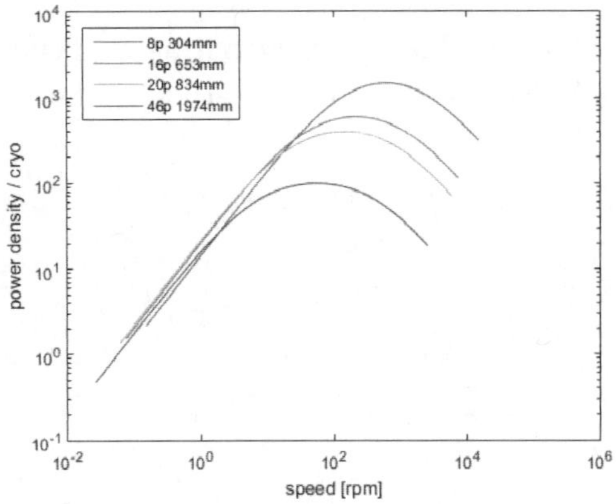

Figure 5: Log-log plot of the power-density-to-cryo ratio vs rotor speed for the four machine designs at 10 MW

With this metric of power-density-to-cryo (PDC), the answer to the initial question investigating the feasibility of fully electric machines for aircraft can be attempted. Designs for the 8-pole and 46-pole cases were compared for both full and partial SC machines, and resulting numbers are presented in Table 2.

Table 2: Comparison of partial and fully SC design parameters and results

Parameter	Units	Fully SC Design 1	Fully SC Design 2	Partial SC Design 1	Partial SC Design 2
Pole count		8	46	8	46
Speed	[rpm]	8000	10	8000	10
Slot depth	[mm]	10	10	25.4	100
Active length	[m]	0.158	0.155	0.157	1.25
Volume	[m^3]	0.09	2.0	0.088	19.75
Power density	[MW/m^3]	110	4.6	110	0.5
Cryo power	[kW]	2208	58	16	92
Efficiency	[%]	81.9	99.8	99.8	99.0
PDC		50	79	7080	5.6

The PDC values of the partial SC designs from Table 2 are shown in Fig. 6. Partial SC Design 1, shown in red, is an 8 pole design with an extremely high ratio of power density to required cooling. The Fully SC Design 1, which was electromagntically sized to match the power density of the partial version, still underperforms in both PDC and efficiency, making it unfeasible for its intended application. This is to be expected, since the only sources of loss for the partial SC machine are in the armature copper and the dc field coils; this is far lower than the losses in a fully SC machine, which are compounded by the high electrical frequency.

By contrast, Partial SC Design 2, shown in green, has a lower PDC ratio than its fully SC

IOP Conf. Series: Materials Science and Engineering **279** (2017) 012026 doi:10.1088/1757-899X/279/1/012026

counterpart. The fully SC machine in this case is operating at such a low frequency that the AC losses are not as significant despite the increased electrical loading. Additionally, minor losses in the field coils are not as drastic for the fully SC design due to the higher operating temperature. In this case, the fully SC design prevails.

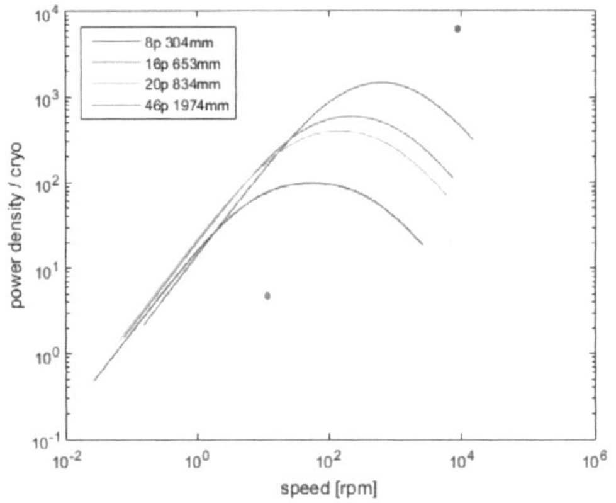

Figure 6: Log-log plot of the power-density-to-cryo ratio vs rotor speed, compared with Partial SC Design 1 (red, upper point) and Partial SC Design 2 (green, lower point)

6. Conclusion and future work

Given current conductor capability, fully SC machines appear to be only suitable for large, slow applications, with some movement towards 100-500rpm range. This is not unexpected, as high frequencies, strong magnetic field and significant electrical loading all contribute to unmanageable AC loss in higher speed applications. With the described motor model, superconducting wire model, and cooling assumptions, the 8-pole motor is infeasible as a fully SC design above 2000 rpm (132 Hz electrical frequency). Above this speed, it is outclassed by the partially SC version in both efficiency and power density. The 46-pole machine encounters this point around 100 rpm (38 Hz).

While these conclusions may be intuitively obvious, we believe the quantitative results and the analytical method provide a framework to rigorously assess the trade-offs as enabling technologies evolve. For instance, a possible aircraft-based cryocooling solution posited by Berg et al. [8] using liquid CH_4 as a heat sink suggests a cooling power of 15 W per 1 W extraction at 20 K; the impact of this improvement in cooling capability for fully SC machine designs can be quickly assessed with the described method, raising the previously described limiting speeds.

One area for future work is in experimental validation of the analytical equations for machine-like conditions. While the AC loss equations have been tested for various isolated conditions, the specific combination of cases for wires within a motor have not been experimentally validated. At the very least, conditions of strong ac magnetic field ($\beta \gg 1$) and simultaneous alternating current for circular wires should be investigated. Testing rotating fields and phase shift between the field and current would provide additional data for model validation.

Tapes, which are likely contenders for higher-temperature SC machines, have been more thoroughly investigated under machine-like conditions, though still only for weak fields and currents. However, analytical models for tapes can be quite complex with dependencies on orientation, tape construction, coil construction, and other factors not necessarily present with circular wires. Integrating established ac loss models for tapes with the method described in this paper is a logical next step for further study.

7. References

[1] G. Snitchler, B. Gamble, and S. S. Kalsi, "The performance of a 5 MW high temperature superconductor ship propulsion motor," *IEEE Transactions on Applied Superconductivity*, vol. 15, no. 2 PART II, pp. 2206–2209, 2005.

[2] T. A. Keim, T. E. Laskaris, J. A. Fealey, and P. A. Rios, "Design and manufacture of a 20 mva superconducting generator," *IEEE Power Engineering Review*, vol. PER-5, no. 6, pp. 56–56, June 1985.

[3] S. S. Kalsi *et al.*, "Development status of rotating machines employing superconducting field windings," *Proceedings of the IEEE*, vol. 92, no. 10, pp. 1688–1704, Oct 2004.

[4] K. S. Haran, D. Loder, T. O. Deppen, and L. Zheng, "Actively shielded high-field air-core superconducting machines," *IEEE Transactions on Applied Superconductivity*, vol. 26, no. 2, pp. 98–105, March 2016.

[5] M. Armstrong *et al.*, "Architecture, Voltage, and Components for a Turboelectric Distributed Propulsion Electric Grid," NASA CR-2015-218440, Tech. Rep., 2015.

[6] P. Gemin *et al.*, "Architecture, Voltage, and Components for a Turboelectric Distributed Propulsion Electric Grid," NASA CR-2015-218713, Tech. Rep., 2015.

[7] G. V. Brown, "Weights and efficiencies of electric components of a turboelectric aircraft propulsion system," *49th AIAA aerospace sciences meeting including the new horizons forum and aerospace exposition*, 2011.

[8] F. Berg *et al.*, "Cryogenic system options for a superconducting aircraft propulsion system," *IOP Conference Series: Materials Science and Engineering*, vol. 101, no. 1, p. 012085, 2015. [Online]. Available: http://stacks.iop.org/1757-899X/101/i=1/a=012085

[9] D. Loder *et al.*, "A conduction cooled nb3sn racetrack coil: Design, construction, and testing," in *2016 IEEE Power and Energy Conference at Illinois (PECI)*, Feb 2016, pp. 1–8.

[10] K. Haran *et al.*, "Superconducting rotating machines - development status and technology roadmap," *Superconductor Science and Technology*, in process 2017.

[11] W. T. Norris, "Calculation of hysteresis losses in hard superconductors carrying ac: isolated conductors and edges of thin sheets," *Journal of Physics D: Applied Physics*, vol. 3, no. 4, pp. 489–507, 1970.

[12] W. J. Carr, "Electromagnetic theory for filamentary superconductors," *Physical Review B*, vol. 11, no. 4, pp. 1547–1554, 1975.

[13] M. Wilson, *Superconducting Magnets*. Clarendon Press Oxford, 1983.

[14] W. Carr, "Ac loss from the combined action of transport current and applied field," *IEEE Transactions on Magnetics*, vol. 15, no. 1, pp. 240–243, Jan 1979.

[15] J. J. Rabbers, D. C. van der Laan, B. ten Haken, and H. H. J. ten Kate, "Magnetisation and transport

current loss of a BSCCO/Ag tape in an external ac magnetic field carrying an ac transport current," *IEEE Transactions on Applied Superconductivity*, vol. 9, no. 2, pp. 1185–1188, June 1999.

[16] L. Prigozhin, "The bean model in superconductivity: Variational formulation and numerical solution," *Journal of Computational Physics*, vol. 200, pp. 190–200, 1996.

[17] D. N. Nguyen *et al.*, "Experimental and numerical studies of the effect of phase difference between transport current and perpendicular applied magnetic field on total ac loss in Ag-sheathed (Bi , Pb) 2 Sr 2 Ca 2 Cu 3 O x tape Experimental and numerical studies of the effect ," vol. 073902, 2005.

[18] E. Pardo, F. Gomory, J. Souc, and J. M. Ceballos, "Current distribution and ac loss for a superconducting rectangular strip with in phase alternating current and applied field," *Superconductor Science and Technology*, vol. 20, no. 4, p. 21, 2005. [Online]. Available: http://arxiv.org/abs/cond-mat/0510314

[19] F. Wan *et al.*, "Architecture and transport properties of multifilamentary MgB2 strands for MRI and low ac loss applications," *IEEE Transactions on Applied Superconductivity*, vol. 27, no. 4, pp. 1–5, 2017.

[20] D. Loder *et al.*, "Demonstration of a practical Nb_3Sn coil for an actively shielded generator," *IEEE Transactions on Applied Superconductivity*, vol. 27, no. 4, pp. 1–7, June 2017.

IOP Conf. Series: Materials Science and Engineering **279** (2017) 012027 doi:10.1088/1757-899X/279/1/012027

Growth rate of YBCO-Ag superconducting single grains

J V J Congreve, Y H Shi, A R Dennis, J H Durrell and D A Cardwell
Department of Engineering, University of Cambridge, Cambridge, UK CB2 1PZ
Email: jvjc2@cam.ac.uk

Abstract
The large scale use of (RE)$Ba_2Cu_3O_7$ bulk superconductors, where RE=Y, Gd, Sm, is, in part, limited by the relatively poor mechanical properties of these inherently brittle ceramic materials. It is reported that alloying of (RE)$Ba_2Cu_3O_7$ with silver enables a significant improvement in the mechanical strength of bulk, single grain samples without any detrimental effect on their superconducting properties. However, due to the complexity and number of inter-related variables involved in the top seeded melt growth (TSMG) process, the growth of large single grains is difficult and the addition of silver makes it even more difficult to achieve successful growth reliably. The key processing variables in the TSMG process include the times and temperatures of the stages within the heating profile, which can be derived from the growth rate during the growth process. To date, the growth rate of the $YBa_2Cu_3O_7$-Ag system has not been reported in detail and it is this lacuna that we have sought to address. In this work we measure the growth rate of the YBCO-Ag system using a method based on continuous cooling and isothermal holding (CCIH). We have determined the growth rate by measuring the side length of the crystallised region for a number of samples for specified isothermal hold temperatures and periods. This has enabled the growth rate to be modelled and from this an optimized heating profile for the successful growth of YBCO-Ag single grains to be derived.

1. Introduction

Single grain (RE)$Ba_2Cu_3O_7$ bulk high temperature superconductors (HTS), where RE=Y, Gd or Sm, have the ability to trap large magnetic fields, typically of more than an order of magnitude larger than those generated by conventional permanent magnets [1, 2]. This potentially enables a wide range of applications including Maglev trains, energy storage flywheels, rotating electrical machines and trapped flux devices [3-5]. However these ceramic materials are inherently brittle [6, 7] and in many practical cases, the mechanical, rather than superconducting, properties limit their potential for application. Numerous authors have reported that the addition of silver is able to significantly improve the mechanical properties [8, 9] of single-grain bulk YBCO by increasing the fracture toughness [10] and bending strength [9, 11] of the samples without degrading their superconducting properties [8, 12, 13].

Practical applications require bulk (RE)BCO to be fabricated in the form of large single grains, since the presence of high-angle grain boundaries in (RE)$Ba_2Cu_3O_7$ superconductors prevent the flow of current. The presence of high angle grain boundaries within the bulk results in a significant reduction in the magnitude of the trapped field generated by the sample [14-17]. It is important to ensure the manufacturing process is reliable as the cost of the raw materials are high and the fabrication process takes a relatively long period of time. A carefully considered heating profile is crucial to the successful growth of single grain bulk superconductors for the top seeded melt growth (TSMG) processing technique, in particular [18, 19].

The fabrication of large, single grain bulk samples by batch processing has been achieved successfully for $YBa_2Cu_3O_7$ (Y-123) and both the $GdBa_2Cu_3O_7$-Ag (GdBCO-Ag) and $SmBa_2Cu_3O_7$-Ag (SmBCO-Ag) systems [20] although successful, reliable batch processing of $YBa_2Cu_3O_7$-Ag (YBCO-Ag) has yet to be achieved due to the greater complexity of the YBCO-Ag system. The addition of greater than 5 wt% Ag to the melt causes a reduction in the peritectic temperature [7] of the YBCO system by approximately 30 °C [21]. This narrows considerably the temperature window available for successful single-grain growth and so a substantially different heating profile is required to that used to melt process YBCO without silver. The reduction in maximum temperature affects many aspects of the growth process, including initial decomposition [22], solute diffusion and interface kinetics and thus reduces the growth rate of YBCO-Ag in comparison to Ag-free Y-123. [7] There is also an increase in the likelihood of spontaneous nucleation of secondary grains due to the reduction in processing temperature [7], and further complications arise due to the low solubility of silver within the melt. [21] In addition, the diffusion of Y to the growth front is much slower than that of Gd or Sm, so further reduction in the diffusion speed due to the decrease in processing temperature reduces the growth rate yet further and makes it much more difficult to fabricate single grains of YBCO-Ag compared to single grains of GdBCO-Ag or SmBCO-Ag.

A number of studies have been carried out on the growth rate of the YBCO system fabricated without silver [2, 23]. It has been established that the relationship between growth rate and undercooling can be modelled by:

$$R_a = 4.5 \times 10^{-7} \Delta T^{1.9} \qquad (1)$$

$$R_c = 2.8 \times 10^{-6} \Delta T^{1.3} \qquad (2)$$

where R_a is the growth rate of the *ac*-plane in *mm/second*, R_c is the growth rate of the *ab*-plane in *mm/second* and ΔT is the undercooling. Experimental results have been found to be in good agreement with these relationships [23].

No growth rate data has yet been reported in detail for the YBCO-Ag system. The measurement of the growth rate for (RE)BCO systems investigated to date have involved a single period of simple isothermal growth [2, 23, 24] and are based on assuming that the growth length of a single grain is proportional to the holding time. However, for the YBCO-Ag system isothermal growth is unreliable and it has proven difficult to grow single grains in this way.

Here, we report measurements of the growth rate using a new, two-stage method, based on continuous cooling and isothermal hold (CCIH), in which a period of slow cooling is followed by a holding period at constant temperature. The slow-cooling stage enables seeding and initialization of the growth for the next stage, which is particularly important in the YBCO-Ag system, since it is difficult to incorporate silver into the lattice, which must take place to ensure that a single grain results. Previous studies involving a single isothermal hold period have shown the growth rate can be modelled by the expression:

$$R = \alpha(\Delta T)^{\beta}; \qquad (3)$$

where R is the growth rate in *mm/hour*, ΔT is the undercooling and α and β are constants that vary with composition of the precursor powder. In the diffusion limited crystal growth model, β can be approximated to two for the YBCO system, assuming that YBCO crystal growth is controlled by the diffusion rate of the Y element in the liquid phase.

The growth rate in the thermal profile used in CCIH is not fully defined by the growth rate model for an isothermal growth process. In this case, the expression required to model the growth rate in the YBCO-Ag system becomes more complex because the additional slow cooled period must also be considered. The definition of growth rate must be considered in order to calculate the parameter α from these data:

$$R = \frac{dL}{dt}; \qquad (4)$$

where L is the growth length and t is the time over which growth occurs. The CCIH growth process is divided into two parts; the continuous cooling period and the isothermal hold period so that the growth length, L, can be related to the growth rate by:

$$L = \int_{t_s}^{t_e} R_{cc}\, dt + R_{ih} t_{ih} \qquad (5)$$

Using equation 3 and noting in the continuously cooled region:

$$\Delta T = -0.5t \qquad (6)$$

$$L = \int_{t_s}^{t_e} \alpha(-0.5t)^{\beta}\, dt + \alpha(\Delta T)^{\beta} t_{ih}; \qquad (7)$$

where t_s is the time at the start of the continuous cooling period, t_e is the time at the end of the continuous cooling period, R_{cc} is the growth rate during continuous cooling, R_{ih} is the growth rate during the isothermal hold period, t_{ih} is the time period of isothermal hold, ΔT is the undercooling below 970°C, L is the side length of the single grain in *mm*, $\propto$ and β are growth parameters to be calculated and t is the time period in *seconds*.

The parameters α and β can be found for each growth condition. When the undercooling is small a significant proportion of the slow cooling and isothermal hold time will be taken up by the seeding and growth initiation period, whereas for large undercooling the majority of this time period will be taken up by growth.

The liquid-phase enrichment technique [25] has been used in the present work to enable successful fabrication of single grains of YBCO-Ag. The growth rate was measured by a new method in which thermal processing, comprising continual cooling and isothermal holding, was followed by measurement of the side lengths of the crystal. The time periods of continuous cooling and isothermal hold and the isothermal hold temperature have been varied for a series of YBCO-Ag samples. Measurement of the growth rate of small samples has provided a large amount of information on the growth process at a low relative cost, because the samples used in the study are small, the time required for growth is short. These data have enabled the growth rate for each condition to be calculated based upon the diffusion-limited growth model and have subsequently enabled a suitable temperature profile to be determined for the successful growth of 20 *mm* as-pressed diameter YBCO-Ag single grains.

2. Method

Nine samples were prepared by the TSMG [14, 26] process with liquid-phase enrichment [25]. Precursor powder was prepared from 99.9 % purity powders in the weight ratio Y-123 : Y-211 : CeO_2 : Ag_2O of 50 : 150 : 1 : 22.2 in order that excess 10 wt% Ag_2O was added to each sample. Additional liquid-phase rich powder was prepared with the composition Yb_2O_3 : $Ba_3Cu_5O_8$: BaO_2 in a molar ratio of 5.0 : 5.6 : 1.0, which was subsequently calcined once at 850°C for 5 hours.

Green compacts were pressed uniaxially in a 20 *mm* diameter cylindrical die with 1.5 g of liquid-phase-rich powder below 10 g of precursor powder. A layer of Yb_2O_3 paste was painted onto the base of each compact to avoid sub-grain formation at the base. Each compact was assembled on ZrO_2 rods above a ceramic plate. A buffer pellet [27-31] of mass 0.15 g was pressed uniaxially from the standard precursor powder without Ag_2O to a cylinder of diameter 5 *mm* and placed centrally above the precursor powder pellet. A generic seed [32] was placed at the center of the buffer pellet prior to melt processing.

Samples were heated above the peritectic temperature in a box furnace. The peritectic temperature was measured to be 989 °C by differential thermal analysis (DTA) for the precursor composition with 10 wt% Ag_2O. The samples were held above this temperature to enable complete decomposition to occur. This was followed by rapid cooling to 970 °C, then slower cooling at 0.5 °C/*hour*, to enable seeding and growth initiation to take place, and then to the isothermal hold temperature (T_i) in the range 970 °C and 952 °C. The sample was then held isothermally for a given time period (t_i) of between 45 minutes and 10 hours. The isothermal hold time was varied to ensure the side length of the partially grown single grain could be measured. Each sample was quenched to ambient temperature at the end of the isothermal hold period. It was assumed that heat dissipation at the growth front is large in comparison to the growth rate, so quenching will enable the growth dimensions and microstructure at the end of the isothermal hold period to be preserved. A schematic of the heating profile can be seen in Fig. 1, which represents the heating profile used conventionally to measure growth rate with a single isothermal hold period superimposed onto the figure.

IOP Conf. Series: Materials Science and Engineering **279** (2017) 012027 doi:10.1088/1757-899X/279/1/012027

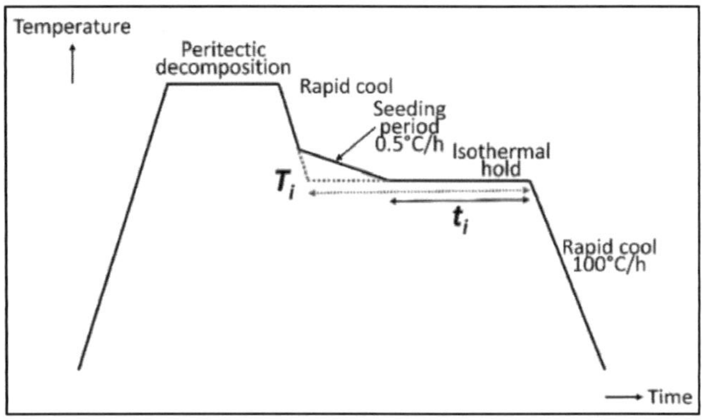

Figure 1: Heating profile for CCIH of YBCO-Ag single grains. The superimposed dotted line represents a heating profile with a single isothermal hold stage used traditionally to measure growth rates of bulk (RE)BCO single grain superconductors.

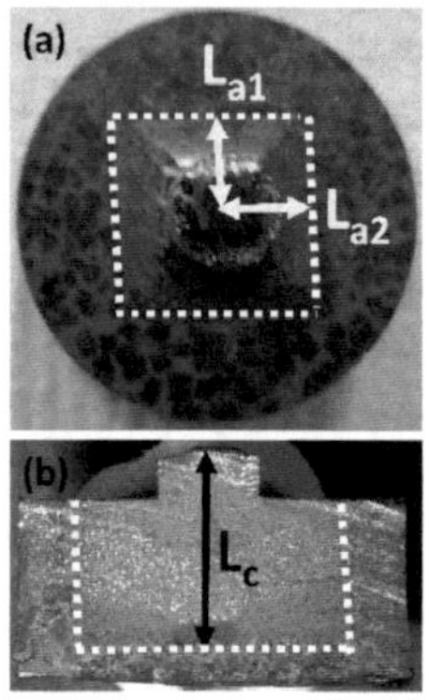

Figure 2: (a) Top view of the single grain where the two side lengths are averaged and (b) side cross-section. The dashed lines mark the edge of the single grained region.

The side length of the square single grain was measured on the top surface of each sample and the average of the two perpendicular side lengths recorded as grain growth in the *a/b*-axis direction. Each sample was then cut in half to enable the extent of growth in the *c*-axis direction to be measured as the vertical distance from the top of the buffer pellet to the edge of the single grained region observed in the rectangular central cross section at the center of the sample. The positions of the measurements taken on each sample cross-section are illustrated in Fig. 2.

3. Results and Discussion

All nine samples were partially grown successfully, with their top surfaces and cross sections shown in Fig. 3, the edge of the single grained region is marked by the yellow dotted line. Measurements of the size of the single grain region enabled the growth rate parameters, $\propto$ and β, to be calculated to give the following models for growth of the YBCO-Ag system:

$$R_{a/b} = 1.4 \times 10^{-3}(\Delta T)^{1.76} + 0.035 \qquad (8)$$

$$R_c = 4.5 \times 10^{-3}(\Delta T)^{1.42} + 0.035 \qquad (9)$$

where $R_{a/b}$ is the growth rate in the *a/b*-axis direction in *mm/hour*, R_c is the growth rate in the *c*-axis direction and ΔT is the undercooling below 970 °C.

These models with an appropriate off-set are a good fit to the actual data, as seen in Fig. 4, with R^2 fit parameters of 0.9984 and 0.9966 for the *a/b*-axis and the *c*-axis direction for an undercooling between 0 and 18 °C. A greater magnitude of undercooling leads to a higher growth rate. However, although a large undercooling enables more rapid growth, the magnitude of undercooling must be limited because the probability of random nucleation of secondary grains increases to a critical level when the undercooling is sufficiently large.

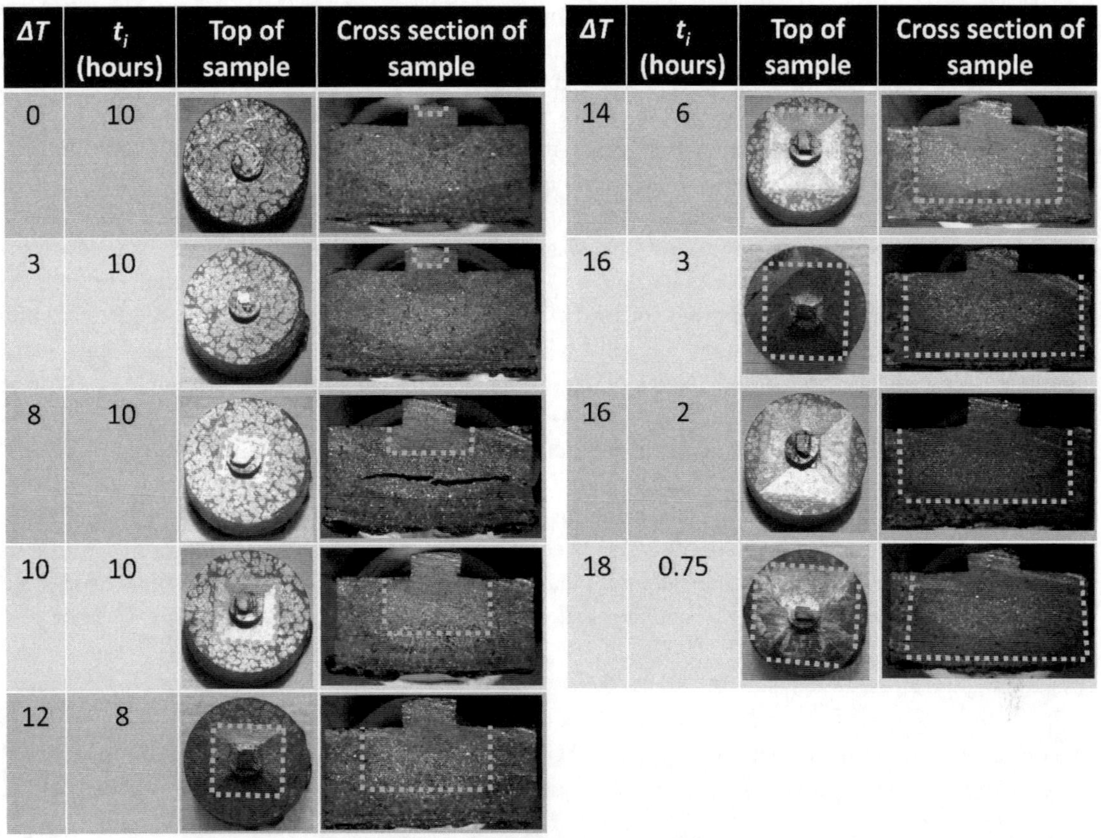

Figure 3: Photographs of the samples grown for the calculation of growth rate. The dashed lines show the edge of the single grained region.

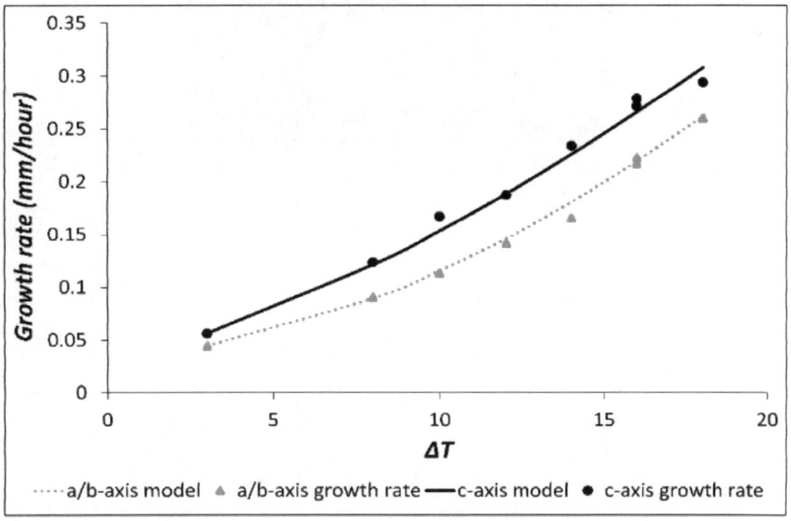

Figure 4: Plot of the actual and modelled growth rates for both the a/b-axis and c-axis directions of the single grain YBCO-Ag samples.

The requirement for an additional continuous slow cooling stage before the isothermal hold period has led to a requirement to develop a revised method to calculate the growth rate, since some growth will occur during the slow cooling period, of which account must be taken. The need for the slow cooling period is evident both from the structure of the bulk samples and the growth rate models. The buffer is clearly evident in the cross sections of all samples, with a region immediately below this that appears to be deficient in silver compared to the central region, as shown in Fig. 5. In addition to lower silver content, the size of the agglomerates present in the silver-deficient region are much smaller in size relative to those in the adjacent region towards the center of the sample, which appears to be silver-rich. Silver is forced to the growth front during the initial stages of growth, since it cannot be easily incorporated into the lattice. As the growth-front advances, too much silver is present to be pushed with the growth front and it is therefore forced to move into the pores already present within the solidifying single grain structure. This discontinuity can be interpreted as the change from a silver deficient region to a silver-rich region, where the silver agglomerates are much larger and fill some of the regions that were initially pores. In addition, the growth rate model required an offset of 0.035 *mm/hour* to fit a power law model. This provides further evidence for the importance of the seeding and growth initiation period required for successful nucleation and growth of the YBCO-Ag system. This offset value is the same for both directions within the sample, suggesting that the slow cooling period is critical to 3D growth and affects both directions to the same extent.

The growth rate model can be compared with that derived for isothermal growth for the YBCO system containing Pt in equations 1 and 2 [23] and the isothermal growth of YBCO of composition 75 wt% Y-123, 25 wt% Y-211, 1 wt% CeO_2: [2]

$$R_{a/b} = 1.6 \times 10^{-3}(\Delta T)^{2.12} \ (mm/hour) \qquad (10)$$

$$R_c = 2.4 \times 10^{-3}(\Delta T)^{2.12} \ (mm/hour) \qquad (11)$$

The differences between the coefficients in these models are due to a combination of factors, including the presence of silver agglomerates, the use of additional liquid-phase and differences in the heating profiles.

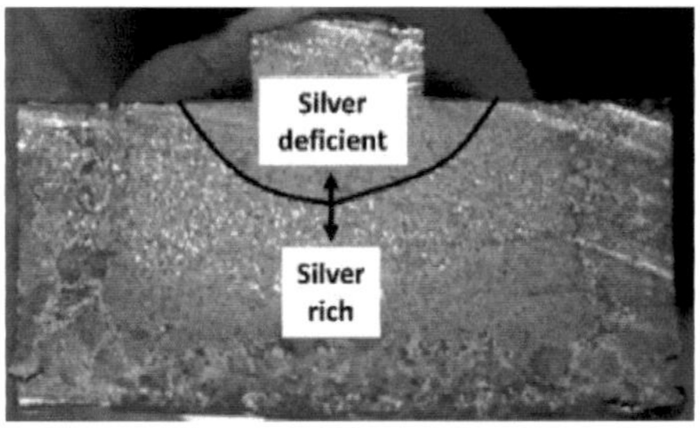

Figure 5: The locations of the silver deficient and silver-rich regions observed in the YBCO-Ag samples.

The dependence of growth rate at various values of undercooling and peritectic temperature, measured using DTA, have enabled the derivation of a suitable heating profile to enable the successful growth of single-grain 20 *mm* as-pressed diameter samples. The optimum heating profile derived from this investigation is shown in Fig. 6, with a sample fully grown using this heating profile shown in the inset.

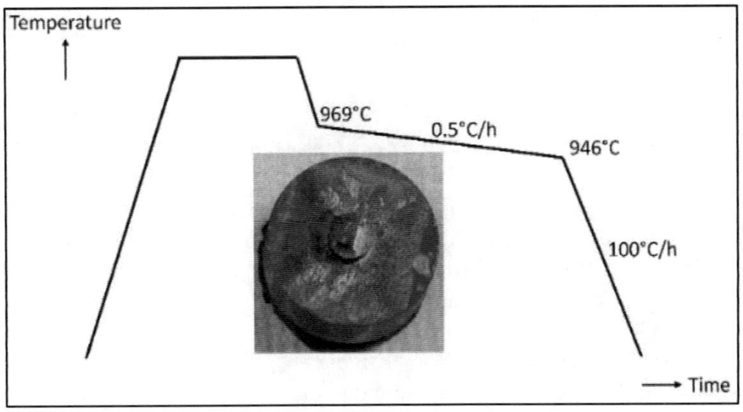

Figure 6: Optimised heating profile for the successful growth of a single grain of 20 mm as-pressed diameter YBCO-Ag sample derived from the growth rate data. A fully grown sample is shown in the inset.

4. Conclusions

The growth rate of single grains in the YBCO-Ag system has been studied in detail using a new method based on continuous cooling and isothermal hold (CCIH), which differs from previous studies in that it provides a seeding and growth initiation period prior to the isothermal growth stage. We have developed a model for the growth rate of YBCO-Ag in both the *a/b*- and *c*-axis directions based on this technique. In addition to the limitations on successful growth and growth rate due to the diffusion of Y species in the melt, the presence of silver further limits the growth rate of this system. The growth rate model has enabled an optimised heating profile to be derived to successfully grow single grains of 20 *mm* as-pressed diameter of YBCO-Ag. We are now able to grow large single grains of YBCO-Ag reliably.

5. References

1 Durrell, J.H., Dennis, A.R., Jaroszynski, J., Ainslie, M.D., Palmer, K.G.B., Shi, Y.H., Campbell, A.M., Hull, J., Strasik, M., Hellstrom, E.E., and Cardwell, D.A.: 'A trapped field of 17.6T in melt-processed, bulk Gd-Ba-Cu-O reinforced with shrink-fit steel', Supercond. Sci. Technol., 2014, 27, (8)

2 Zhai, W., Shi, Y.H., Durrell, J.H., Dennis, A.R., and Cardwell, D.A.: 'The Influence of Y-211 Content on the Growth Rate and Y-211 Distribution in Y-Ba-Cu-O Single Grains Fabricated by Top Seeded Melt Growth', Crystal Growth & Design, 2014, 14, (12), pp. 6367-6375

3 Campbell, A.M., and Cardwell, D.A.: 'Bulk high temperature superconductors for magnet applications', Cryogenics, 1997, 37, (10), pp. 567-575

4 Werfel, F.N., Floegel-Delor, U., Rothfeld, R., Riedel, T., Goebel, B., Wippich, D., and Schirrmeister, P.: 'Superconductor bearings, flywheels and transportation', Supercond. Sci. Technol., 2012, 25, (1)

5 Li, B.Z., Zhou, D.F., Xu, K., Hara, S., Tsuzuki, K., Miki, M., Felder, B., Deng, Z.G., and Izumi, M.: 'Materials process and applications of single grain (RE)-Ba-Cu-O bulk high-temperature superconductors', Physica C, 2012, 482, pp. 50-57

6 Nishio, T., Itoh, Y., Ogasawara, F., Suganuma, M., Yamada, Y., and Mizutani, U.: 'SUPERCONDUCTING AND MECHANICAL-PROPERTIES OF YBCO-AG COMPOSITE SUPERCONDUCTORS', Journal of Materials Science, 1989, 24, (9), pp. 3228-3234

7 Iida, K., Babu, N.H., Shi, Y.H., Miyazaki, T., Sakai, N., Murakami, M., and Cardwell, D.A.: 'Single domain YBCO/Ag bulk superconductors fabricated by seeded infiltration and growth', in Hoste, S., and Ausloos, M. (Eds.): '8th European Conference on Applied Superconductivity' (2008)

8 Azambuja, P.d., Rodrigues JÃºnior, P., Jurelo, A.R., Serbena, F.C., Foerster, C.E.n., Costa, R.n.M., Souza, G.B.d., Lepienski, C.M.c., and Chinelatto, A.L.: 'Effects of Ag addition on some physical properties of granular YBa2Cu3O7-´ superconductor', Brazilian Journal of Physics, 2009, 39, pp. 638-644

9 Yeh, F., and White, K.W.: 'FRACTURE-TOUGHNESS BEHAVIOR OF THE YBA2CU3O7-X SUPERCONDUCTING CERAMIC WITH SILVER-OXIDE ADDITIONS', Journal of Applied Physics, 1991, 70, (9), pp. 4989-4994
10 Diko, P., Krabbes, G., and Wende, C.: 'Influence of Ag addition on crystallization and microstructure of melt-grown single-grain YBa2Cu3O7 bulk superconductors', Supercond. Sci. Technol., 2001, 14, (7), pp. 486-495
11 Nakamura, Y., Tachibana, K., and Fujimoto, H.: 'Dispersion of silver in the melt grown YBa2Cu3O6+x crystal', Physica C: Superconductivity, 1998, 306, (3–4), pp. 259-270
12 Mendoza, E., Puig, T., Varesi, E., Carrillo, A.E., Plain, J., and Obradors, X.: 'Critical current enhancement in YBCO–Ag melt-textured composites: influence of microcrack density', Physica C: Superconductivity, 2000, 334, (1–2), pp. 7-14
13 Matsui, M., Sakai, N., and Murakami, M.: 'Effect of Ag 2 O addition on the trapped fields and mechanical properties of Nd–Ba–Cu–O bulk superconductors', Superconductor Science and Technology, 2002, 15, (7), pp. 1092
14 Cardwell, D.A.: 'Processing and properties of large grain (RE)BCO', Materials Science and Engineering B-Solid State Materials for Advanced Technology, 1998, 53, (1-2), pp. 1-10
15 Dimos, D., Chaudhari, P., Mannhart, J., and Legoues, F.K.: 'ORIENTATION DEPENDENCE OF GRAIN-BOUNDARY CRITICAL CURRENTS IN YBA2CU3O7-DELTA BICRYSTALS', Physical Review Letters, 1988, 61, (2), pp. 219-222
16 Todt, V.R., Zhang, X.F., Miller, D.J., StLouisWeber, M., and Dravid, V.P.: 'Controlled growth of bulk bicrystals and the investigation of microstructure-property relations of YBa2Cu3Ox grain boundaries', Applied Physics Letters, 1996, 69, (24), pp. 3746-3748
17 Durrell, J.H., Hogg, M.J., Kahlmann, F., Barber, Z.H., Blamire, M.G., and Evetts, J.E.: 'Critical current of YBa2Cu3O7-delta low-angle grain boundaries', Physical Review Letters, 2003, 90, (24)
18 Lo, W., Cardwell, D.A., Dung, S.-L., and Barter, R.G.: 'Processing of bulk YBa2Cu3O7−δ ceramics prior to peritectic solidification', Journal of Materials Science, 1995, 30, (16), pp. 3995-4002
19 Volochova, D., Diko, P., Radusovska, M., Piovarci, S., Antal, V., Zmorayova, K., and Sefcikova, M.: 'Parameters of TSMG Process for Y1.5Ba2Cu3Ox System with CeO2 Addition', J. Supercond. Nov. Magn, 2013, 26, (4), pp. 885-889
20 Shi, Y., Babu, N.H., Iida, K., Yeoh, W.K., Dennis, A.R., Pathak, S.K., and Cardwell, D.A.: 'Batch-processed GdBCO-Ag bulk superconductors fabricated using generic seeds with high trapped fields', Physica C, 2010, 470, (17-18), pp. 685-688
21 Reddy, E.S., Goodilin, E.A., Tarka, M., Zeisberger, M., and Schmitz, G.J.: 'Low temperature processing of single domain YBa2Cu3Oy thick films from Y2O3 fabrics on Ag–Pd alloy substrates', Physica C: Superconductivity, 2002, 372–376, Part 2, pp. 1200-1203
22 Cai, C., Tachibana, K., and Fujimoto, H.: 'Study on single-domain growth of Y 1.8 Ba 2.4 Cu 3.4 O y /Ag system by using Nd123/MgO thin film as seed', Superconductor Science and Technology, 2000, 13, (6), pp. 698
23 Endo, A., Chauhan, H.S., Nakamura, Y., and Shiohara, Y.: 'Relationship between growth rate and undercooling in Pt-added Y1Ba2Cu3O7-x', Journal of Materials Research, 1996, 11, (5), pp. 1114-1119
24 Shi, Y., Babu, N.H., Iida, K., and Cardwell, D.A.: 'Growth rate and superconducting properties of Gd-Ba-Cu-O bulk superconductors melt processed in air', IEEE Trans. Appl. Supercond., 2007, 17, (2), pp. 2984-2987
25 Congreve, J.V.J., Shi, Y.H., Dennis, A.R., Durrell, J.H., and Cardwell, D.A.: 'Improvements in the processing of large grain, bulk Y-Ba-Cu-O superconductors via the use of additional liquid phase', Supercond. Sci. Technol., 2017, 30, (1), pp. 11
26 Li, G.Z., Li, D.J., Deng, X.Y., Deng, J.H., and Yang, W.M.: 'Infiltration Growth and Crystallization Characterization of Single-Grain Y-Ba-Cu-O Bulk Superconductors', Crystal Growth & Design, 2013, 13, (3), pp. 1246-1251

27 Li, T.Y., Cheng, L., Yan, S.B., Sun, L.J., Yao, X., Yoshida, Y., and Ikuta, H.: 'Growth and superconductivity of REBCO bulk processed by a seed/buffer layer/precursor construction', Supercond. Sci. Technol., 2010, 23, (12)
28 Kim, C.J., Lee, J.H., Park, S.D., Jun, B.H., Han, S.C., and Han, Y.H.: 'Y2BaCuO5 buffer block as a diffusion barrier for samarium in top seeded melt growth processed YBa2Cu3O7-y superconductors using a SmBa2Cu3O7-d seed', Supercond. Sci. Technol., 2011, 24, (1)
29 Zhou, D.F., Xu, K., Hara, S., Li, B.Z., Deng, Z.G., Tsuzuki, K., and Izumi, M.: 'MgO buffer-layer-induced texture growth of RE-Ba-Cu-O bulk', Supercond. Sci. Technol., 2012, 25, (2)
30 Shi, Y.H., Dennis, A.R., and Cardwell, D.A.: 'A new seeding technique for the reliable fabrication of large, SmBCO single grains containing silver using top seeded melt growth', Supercond. Sci. Technol., 2015, 28, (3)
31 Kumar, N.D., Shi, Y.H., Zhai, W., Dennis, A.R., Durrell, J.H., and Cardwell, D.A.: 'Buffer Pellets for High-Yield, Top-Seeded Melt Growth of Large Grain Y-Ba-Cu-O Superconductors', Crystal Growth & Design, 2015, 15, (3), pp. 1472-1480
32 Shi, Y., Babu, N.H., and Cardwell, D.A.: 'Development of a generic seed crystal for the fabrication of large grain (RE)-Ba-Cu-O bulk superconductors', Supercond. Sci. Technol., 2005, 18, (4), pp. L13-L16

Acknowledgements
The authors would like to acknowledge the support of the Engineering and Physical Sciences Research Council (EPSRC grant ref. EP/K02910X/1) for financial support.

IOP Conf. Series: Materials Science and Engineering **279** (2017) 012028 doi:10.1088/1757-899X/279/1/012028

Present status of PIT round wires of 122-type iron-based superconductors

T Tamegai[1], T Suwa[1], S Pyon[1], H Kajitani[2], K Takano[2], N Koizumi[2], S Awaji[3] and K Watanabe[3]

[1]Department of Applied Physics, The University of Tokyo, 7-3-1 Hongo, Bunkyo-ku, Tokyo 113-8656, Japan
[2] Naka Fusion Institute, National Institutes for Quantum and Radiological Science and Technology (QST), 801-1 Mukoyama, Naka-shi, Ibaraki 311-0193, Japan
[3]High Field Laboratory for Superconducting Materials, Institute for Materials Research, Tohoku University, Sendai 980-8577, Japan

E-mail: tamegai@ap.t.u-tokyo.ac.jp

Abstract. Outstanding characteristics with high T_c and H_{c2} and small anisotropy in iron-based superconductors (IBSs) have triggered the development of superconducting wires and tapes using these novel superconductors. In this short article, developments and present status of round wires of 122-type IBSs are reviewed. By introducing hot-isostatic pressing (HIP) technique, J_c in round wires of 122-type IBSs has been improved significantly. Further improvements have been realized by refining the fabrication process of the core material and introducing partial texturing of the wire core. The largest transport J_c for round wires at 4.2 K at self-field and 100 kOe are $2.0x10^5$ A/cm^2 and $3.8x10^4$ A/cm^2, respectively. We also compare the J_c characteristics of wires and tapes processed by HIP.

1. Introduction

Iron-based superconductors (IBSs) have very promising characteristics for high-field applications with their reasonably high T_c, H_{c2}, and J_c, and small anisotropy [1,2]. Among various IBSs, 122-type compounds are most extensively studied due to the fact that stable single crystals with good characteristics can be easily obtained. J_c close to $1.0x10^6$ A/cm^2 at low temperatures is realized in single crystals of 122-type IBSs [3,4] and further enhancement of J_c by irradiation has also been demonstrated [5-7]. In particular, (Ba,K)Fe_2As_2 single crystal has a large J_c, and it can be enhanced over $1x10^7$ A/cm^2 (4.2 K, self-field) after irradiation [8-10]. Several techniques have been applied to fabricate superconducting wires and tapes of IBSs [1]. Among them, round wires fabricated using the powder-in-tube (PIT) method is the simplest ones and the first wires using 1111-type IBS was fabricated using this technique [11]. In addition, round wires produced by PIT technique is most suitable for winding high-field magnets. Up to now, using 122-type IBSs, (Ba,K)Fe_2As_2 and (Sr,K)Fe_2As_2 PIT round wires have been fabricated, and reasonably high in-field J_c has been reported [8, 12-19]. On the other hand, J_c characteristics in tapes are better than round wires, in particular, in pressed tapes. The practical level of J_c of $1x10^5$ A/cm^2 has been achieved even at 100 kOe in pressed tapes [20,21]. This is due to the high degree of texturing and high density in the pressed tape.

There are three kinds of sites where we can modify physical properties of 122-type IBSs, namely

$(Ba,A)Fe_2As_2$ (*A*: Alkali earths), $Ba(Fe,M)_2As_2$ (*M*: transition metals), and $BaFe_2(As,P)_2$. Among these three kinds of materials, $BaFe_2(As,P)_2$ is least extensively studied. Although reasonably high-quality bulk polycrystalline materials are reported, their J_c characteristics are not well known [22]. Actually, when the PIT wires are fabricated using $BaFe_2(As,P)_2$, only a small value of J_c~1,000 A/cm^2 (4.2 K, self-field) is realized possibly due to the presence of minor ferromagnetic impurity phase Fe_2P. The situation is similar in the case of $Ba(Fe,Co)_2As_2$. The largest J_c reported for wires of $Ba(Fe,Co)_2As_2$ is ~ 1x10^4 A/cm^2 at 4.2 K and self-field [14,19,23].

In this short article, present status of J_c characteristics of PIT round wires of 122-type materials are reviewed focusing on $(Ba,K)Fe_2As_2$. We report the latest result of the highest J_c of 3.8x10^4 A/cm^2 at 4.2 K under 100 kOe in $(Ba,K)Fe_2As_2$ PIT wire. Characterizations of the PIT wires including magneto-optical imaging and X-ray analyses are reported. We also compare the J_c characteristics of wires and tapes, both processed under high pressure.

2. Experiments

All the superconducting wires of 122-type IBSs were fabricated using powder-in-tube (PIT) method. For the last couple of years, we have constantly improved the fabrication processes. Five kinds of PIT wires of $Ba_{0.6}K_{0.4}Fe_2As_2$ are reported in this article [15-18]. Typical characteristics of these wires are summarized in Table 1.

Table 1. Characteristics of round wires of $Ba_{0.6}K_{0.4}Fe_2As_2$, #1 [15], #2 [16], #3 [17], #4 [18], and new wire #5. "G" and "D" in the drawing method mean "groove-roller" and "dies", respectively.

Wires	Drawing method	Sintering T (C^o)	Sintering P (MPa)	Sintering time (h)	Diameter (mm)	J_c (4.2 K, sf) (A/cm^2)	J_c (4.2 K, 10 T) (A/cm^2)
#1	G	900	0.1	36	0.6	1.0x10^4	500
#2	G	600	120	4	1.2	3.8x10^4	3.0x10^3
#3	G	700	175	0.5	1.2	1.2x10^5	9.0x10^3
#4	G	700	175	4	1.2	1.8x10^5	2.0x10^4
#5	G/D	700	175	4	1.2	2.0x10^5	3.8x10^4

In the first generation PIT wire #1, starting raw materials were mixed in a glove box filled with Ar gas but with insufficient purity, and the mixing process was performed by hand using agate mortar [15]. After the introduction of glove box with a gas purification function and planetary ball milling machine for pulverizing raw materials for wire #2-#5, J_c was improved significantly. We also introduced hot-isostatic pressing (HIP) process after wire #2 [16]. In the HIP process, starting powders are first filled in a Ag tube with inner diameter of 3 mm, and then drawn down to a diameter of 1.2 mm. After this, the drawn wire was put in a 1/8" outer diameter Cu tube and drawn to a final diameter of ~1.2 mm. After the drawing process, both ends were sealed by using arc welding. Details of wire fabrication conditions for each wire are described in the corresponding reference, except for the #5. In the latest wire fabrication #5, wire-drawing dies were introduced to make the wires into more perfect round shape, which also helped to make the cross section of the core more circular. Unfortunately, however, we had to introduce groove-roller in the last stage of the wire drawing, since wires were often broken in the drawing process using dies when the diameter became small.

Phase purity of the wire was confirmed by using X-ray diffraction. Shielding characteristics and magnetic J_c were characterized by using a commercial SQUID magnetometer (MPMS-XL5, Quantum Design). Transport J_c was evaluated at IMR, Tohoku university by using 15 T superconducting magnet. Electrical contacts to the wire were made by soldering, and the wire was immersed in liquid He to minimize the effect of heating. Magneto-optical (MO) imaging of the wires was performed after cutting the wire to reveal the specific cross section of the wire core and polishing them using lapping films. Magnetic induction distribution was visualized by using a garnet film. MO imaging was performed using a continuous-flow type of cryostat (Microstat-HR, Oxford Instruments), and the

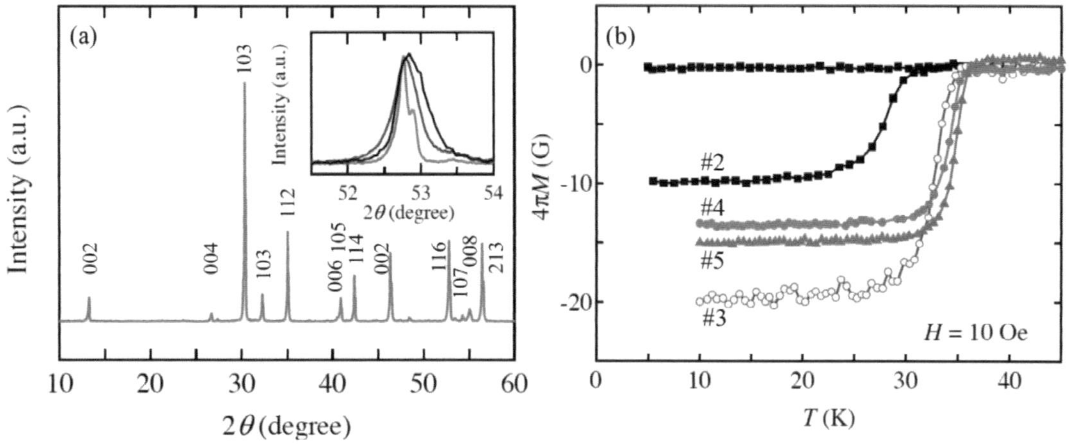

Figure 1. (a) X-ray diffraction pattern of (Ba,K)Fe_2As_2 powder for wire #4. Inset shows blow-ups of 116 peaks for wires #2 (black), #3 (blue), and #4 (red). (2) Magnetization as a function of temperature at 10 Oe for wires #2 (closed squares) #3 (open circles), #4 (closed circles), and #5 (closed triangles). As the fabrication process is refined, T_c increases and the transition width becomes narrower.

image was captured by using a cooled-CCD camera (Orca-ER, Hamamatsu).

3. Results

Figure 1(a) shows X-ray diffraction pattern of the final powder samples used for the wire #4 [18]. All peaks can be indexed by α-$ThCr_2Si_2$-type crystal structure, and no impurity peaks are identified. In the inset of Fig. 1(a), profiles of (116) diffraction peaks of powders used for the wire #2, #3, and #4 are compared. It is obvious that the new powder for wire #4 shows narrow peak profile and clear separation of $K_{\alpha 1}$ and $K_{\alpha 2}$ peaks is observed. This narrow diffraction peak also guarantees homogeneity of the grains in the wire core. In Fig. 1(b), the temperature dependence of magnetization is compared for the wire #2, #3, #4, and #5. As we successively introduced new processes for the improvement of wires, the onset of diamagnetism of the wire shifts to higher temperatures and the transition width becomes narrower. It should be noted that IBSs show strong sensitivity to strains, and T_c and volume fraction of superconductivity of the as-drawn wires are severely suppressed [17]. Only after proper sintering process, the value of T_c and volume fraction recover. In this sense, T_c as high as 36 K with very narrow transition width in the fabricated wire is not so obvious.

Figures 2(a), (b), (c), (d), (e) show optical micrographs of the cross section of the wires #1, #2, #3, #4, and #5, respectively. The corresponding MO images are shown in Fig. 2(f), (g), (h), (i), and (j), respectively. Obviously, the first generation wire #1 without HIP process shown in Fig. 2(f) is granular and low J_c is expected. Actually, transport J_c at 4.2 K and self-field for the wire #1 is only 1.0×10^4 A/cm^2 [15]. After introducing Ar-gas purification system, planetary ball milling, and high pressure sintering (HIP process) at 120 MPa, obvious granularity has been eliminated as shown in the image for wire #2 (Fig. 2(g)). Introduction of better-quality starting powders sustained the suppression of granularity as shown in Fig. 2(h). MO images keep to demonstrate large intergranular current after introducing refined powders by eliminating outer skin part of reacted materials in wire #4 (Fig. 2(i)). Now, the latest round wire partly drawn using dies has better geometrical symmetry of the core compared with earlier wires as shown in Fig. 2(j). One may notice some darker linear defects in some of MO images, as shown in upper half of Fig. 2(h). These are related to the cracks in the core of the wire which were introduced in the process of drawing. Similar crack formations have been reported in pressed tapes of 122-type IBSs [24]. The presence and direction of the cracks with respect to the current flow is responsible for the discrepancy between the J_c values evaluated by transport method and those evaluated magnetically [25].

Transport J_c as a function of magnetic field up to 140 kOe for wire #2 (open squares), #3 (closed

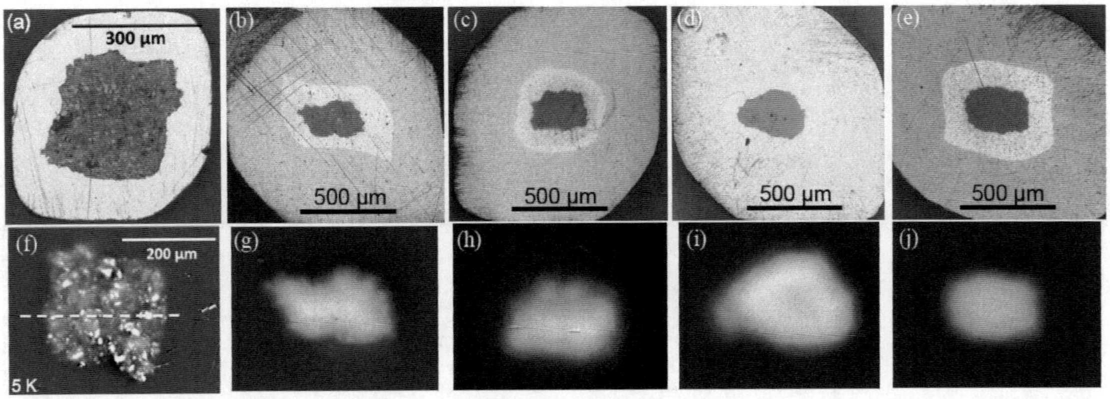

Figure 2. Optical micrographs of PIT wires (a) #1, (b) #2, (c) #3, (d) #4, and (e) #5. Corresponding MO images of PIT wires in the remanent state at 5 K (f) #1, (g) #2, (h) #3, (i) #4, and (j) #5.

squares), #4 (open circles), and #5 (closed circles) are shown in Fig. 3(a). It is obvious that transport J_c increases steadily as the fabrication process is refined. After the introduction of HIP process, the low-field J_c exceeded $1.0x10^5$ A/cm^2, and it reached $2.0x10^5$ A/cm^2 in wire #5. The transport J_c at 4.2 K under 100 kOe has also been enhanced steadily and reached $3.8x10^4$ A/cm^2 in wire #5. In Fig. 3(b), transport J_c as a function of magnetic field for HIP tapes fabricated in parallel with HIP wires #4 (#4': open triangles) and #5 (#5': closed triangles) are shown in Fig. 3(b) [18]. It should be noted that HIP tape has better J_c characteristics compared with HIP wires especially at lower fields even if they are fabricated from the same starting powder.

4. Discussion

Let us discuss the reason why we obtained almost twice as large as J_c in wire #5 compared with wire #4 at high fields. Our initial intention of using wire-drawing dies was to make the shape of the core more symmetric and its density higher. However, it is well known that grains in the core of the PIT round wires of quasi-two-dimensional materials such as cuprate or MgB_2 are textured [26]. This is due to the plate-like shape of the grains. In order to confirm if grains in wire #5 are textured to some extent or not, we took X-ray diffraction pattern for two orthogonal cross sections shown in Figs. 4(e) and (f). In the case of X-ray diffraction from the longitudinal cross section (Fig. 4(c)), the intensity ratio of

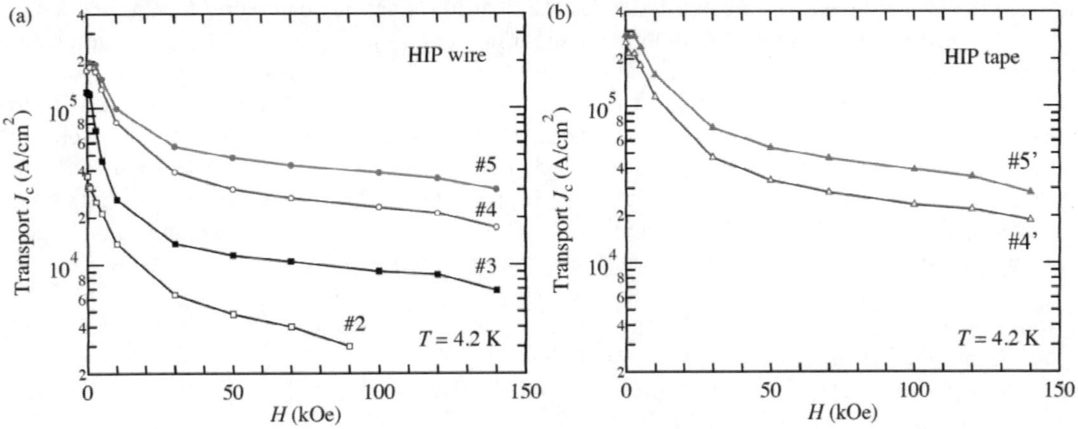

Figure 3. (a) Transport J_c as a function of field up to 140 kOe for wire #2 (open squares), #3 (closed squares), #4 (open circles), and #5 (closed circles). (b) Transport J_c as a function of magnetic field for HIP tapes fabricated in parallel with HIP wires #4 (#4': open triangles) and #5 (#5': closed triangles).

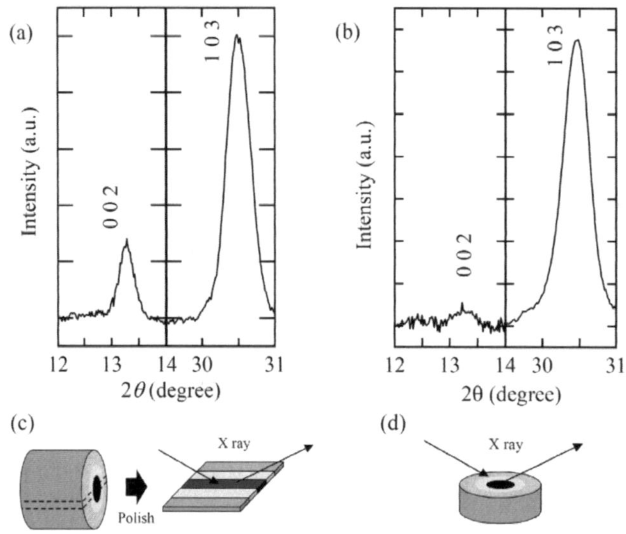

Figure 4. (a) X-ray diffraction profiles of (002) and (103) peaks taken for the longitudinal cross section shown in (c). The same taken for the transverse cross section shown in (d).

(002) and (103) peaks $I(002)/I(103) \sim 0.28$. On the other hand, in the case of transverse cross section (Fig. 4(d)), the same ratio $I(002)/I(103)$ is less than 0.06. The weaker $I(002)$ in the case of the transverse cross section indicates that the fraction of *c*-axis oriented grain on this surface is small. On the other hand, modest value of $I(002)/I(103)$ for the longitudinal cross section supports that fraction of *c*-axis oriented grains on this surface is not negligible. Such *c*-axis-oriented grains can couple along the direction of the wire and at least partly suppress weak-links. In the ideal case of round wires made out of quasi-two dimensional material with tetragonal symmetry, all grains are expected to align so that they become axisymmetric with respect to the center of the wire. To quantitatively evaluate the degree of texturing in the wire core, we may need to investigate the evolution of the X-ray diffraction intensity by successively exposing different surfaces at different depths or X-ray with much shorter wavelength may be necessary.

5. Summary

Present status of the development of round wires using 122-type iron-based superconductors is reviewed. Transport J_c has been enhanced by successive introductions of several techniques such as high-pressure sintering, control of drawing and sintering condition, promotion of the reaction of raw materials by densification, and introduction of dies in the drawing process, which enhanced the texturing inside the wire core. The present records of transport J_c of round wires are realized for that processed at 175 MPa with values at 4.2 K under self-field and 100 kOe being 2.0×10^5 A/cm^2 and 3.8×10^4 A/cm^2, respectively. HIP tapes processed in the same condition also demonstrated good performance with its transport J_c at self-field and 100 kOe being 2.8×10^5 A/cm^2 and 3.8×10^4 A/cm^2, respectively. We expect that further enhancement of J_c can be realized by enhancing the degree of texturing and further densification of the core using higher pressure.

6. References

[1] Ma Y 2012 *Supercond. Sci. Technol.* **25**, 113001
[2] Putti M *et al.* 2010 *Supercond. Sci. Technol.* **23**, 034003
[3] R Prozorov, Ni N, Tanatar M A, Kogan V G, Gordon R T, Martin C, Blomberg E C, Prommapan P, Yan J Q, Bud'ko S L and Canfield P C 2008 *Phys. Rev.* B **78**, 224506
[4] Nakajima Y, Taen T and Tamegai T 2009 *J. Phys. Soc. Jpn.* **78**, 023702

[5] Nakajima Y, Tsuchiya Y, Taen T, Tamegai T, Okayasu S and Sasase M 2009 *Phys. Rev.* B **80**, 012510
[6] Tamegai T, Taen T, Yagyuda H, Tsuchiya Y, Mohan S, Taniguchi T, Nakajima Y, Okayasu S, Sasase M, Kitamura H, Murakami T, Kambara T and Kanai Y 2012 *Supercond. Sci. Technol.* **25**, 084008
[7] Taen T, Nakajima Y, Tamegai T and Kitamura H 2012 *Phys. Rev.* B **86**, 094527
[8] Pyon S, Taen T, Ohtake F, Tsuchiya Y, Inoue H, Akiyama H, Kajitani H, Koizumi N, Okayasu S and Tamegai T 2013 *Appl. Phys. Express* **6**, 123101
[9] Ohtake F, Taen T, Pyon S, Tamegai T, Okayasu S, Kambara T and Kitamura H 2015 *Physica* C **518**, 47
[10] Taen T, Ohtake F, Pyon S, Tamegai T and Kitamura H 2015 *Supercond. Sci. Technol.* **28**, 085003
[11] Gao Z S, Wang L, Qi Y P, Wang D L, Zhang X P and Ma Y M 2008 *Supercond. Sci. Technol.* **21**, 105024
[12] Qi Y P, Wang L, Wang D L, Zhang Z Y, Gao Z S, Zhang X P and Ma Y W, *Supercond. Sci. Technol.* **23**, 055009
[13] Togano K, Matsumoto A and Kumakura H 2011 *Appl. Phys. Express* **4** 043101
[14] Weiss J D, Tarantini C, Jiang J, Kametani F, Polyanskii A A, Larbalestier D C and Hellstrom E E 2012 *Nat. Mater.* **11**, 682
[15] Ding Q P, Prombood T, Tsuchiya Y, Nakajima **Y** and Tamegai T 2012 *Supercond. Sci. Technol.* **25**, 035019
[16] Pyon S, Tsuchiya Y, Inoue H, Kajitani H, Koizumi N, Awaji S, Watanabe K and Tamegai T 2014 *Supercond. Sci. Technol.* **27**, 095002
[17] Pyon S, Yamasaki Y, Kajitani H, Koizumi N, Tsuchiya Y, Awaji S, Watanabe K and Tamegai T 2015 *Supercond. Sci. Technol.* **28**, 125014
[18] Pyon S, Suwa T, Park A, Kajitani H, Koizumi N, Tsuchiya Y, Awaji S, Watanabe K and Tamegai T 2016 *Supercond. Sci. Technol.* **29**, 115002
[19] Tamegai T, Pyon S, Ding Q P, Inoue H, Kobayashi H, Tsuchiya Y, Kajitani H and Koizumi N 2014 *J. Phys. Conf. Ser.* **507**, 022041
[20] Gao G S, Togano K, Matsumoto A and Kumakura H 2014 *Sci. Rep.* **4**, 4065
[21] Zhang X P, Yao C, Lin H, Cai Y, Chen Z, Li J Q, Dong C H, Zhang Q J, Wang D L, Ma Y W, Oguro H, Awaji S and Watanabe K 2014 *Appl. Phys. Lett.* **104**, 202601
[22] Adachi S, Murai Y, Tanabe K 2014 *Physica* C **483**, 67
[23] Inoue H, Tsuchiya Y, Tada S, Pyon S, Kajitani H, Koizumi N and Tamegai T 2014 *Physica* C **504**, 73
[24] Lin H, Yao C, Zhang X P, Zhang H T, Wang D L, Zhang Q J, Ma Y Wa, Awaji S and Watanabe K 2015 *Sci. Rep.* **4**, 4465
[25] Pyon S, Mine A, Suwa T and Tamegai T 2015 *Physica* C **530**, 76
[26] Bjoerstad R, Scheuerlein C, Rikel M O, Ballarino A, Bottura L, Jiang J, Matras M, Sugano M, Hudspeth J and Michiel M D 2015 *Supercond. Sci. Technol.* **28**, 062002

Acknowledgements
This work was partially supported by a Grant-in-Aid for Scientific Research (A) (17H01141), a Grant-in-Aid for Young Scientists (B) (16K17745), and the Japan-China Bilateral Joint Research Project by the Japan Society for the Promotion of Science (JSPS). We thank Z. Gao, K. Togano and H. Kumakura for valuable comments for the improvement of the quality of polycrystalline sample.

IOP Conf. Series: Materials Science and Engineering **279** (2017) 012029 doi:10.1088/1757-899X/279/1/012029

A high field and cryogenic test facility for neutron irradiated superconducting wire

A Nishimura[1], H Miyata[2], M Yoshida[3], M Iio[3], K Suzuki[4], T Nakamoto[4], M Yamazaki[5] and T Toyama[5]

[1] National Institute for Fusion Science, Toki, Gifu 509-5292 Japan
[2] Japan Superconductor Technology, Inc. Kobe, Hyogo 651-2271 Japan
[3] Japan Proton Accelerator Research Complex, Tokai, Ibaraki 319-1195 Japan
[4] High Energy Accelerator Research Organization, Tsukuba, Ibaraki 305-0801 Japan
[5] Tohoku University, Oarai, Ibaraki 311-1313 Japan

nishi-a@nifs.ac.jp

Abstract. A 15.5 T superconducting magnet and a variable temperature insert (VTI) system were installed at a radiation control area in Oarai center in Tohoku University to investigate the superconducting properties of activated superconducting materials by fast neutron. The superconductivity was measured at cryogenic temperature and high magnetic field. During these tests, some inconvenient problems were observed and the additional investigation was carried out. The variable temperature insert was designed and assembled to perform the superconducting property tests. without the liquid helium. To remove the heat induced by radiation and joule heating, high purity aluminum rod was used in VTI. The thermal contact was checked by FEM analysis and an additional support was added to confirm the decreasing the stress concentration and the good thermal contact. After the work for improvement, it was affirmed that the test system works well and all troubles were resolved. In this report, the improved technical solution is described and the first data set on the irradiation effect on Nb_3Sn wire is presented.

1. Introduction

Large scale plasma experimental devices aimed to realize deuterium-tritium or deuterium-deuterium reaction have been planned. ITER and JT-60SA are under construction [1,2] and design activities of DEMO or a prototype reactor are ongoing. Since some neutrons will stream out of the plasma vacuum vessel and pass through the shielding blankets, magnets in these devices will be irradiated by fast neutrons [3,4]. Also, high energy particle physics requires excellent performance of superconducting magnets under high energy particles irradiation, such as International Linear Collider and LHC HL upgrade project [5]. These projects are expected to achieve ultrahigh level nuclear reactions and generate a lot of high energy neutrons and particles in GeV order. Therefore, it calls the knowledge of neutronics in the magnet community and the investigations on the effect of the neutron irradiation on the superconducting properties.

The apparatus of the VTI is shown in Figure 1 (a) and the assembly drawing is shown in Figure 1 (b). And the sample holder part is enlarged in Figure 1 (c). The VTI has two sets of G-M refrigeration and the second stage of the refrigerator is connected to a high purity aluminum rod with a copper plate. Superconducting wire is set on the sample holder by soldering. The heat of radiation and joule heating is transferred to G-M refrigerators by the conduction cooling with the high purity aluminum rod. The contact part of the copper plate with the high purity aluminum rod is described in Figure 2. The heat comes up from the bottom of the VTI and transferred to the copper plate passing through the contact surface.

Some activities on the commissioning and studies on the activated superconducting strands have been performed and inconvenient troubles were recognized. In this paper, the typical troubles found during commissioning will be presented and the work done for improvement will be described. In addition, the experimental results of irradiated Nb_3Sn wire up to 1.0 x 10^{22} n/m^2 of over 0.1 MeV neutron will be reported as the first report.

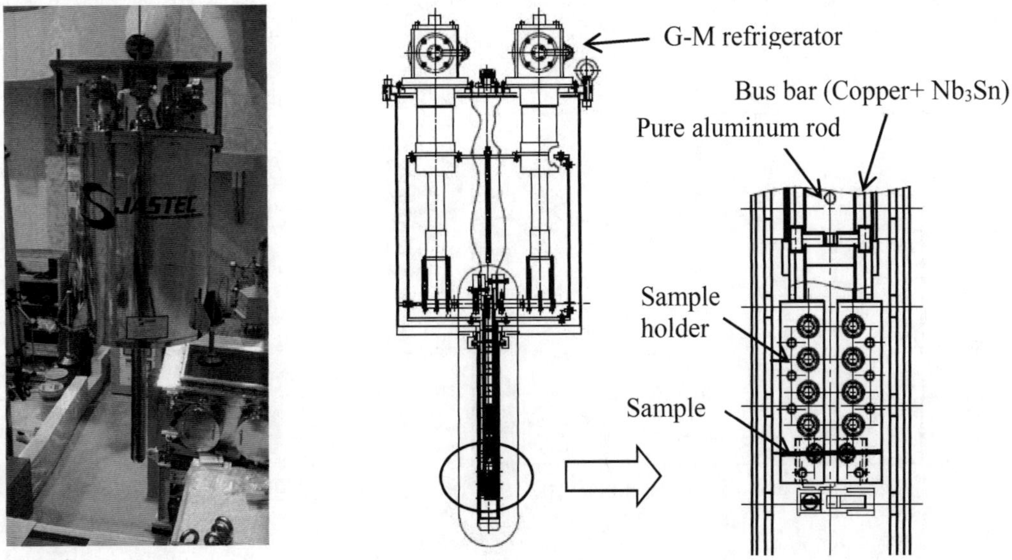

(a) Apparatus of VTI. (b) Illustration of VTI. (c) Sample holder.

Figure 1. Variable temperature insert.

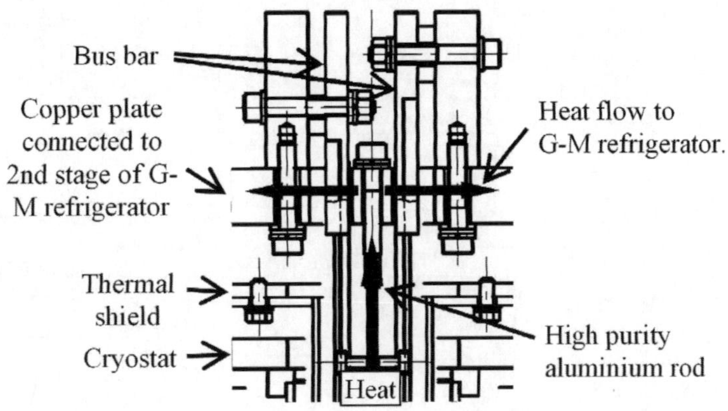

Figure 2. Connection part of copper plate with high purity aluminum rod.

2. Example of trouble

As part of commissioning tests, the current test at high magnetic field was carried out. One result tested at 15.5 T is shown in Figure 3. The sample was an unirradiated Nb_3Sn wire and three CERNOX sensors were attached. All three readings increased as the current was swept significantly. As the conduction cooling was adopted, some temperature rise was expected, but the result was too large of a temperature rise. From the results, it is recognized that the heat removal ability is less than the designed one. Also, it is noticed that the plus and minus electrodes temperatures are not the same and the sample temperature is almost the average of two electrodes temperatures. Since the heat removal ability depends on the thermal contact, both contact conditions of the bus bars would not be same.

Another result is shown in Figure 4. The current test was performed at 0 T, 2.5 T, 5.0 T and 7.5 T with an unirradiated Nb_3Sn wire. The current was swept up to 500 A in 20 seconds. A small temperature rise was observed at 0 T, 2.5 T and 5.0 T. The temperature rise became larger as the magnetic field

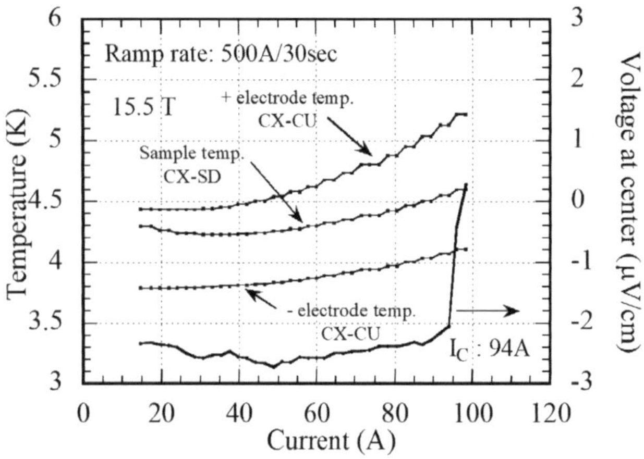

Figure 3. Current test results of unirradiated Nb_3Sn wire at 15.5 T.

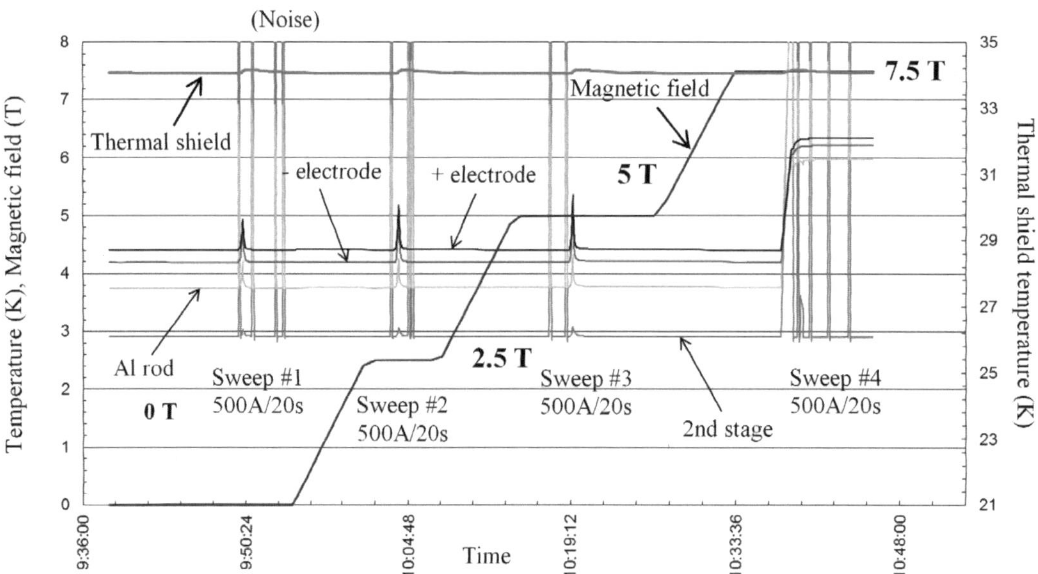

Figure 4. Temperature rise observed at high magnetic field with unirradiated Nb_3Sn wire.

increased. This seems to be caused by the effect of magnetic resistance. At 7.5 T, the temperatures of plus, minus electrodes and the aluminum rod did not return to the original temperature after the current test. Since the second stage temperature came back to the original position and the aluminum rod temperature did not return, it is considered that the heat transfer on the interface between the copper plate and the high purity aluminum rod was degraded by the electro-magnetic force.

3. Analysis of VTI and reassembled VTI

An illustration of the rod of VTI is shown in Figure 5. There is a thick copper plate connecting to the second stage of G-M refrigerator and a high purity aluminum rod which is sandwiched by a negative and a positive bus bars. The superconducting wire sample is soldered on the sample holder which locates at the top of the VTI rod. (See Figure 1) When the current is put into the sample wire under the high magnetic field, the electromagnetic in the vertical direction to the sample current acts on the sample, and the rod is bended like a cantilever beam. The expected maximum electromagnetic force is about 170 kN. (500 A x 15.5 T x 22.6 mm (span length of the sample wire) / 1000 mm) The fixed point becomes the thermal contact point. Therefore, it was considered that the contact point was deformed by the bending moment resulting in the degradation of the heat transfer. To confirm this consideration, the stress and deformation analysis was carried out with and without an additional support.

The results of the analysis are summarized in Figure 6. Two additional supports are considered. The first one is a fixer (attachment of copper) which pushes the contact surface of the high purity aluminum rod to the copper plate and keeps the gap between the copper plate and the aluminum less than 0.001 mm. The second is the L-type support made of 304 stainless steel. By adding these supports, the stress concentration at the corner of the aluminum rod decreased from about 40 MPa to about 17 MPa, and the stress concentration at the corner of the L-type support is about 100 MPa. The permanent deformation

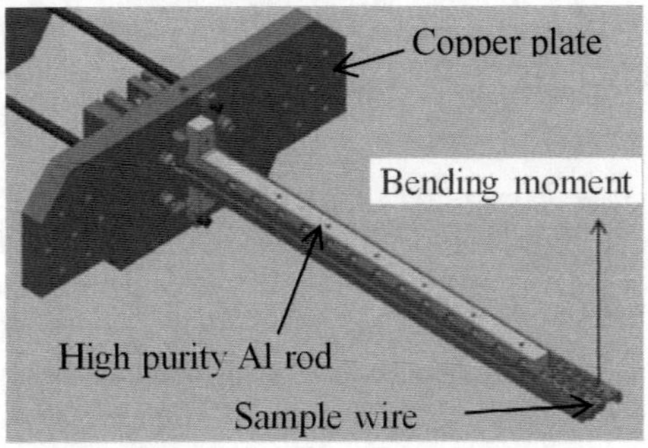

Figure 5. Structure of the rod of VTI.

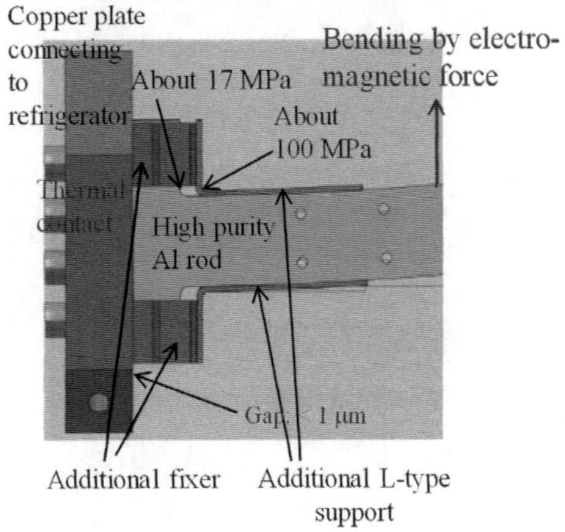

Figure 6. Results of stress and deformation analysis.

IOP Conf. Series: Materials Science and Engineering **279** (2017) 012029 doi:10.1088/1757-899X/279/1/012029

at the corner of the aluminum rod would be reduced by this support and the aluminum rod would be pushed to the copper plate more tightly.

A photo of the reassembled contact part of the VTI is shown in Figure 7. The aluminum rod was attached to the copper plate firmly. On the side of the copper bus bar, the soldered part is able to be seen. The Nb_3Sn wire was embedded to reduce the joule heating.

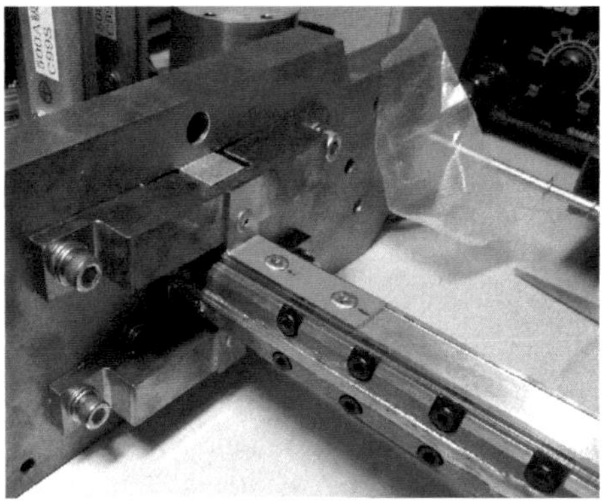

Figure 7. Reassembled contact part of high purity aluminum rod and copper plate.

4. Results of commissioning test

After the analytical activity, the improvement work has been performed in the company. The work was confirmed at the company and then the VTI was installed again at Oarai center of Tohoku University. The commissioning test was carried out using the 15.5 T superconducting magnet and the neutron irradiated Nb_3Sn wire (irradiated up to 1 x 10^{22} n/m^2 (> 0.1 MeV) at JRR-3, Japanese fission reactor for research) was set on the sample holder. The cooling of the VTI was performed without any troubles and the current test was carried out at cryogenic temperature. The Nb_3Sn wire was fabricated at Furukawa Electric Corporation Ltd [6,7].

Obtained data at 10 T is shown in Figure 8. A ramp rate was 250 A/ 10 s and the sampling rate of the current and the voltage was 10 Hz (Figure 8 (a)) and that of the temperature was 0.2 Hz (Figure 8 (b)). The maximum current was about 227 A and shut down successfully. The temperatures of negative and positive electrodes rose to about 14 K and cooled down again to the same temperatures as those before test within a couple of minutes. This temperature rise was caused by the joule heating at sample holder made of copper. The heat would be transferred to the CERNOX sensor and heat up the sensor. Therefore, there is a few seconds delay and cooled again by the heat conduction. The 500 A current test at 15.5 T

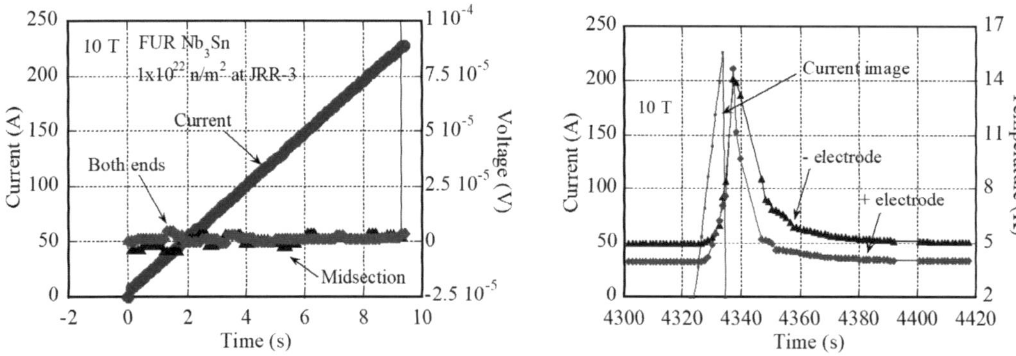

(a) Current and voltage vs time diagram.

(b) Current and temperature vs time diagram.

Figure 8. Commissioning test results at 10 T (Nb_3Sn irradiated to 1 x 10^{22} n/m^2).

will be performed in neat future. From these results, it is recognized that the heat transfer on the interface of the cooper plate to the high purity aluminum rod was improved, and that the VTI works well.

Figure 9 shows the effect of neutron irradiation on the critical current of low temperature superconducting wires [6]. Star symbols show the results obtained in this study and all other data was from samples irradiated by 14 MeV pure neutron at Fusion Neutron Source in JAEA (shut down in March 2017.) and the critical currents were measured at High Field Laboratory for Superconducting Materials in Tohoku University using liquid helium. The samples were immersed in the liquid helium and the sample temperature was kept at 4.2 K. As shown in Figure 8 (b), the negative electrode temperature in this study was around 5 K. So, the critical temperature might be a little bit low. But even if this consideration is included, there is a big difference in the critical temperatures of the irradiated samples. Namely, the star symbols are plotted around the solid round symbols. It suggents that the 14 MeV neutron irradiation would have a stronger effect than the fission neutron irradiation on the change in the critical current of Nb_3Sn wire.

Further investigations are needed. The critical current measurement of the neutron irradiated low temperature superconducting wires will be continued with the improved VTI system and the mechanism of the neutron irradiation effect would be clarified in near future.

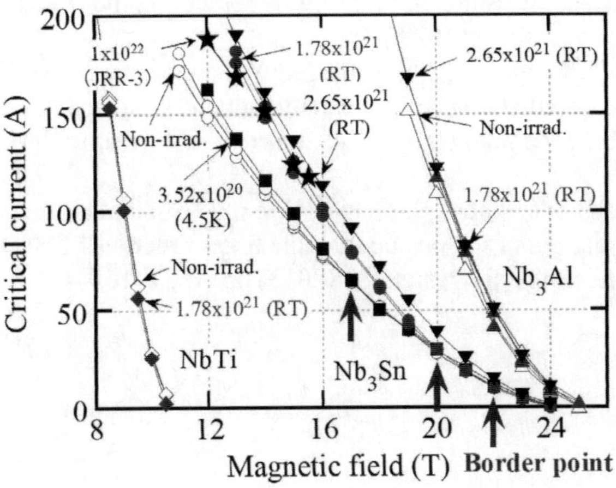

Figure 9. Comparison of critical currents after neutron irradiation.

5. Summary

During the several tests for the critical current measurement with the VTI system, the unexpected temperature rise was observed and the reason for the trouble was investigated. Then the improvements to the thermal connections of VTI system were implemented that allowed the system to function properly and the following conclusions to be made.

The interface between the copper plate and the high purity aluminum rod works as a heat transfer path and the thermal contact is very sensitive on the electromagnetic force. To keep the good contact condition, the mechanical support is important and the additional supports considered in the analytical investigation improved the performance of the VTI. The designed electromagnetic force has not been applied to the sample holder, but such commissioning test will be carried out in near future.

The neutron irradiated Nb_3Sn wire was investigated using the improved VTI and is was noticed that the fission neutron irradiation would have a weak effect than 14 MeV neutron irradiation on the change in the critical current. This is very new data and the following experiments would be performed continuously. The mechanism of the neutron irradiation effect on the critical current will be clarified and it would contribute to the development of the new low temperature superconductor.

IOP Conf. Series: Materials Science and Engineering **279** (2017) 012029 doi:10.1088/1757-899X/279/1/012029

Acknowledgments
The authors would like to express the thanks to Dr. T. Shikama at Hachinohe Institute of Technology, Drs. T. Takeuchi, G. Nishijima at NIMS, Y. Hishinuma, T. Tanaka, T. Muroga at NIFS, H. Oguro at Tokai University and K. Ochiai at QST for their contribution to perform this study, and to Japan Superconductor Technology, Inc. for improving the system.

This work was performed at International Research Center for Nuclear Materials Science and High Field Laboratory for Superconducting Materials in Institute for Metals Research, Tohoku University.

References
[1] ITER home page, http://www.iter.org/.
[2] JAEA home page, http://www.naka.jaea.go.jp/.
[3] T. Tanaka, A. Sagara, M.Z. Youssef: "Development of three-dimensional neutronics calculation system for design studies on helical reactor FFHR," Fusion Engineering & Design 81 (2006) pp2761–2766.
[4] S. Sato, H. Iida, and T. Nishitani: "Evaluation of shutdown gamma-ray dose rates around the duct penetration by three-dimensional Monte Carlo decay gamma-ray transport calculation with variance reduction method," Journal of Nuclear Science and Technology 39 (2002) 1237-1246.
[5] F. Cerutti, M. Mauri, A. Mereghetti, E. Todesco, E. Wildner: "Energy Deposition in the New IRs," http://indico.cern.ch/ conferenceDisplay.py?confId=36286
[6] A. Nishimura, T. Takeuchi, S. Nishijima, K. Ochiai, G. Nishijima, K. Watanabe, M. Narui, H. Kurishita, T. Shikama; "Study on neutron irradiation effect of superconductors and installation of 15.5 T magnet in hot laboratory at IMR in Tohoku University", Journal of Nuclear Materials, 417 (2011) pp. 842–845.
[7] A. Nishimura, T. Takeuchi, G. Nishijima, K. Watanabe, K. Ochiai, D. Li, T. Shikama, "A new facility for investigation on neutron irradiation effect on superconducting properties of Nb_3Sn strand for fusion magnet", Fusion Engineering and Design, 88 (2013) pp. 1551–1554.

IOP Conf. Series: Materials Science and Engineering **279** (2017) 012030 doi:10.1088/1757-899X/279/1/012030

Towards Isotropic Vortex Pinning in YBCO Films with Double-doping BHO-Y_2O_3 and BZO-Y_2O_3 Artificial Pining Centers

Bibek Gautam[*1], Mary Ann Sebastian[2], Shihong Chen[1], Timothy Haugan[2], Yanbin Chen[3], Zhongwen Xing[4], Joseph Prestigiacomo[5], Mike Osofsky[5] and Judy Wu[*1]

[1]*Department of Physics and Astronomy, University of Kansas, Lawrence, Kansas 66045, USA*
[2]*U.S. Air Force Research Laboratory, Propulsion Directorate, WPAFB, OH 45433 USA*
[3]*Physics Department at Nanjing University, Nanjing, Jiangsu 210093, China*
[4]*College of Engineering and Applied Science, Nanjing University, Nanjing, Jiangsu 210093, China*
[5]*US Naval Research Laboratory, 4555 Overlook Ave, SW Washington, DC 20375, USA*

Email: gautbibe@ku.edu, jwu@ku.edu

Abstract. Strong and isotropic vortex pinning landscape is demanded for high field applications of $YaBa_2Cu_3O_{7-x}$ (YBCO) epitaxial thin films. Double-doping (DD) of artificial pinning centers (APCs) of mixed morphologies has been identified as a viable approach for this purpose. This work presents a comparative study on the transport critical current density $J_c(H, \theta)$ of 3.0 vol.%Y_2O_3+2.0 (or 6.0) vol.% $BaZrO_3$ (BZO DD) and 3.0 vol.%Y_2O_3+ 2.0 (or 6.0) vol.% $BaHfO_3$ (BHO DD) films. Based on the elastic strain model, $BaHfO_3$ (BHO) nanorods have lower rigidity than their $BaZrO_3$ (BZO) counterparts, which means their c-axis alignment is more susceptible to the local strain generated by the secondary dopant of Y_2O_3. Considering the increasing strain field with higher BZO (or BHO doping), the higher susceptibility may result in a large portion of the BHO APCs moving away from perfect c-axis alignment and enhancing isotropic pinning with respect to the H orientation. This is confirmed since the BHO DD films illustrate a less pronounced J_c peak at H//c-axis and hence more isotropic $J_c(\theta)$ than their BZO DD counterparts. At 9.0 T, the variation of the J_c across the entire θ range (0-90 degree) is less than 18% for the BHO DD film, in contrast to about 100% for the 2.0 vol.% BZO DD counterpart. At the higher BHO concentration of 6.0 vol.%, this higher tunability of the Y_2O_3 leads to increased ab-plane aligned BHO APCs and hence enhanced J_c at *H*//ab-plane.

1. Introduction

Artificial pinning centers (APCs) created by inserting impurity phase materials like $BaZrO_3$ (BZO), $BaSnO_3$ (BSO), $BaHfO_3$ (BHO), and $YBa_2(Nb/Ta)O_6$ in the film matrix have shown to provide an effective approach to enhance the critical current density J_c in applied magnetic fields (H) [1-7]. This has motivated efforts in controlling the APC morphology, dimension, orientation, and concentration for optimized $J_c(H, \theta, T)$. Double doping (DD) has been recently employed to generate APCs of mixed morphology such as 1D+3D APCs in BZO+Y_2O_3 [1, 8], BSO+Y_2O_3 [9, 10], BHO+Y_2O_3 [11], $BaTiO_3$+Y_2O_3 [12] and $Ba_2Y(Nb/Ta)O_6$ [13, 14] doped RE-123 matrices. The c-axis aligned 1D APCs can provide strong correlated pinning to address the weak pinning due to the layered structure of the RE-123. The 3D APCs have been found to further reduce the angular dependence of the J_c with respect to the H orientation. Among others, the BSO+Y_2O_3 (or Y211) doped YBCO thin films and conductors (BSO DD) have received most intensive study since BSO 1D APCs (or nanorods) have a relatively larger diameter of 7-10 nm [10] than the most other 1D APCs' and therefore remain their c-axis alignment in the BSO DD films in the moderate doping range [9, 11, 15]. 1D APCs of smaller diameters, such as BZO (5-6 nm) [3], BHO (4-5 nm) [2] are expected to have less rigidity and therefore are more susceptible to the local nonuniform strain field, induced by seeded growth, or DD, or vicinal growth, resulting in shorter segments of 1D APCs or splayed around the c-axis [7, 8, 16] An unique benefit of the more disordered APC landscape study is the reduced strain field overlap at high APC concentration and hence reduced T_c degradation and improved J_c at 77 K. In DD format, 1D APCs of smaller diameters may allow high concentration doping for enhanced isotropic pinning.

Motivated by this, we investigate transport $J_c(H,\theta)$ in 2 and 6 vol.% BZO or BHO + 3 vol.% Y_2O_3 (BZO or BHO DD) in YBCO thin films. Although having comparable diameters of ~5-6 nm [17, 18], BZO 1D APCs have higher rigidity than their BHO counterparts according to an elastic strain model [19] with consideration of the lattice mismatch of APC and YBCO along with the elastic properties of both. The comparison of the 2 vol.% BZO and BHO DD films has revealed promising trend anticipated from the theory. At 9.0 T, for example, the variation of the J_c across the entire θ range (0-90 degree) is less than 18% for the BHO DD film, in contrast to about 100% for the 2 vol.% BZO DD counterpart. This work intends to further this study by including the 6 vol.% BZO and BHO DD films in this comparative study of the transport $J_c(H,\theta)$ at different temperatures (77 K and 65 K) and field up to 9.0 T at different orientation at $\theta = 0^o$ (*H*//c-axis), $\theta = 45^o$, and $\theta = 90^o$ (*H*//ab-plane). A strong and isotropic pinning can be achieved in the low concentration (2 vol. %) BHO DD film, a significant orientation switch of BHO APCs from c-axis aligned to ab-plane aligned occurs at 6 vol.% and hence improves J_c at $\theta=90^o$. This is in contrast to the well mixed APC morphology in 6 vol% BZO DD case.

2. Sample preparation and experiment

A fixed 3 vol.%Y_2O_3 and variable 2 and 6 vol.% $BaZrO_3$ or $BaHfO_3$ mixed with YBCO targets were used to fabricate two sets of BZO DD and BHO DD thin films on (100) $SrTiO_3$ (STO) single crystal substrates at their optimum growth temperatures 825 °C and 810 °C, respectively, using pulse laser deposition (PLD). PLD system utilizes a Lambda Physik LPX 300 KrF excimer laser of wavelength (λ) =248nm and a fluence of approximately 1.6 J/cm^2 for the depositions. All the films were deposited at laser energy of 450 mJ and a repetition rate of 8 Hz, within the 300 mTorr oxygen pressure which were annealed for 30 minutes at 500 °C within an oxygen environment after deposition. The thicknesses of the film were around 140 nm and 160 nm for BZO DD and BHO DD thin films, respectively.

For electrical transport measurement, two parallel microbridges of the width ~20 μm or ~40 μm and the length ~500 μm were patterned respectively using photolithography. Quantum Design Evercool II Physical Properties Measurement System (PPMS) with a vibrating sample magnetometer (VSM) was utilized for measuring the critical temperature (T_c) and the transport J_c (*H*) at H orientations of $\theta = 0^o$ (*H*//c-axis), 45°, and 90° (*H*//ab plane) in the plane where H is always perpendicular to J_c. Angular dependence $J_c(\theta)$ at different field was also measured for understanding isotropic pinning properties. To minimize sample heating effect short current pulses of width of 50 ms and pulse interval of 3.0 s were employed in the I-V measurement.

3. Results and discussions

Figures 1a-d compare the X-ray diffraction (XRD) θ-2θ spectra of 2 vol.% and 6 vol.% BZO DD or BHO DD films. The major diffraction (00l) peaks for all films indicate they are highly crystalline c-axis oriented. The c-axis lattice constants are 1.172 nm and 1.177 nm for 2 vol.% BZO and BHO DD respectively, and 1.173 nm for 6 vol. % BZO or BHO DD films. Some weak peaks are also observed which are attributed to either crystalline BZO (or BHO) or Y_2O_3 APCs. A distinctly larger full width at a half maximum (FWHM) of 2 vol. % BZO DD (0.839 deg.) compared to FWHM of 2 vol. % BHO DD (0.353 deg.) implies a more disturbance of lattice in former as compared to the latter. However, the comparable FWHM of 0.489 deg. and 0.567 deg., respectively for 6 vol. % BZO DD and BHO DD films suggests the difference in the YBCO lattice in the two cases reduces considerably at higher BZO (or BHO) concentrations.

Figures 2a-c compare the $J_c(H)$ curves in the field up to 9.0 T at *H* orientations of $\theta = 0^o$ (*H*//c-axis), θ=45°, and θ=90° (*H*//ab plane) for 2 vol.% BZO DD (circle) and BHO DD (inverted triangle) films measured at 65 K (open) and 77 K (solid and half filled). At 65 K, J_c(*H*//c-axis) is slightly higher for 2 vol.% BZO DD film at all field range up to 9.0 T while the lower $J_c(H)$ value of this sample at 77 K may be due to the lower T_c ~ 85.71 K for BZO DD film as compared to the T_c ~ 87.08 K for 2 vol.% BHO DD film. This same T_c effect may explain the lower $J_c(H)$ value of the 2 vol.% BZO DD film at θ=45°, and θ=90° (Figures 2b-c). Based on the comparison at 65 K, both 2 vol.% BZO DD and BHO DD samples have high-concentration c-axis aligned 1D APCs as indicated by the high $J_c(H)$ at *H*//c-axis. These 1D APCs act

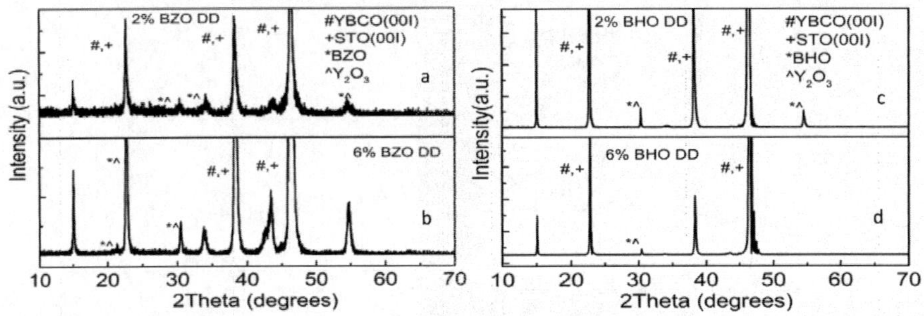

Figure 1. XRD θ-2θ scan for (a) 2 vol.%, (b) 6 vol.% BZO + 3 vol.% Y_2O_3 (BZO DD), and (c) 2 vol.%, (d) 6 vol.% BHO + 3 vol.% Y_2O_3 (BHO DD) doped YBCO thin films on STO substrates. Indices follow the same for the 2 and 6% BZO DD, and 2 and 6% BHO DD respectively.

as strong correlated pinning centers up to the so called matching field $H=n*\phi_0$, where n is the areal density of 1D APCs in c-axis and ϕ_0 is the flux quantum. Quantitatively, this enhanced pinning can be described by the α value through fitting of $J_c(H)\sim H^{-\alpha}$. For reference, YBCO films without APCs, α~0.5-0.6 [20]. The enhanced pinning in the 2 vol.% DD samples is illustrated in the much reduced α value ~ 0.17 and 0.16 for the 2 vol.% BZO DD and BHO DD respectively. At 65 K, the trend at θ=45°, indicates the enhanced pining at wide angular range is probably due to misaligned 1D APCs rather by splayed BHO 1D APCs which probably could pin the vortices in small angular range (Figure 2b). Interestingly, at 65 K, J_c (*H* // ab-plane) overcomes for 2 vol. % BHO DD film at higher field (>2.0 T) which can be seen as the crossover of the J_c at round 2.0 T (Figure 2c). This crossover trend of J_c(H) value at θ=45° and θ=90° is an indication of strong pinning ability of BHO APCs not only along ab-plane but also the angle between c-axis and ab-plane. Such enhancement at wide angular range may possibly due to the formation of misaligned and short BHO APCs in the film by influence of kinematic diffusion of BHO along c-axis in the presence of Y_2O_3.

In parallel, the $J_c(H)$ curves of the 6 vol.% BZO DD (circle) and BHO DD (inverted triangle) measured at the same conditions are compared in Figures 2d-f. With the increasing BZO and BHO doping to 6 vol.%, the overall J_c(*H*//c-axis) is higher for BZO DD film for the entire field range 0-9.0 T, and at both temperature 77 K and 65 K. Interestingly, the T_c values for 6 vol.% BHO DD film reduces to 85.8 K in contrast to T_c ~ 87.8 K for 6 vol.% BZO DD films. Despite of similar PLD sample fabrication process was used to these films, 3 vol.% Y_2O_3 influenced differently in pinning landscape. The increase T_c for 6 vol.% BZO DD film is probably through reducing strain field overlap, is also observed in 6 vol.% BZO single doped YBCO film on vicinal substrate [3]. Since $J_c(H)$~effective length of APCs, the higher J_c(*H*//c-axis) at 65 K for 6 vol.% BZO DD possibly due to longer effective c-axis aligned 1D APCs. The similar trend with comparable J_c(*H* at θ=45°) for both 6 vol.% BZO and BHO DD films observed at 65 K, where T_c effect is negligible, implies comparable pinning strength of 1D APCs (Figure 2e). Interestingly, at 65 K, J_c (*H* // ab-plane) overcomes for 6 vol.% BHO DD film at higher field (>2.0 T) same as the 2 vol.% BHO DD film. The slow decrease of J_c at high field when *H* // ab-plane or about 1.7 times higher J_c(*H* // ab-plane) in average for the entire field upto 9.0 T compared to J_c(*H* // c-axis) at 65 K for 6 vol.% BHO DD indicates the strong pinning possibly due to ab-aligned BHO 1D APCs.

However, the reverse results are observed in 6 vol.% BZO DD film in which higher $J_c(H)$ along c-axis compared to ab-plane. The higher J_c (H // ab-plane) may be due to strong intrinsic pinning of layered structure or intrinsic pining plus 1D APCs aligned along ab-plane. The former may cause sharp $J_c(\theta)$ at θ=90° while latter may cause wider peak at around θ=90°. The latter may be more convincing in this case as we have observed wider $J_c(\theta=90°)$ [details in later section].

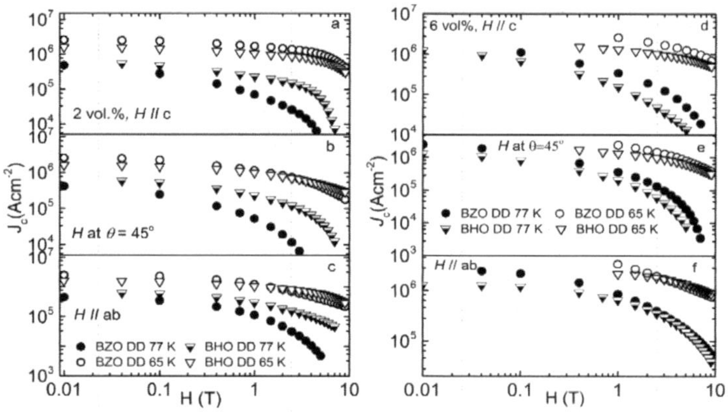

Figure 2. J_c vs. H curves measured on 3% Y_2O_3+2% BZO or BHO doped YBCO (BZO (circle) or BHO (inverted triangle) DD) nanocomposite films at (a) θ =0° (H//c-axis), (b) θ = 45°, and (c) θ = 90° (H//ab-plane) and 3% Y_2O_3+6% BZO or BHO doped YBCO (BZO or BHO DD) nanocomposite films at (d) θ =0° (H//c-axis), (e) θ = 45°, and (f) θ = 90° (H//ab-plane) at 77 K (solid and half filled) and 65 K (open) respectively.

Figures 3a-c compare the field dependence of the pining force density F_p=J x H at 77 K and 65 K at field ordination θ=0° (H // c-axis), θ=45° and θ=90° (H // ab-plane) for 2 vol.% BZO or BHO DD film. Except H at θ=90° for 2 vol.% BZO DD film, F_p vs H curve show the typical inverted bell shape with different F_p peak location (H_{max}) and peak value $F_{p,max}$. At H // c-axis (Figure 3a), the H_{max}, representing so called matching field, are comparable for BZO and BHO APCs at 65 K, explore the comparable number density of APCs are effectively acting along c-axis. Noticeably high $F_{p,max}$ (~ 1.5 times) for BZO DD compared to BHO DD films at 65 K, where T_c effect is negligibly small, support the argument of strong F_p (=n*f_p, n is the effective no density of APCs and f_p is the pinning force per unit length of APCs) through longer BZO 1D APCs aligned to c-axis. Interestingly, the same trend is not followed when H at θ=45° and H // ab-plane (Figures 3b and 3c) at 65 K. At both temperatures, the $F_{p,max}$ and H_{max} are higher for BHO DD films especially the cross-over of $F_{p,max}$ at 4.5 T and 2.5 T respectively, is an indication of increasing effectiveness of the BHO 1D APCs. In addition, at 65 K, and H at θ=45°, $F_{p,max}$~ 26.71 GN/m^3, and H_{max}~8.5 T are about 1.14 and 2.0 times higher respectively for BHO DD film compared to its counterpart. The similar trend of $F_{p,max}$ and H_{max} are observed at H // ab-plane (Figure 3c). It further implies that the addition of 3 vol.% Y_2O_3 influence more on orientation and morphology of BHO APCs compared to BZO APCs due to higher tunability of BHO as predicted theoretically in our other paper [21].

In addition, Figures 3d-f compare the F_p vs H curves measured at the same condition for 6 vol.% BZO or BHO DD film respectively. The higher $F_{p,max}$ are observed for BZO DD films except when H // ab-plane, at which $F_{p,max}$ overtakes for BHO DD film. At 65 K, when H // c-axis, $F_{p,max}$ ~ 67.99 GN/m^3 for BZO DD is about 1.5 times in contrast to BHO DD $F_{p,max}$~ 45.78 GN/m^3 but lower H_{max}~7.0 T for former compared to H_{max}~8.5 T latter. It means weaker pinning strength (measured through H_{max}) of

BZO APCs compared to BHO APCs along c-axis. Interestingly, $F_{p,max}$ gap decreases when H at θ=45° (Figure 3e) with $F_{p,max}$~40.65 GN/m³ for BZO DD. It is only about 1.3 times higher than BHO DD.

But the higher H_{max}~6.0 T for BHO DD compared to 4.5 T for its counterpart. It is indicative of higher pinning strength possibly due to high effective areal density of misaligned BHO APCs. At 65 K, the constant difference of H_{max} (1.5 T) when H // c-axis and H at θ=45° for both films suggest the equable pining strength of BZO and BHO APCs along those directions. However, the same trend is not followed for the field (H) // ab-plane (Figure 3f). The higher $F_{p,max}$>78.63 GN/m³for BHO DD is an indicative of the strong pinning by BHO APCs in contrast to BZO APCs ($F_{p,max}$ ~68.72 GN/m³). More interestingly, the constant gap of $F_{p,max}$ beyond 6.0 T, at H // ab-plane explores the fact that the almost an equal number of effective 1D APCs are involved in vortex pinning in both cases. The bell shaped F_p vs H curve for 2 vol.% BHO DD film has changed to constantly increasing $F_{p,max}$ at H>6.0 T for 6 vol.% BHO DD also suggests that BHO APCs has more tunability in the presence of Y_2O_3 in contrast to BZO APCs. It further suggests that the orientation of 1D APCs more likely changes from c-axis alignment to ab-plane alignment with increasing BHO concentration to 6 vol.% compared to similarly doped BZO film (Figures 3c and 3f). This result is different from BZO DD case in which 6 vol.% BZO DD film shows more isotropic at low temperature (≤ 65 K) in compared to lower BZO concentration [1]. This indicates that the BZO APCs in DD film may provide effective APCs along with Y_2O_3 improving vortex pinning for the field orientation other than c-axis.

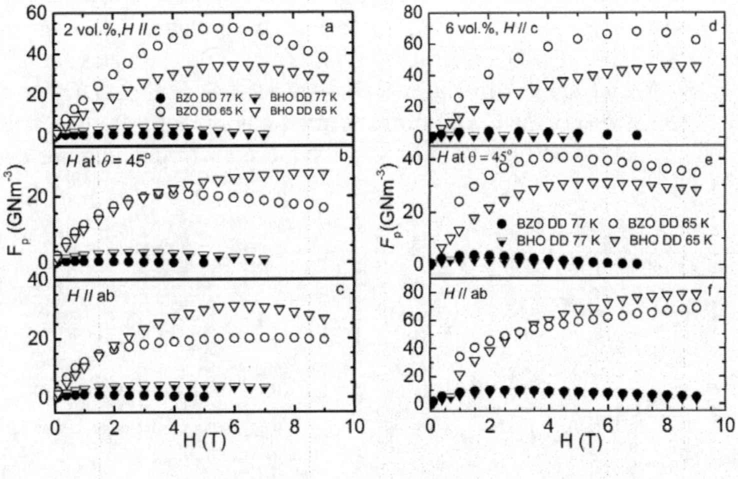

Figure 3. F_p vs. H curves measured on 3% Y_2O_3+2% BZO or BHO doped YBCO (BZO (circle) or BHO (inverted triangle) DD) nanocomposite films at (a) θ=0° (H//c-axis), (b) θ= 45°; and (c) , θ= 90° (H//ab-plane) and 3% Y_2O_3+6% BZO or BHO doped YBCO (BZO or BHO DD) nanocomposite films at (d) θ=0° (H//c-axis), (e) θ= 45°; and (f) , θ= 90° (H//ab-plane) at 77 K (solid and half filled) and 65 K (open) respectively.

Based on the discussion so far, the schematic of the BZO, BHO and Y_2O_3 nanostructures morphology and orientation in BZO DD (a,c) and BHO DD (b,d) films can be schematically represented as in Figures 4a-d. At low concentration (2 vol.%), the orientation of 1D-APCs may be perpendicular to the surface (c-axis of the film), but they may form more splayed and short APCs in BHO DD compared to BZO DD film. However, at higher concentration (6 vol.%), 1D APCs may noticeably be either parallel or perpendicular to the surface (ab-plane or c-axis) of the films. Due to the higher rigidity of BZO compared to BHO, the majority of the BZO 1D APCs may form along c-axis of the film in contrast to BHO 1D APCs along ab-plane (Figures 4c and 4d).

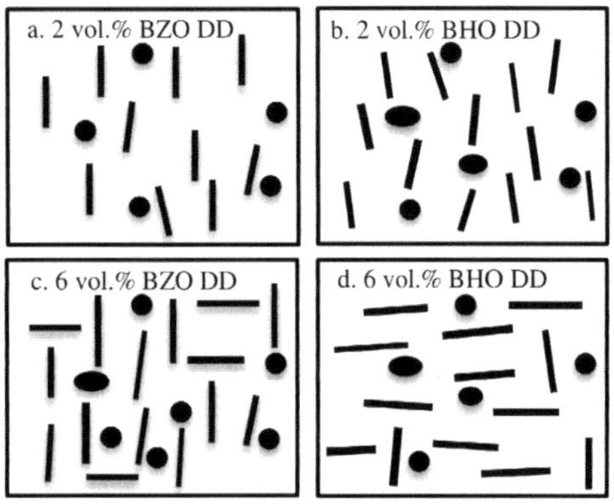

Figure 4. Schematic of the distribution of the 1D APCs (rod shape) and 3D APCs (spherical or oval shape) nanostructures in the YBCO films with (a) 2 vol.% and (c) 6 vol.% BZO DD films, (b) 2 vol.% and (d) 6 vol.% BHO DD films.

Figures 5a-b compare the $J_c(\theta)$ curves at 65 K, for 2 vol.% and 6 vol.% BZO and BHO DD films respectively. The differences in $J_c(\theta)$ curves of 2 vol.% BZO and BHO DD films (Figure 5a) are the lower $J_{c,max}$ (J_c peak value) and higher $J_{c,min}$ (the lowest J_c value of the $J_c(\theta)$ curve) for the entire angular range θ=0°-90° in BHO DD film compared to BZO DD film at 5.0 and 9.0 T. The reducing value of $J_{c,min}$ indicates weaker vortex pinning in that angular range while increasing $J_{c,max}$ is indicative of strong pinning by parallel 1D APCs aligned either along the c-axis or ab-plane. The J_c anisotropy (($J_{c,max}$ - $J_{c,min}$)/$J_{c,min}$) is about 129% for the BZO DD film at 5.0 T in contrast to 47% for its counterpart BHO DD (Figure 5a). J_c anisotropy further reduces to 100% at 65 K and 9.0 T for BZO DD film which is still 5 times higher than the J_c anisotropy of BHO DD film at same field and temperature. This indicates that at lower concentration (2 vol.%) BHO DD film shows more isotropic pinning compared to its counterpart BZO DD film.

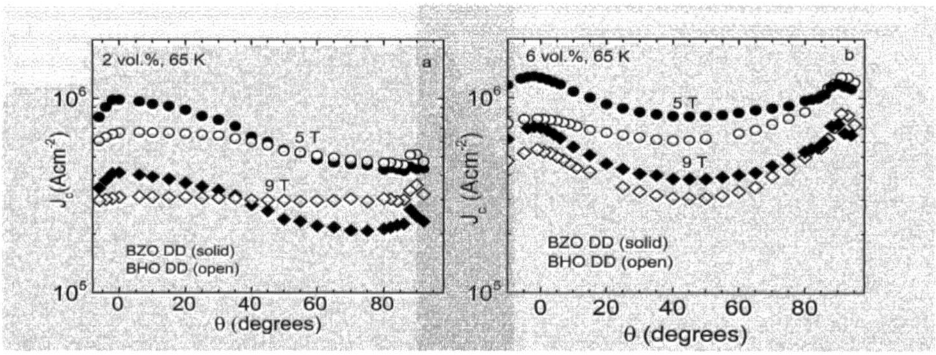

Figure 5. Angular dependence of J_c measured on (a) 3% Y_2O_3+2 vol% BZO or BHO doped YBCO (BZO(solid) or BHO (open) DD) (b) 3% Y_2O_3+6 vol% BZO or BHO doped YBCO (BZO or BHO DD) nanocomposite films measured at 5.0 T (circle), and 9.0 T (diamond) at 65 K.

However, this trend doesn't follow for the 6 vol. % DD films. At 65 K, the overall J_c is higher for 6 vol% BZO DD film for the entire angular range as defined previously at 5.0 T and 9.0 T. Thus, an increase of $J_c(\theta)$ value for wide angular range for BZO DD film is indicative of strong and effective pinning by 1D APCs along with 3D APCs for the field other than c-axis compared to BHO DD film. Comparing 2 vol.% DD films, J_c peaks are observed at *H* // c-axis and *H* // ab-plane for both 6 vol.% DD films. There is an absence of such peak at *H* // c-axis in 2 vol.% BHO DD film appears at 6 vol.% BHO DD films. The J_c anisotropy of 6 vol.% BZO DD at 5.0 T is about 60% compared to 111% for BHO DD film. In addition, at 9.0 T, J_c anisotropy is about 92% for the BZO DD film which is almost half of the BHO DD film (173%). The increasing J_c anisotropy in BHO DD film may be due to increasing J_c peak towards the ab-plane rather than J_c peak at c-axis which is different from BZO DD film at which J_c peaks appear at around c-axis. It means at 6 vol.% BZO DD film, more isotropic pinning is observed in contrast to BHO DD film. This illustrates that the influence of Y_2O_3 in BZO and BHO 1D-APCs is different, more in the latter compared to former.

4. Conclusion

A comparative study of the transport J_c (H, θ) of the BZO DD and BHO DD films has been carried out at 77 K and 65 K, in magnetic field up to 9.0 T. A secondary Y_2O_3 3D APC doping was employed to probe the susceptibility of the c-axis alignment of BZO and BHO nanorods to the local strain generated by the Y_2O_3 3D APC. At a fixed 3.0 vol.% of Y_2O_3 3D APC doping, the concentration of the BZO and BHO nanorods was varied from 2.0 vol.% to 6.0 vol.%. At the 2.0 vol.% concentration, the BHO DD film shows more isotropic pinning in the entire field range at 65 K, reducing J_c anisotropy. At the higher concentration of 6.0 vol.%, the overall higher $J_c(\theta)$ values are observed in BZO DD samples for wide angular range from *H*//c-axis at 65 K except around H//ab-plane. This may be attributed to the switch of a larger proportion of the BHO APCs to ab-plane alignment at the higher strain field overlap as compared to the BZO counterparts. Consequently, the J_c anisotropy increased from 18 % at 2.0 vol.% to 173% at 6.0 vol.% BHO concentration in BHO DD films at 9.0T and 65 K. This is in contrast to the opposite trend of the J_c anisotropy of 100% to 92% at the same condition in BZO DD films as the BZO concentration was increased from 2.0 vol.% to 6.0 vol %. This result illustrates the importance of understanding the effect of the strain field on the APC morphology via DD and other approaches.

5. References

[1] S. Chen, M. Sebastian, B. Gautam, J. Wilt, T. Haugan, Z. Xing, *et al.*, "Enhancement of Isotropic Pinning Force in YBCO Films With $BaZrO_3$ Nanorods and Y_2O_3 Nanoparticles," *IEEE Transactions on Applied Superconductivity,* vol. 27, pp. 1-5, 2017.

[2] S. Miura, Y. Yoshida, Y. Ichino, Q. Xu, K. Matsumoto, A. Ichinose, *et al.*, "Improvement in J c performance below liquid nitrogen temperature for SmBa2Cu3O y superconducting films with BaHfO3 nano-rods controlled by low-temperature growth," *APL Materials,* vol. 4, p. 016102, 2016.

[3] J. Wu, J. Shi, F. Baca, R. Emergo, J. Wilt, and T. Haugan, "Controlling $BaZrO_3$ nanostructure orientation in $YBa_2Cu_3O_7$ films for a three-dimensional pinning landscape," *Superconductor Science and Technology,* vol. 28, p. 125009, 2015.

[4] M. Erbe, J. Hänisch, R. Hühne, T. Freudenberg, A. Kirchner, L. Molina-Luna, *et al.*, "BaHfO3 artificial pinning centres in TFA-MOD-derived YBCO and GdBCO thin films," *Superconductor Science and Technology,* vol. 28, p. 114002, 2015.

[5] M. A. P. Sebastian, J. N. Reichart, J. L. Burke, and L. B. Brunke, "Optimizing Flux Pinning of YBCO Superconductor with $BaSnO_3$+ Y2O3 Dual Mixed Phase Additions," *Optimizing Flux Pinning of YBCO Superconductor with $BaSnO_3$+ Y2O3 Dual Mixed Phase Additions,* 2013.

[6] X. Obradors, T. Puig, S. Ricart, M. Coll, J. Gazquez, A. Palau, *et al.*, "Growth, nanostructure and vortex pinning in superconducting YBa2Cu3O7 thin films based on trifluoroacetate solutions," *Superconductor Science and Technology,* vol. 25, p. 123001, 2012.

[7] F. Baca, P. Barnes, R. Emergo, T. Haugan, J. Reichart, and J. Wu, "Control of $BaZrO_3$ nanorod alignment in $YBa_2Cu_3O_{7-x}$ thin films by microstructural modulation," *Applied Physics Letters,* vol. 94, p. 102512, 2009.

[8] B. Maiorov, S. A. Baily, H. Zhou, O. Ugurlu, J. A. Kennison, P. C. Dowden, *et al.*, "Synergetic combination of different types of defect to optimize pinning landscape using $BaZrO_3$ doped $YBa_2Cu_3O_7$," *Nature materials,* vol. 8, pp. 398-404, 2009.

[9] A. K. Jha, K. Matsumoto, T. Horide, S. Saini, P. Mele, Y. Yoshida, *et al.*, "Systematic Variation of Hybrid APCs Into YBCO Thin Films for Improving the Vortex Pinning Properties," *IEEE Transactions on Applied Superconductivity,* vol. 25, pp. 1-5, 2015.

[10] P. Mele, K. Matsumoto, A. Ichinose, M. Mukaida, Y. Yoshida, S. Horii, *et al.*, "Systematic study of $BaSnO_3$ doped $YBa_2Cu_3O_{7-x}$ films," *Physica C: Superconductivity,* vol. 469, pp. 1380-1383, 2009.
[11] M. A. P. Sebastian, J. N. Reichart, M. M. Ratcliff, T. J. Bullard, J. L. Burke, C. R. Ebbing, *et al.*, "Study of the Flux Pinning Landscape of YBCO Thin Films With Single and Mixed Phase Additions BaMO3+ Z: M= Hf, Sn, Zr and Z= Y2O3, Y211," *IEEE Transactions on Applied Superconductivity,* vol. 27, pp. 1-5, 2017.
[12] W. Hong-Yan, D. Fa-Zhu, G. Hong-Wei, and Z. Teng, "Strongly enhanced flux pinning in the YBa2Cu3O7− x films with the co-doping of BaTiO3 nanorod and Y2O3 nanoparticles at 65 K," *Chinese Physics B,* vol. 24, p. 097401, 2015.
[13] M. Sieger, P. Pahlke, M. Lao, M. Eisterer, A. Meledin, G. Van Tendeloo, *et al.*, "Tailoring Microstructure and Superconducting Properties in Thick BaHfO3 and Ba2 Y (Nb/Ta) O6 Doped YBCO Films on Technical Templates," *IEEE Transactions on Applied Superconductivity,* vol. 27, pp. 1-7, 2017.
[14] F. Rizzo, A. Augieri, A. Angrisani Armenio, V. Galluzzi, A. Mancini, V. Pinto, *et al.*, "Enhanced 77 K vortex-pinning in Y Ba2Cu3O7− x films with Ba2Y TaO6 and mixed Ba2Y TaO6+ Ba2Y NbO6 nano-columnar inclusions with irreversibility field to 11 T," *APL Materials,* vol. 4, p. 061101, 2016.
[15] T. Horide, T. Kawamura, K. Matsumoto, A. Ichinose, M. Yoshizumi, T. Izumi, *et al.*, "Jc improvement by double artificial pinning centers of $BaSnO_3$ nanorods and Y_2O_3 nanoparticles in $YBa_2Cu_3O_7$ coated conductors," *Superconductor Science and Technology,* vol. 26, p. 075019, 2013.
[16] R. L. S. Emergo, F. J. Baca, J. Z. Wu, T. J. Haugan, and P. N. Barnes, "The effect of thickness and substrate tilt on the BZO splay and superconducting properties of $YBa_2Cu_3O_{7-x}$ films," *Superconductor Science and Technology,* vol. 23, p. 115010, 2010.
[17] H. Tobita, K. Notoh, K. Higashikawa, M. Inoue, T. Kiss, T. Kato, *et al.*, "Fabrication of $BaHfO_3$ doped $Gd1Ba_2Cu_3O_{7-\delta}$ coated conductors with the high Ic of 85A/cm-w under 3T at liquid nitrogen temperature (77 K)," *Superconductor Science and Technology,* vol. 25, p. 062002, 2012.
[18] J. F. Baca, T. J. Haugan, P. N. Barnes, T. G. Holesinger, B. Maiorov, R. Lu, *et al.*, "Interactive Growth Effects of Rare - Earth Nanoparticles on Nanorod Formation in $YBa_2Cu_3O_x$ Thin Films," *Advanced Functional Materials,* vol. 23, 2013.
[19] J. J. Shi and J. Z. Wu, "Micromechanical model for self-organized secondary phase oxide nanorod arrays in epitaxial $YBa_2Cu_3O_{7-\delta}$ films," *Philosophical Magazine,* vol. 92, pp. 2911-2922, 2012.
[20] V. Matsui, V. Flis, V. Moskaliuk, A. Kasatkin, N. Skoryk, and V. Svechnikov, "Current-carrying abilities of nano-structured HTS thin films," *J. Nanoscience Nanoengineering,* vol. 1, pp. 38-43, 2015.
[21] B. Gautam, M. A. Sebastian, S. Chen, J. Shi, T. Haugan, Z. Xing, *et al.*, "Transformational dynamics of BZO and BHO nanorods imposed by Y2O3 nanoparticles for improved isotropic pinning in YBa2Cu3 O 7-δ thin films," *AIP Advances,* vol. 7, p. 075308, 2017.

Acknowledgements
This research was supported in part by NSF contracts Nos. NSF-DMR-1337737 and NSF-DMR-1508494, the AFRL Aerospace Systems Directorate, and the Air Force Office of Scientific Research (AFOSR), the US National Science Foundation (DMR-1565822) for TEM characterization

IOP Conf. Series: Materials Science and Engineering **279** (2017) 012031 doi:10.1088/1757-899X/279/1/012031

Comparison study of the flux pinning enhancement of YBCO superconductor with BZO and BZO + Y_2O_3 mixed phase additions

M A Sebastian[1,2], C Ebbing[2], W Zhang[3,4], J Huang[3,4], H Wang[3,4], S Chen[5], B Gautum[5], J Wu[5] and T Haugan[1]

[1] U.S. Air Force Research Laboratory, Aerospace Systems Directorate, WPAFB, OH 45433 USA
[2] UDRI, University of Dayton, OH 45469 USA
[3] Materials Science and Engineering Program, Texas A&M, College Station, TX 77843 USA
[4] School of Materials Engineering, Purdue University, West Lafayette, IN 47907 USA
[5] Department of Physics & Astronomy, the University of Kansas, Lawrence, Kansas 66045, USA
[6] Author to whom correspondence should be addressed: sebastianm1@udayton.edu
[7] This research is funded by AFRL/ Aerospace Systems Directorate and the Air Force Office of Scientific Research (LRIR No. 14RQ08COR), and by National Science Foundation DMR-1565822 for TEM courtesy of H. Wang's research group, and (ARO contract No. ARO-W911NF-16-1-0029, and NSF contracts Nos. NSF-DMR-1337737 and NSF-DMR-1508494 for angular current density measurements courtesy of J. Wu's research group.

Abstract. Adding nanophase defects to $YBa_2Cu_3O_{7-\delta}$ (YBCO) superconductor thin films is well-known to enhance flux pinning, resulting in an increase in current density (J_{ct}). While many previous studies focused on single phase additions, the addition of several phases simultaneously shows promise in improving current density by combining different pinning mechanisms. This paper compares the effect of the addition of two insulating, nonreactive phases of barium zirconium oxide (BZO) and yttrium oxide Y_2O_3, both as a single addition of BZO and as a double addition in conjunction with Y_2O_3. Processing parameters vary the target composition volume percent of BZO from 2-6 vol. %, while maintaining 3 vol. % Y_2O_3, and the remaining vol. % YBCO. Pulsed laser deposition produced thin films on $LaAlO_3$ (LAO) and $SrTiO_3$ (STO) substrates at various deposition temperatures. Comparison of strong and weak flux pinning mechanisms, current densities, critical temperatures, and microstructures of the resulting films will be presented. Films produced from the 2 vol. % BZO + 3 vol. Y_2O_3 doped YBCO target at a deposition temperature of 825 °C attained the highest current density. The addition of second phase Y_2O_3 impacted the film microstructure, resulting in more isotropic behavior when compared to the YBCO films doped with only BZO.

…

1. Introduction

Addition of nano-sized insulating phases to $YBa_2Cu_3O_{7-\delta}$ (YBCO) thin films enhances flux pinning, resulting in higher current densities. Previous research has focused on combining the artificial pinning centers resulting from BZO nanorods [1], [2], [3] and Y_2O_3 nanoparticles [4]. Liu utilized a 1.5 vol. % BZO and YBCO target to produce films on LAO decorated with Y_2O_3 nano-islands [5]. Optimization of film thickness for films consisting of YBCO with BZO and Y_2O_3 has also been studied, utilizing a 5 mol. % BZO and 5 mol. % Y_2O_3 target [6]. It is hypothesized that the combination of BZO and Y_2O_3 with YBCO influences the lattice strain in the film. This research seeks to explore the effects on current density of doping a YBCO target with 2, 4, and 6 volume percent BZO along with 3 vol. percent Y_2O_3, with the remaining volume percent YBCO.

2. Experimental Procedures

2.1. Target Preparation

Targets were produced via solid state processing of commercial powders of YBCO, $BaZrO_3$, and Y_2O_3, which were dried in a box furnace for 8 hours at 450 °C. The dried powders were measured and mixed with an agate mortar and pestle to comprise compositions listed in Table 1. The mixtures were then individually pressed utilizing a Carver die press with 1.25 inch and 1.125 inch diameter dies and a pressure of 6.89 MPa. The targets were then sintered at 850 °C for 60 hours and 920 °C for 156 hours. Final densities of the three targets ranged between 86% and 93%. A YBCO - only target was prepared by similar conditions, with final density 93%.

Table 1. PLD Target Compositions.

Target	YBCO Volume %	BZO Volume %	Y_2O_3 Volume %
TS199	100.0		
TS196	98.0	2.0	
TS201	95.0	2.0	3.0
TS202	93.0	4.0	3.0
TS203	91.0	6.0	3.0

2.2. Thin Film Production via Pulsed Laser Deposition (PLD)

Thin films were produced using pulsed laser deposition (PLD) on $SrTiO_3$ (STO) single crystal substrates, having dimensions of 3.2 mm x 3.2 mm x 0.5 mm and 4.0 mm x 10.0 mm x 0.5 mm. Substrates were cleaned via an ultrasonic cleaner for five minutes with acetone, followed by five minutes with isopropyl alcohol. The substrates were mounted to a YBCO coated heater block with colloidal silver paint. Films were produced via PLD with a Lambda Physik LPX 300 Eximer KrF laser using an energy of 450 mJ, a repetition rate of 8Hz, and a 40Pa O_2 atmosphere, followed by annealing at 500 °C in an O_2 atmosphere for 30 minutes.

2.3. Film Characterization

Film thicknesses were determined by etching the thin films with nitric acid and measuring the stair step thickness with a KLA Tencor D-120. Resulting film thicknesses were 290, 247, and 256 nm for the 2, 4, and 6 vol. % double doped BZO + Y_2O_3 respectively. Single doped BZO films were 135 nm thick. Magnetic current density (J_{cm}) and critical transition temperature (T_c) were measured with a Quantum Design Physical Properties Measurement System (PPMS) with a vibrating sample magnetometer (VSM) probe. The VSM data was attained for conditions of 77 K, 65 K, 20 K, and 5 K with an applied field varied from 0-9 T for H // c-axis of the films. $T_{c\text{-}onset}$ temperatures of the films were attained from zero-field-cooled measurements. The measured magnetic data was used to determine the current densities by utilizing a simplified form or the Bean Critical State Model [7,8,9]. Transport current density measurements were conducted through collaboration with Dr. Judy Wu's research group at University of Kansas. Silver contacts were sputtered through a metal shadow mask on the films, to reduce the

IOP Conf. Series: Materials Science and Engineering **279** (2017) 012031 doi:10.1088/1757-899X/279/1/012031

contact resistance to the microbridges. Microbridges of ≈ 20 μm width and ≈500 μm length were patterned utilizing photolithography. Film surfaces were spin coated with 351 photoresist, exposed at 500W UV for 70 s., developed for 50 s., followed by etching for 2 min. with 0.05% HNO_3. Platinum wires were attached to the microbridges with indium for the electrical connections. A Keithley 2430 1KW pulsed current source meter and HP 34420 nanovoltmeter were used to attain the transport current density. LabVIEW was used to control the increasing amplitude of the current pulse train and synchronize the resulting voltage detected across the film. A short 50 ms pulse width and a long time interval of 3000 – 4000 ms minimized heating. The current density was determined using a 1 μV/cm criteria. A Quantum Design Evercool II Physical Property Measurement System (PPMS) was employed to measure the transport current density as a function of temperature, magnetic field (0-9T), and angle θ between the field and the c-axis of the film, maintaining the field perpendicular to the current. Transport current densities were measured at θ = 0° (corresponding to *H* // c), 45°, and 90° (corresponding to *H*// ab).
A FEI Sirion Scanning Electron Microscope (SEM) with ultra-high resolution at 5 kV and a spot size of 3 was used to obtain images of the films' surface. Cross-sectional Transmission Electron Microscopy (XTEM) images and Scanning Transmission Electron Microscopy image (STEM) were obtained using a FEI Tecnai F20 analytical microscope under the acceleration voltage of 200KV. TEM work was courtesy of H. Wang's group previously located at Texas A & M University, and currently located at Purdue University. Lattice parameters were determined by x-ray diffraction analysis utilizing the Bruker D8 Discover diffractometer with the following operation conditions: high resolution, Cu tube 1.5418 Å, Power: 40kV, 40mA; 2T: 5-70°; rocking curves for YBCO *(005)*.

3. Results and Discussion

The experimental results in figure 1 and figure 2 show the resulting current densities verses applied field at 77, 65, 20, and 5K for YBCO films with varying 2, 4, and 6 volume percent BZO and 3 volume percent Y_2O_3, at two different deposition temperatures of 825°C and 810°C. Looking at figure 2, the 2 vol. % double doped BZO films (810 °C deposition temperature) attained a higher current density than YBCO films at 65K, but not at the lower temperatures of 20 and 5K. In contrast in figure 1, the 2 vol. % BZO double doped films grown at 825 °C attained higher current densities than YBCO at both high and low temperatures. Comparing these graphs shows that for all of the temperatures 77K-5K, the highest current density was attained with the YBCO film doped with 2 vol. percent BZO + 3 vol. percent Y_2O_3 at a deposition temperature of 825 °C. The associated pinning force curves for this film is shown in Fig. 3(b). Interestingly, in figure 3(a), the double doped 2 vol. % BZO with Y_2O_3 outperforms the single doped 2 vol. % BZO when applied fields exceed 4 tesla. The single doped 2 vol. % film has a matching field of 3 T for 65K, as indicated by the *. The flattening of the current density curve for the double doped film verses the single doped film signifies an increased isotropic behavior, which is confirmed with angular current density measurements in Fig. 3(c)-(d). In the angular current density graphs shown in Fig. 3, Theta = 0° corresponds to the field *H* applied parallel to the c direction, and Theta = 90°corresponds to the field *H* applied parallel to the a-b plane. As the temperature decreases and the applied field increases, the magnitude of the strong pinning due to the dopants at 0°surpasses the magnitude of the intrinsic pinning at 90°. Collaborative research with Dr. Judy Wu's group also showed this benefit of increased isotropic behavior in current density resulting from the utilization of double doping with BZO and Y_2O_3 [10]. Figure 4 shows the variation in critical temperature verses vol. % BZO dopants in YBCO films, also doped with 3 vol. % Y_2O_3, at deposition temperatures of 810 °C and 825 °C. The critical temperatures were degraded at the higher deposition temperature, and for the optimized 2 vol. % double doped BZO film. This corresponds to increased stacking faults seen as deposition temperature increases and is also typical of a tradeoff between increased current densities and lower critical temperatures [11].

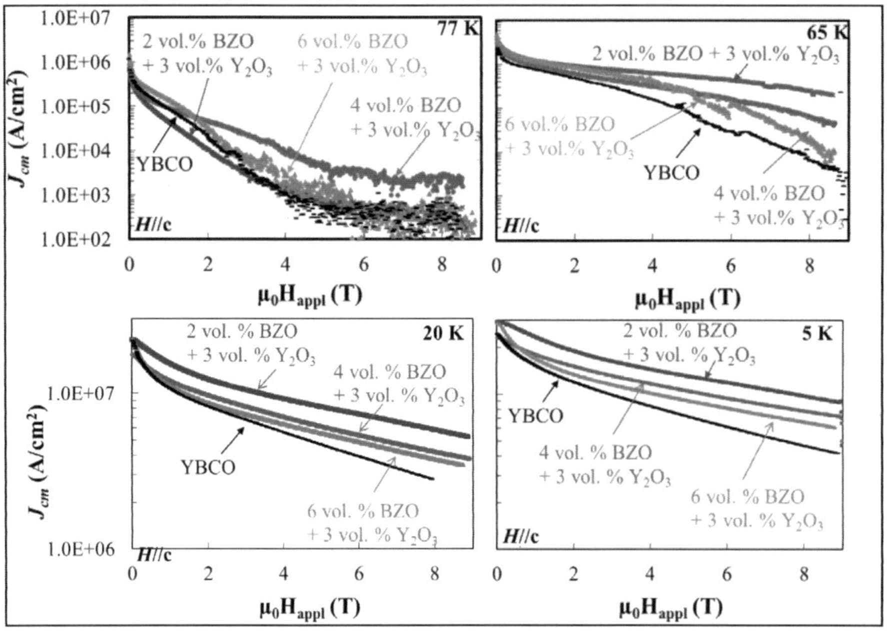

Figure 1. Current density as a function of applied field parallel to the c-direction, measured at 77, 65, 20, and 5K for $YBCO_{100-(x+3)}BZO_x$ (x = 2, 4, 6 vol. %) Y_2O_3 = 3 vol. % at deposition temperature of 825 °C.

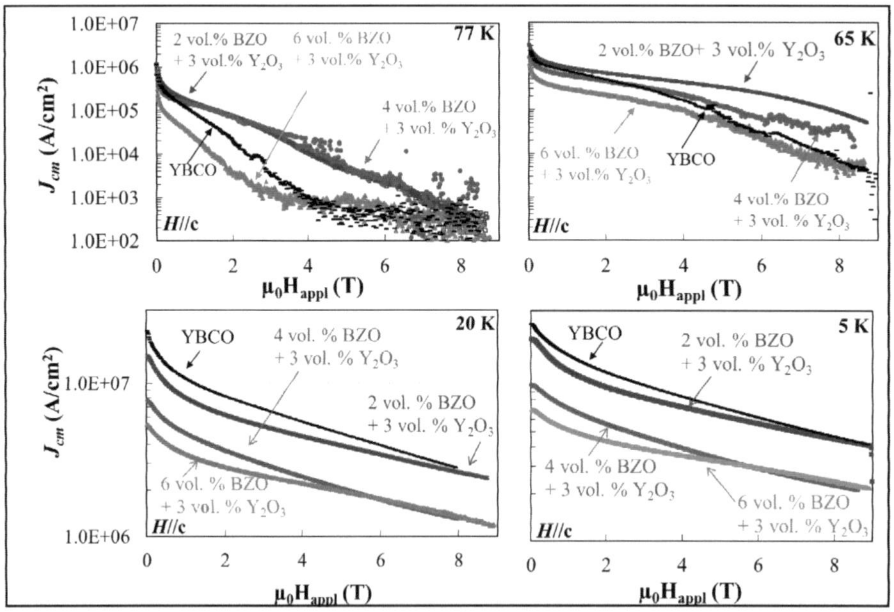

Figure 2. Current density as a function of applied field parallel to the c-direction, measured at 77, 65, 20, and 5K for $YBCO_{100-(x+3)}BZO_x$ (x = 2, 4, 6 vol. %) Y_2O_3 = 3 vol. % at a deposition temperature of 810 °C.

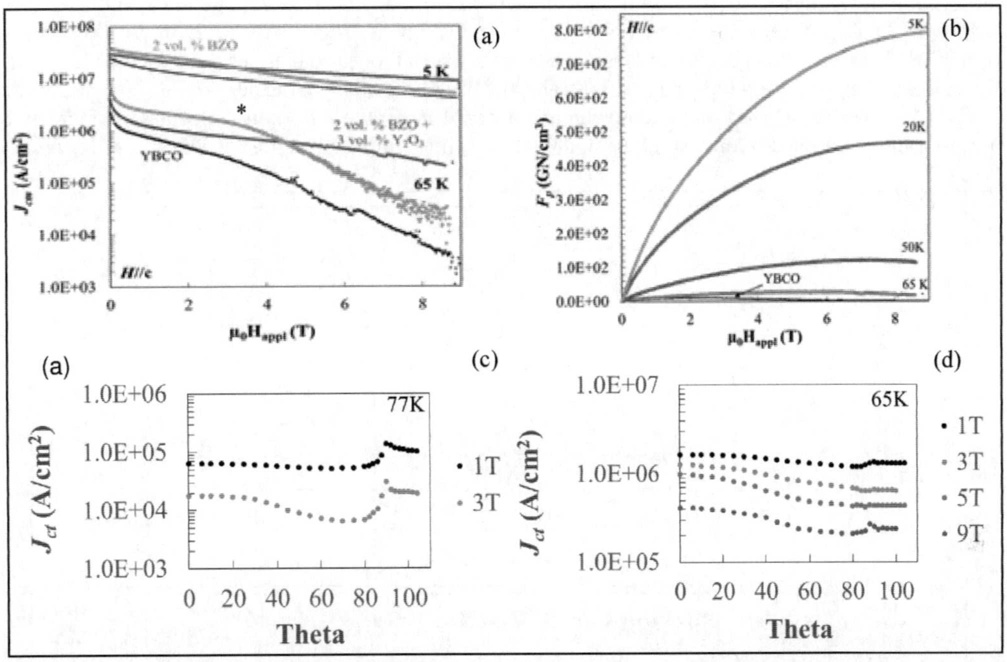

Figure 3. a) Current density as a function of applied field measured at 65 and 5 K for single doped 2 vol. % BZO and YBCO film and double doped 2 vol.% BZO + 3 vol.% Y_2O_3 YBCO film (STO substrate and 825°C deposition temperature). Note * signifies matching field for single doped film. (b) corresponding pinning force curves for optimized 2 vol. % BZO + 3 vol. % Y_2O_3 doped YBCO film at 65, 50, 20, and 5K deposited at 825 °C and for YBCO film shown as black curve. Angular dependence of current density measured at (c) 77K and (d) 65K at 1T, 3T, 5T, and 9T for 2 vol. % BZO + 3 vol. % Y_2O_3 doped YBCO film deposited at 825 °C.

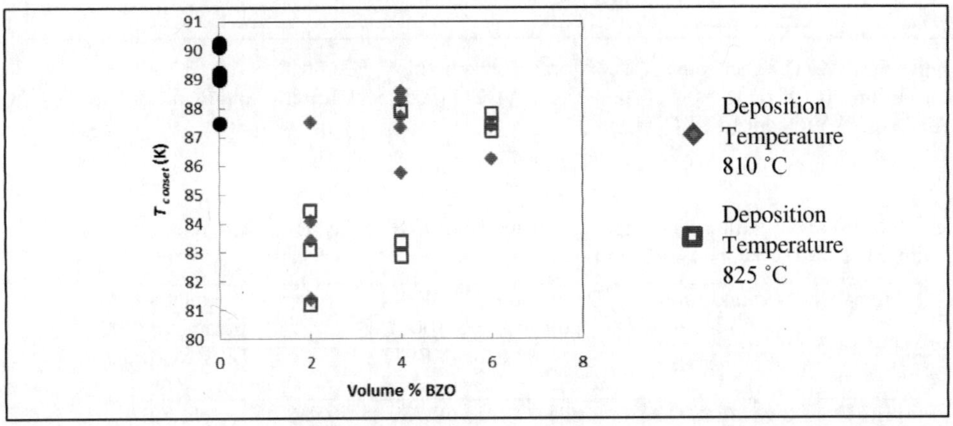

Figure 4. Critical temperatures (T_c) for 2, 4, and 6 vol. % BZO with 3 vol. % Y_2O_3 doped YBCO films, deposited at 810 °C and 825 °C. Reference T_c for undoped YBCO films shown as black circles.

Microstructure studies involved xrd analysis, along with SEM and TEM. From figure 5(a), the 2 Theta omega scan clearly shows nice epitaxial film growth, and confirms BZO c-axis alignment and the presence of Y_2O_3 for the double doped YBCO films TJ2734, TJ2712 and TJ2768 (2, 4, and 6 vol. % BZO with 3 vol. % Y_2O_3 respectively). Note that the Y_2O_3 peak is absent from the 2 vol. % BZO doped film TJ2856. The rocking curves were measured for the YBCO *(005)* peak for each of the double doped films. Recall that peak shift indicates macrostrain, and peak broadening indicates non-uniform strain. A FWHM > 0.2 defines a broad peak. Referring to figure 5(b) and Table 2, the non-uniform strain increased as the volume percent BZO increased the double doped films, until some stress relief occurred at 6 vol. % BZO.

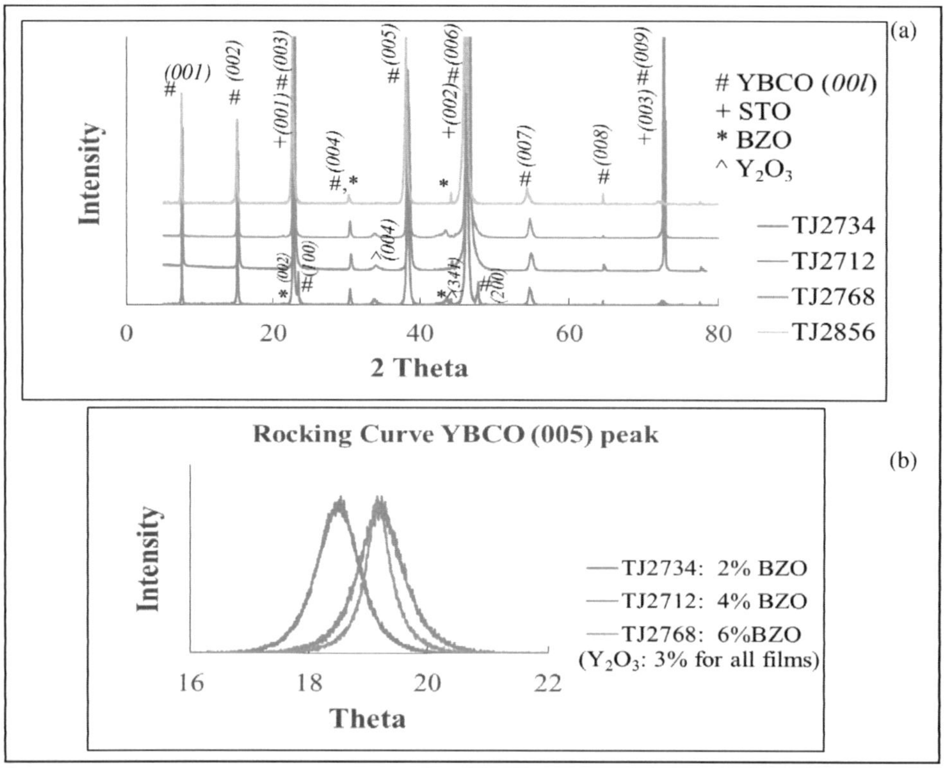

Figure 5. (a) XRD 2Theta omega scans for 2, 4, and 6 vol. % BZO with 3 vol. % Y_2O_3 doped YBCO films. (b) XRD rocking curve scans of YBCO (005) peak for 2, 4, and 6 vol. % BZO with 3 vol. % Y_2O_3 doped YBCO films.

Table 2. FWHM and c-lattice parameters calculated from XRD analysis for 2, 4, and 6 vol. % BZO with 3 vol. % Y_2O_3 doped YBCO films.

Film #	Composition	Deposition Temp. °C	FWHM (005) YBCO	β	c lattice parameter (005) YBCO peak (Å)
TJ2734	2 vol.% BZO + 3% Y_2O_3	825 °C	0.839	0.988	11.727
TJ2712	4 vol.% BZO + 3% Y_2O_3	825 °C	0.842	1.026	11.693
TJ2768	6 vol.% BZO + 3% Y_2O_3	825 °C	0.489	0.665	11.732

From the SEM micrographs in figure 6, one can see how the surface of the film doped with 2 vol. % BZO + 3 vol. % doped YBCO (c,d) contains an increased number of pores and defects in comparison to the YBCO (a,b) film. The effect of the addition of 3 vol. % Y_2O_3 also has a remarkable impact on the film surface, which can be seen by comparing figure 6 (e,f), the double doped 2 vol. % BZO to figure 6 (g,h), the single doped 2 vol. % BZO film. Interestingly in figure 6(h), within the pores of the film, square shapes are imaged, and are apparently the cross sections of $BaZrO_3$ nanorods (inset), which have a cubic unit cell perovskite structure [12]. The number of pores and defects also increases with increasing deposition temperature, due to the increased adatom mobility (figure 6 e,f). This effect of temperature is also shown with the 4 vol. % BZO double doped film in figure 7(a-d). In comparing figures 6, and 7, the defects and pores and grains also increase with the composition of volume % BZO, with evidence of a-axis film growth associated with the basket weave pattern shown in figure 7(a) [13]. TEM low magnification and high magnification images in figure 8, shows the presence of the second phase dopants, with their density increasing as the volume % BZO of the target utilized increases from 2, 4, and 6 volume % for the double doped BZO films. Due to the proximity of Y and Zr on the periodic table, EDS cannot distinguish between the two with elemental analysis. In figure 8(c) the impact of increased concentration of dopants on the microstructure is evidenced in the degraded film structure and the impact on the BZO nanrod dimensions [10] of the 6 vol. % double doped BZO film on the right compared to the 4 vol. % double doped BZO film on the left.

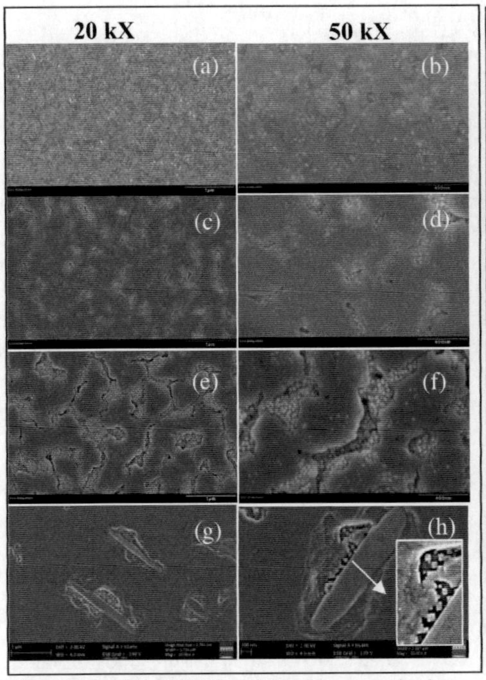

Figure 6. SEM 20kX and 50kX respectively: (a), (b) YBCO (c) (d) 2 vol.% BZO + 3 vol. % Y_2O_3 doped YBCO on STO substrate at 810 °C deposition temperature; (e), (f) 2 vol.% BZO + 3 vol. % Y_2O_3 doped YBCO on STO substrate at 825 °C deposition temperature; (g), (h) 2 vol. % BZO doped YBCO on STO substrate at 825 °C deposition temperature.

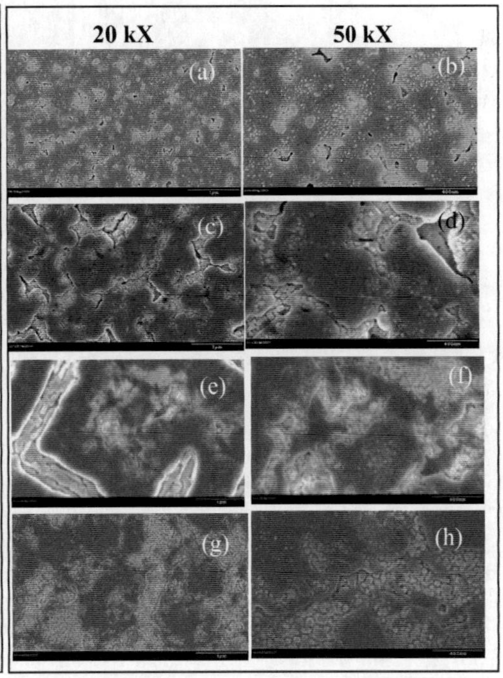

YBCO on STO substrate at 810 °C deposition temperature. (c), (d) 4 vol. % BZO + 3 vol. % Y_2O_3 doped YBCO on STO substrate at 825 °C deposition temperature. SEM 20kX and 50kX respectively: (a), (b) 6 vol.% BZO + 3 vol.% Y_2O_3 doped YBCO on STO substrate at 810 °C deposition temperature; (c), (d) 6 vol. % BZO + 3 vol. % Y_2O_3 doped YBCO on STO substrate at 825 °C deposition temperature.

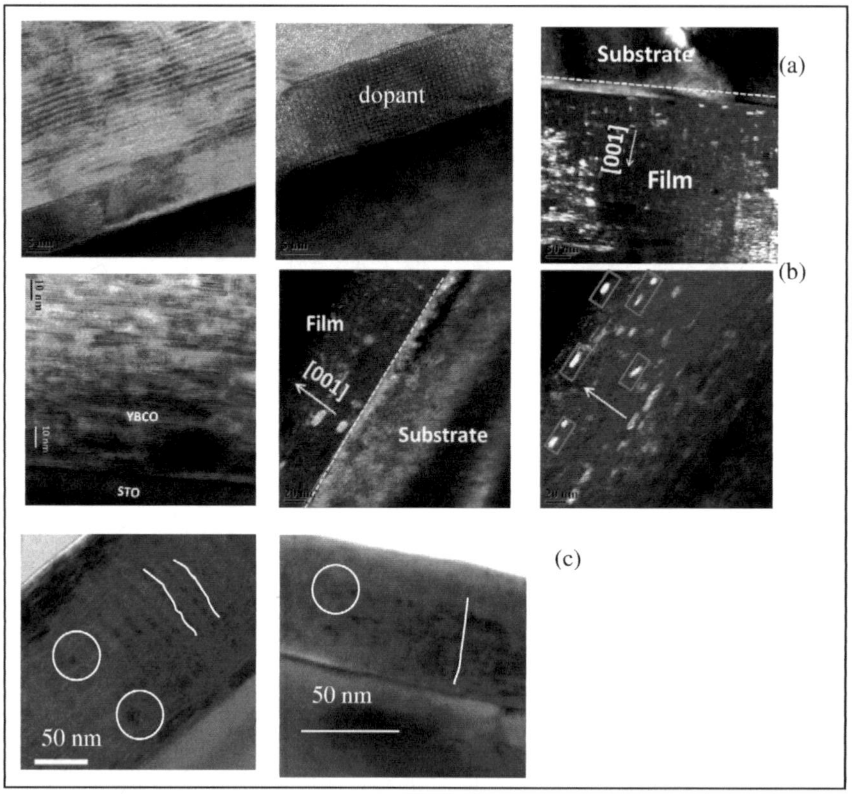

Figure 8. TEM intermediate and high resolution images respectively: (a) 2 vol. % BZO + 3 vol. % Y_2O_3 doped YBCO on STO substrate at 825 °C deposition temperature. (b) 4 vol. % BZO + 3 vol. % Y_2O_3 doped YBCO on STO substrate at 825 °C deposition temperature. Courtesy of H. Wang's research group Texas A & M University & Purdue University (c) 4 vol. % BZO + 3 vol. % Y_2O_3 doped YBCO and 6 vol. % BZO + 3 vol. % Y_2O_3 doped YBCO on STO substrate at 825 °C deposition temperature. Courtesy of Kansas University.

4. Conclusion

This research compares the effect of the addition of two insulating, nonreactive phases of barium zirconium oxide (BZO) and yttrium oxide Y_2O_3, both as a single addition of BZO and as a double addition in conjunction with Y_2O_3. Processing parameters vary the target composition volume percent of BZO from 2-6 vol. %, while maintaining 3 vol. % Y_2O_3, and the remaining vol. % YBCO. Pulsed laser deposition produced thin films on $LaAlO_3$ (LAO) and $SrTiO_3$ (STO) substrates at various deposition temperatures. XRD confirmed epitaxial, c-axis aligned YBCO films, the presence of the BZO and Y_2O_3 dopant phases, and the resulting stress present in the films. Films produced from the 2 vol. % BZO + 3 vol. % Y_2O_3 doped YBCO target at a deposition temperature of 825 °C attained the highest current density. The addition of second phase Y_2O_3 impacted the film microstructure, resulting in more isotropic behavior when compared to the YBCO films doped with only BZO. This was evidenced in the angular current density measurements, and confirmed by XRD, SEM, and TEM analysis. The 2 vol. % BZO double doped with Y_2O_3 YBCO films achieved higher current density than YBCO films doped only with 2 vol. % BZO at fields greater than 5T at 65K and 4T at 5K, providing insight on tailoring the microstructure to enhance current density at higher fields.

IOP Conf. Series: Materials Science and Engineering **279** (2017) 012031 doi:10.1088/1757-899X/279/1/012031

References

[1] Haugan TJ, Barnes PN, Campbell TA, Pierce NA, Baca FJ and Maartense I 2007 Flux pinning of Y-Ba-Cu-O films doped with $BaZrO_3$ nanoparticles by multilayer and single target methods *IEEE Trans on Appl Supercond* **17**(2) 3724-3728

[2] Maiorov B, Baily SA, Zhou H, Ugurlu O, Kennison JA and Dowden PC, et al. 2009 Synergetic combination of different types of defect to optimize pinning landscape using BaZrO3-doped $YBa_2Cu_3O_7$ *Nature Materials* **8**(5) 398-404

[3] Gutiérrez J, Llordés A, Gázquez J, Gibert M, Romà N and Ricart S, et al. 2007 Strong isotropic flux pinning in solution-derived $YBa_2Cu_3O_{7-x}$ nanocomposite superconductor films *Nature Materials* **6** 367-373

[4] Baca FJ, Haugan TJ, Barnes PN, Holesinger TG, Maiorov B, and Lu R et al. 2013 Interactive growth effects of rare-earth nanoparticles on nanorod formation in $YBa_2Cu_3O_x$ thin films *Adv Funct Mater* **23**(38)4826-4831

[5] Liu Y and Du G 2011 Preparation and flux-pinning properties of multilayered yttrium barium copper oxide thin films containing alternating barium zirconate and yttria nanostructures *J Electron Mater* **40**(7) 1512-1516

[6] Zhou H, Maiorov B, Baily SA, Dowden PC, Kennison JA and Stan L, et al. 2009 Thickness dependence of critical current density in YBa2Cu3O7−δ films with $BaZrO_3$ and Y_2O_3 addition *SuScT* **22**(8) 085013-1

[7] Kell JW, Haugan TJ, Locke MF and Barnes PN 2005 Tb and Ce doped Y123 films processed by pulsed laser deposition *IEEE Trans Appl Supercond* **15**(2) 3726-3729

[8] Bean CP 1964 Magnetization of high-field superconductors *Reviews of Modern Physics* **36**(1) 31-39

[9] Bean CP 1962 Magnetization of hard superconductors *Phys Rev Lett* **8**(6) 250-253

[10] Chen S, Sebastian MA, Gautam B, Wilt J, Haugan T and Xing Z, et al. 2017 Enhancement of isotropic pinning force in YBCO films with $BaZrO_3$ nanorods and Y_2O_3 nanoparticles *IEEE Trans Appl Supercond* **27**(4) 1-5

[11] Wang J, Kwon J, Yoon J, Wang H, Haugan T and Baca F, et al. 2008 Flux pinning in $YBa_2Cu_3O_{7-delta}$ thin film samples linked to stacking fault density *Appl Phys Lett* **92**(8) 2507

[12] Macario LR, Moreira ML, Andres J and Longo E. 2010 An efficient microwave-assisted hydrothermal synthesis of $BaZrO_3$ microcrystals: growth mechanism and photoluminescence emissions *CrystEngComm* **12**(11) 3612-3619

[13] Chang CC, Wu XD, Ramesh R, Xi XX, Ravi TS and Venkatesan T, et al. 1990 Origin of surface roughness for c-axis oriented Y-Ba-Cu-O superconducting films *Appl Phys Lett* **57**(17) 1814

IOP Conf. Series: Materials Science and Engineering **279** (2017) 012032 doi:10.1088/1757-899X/279/1/012032

Influence of field penetration ratios and filamentation on end-effect related hysteretic loss reductions for superconducting strips

K. Dong[1,2], M. Sumption[1], E. W. Collings[1], M. Majoros[1], H.Yu[2], M. Hu[2]

[1]CSMM, Materials science Department, The Ohio State University, Columbus, OH 43210

[2]School of Electrical Engineering, Southeast University, Nanjing, China, 210096

Abstract. There are a few key conductor-specific factors which influence the power loss of superconductors; these include critical current, geometry, and normal metal resistivity. This paper focuses on the influence of sample geometry on the power loss of superconducting strips and the effect of filamentation and sample length as a function of the field penetration state of the superconductor. We start with the analytical equations for infinite slabs and strips and then consider the influence of end effects for both unstriated and striated conductor. The loss is then calculated and compared as a function of applied field for striated and unstriated conductors. These results are much more general than they might seem at first glance, since they will be important building blocks for analytic loss calculations for twisted geometries for coated conductors, including helical (Conductor on Round Core, CORC), and twisted (e.g., twist stack cables) geometries. We show that for relatively low field penetration, end effects and reduced field penetration both reduce loss. In addition, for filamentary samples the relevant ratio of length scales becomes the filament width to sample length, thus modifying the loss ratios.

1. Introduction

Understanding and reducing the AC loss of coated conductor and the cables wound from superconducting strips is important for enabling superconducting AC machines. Superconducting machines of interest include motors, generators, fault current limiters, fast ramping magnets, and a number of other devices. YBCO coated conductors of increasing interest for these applications. Based on the critical state model [1-3], loss expressions for semi-infinite superconducting slabs are findable in most textbooks on superconductivity, e.g., [4-6], while the expressions for loss in simple YBCO tapes are well known from the work of Brandt [7, 8] and later Muller [9].

While the losses for monofilaments or even multifilamentary coated conductor tapes is understood for the simple case of a flat tape in a perpendicular applied field, the loss for coated conductors in helical or twisted geometries is not well known. An approach which considers the applied field to vary as a trigonometric function along the sample length suggests a simple rule that the loss of a twisted or helical sample should be reduced by a factor of $2/\pi$. This treatment is correct for long pitches, but a consideration of end effects – i.e. the reduction in magnetization and loss for samples where the sample length is larger but not very much greater than the sample width – should be important to consider for tighter pitches.

Below we first remind the reader of the loss expressions for slabs and strips. We then describe the basic geometry of helical and twisted coated conductor samples, and describe the simple approach of averaging a sinusoidal spatial for larger pitches. We then consider a helical conductor as a series of segments where an end effect or termination effect is present, but one where an additional spatial field variation also exists. The difficulty of this simple problem leads us back to a description of the simpler case where the field is uniform, and a termination is present, but both coated conductor geometries and filament striation modify the termination effect.

1.1. Superconducting semi-infinite slab with $t \gg w$

Consider the critical state of a slab ($t \gg w$, and infinitely long) of superconductor of width $w = 2a$ in a field H, applied along $t = 2t_h$ (the y direction) as shown in Figure 1. For low applied field the sample is only partially penetrated, and the loss per cycle, per volume (Q) is [4]

$$Q = \frac{2\mu_0 H_0^3}{3H_p} \tag{1}$$

where the penetration field, H_p is given by $H_p = J_c\, w/2$.

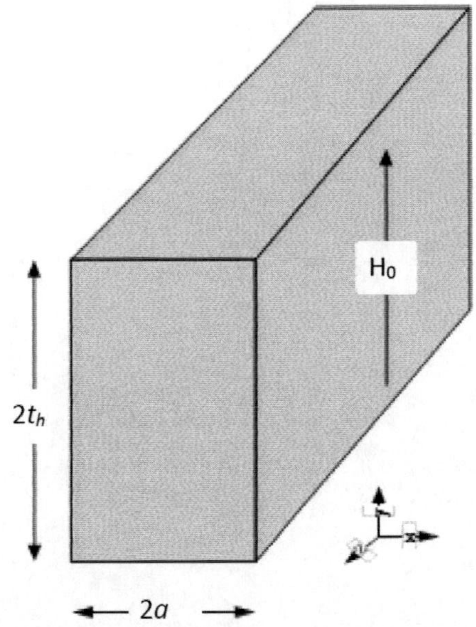

Figure 1. A superconducting sample whose cross section is rectangular.

When $H_0 \geq H_P$, the slab becomes fully-penetrated and the per cycle, per volume loss is given by [4]

$$Q = 2\,\mu_0 J_c a H_0\left(1 - \frac{2}{3}\frac{J_c a}{H_0}\right) = 2\mu_0 H_0^2\left(\frac{H_P}{H_0} - \frac{2}{3}\left(\frac{H_P}{H_0}\right)^2\right) \tag{2}$$

Comparing equations (1) and (2), the loss initially increases rapidly as H_0^3 until $H_0 > H_p$, after which

$$Q = 2\mu_0 H_0^2 \tau(\beta) \tag{3}$$

where $\beta = H_0/H_P$ and $\tau(\beta)$ given by

$$\tau(\beta) = \begin{cases} \dfrac{\beta}{3} & \beta \leq 1 \\ \dfrac{1}{\beta} - \dfrac{2}{3\beta^2} & \beta \geq 1 \end{cases} \tag{4}$$

Thus when the applied field is just above the penetration field both linear and quadratic loss term are seen. But as $H_0 \gg H_p$, $Q \approx 2\mu_0 H_0 H_P = \mu_0 H_0 J_c w$

1.2. An infinitely long superconducting strip

The result for semi-infinite slabs which are thick along the direction of the applied field but have zero demagnetization are well described by Eq (1)-(4). However, if the sample thickness, t, becomes much thinner that the sample width ($w = 2a$), the situation will be different. For $H_0 >> H_p$, the loss equations are identical to those of the semi-infinite slab (Eq(1)-(4)). However, as H_0 drops below the penetration field, which is itself modified, the loss expressions are significantly modified by a kind of demagnetization effect. Brandt [7, 8] and Muller [9] showed that

$$Q=N\mu_0 H_0 J_c w \tag{5}$$

Where

$$N = \left(\frac{H_0}{H_d}\right) g\left\{\frac{H_0}{H_d}\right\}$$

$$g\left\{\frac{H_0}{H_d}\right\} = \frac{H_d}{H_0}\left[\frac{2H_d}{H_0}\ln\left(\cosh(\tfrac{H_0}{H_d})\right) - \tanh(\tfrac{H_0}{H_d})\right] \text{ and } H_d = 0.4J_c t \tag{6}$$

The penetration field is given by

$$H_p = \frac{J_c t}{\pi}\left[Ln\left(\frac{w}{t}+1\right)\right] = \frac{5}{2\pi}H_d\left[Ln\left(\frac{w}{t}+1\right)\right] \tag{7}$$

We note that, from Ref [10], that

$$N \approx 1 - 2\left(\frac{H_d}{H_0}\right)Ln(2) \tag{8}$$

Therefore, when $H_0 >> H_d$, the loss strip is the same as that of a slab with the same width and J_c. At lower fields, the loss is modified by N.

1.3. Twisted and Helical Geometries

Figure 2 shows the geometry of a twisted sample (left) and a helical wrap coated conductor (right). We might imagine a simple loss treatment which is an extension of either the slab or the strip model, but modified by the space varying field. Our task is simplified if we take fields well above the penetration field, where the slab and strip model are equivalent.

Figure 2. Coated Conductor in a twisted geometry (left) and a helical geometry (right).

Let us consider Eq 5 with $N = 1$, but take $H = H_0 \sin(2\pi z/L_p)\sin(\omega t)$ as the field applied perpendicular to the twisted strand (left), or to the helical wrap strand (right). Here z is the position along the length of the axis of symmetry, and L_p is the twist pitch in the case of the twisted strand, and for the case of the helix an effective twist pitch

$$L_{peff} = \sqrt{L_h^2 + \pi^2 D_h^2} \tag{9}$$

where L_h is the length of one helix period as taken along the length (z-axis) of the helix. Following Carr [11], we can integrating the average of Eq (5) over a spatial field cycle, such that

$$Q = \frac{N\mu_0 J_c w H_0}{L_p/_2}\int_0^{\pi} sin\left(\frac{2\pi z}{L_p}\right)dz = \frac{N\mu_0 J_c w H_0}{L_p/_2}\frac{L_p}{2\pi}(2) = \left(\frac{2}{\pi}\right)N\mu_0 J_c w H_0 = \left(\frac{2}{\pi}\right)Q_0 \quad (10)$$

Where Q_0 is the loss for a slab or strip where the field is a field that is time varying and spatially uniform of maximum amplitude H_0. For $L_p \gg w$, this leads to a proper result. However, we have neglected to account for the details of the current paths, which will be important as pitch is reduced. The currents generally flow along the length of the tape, but as the polarity of H is changing over the twist pitch, the sense of the shielding (or trapping) currents do too, requiring them to flow across the tape at periods like $L_p/2$ (or $L_{peff}/2$ for a helix). This can be seen more clearly if we consider a finite sample and its end effects.

1.4. Influence of Sample End Effects

As shown by a number of authors [12-14], finite length superconductors in applied fields have flux gradients which penetrate into the sample in multiple dimensions. If we consider, paralleling the case of the infinite slab, a sample which is infinitely thick (y-direction), and has a length L (along z) and a half width $a=w/2$ (along x), and field is applied along y, as shown in Figure 3, flux will penetrate in the x and z directions. Because the sample is longer than it is wide, the flux penetrates most effectively in the width (x) direction, and over most of the length of the sample the shielding (or trapping) currents flow along the length of the sample. However, Kirchhoff's law requires that they do not diverge and thus must flow across the width of the sample at the ends of the sample, leading to the current flow regions shown in Figure 3. The end effect regions are $2a$ wide, and penetrate a maximum depth of $L_t = a$ assuming that J_c is isotropic.

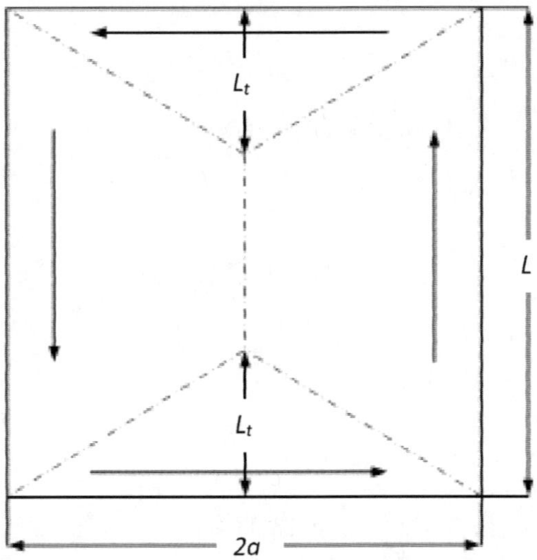

Figure 3. The transferring path of induced current with the consideration of terminal effect

More generally

$$L_t = a\frac{J_{cz}}{J_{cy}} \quad (11)$$

where J_{cz} and J_{cy} are the critical currents in z and y directions respectively. Here we assume the superconducting sample is isotropic ($J_{cz}/J_{cy} = 1$) and the loss for a slab becomes

$$Q = \mu_0 J_c H_0 w\left(1 - \frac{w}{3L}\right) \tag{12}$$

A straightforward extension of this to a strip should be possible by the addition of the prefactor N, such that

$$Q = N\mu_0 J_c H_0 w\left(1 - \frac{w}{3L}\right) \tag{13}$$

Let us explore the behaviour of expression (13) by inserting some specific and relevant values for a coated conductor and then plotting Q/H_0 as a function of applied field. Using w = 1.2 cm, $J_c = 10^{10}$ A/m^2, and $L = (4w, 5w, 10w, 20w, 50w, 100w)$ we plot Q/H_0 vs H_0/H_d in Figure 4 (a). Using Eq (6) we find that $B_d = \mu_0 H_d$ = 5 mT, and using Eq (7) we find that $B_p = \mu_0 H_p$ = 37.6 mT. In Figure 4(a) we see that Q/H_0 tends to 150 J/(m^{2}A) for samples with $L \gg w$ and $H_0 \gg H_p$ agreeing with the value for the infinite slab (or infinite strip for fields well above penetration). As L/w is reduced, the saturation value for a given L/w ratio is reduced. If $H_0 \le H_p \approx 8H_d$, loss is further reduced by field penetration effects related to the coated conductor's highly aspected geometry.

2. Effect of Filamentarization

The loss of coated conductors can be reduced by striation (cutting filaments into the superconducting strip). The per cycle loss per unit volume loss is then given by

$$Q = N\mu_0 H_0 \frac{J_c w}{N_f}\left(1 - \frac{w}{3LN_f}\right) \tag{14}$$

where N_f is the number of filaments. Using the same parameters as above, and setting L/w = 4, QN_f/H_0 can be plotted vs H_0/H_d for tapes with various numbers of filaments, as shown in Fig. 4 (b).

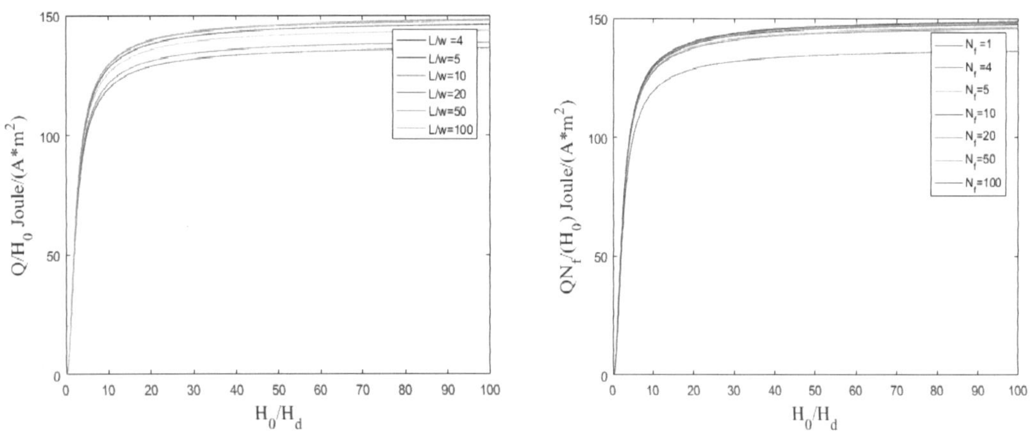

Figure. 4. (a) Q/H_0 vs H_0/H_d for an unstriated coated conductor strip for w = 1.2 cm, J_c = 10^{10} A/m^2, for various L/w ratios, (b) Q/H_0 vs H_0/H_d for a striated coated conductor strip for w = 1.2 cm, J_c = 10^{10} A/m^2, for various L/w = 4.

As we can see in Figure 4 (b), for a one filament (unstriated) strand with L/w = 4, Q/H_0 is about 135 J/Am^2, reduced from its value for infinite length (150 J/Am^2) by about 10% due to end effects. As we introduce more filaments, the loss is reduced proportional to the number of filaments, but by plotting Q/H_0N_f, we can see instead the effect of filamentation on sample length. We see that as the filament number reaches 100 the sample end effect based loss reduction is removed.

Conclusions

This paper investigates the influence of sample geometry on the loss of superconducting strips and the effect of filamentation and sample length as a function of the field penetration state of the superconductor. The loss for finite segments is calculated as a function of applied field for striated and unstriated conductors. These results are much more general than they might seem at first glance, since they will be important building blocks for analytic loss calculations for twisted geometries for coated conductors, including helical (Conductor on Round Core, CORC), and twisted (e.g., twist stack cables) geometries. For high levels of flux penetration, the end effects are those for superconducting slabs, while for relatively low field penetration, end effects and reduced field penetration both reduce loss. However, for filamentary samples the ratio of length scales becomes filament width to sample length, thus modifying the loss ratios. This leads to an apparent reduction in the end effects, since the relevant ratio which controls them is the filament width to sample length, rather than the whole conductor width to sample length.

Acknowledgments

Funding was provided by the U.S. Department of Energy, Office of High Energy Physics, under Grants No. DE-SC0011721 and the Chinese Scholarship Council.

References

[1] Bean C P 1962 *Phys. Rev Lett* **8** 250
[2] London H 1963 *Phys. Lett.* **6** 162
[3] Bean C P 1964 Magnetization of High-Field Superconductors *Reviews of Modern Physics* **36** 31–39
[4] Wilson M N 1983 Superconducting magnets
[5] Campbell A 1998 Hysteresis Loss in Superconductors *in Handbook of Applied Superconductivity* (IOP Publishing, London) Ed: Seeber 186-204.
[6] Orlando T P and Delin K A *Foundations of Applied Superconductivity* (Addison-Wesley) 1991.
[7] Brandt E H and Indenbom M 1993 Type-II-superconductor strip with current in a perpendicular magnetic field *Physical review B* **48** 12893
[8] Brandt E H 1996 Superconductors of finite thickness in a perpendicular magnetic field: Strips and slabs *Physical review B* **54** 4246
[9] Müller K-H 1997 AC power losses in flexible thick-film superconducting tapes *Physica C: Superconductivity* **281** 1-10
[10] Sumption M, Lee E, Cobb C B, Barnes P N, Haugan T J, Tolliver J, Oberly C E and Collings E 2003 Hysteretic loss vs. filament width in thin YBCO films near the penetration field *IEEE transactions on applied superconductivity* **13** 3553-6
[11] Carr W J Jr 2001 AC Loss and Macrosciopic Theory of Superconductors 2nd Edition (London: Taylor and Francis)
[12] E. M. Gyorgy, R. B. vanDover, K. A. Jackson et al., Appl. Phys. Lett 55,283 (1989).
[13] F. M. Sauerzopf, H. P. Wiesinger and H. W. Weber, Cryogenics 30,650 (1990).
[14] S. Hq H. Hojaji, A. Barkatt et aL, Phys Rev. B. 43, 2878 (1991).

IOP Conf. Series: Materials Science and Engineering **279** (2017) 012033 doi:10.1088/1757-899X/279/1/012033

Design and Preparation of Two ReBCO-CORC® Cable-In-Conduit Conductors for Fusion and Detector Magnets

T Mulder[1,2], D van der Laan[3,4], J D Weiss[3,4], A Dudarev[2], M Dhallé[1] and H H J ten Kate[2]

[1] University of Twente, Enschede, the Netherlands
[2] CERN, Geneva, Switzerland
[3] Advanced Conductor Technologies, Boulder, CO, USA
[4] University of Colorado, Boulder, CO, USA

E-mail: tim.mulder@cern.ch

Abstract. Two new ReBCO-CORC® based cable-in-conduit conductors (CICC) are developed by CERN in collaboration with ACT-Boulder. Both conductors feature a critical current of about 80 kA at 4.5 K and 12 T. One conductor is designed for operation in large detector magnets, while the other is aimed for application in fusion type magnets. The conductors use a six-around-one cable geometry with six flexible ReBCO CORC® strands twisted around a central tube. The fusion CICC is designed to be cooled by the internal forced flow of either helium gas or supercritical helium to cope with high heat loads in superconducting magnets in large fusion experimental reactors. In addition, the cable is enclosed by a stainless steel jacket to accommodate with the high level of Lorentz forces present in such magnets. Detector type magnets require stable, high-current conductors. Therefore, the detector CORC® CICC comprises an OFHC copper jacket with external conduction cooling, which is advantageous due to its simplicity. A 2.8 m long sample of each conductor is manufactured and prepared for testing in the Sultan facility at PSI Villigen. In the paper, the conductor design and assembly steps for both CORC® CICCs are highlighted.

1. Introduction

Magnets for large scale particle detectors, magnets for fusion experiments, and their bus lines require high currents in high magnetic fields at a wide range of operating temperatures. The operating limits of NbTi and Nb_3Sn conductors in terms of magnetic field, current density and temperature warrant the development of cables made from high temperature superconductors (HTS).

HTS materials such as ReBCO (Re = Rare Earth), BSCCO or MgB_2 greatly broaden the operating spectrum of temperature, current and magnetic field ranges compared to cables using traditional Nb-based low temperature superconductors. CERN is exploring several HTS options, including the use of MgB_2 for superconducting links that power parts of the Large Hadron Collider [1], as well as ReBCO based conductors using either Roebel [2] or Conductor On Round Core (CORC)® cable variants [3].

The CORC® cable is a round conductor that offers isotropic flexibility in combination with high current densities, ranging from 100 to 600 A/mm^2 at 4 K and 10 T, and high electrical

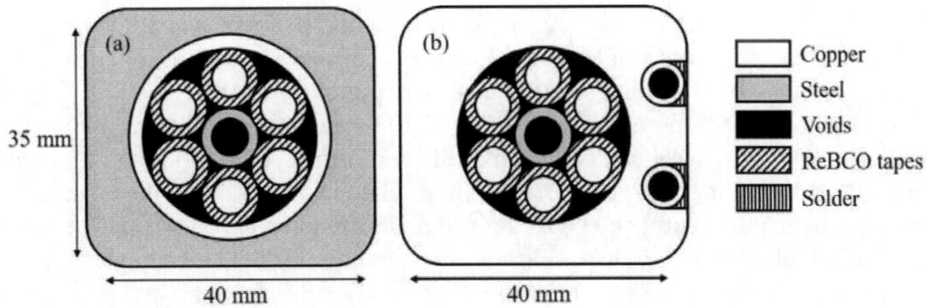

Figure 1. CORC® Cable-In-Conduit Conductor cross-sections with (a) stainless steel jacket for fusion magnets and (b) copper jacket for detector magnets and bus bars, comprising two additional cooling channels for conduction cooling.

and thermal stability. Several CORC® strands can be combined within a jacket to make a Cable-In-Conduit Conductor (CICC) designed for carrying high currents up to 100 kA at 4 K and 10 T, which are required for the next generation of magnets for particle detectors and their bus bars.

The first of its kind CORC® six-around-one CICC was manufactured and successfully tested in 2016 [3]. This demonstration conductor comprised 6 CORC® strands in the six-around-one configuration in combination with an aluminium alloy jacket. Two tests were performed: first in liquid helium in an external magnetic field of 9 T, and second in liquid nitrogen in self-field.

Two new and much improved CORC® CICCs are presented here, featuring a higher current rating and internal forced-flow cooling combined with external conduction cooling for one of the samples, as presented in Figure 1. One new conductor is especially designed for use in high magnetic field- and high heat load environments found in tokamak type fusion magnets. The other conductor is designed for high thermal- and electrical stability, as required in detector magnets and superconducting bus bars. Two 2.8 m long conductors were manufactured at CERN in collaboration with Advanced Conductor Technologies (ACT). Both conductors are to be tested in the SULTAN cable test facility at PSI, Villigen, Switzerland in the second half of 2017.

2. New CORC® Cable-In-Conduct Conductors

The two new CORC® CICC conductors are similar in terms of general design, cable configuration, performance and joint terminals. They differ in method of cooling and jacket material.

2.1. ReBCO CORC® Six-Around-One Cable

Both conductors use the six-around-one cabling layout where the cable comprises six CORC® strands, helically twisted around a stainless steel center tube. Each strand contains 42 SCS 4050 ReBCO tapes from Superpower Inc., for a total of 252 tapes in the CICC. The tapes have 5 μm of copper plating on each side which greatly increases the strand current density compared to the tapes in the previous CORC® CICC that had 20 μm of copper plating on each side.

The CORC® strands for each CICC are slightly different in design. The core of the CORC® strands for the fusion-type CICC is a solid copper former approximately 5 mm in diameter while the strands for the detector-type CICC has a solid copper former of 4 mm in diameter that is extended to 5 mm diameter using 5 layers of copper tapes. The thinner solid copper core makes the second type of strands easier to twist. The disadvantage is that after twisting, these strands tend to spring out a bit more than the strands with the thicker, more rigid core.

IOP Conf. Series: Materials Science and Engineering **279** (2017) 012033 doi:10.1088/1757-899X/279/1/012033

The CORC strands consist of 14 layers of ReBCO tape that are wound around the 5 mm copper core followed by two additional layers of copper tapes and a layer of polyester insulation. The outer diameter of each strand is 7.6 mm. The twisting from CORC® strand to six-around-one CORC® conductor was done manually. The twist starts around 150 mm distance from the joint section. The strand pitch length is 400 mm allowing a full pitch in the high magnetic field region of the SULTAN facility. There are 4.5 cable pitches between the joint terminals.

The polyester insulation around the CORC® strands protects the tapes during twisting. This insulation was removed after twisting to allow firstly direct contact between gas/liquid and the tapes, and secondly, electrical contact between CORC® strands. Each six-around-one bundle is wrapped in a layer of 50 μm thick and 40 mm wide copper foil for mechanical protection inside the jacket.

A detailed overview on the conductor properties is presented in Table 1. In addition the table provides details on the conductor that was tested in 2016 at CERN to show the evolution in size and performance of the CORC CICCs.

Table 1. Properties of the new (2017) and old (2016) CORC® Cable-In-Conduit Conductors.

	2016 CICC Test Sample	2017 CICC Fusion Sample	2017 CICC Detector Sample
Layout	Six-Around-One	Six-Around-One	Six-Around-One
Cooling Method	Liquid Bath	Int. Forced-Flow	Int. Forced-Flow + Ext. Conduction
Jacket Material	Aluminium Alloy	Stainless Steel	OFHC Copper
Jacket Dimensions [mm x mm]	30 x 30	40 x 35	40 x 35
Number of Strands	6	6	6
Strand Diameter [mm]	7.6	7.6	7.6
ReBCO Tapes per Strand	38	42	42
ReBCO Layers per Strand	12	14	14
Length [mm]	1700	2777	2777
Cable Pitch Length [mm]	400	400	400
Cable Pitches	1.5	4.5	4.5
Calc. I_c at 4K/12T [kA]	42	80	80
$J_{Stabilizer}$ at 80 kA [A/mm^2]	133	79	232

2.2. Cable Performance

The cable is rated for 80 kA at 12 T and 4 K. The CORC® strands are twisted to form a cable of 23.6 mm in diameter. The outer layers of copper foil increase the diameter to 24.0 mm before jacketing for a total area of 452 mm^2. The cable current density is rated for 176 A/mm^2 at 12 T and 4 K.The critical current is calculated using single tape I_c(B, T) scaling data, in combination with the corresponding length and angle dependence after being wound to a CORC® strand.

The calculated engineering current density of the entire conductor including the jacket sized 35 mm x 40 mm, is 57 A/mm^2. Measurements are scheduled at temperatures ranging from 4 K up to 50 K. The calculated I_c(B) curve of the CICCs for various temperatures is presented in Figure 2.

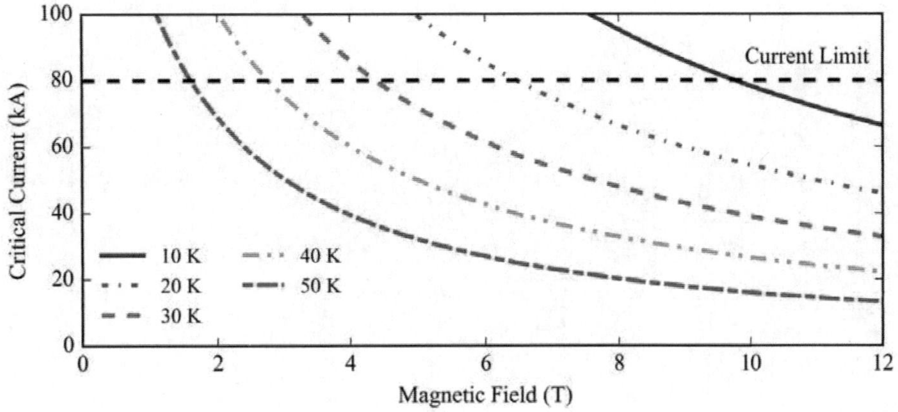

Figure 2. Projected cable critical current versus magnetic field, $I_c(B)$, for various operating temperatures. The measurement station has a current supply limit of 80 kA and therefore it is not possible to measure $I_c(B)$ at low applied magnetic field and low temperatures.

2.3. Conductor Jackets and Cooling layout

The fusion-type CORC® CICC has a stainless steel jacket and internal force flow cooling. High strength materials, such as stainless steel, are needed in modern tokamak magnets for taking up and transferring the enormous accumulated Lorentz forces acting on the superconducting cables during operation. The high heat load from fusion heating on the conductor requires internal flow cooling, as conduction cooling is not effective via the stainless steel jacket. The bore of the stainless steel jacket is 28 mm in diameter. The six-around-one CORC® cable is enclosed by an OFHC copper tube with a wall thickness of 2 mm for a total cable outer diameter of 28 mm. The copper tube is soldered in to the copper joint terminals to provide a parallel path for the current if necessary. The CICC and the six-around-one cable during assembly are shown in Figure 3.

The forces acting on the conductors of detector magnets and their bus lines are much lower compared to the forces in fusion magnets. Therefore, a material with lower strength and higher thermal and electrical conductivity can be used to jacket the six-around-one CORC® cable. In practice, high purity aluminium is commonly used in conductors for detector magnets due to their high radiation transparency. OFHC copper stabilizer is not uncommon in conductors for detector magnets, but better suited for their bus bars. The high thermal conductivity of these materials opens the possibility of a conduction cooled conductor and greatly improves its thermal and electrical stability. Conduction cooling is preferred for its simplicity and practical implementation. The detector-type CORC® CICC sample has an OFHC copper jacket with two helium lines integrated in the jacket for conduction cooling, as well as the option of internal forced-flow cooling via an additional helium line entering the jacket near the joint terminals. The cooling schematic for the CORC® CICC samples is presented in Figure 4.

Copper was chosen over aluminium in the detector-type CICC sample because it more closely matches the thermal expansion of the stainless steel jacket of the fusion-type CICC, since both sample are measured in series in the same sample holder. The copper material of the jacket is cold worked, high tensile strength OFHC copper, since stresses up to 200 MPa are expected in the jacket during the test in Sultan that would exceed the yield strength of fully annealed OFHC copper. In practice, the copper jacket can be fully annealed if this conductor is used in a lower stress environment such as in detector magnets and their bus lines. The peak stress in

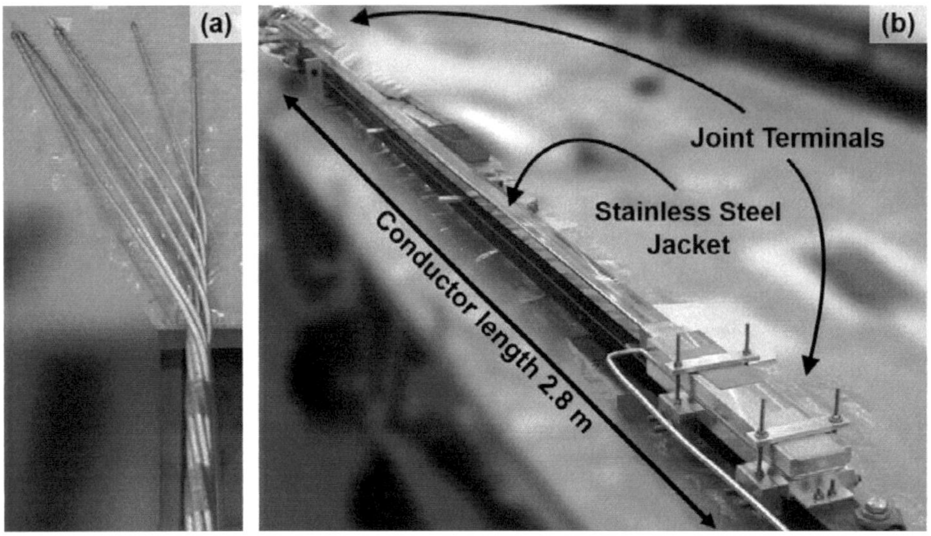

Figure 3. (a) Winding of the six-around-one CORC® cable performed manually around a stainless steel tube. (b) The CORC® CICC with the two joint terminals and the stainless steel jacket during assembly. The conductor is mounted on a stainless steel bar during assembly to ensure that both joint terminals are mounted exactly parallel.

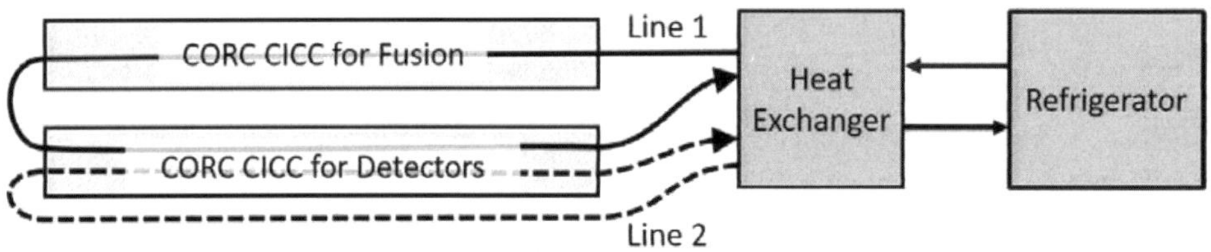

Figure 4. Schematic of the two cooling channels including the heat exchanger. The temperature of the helium gas in each channel is individually regulated by heaters and the heat exchanger. The heat exchanger is not present if the sample is cooled with liquid helium. The mass-flow in each channel can be regulated. Line 1 (solid line) applies internal forced flow cooling to the CORC® CICC within the stainless steel jacket and returns as a conduction cooling line for the CORC® CICC with copper jacket. Line 2 (dashed line) supplies additional forced-flow cooling to the CORC® CICC with copper jacket if necessary.

the stainless steel jacket is well below its yield strength.

Each jacket is divided in two halves lengthwise. The halves of the jackets are mounted and welded using electron beam welding for both the stainless steel jacket and the copper jacket. A bimetallic connection between copper joint terminals and stainless steel jacket is created by brazing a 50 mm long stainless steel adaptor piece to the joint terminals. The adaptors are shown in Figure 5.

The six-around-one CORC® cable was inserted via the adaptor to the terminal. Access to the terminal for soldering was possible via the small gaps between CORC® strands. The CORC® CICC with the copper jacket has the same adaptor pieces to provide symmetry between the two conductors. The adaptor pieces also serve as inlet and outlet points for the internal gas flow cooling.

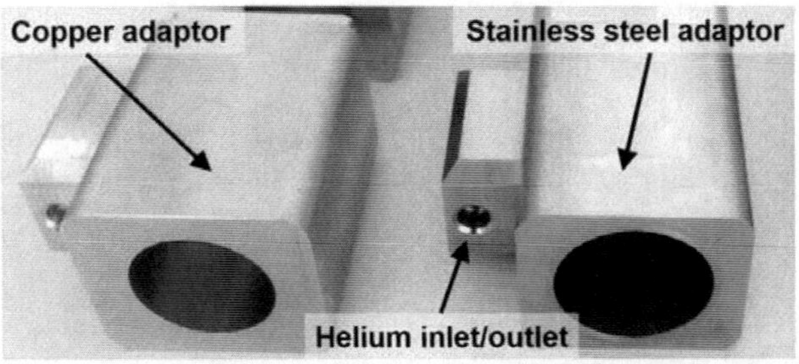

Figure 5. The picture shows the copper and stainless steel adaptor pieces before assembly in the conductor. The adaptor pieces are later brazed to the copper joint terminals and form the link between terminal and jacket. The jackets are welded to the adaptor pieces using electron beam welding. These adaptors also serve as inlet and outlet points for the internal forced-flow cooling.

3. Joint Terminal Layout and Optimization

The joint terminals are designed for a low resistance connection that distributes current equally between CORC® stands. The expected joint terminal resistance is about 2 nΩ per joint terminal at 4 K for a total loop resistance of 8 nΩ. At 50 K, the expected resistance is approximately 6 nΩ per terminal for a total loop resistance of 24 nΩ. The copper joint terminals are 305 mm in length, in which the CORC® strands are inserted over a length of 285 mm.

The connection from CORC® strands to terminals is made in a similar fashion as in the previous (2016) CORC® CICC sample [5]. The entire terminal is heated to about 196±2 °C and filled with eutectic Sn63Pb37 solder. The ends of the CORC® strands are tapered to a staircase like geometry and solder coated. That means several upper layers of the CORC® strands are cut short to facilitate direct, low resistive current injection to the inner ReBCO layers. This termination method has been successfully applied in terminals for single CORC® cables [4] and is explained in more detail in reference [5]. Twisting of the strands within the terminal is hardly possible due to the solder coating on the strand ends, therefore they remain straight in the terminals. The cooling tubes are soldered in a groove on the surface of the joint terminals in the last step of the CICC assembly. Soldering of the cooling tubes is performed with low temperature solder (Sn44In42Cd14), such that the solder in the terminals is not affected.

3.1. Current Distribution in the Cables

The samples are relatively short with 2777 mm end-to-end distance. Current is mainly distributed in the joint terminals, since there is only a relatively low pressure contact between strands. The CORC® strands are twisted such that half a cable pitch difference remains between joint terminals. The half pitch difference ensures a specific strand layout within the terminals, as shown in Figure 6. This layout purposely avoids that one or two strands have a much lower contact resistance compared to the other strands and instead forces a very similar contact resistance to each strand, thereby greatly improving current distribution between strands.

Calculations show that by using this shifted layout, the current distribution between strands differs up to 5% at temperatures around 40 to 50 K compared to 35% in case of a full cable pitch difference [6]. The current distribution in the 5 to 15 K range improves as well by applying this strand layout, however the effect is less pronounced due to the lower resistivity of the copper. The difference in current distribution is reduced from 20% to about 3% in this case.

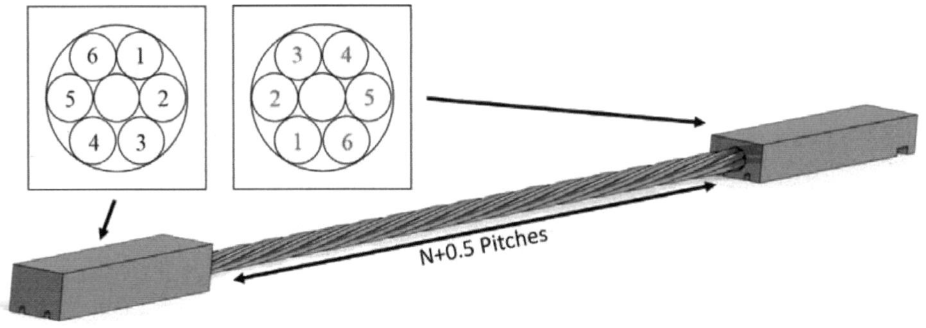

Figure 6. The CORC® six-around-one cable with a terminal connected on each end, including the strand layout inside the two terminals. Current is injected/extracted from the bottom of the terminals. There is half a cable pitch difference between terminals. This shift in cable pitch equalizes the total contact resistance per strand and forces a more homogeneous current distribution in the cable.

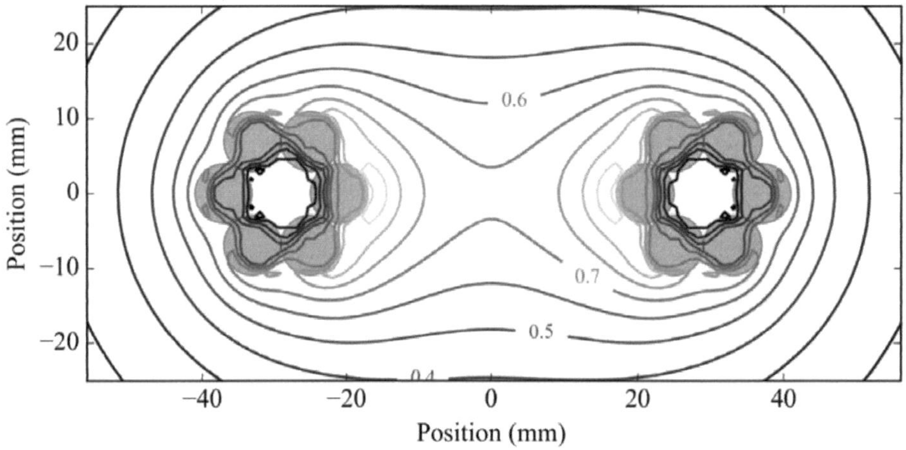

Figure 7. The contour-lines of the self-field of the CORC® CICC conductors in the sample holder for the test at PSI. Both conductors combined produce approximately 1 T per 50 kA peak field on the cable. Since the 6-around-1 cable is twisted, all six strands will experience this peak field at some point within a full cable pitch.

4. Full Sample and Measurement Facility

The SULTAN cable test facility features a 10.8 T split pair solenoid providing the test magnetic field with 2% uniformity over a section of 400 mm [7]. The self-field on the sample is about 1 T per 50 kA in all cable orientations as presented in Figure 7. That results in a maximum magnetic field on the conductor of 12.4 T in the peak field region of the SULTAN magnet system at full background field.

The Sultan test facility offers the possibility to cool the sample with 10 g/s helium gas in the range of 5 to 50 K or with liquid helium at 4.5 K. In case when the sample is cooled by gas, the temperature is regulated with the combination of a heat exchanger and heaters. Both straight sections of CORC® conductors are combined in series to a hairpin test sample with one leg carrying the feed current and the other carrying the returning current through the high field region of the facility.

The forces in the samples are directed outwards. An aluminium alloy clamping structure

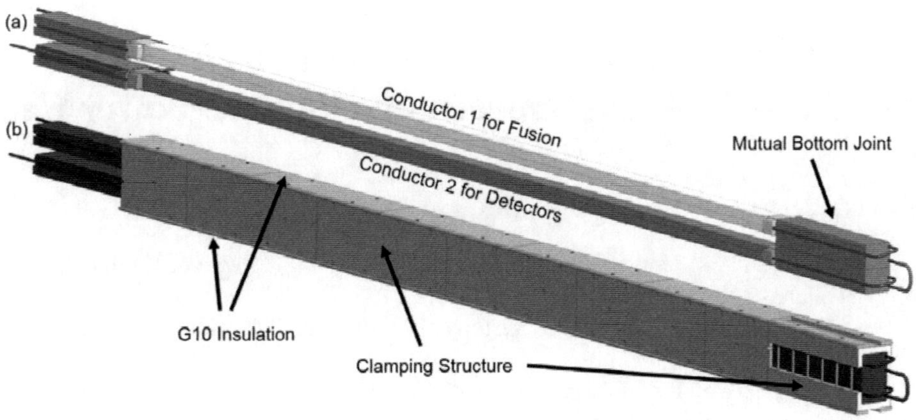

Figure 8. (a) The two bare CICCs jointed together on one side without any of the clamping structure. (b) Both CICCs including the clamping structure and insulation/guiding plates.

keeps both conductors together. G10 plates on the surface of the aluminium alloy clamps serve as a guiding system for inserting the sample into the station and also provide insulation between the cryostat and the sample. The joint terminals of the sample make a clamping contact to the terminals of the facility using squeezed indium wires as the interface to minimize the contact resistance. The terminals on the bottom side of the sample where the current returns feature a soldered contact using indium as soldering material. A schematic drawing of the two conductors in combination with the clamping system in presented in Figure 8.

5. Conclusion and Outlook

Two new ReBCO CORC® Cable-In-Conduit Conductors were manufactured at CERN. One has a stainless steel jacket aimed for use in magnets for fusion experiments and the other CICC has a copper jacket for implementation in detector magnets and superconducting bus bars. Both conductors are rated for 80 kA at 12T/4K.

The joint terminals were produced using a proven method that was developed and tested with the previous CORC® CIC-Conductor [3]. The joint terminal resistance is expected to be about 2 nΩ at 4 K and 6 nΩ at 50 K.

The performance test of both conductors and their joint terminals at various temperatures will take place in the SULTAN cable test facility in August and October of 2017.

Acknowledgements

This work has been supported in-part by the U.S. Department of Energy under contract numbers DE-SC0007660 and DE-SC0014009.

References

[1] Ballarino, A. 2014 Supercond. Sci. Technol. **27** 044024.
[2] Kirby G A *et al.* 2017 IEEE Trans. Appl. Supercond. **27** 4003307.
[3] Mulder T, Fleiter J, Willering G, Dudarev A, Mentink M, Dhalle M and Ten Kate H 2017 IEEE Trans. Appl. Supercond. **27** 4801004.
[4] Mulder T, Dudarev A, van der Laan D, Mentink M, Dhalle M and ten Kate H 2015 IOP Conference Series: Materials Science and Engineering **102** 012026.
[5] Mulder T, Dudarev A, Mentink M, Dhalle M and ten Kate H 2016 IEEE Trans. Appl. Supercond. **26** 4801704.
[6] Mulder T, Presentation at WAMHTS-04 2017, Barcelona, Spain.
[7] Sultan Team 2002 'User Specification for Conductor Samples to be Tested in the SULTAN Facility', CRPP-REPORT-2002-009.

IOP Conf. Series: Materials Science and Engineering **279** (2017) 012034 doi:10.1088/1757-899X/279/1/012034

Technical Challenges and Potential Solutions for Cross-Country Multi-Terminal Superconducting DC Power Cables

A Al-Taie[1,2,3], L Graber[4], S V Pamidi[1,2]

[1] Florida A&M University - Florida State University College of Engineering, Tallahassee, FL 32310, USA
[2] Center for Advanced Power Systems, Florida State University, Tallahassee, FL 32310, USA
[3] University of Technology, Baghdad, 10066, Iraq
[4] Georgia Institute of Technology, Atlanta, GA 30332, USA

aha16b@my.fsu.edu

Abstract. Opportunities for applications of high temperature superconducting (HTS) DC power cables for long distance power transmission in increasing the reliability of the electric power grid and to enable easier integration of distributed renewable sources into the grid are discussed. The gaps in the technology developments both in the superconducting cable designs and cryogenic systems as well as power electronic devices are identified. Various technology components in multi-terminal high voltage DC power transmission networks and the available options are discussed. The potential of ongoing efforts in the development of superconducting DC transmission systems is discussed.

1. Introduction

There is a renewed interest in DC systems in power transmission and distribution. By the start of the last century, an efficient electrical power transmission system was needed to transfer the power over long distances and to start the expansion of the power grid. Changing the voltage levels by the AC power transformers was the settling point for the War of the Currents and the start of utilizing the AC power system by the end of the nineteenth century [1]. Even though AC power became the norm since then, DC power transmission was not totally ignored. This is due to the fact that the DC power technology provides an efficient and effective interconnection between any two independent AC grids to transfer power [2], in addition to adding rigidity to both of them. The conventional solution for this is known as the rotary converter. It is essentially a motor-generator coupled machine. The unit connected to the grid that supplies the power to be transferred operates as a motor, and the other unit operates as a generator and is connected to the grid receiving the power. This method is used in several locations around the world. For instance, in Japan, it is used to connect power networks of adjacent areas with different frequencies, 50 and 60 Hz. Hundreds of MVAs can be transferred using this method [3]. Basically, the motor-generator coupled machine need to be built to the maximum rating of the power transferred, so the coupled machine must be built to GW scale if it is intended to transfer power in that range. However, there is a growing trend towards using of solid-state solutions based back-to-back (B2B) high voltage direct current (HVDC) grids tying systems [4].

Having mostly relied on the AC power transmission and distribution (T & D) systems for the last century, the utilization of the DC systems is being considered and the interest is growing [2]. The new trend of increasing reliance on renewable energy sources, which provide ample amounts of energy usually at remote locations, requires a very efficient method to transmit these large amounts of energy to the urban demand centers. The rapid development in power electronics has contributed in utilizing

HVDC power transmission and making it feasible and controllable. Currently, and after decades of notable success in the HVDC transmission field, DC transmission is preferred over the AC when the power transmission distance exceeds the break-even point. It presents the meeting point for the AC and DC systems' capital and operational cost curves. The break-even point is in the range of 500 to 800 km for overhead lines, but much shorter for underground cables, which is between 20 and 50 km for submarine cables and twice as long for underground cables [5]. The reason behind the lower values for the break-even length in cables is due to the cable's capacitive charging length when transmitting power in AC form [6]. However, this limitation does not exist when transmitting DC power. Additional benefits for the HVDC transmission system over the HVAC are discussed elaborately in the literature [1, 7]. In conclusion, HVDC transmission is an efficient and economical choice for transmitting large amounts of power over long distances.

The next step will be to expand HVDC point-to-point links to a true multi-terminal HVDC (MT-HVDC) system, which has the ability to provide all the above mentioned benefits while adding more flexibility with respect to system control. Recently, enhancing the transmission system capacity and improving the system control are the two major challenges that are facing the power sector in North America [2]. Therefore, considering the MT-HVDC power transmission systems is a prospective innovative solution that can help with addressing these two challenges. MT-HVDC systems support the grid in many aspects, like serving as asynchronous ties between multiple grids/networks and providing a system that can transfer power in a more efficient way over long distances, besides the ability to receive or deliver the power at multiple points necessary for integrating distributed renewable energy sources into the grid.

The realization of the need for the MT-HVDC systems is increasing every day because they offer a full seizing of the technical and economic benefits of the HVDC transmission system [8]. Figure 1 below shows the potential of MT-HVDC systems in decreasing the total capital cost by reducing the converter stations needed. However, switchgears are needed to control the power flow.

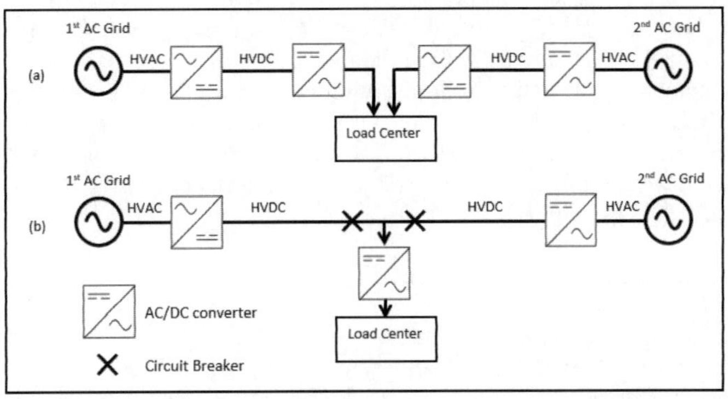

Figure 1. (a) Two point-to-point HVDC links with 4 converters. (b) The same Two DC links connected in a MT-HVDC configuration with 3 converters.

2. Need of Superconducting Cable Technology for the MT-HVDC Systems

The MT-HVDC system involves handling large amounts of power, on GW scale. This will lead to some challenges such as:

i. If the suspended overhead cables and transmission lines were used, they would be vulnerable to weather conditions.

ii. If the underground conventional cables were used, the complexity and size of the project would be substantially high since these cables need to be separated by several meters to avoid the overheating [9]. Besides, due to their size, it is more difficult to obtain right of ways for larger areas for the station footprints, especially when it is within urban areas [2].

All these problems could potentially be addressed by replacing the conventional cables technology with state of art high temperature superconducting (HTS) cables. Underground HTS cables will address the weather conditions concerns. Besides, the Superconducting DC power cables have the attractive "zero loss" state. Although the Superconducting cables are much higher in efficiency when transmitting AC power, compared to the conventional cable technology, they are even better when transmitting DC power due to the absence of the significant negative impact of the AC losses, which will be discussed in the next section. Thus, each Superconducting HVDC cable has the ability of carrying currents up to 100 kA whereas the conventional HVDC line can carry a few kA at most [2], and so, less right-of-way and foot print are needed. In addition, if more cables are needed for the sake of expansion, the Superconducting DC cables can be buried closer to each other unlike the conventional cables due the fact that magnetic field can be completely eliminated outside the superconducting cable with a proper cable design [9]. In summary, the HVDC based HTS cable technology offers the optimum solution to achieve higher performance and power density for the MT-HVDC systems.

In Europe, the vision of using superconducting HVDC for long distance transmission is being explored under a large project aimed at moving towards low-carbon economy called the "Best Paths" program. The program intends to develop technology to increase the capacity of the European interconnected transmission networks [10]. The main goal is to integrate the abundant renewable energies, from the wind farms, smoothly and efficiently into the pan-European power transmission network. Thus, the program intends to achieve the European MT-HVDC based HTS cable power transmission network. The HTS cable for the program intends to use the magnesium diboride (MgB_2) superconductor and cooled with gaseous helium (GHe), and they are investigated to carry currents up to 5 – 10 kA under 200 – 320 kV of high voltage levels to transfer up to 3.2 GW of power [11].

3. AC Losses in Superconducting Cables Due to The Ripple and Harmonics from Converters

AC Superconducting power cables suffer from AC losses. DC superconducting cables do not produce any losses as long as the DC current is clean and free from AC ripple and harmonics that originate from the power converters [12]. The extent of losses due to ripple is small, but it is necessary to eliminate it for long distance superconducting DC cable systems to make them more efficient and to minimize the expensive cryogenic cooling capacity required [13]. The harmonics could also come from reversing the power flows in HVDC based HTS systems. There are a few different converter topologies that can be used in HVDC systems [13-17]. The selection of converter topologies for HTS HVDC systems has to be based on their suitability for high current operation and the need for clean DC power. It has been reported that the parallel hybrid modular multilevel HVDC converter design offers significant reduction in the extent of ripple in HVDC systems. [17]. Long distance cables tend to dampen the ripple in general, but it is important to consider this issue in the design of superconducting cable for HVDC applications for negligible steady state losses [12-16].

4. Technical Challenges and Solutions for the MT-HVDC HTS Cables Technology

4.1. The Superconducting Technology Challenges

HTS technology has the potential for providing an efficient MT-HVDC systems, but has its own technical challenges, cost issues, and lack of standards and industry accepted test protocols.

Regarding to cost, there have been several early and recent research studies for utilizing superconductors to transport large amounts of power over long distances [18, 19]. The electric power research institute (EPRI) conducted research on the economic feasibility of utilizing superconducting cables for transmitting large amounts of power over long distances. The analysis concluded that a superconducting cable system would cost approximately the same as a conventional transmission line if the line is transmitting power in excess of 5 GW, over a long distance of about 1,600 km. Part of that result due to the fact that the cost of the cryogenic system, which is needed to maintain the superconducting properties, will be a small share of the total cost [2]. In addition, while the efficiency of the conventional power transmission lines drops substantially when carrying bulk power over long distances, the high efficiency of superconducting cable systems will not be affected. Furthermore, the power losses in the terminations are constant and not affected by the cable length. These terminations losses are a large fraction of the total losses of the system because they handle the connection between the ambient temperature of the current leads and the cryogenic temperature of the superconducting cable. However, these losses decrease relative to the whole system losses with the increase in cable length to negligible values with respect to the total cost. Furthermore, a drop in the cost of the superconductor and power electronic converter technology is expected in the near future. The increase in manufacturing capacity of volume of HTS materials and cables will bring about a substantial decrease in the cost due economies of scale. On the power electronics side, the device prices are dropping rapidly, which will reflect on the capital cost drop for the converter substations. However, the new technology devices, like silicon carbide (SiC) devices, are still not in the mass production level to allow a practical cost estimation. In addition, the prospective development in the HTS materials and the efficiency of the cryogenic cooling system is expected to further reduce the capital and operational costs. Thus, at this point the technology is not mature enough to conduct a meaningful economic analysis for such systems.

One obstacle for HTS technology is the lack of standardized test methods. It is still an issue for the utility companies to perform testing for the HTS DC cables because there is no a standard test for it, in contrast to the HTS AC cables as was achieved discussed [20]. This presents one of the major challenges for utilizing HTS DC cables widely in power grids. HTS DC cable has its unique challenge in performing a load cycle, contrary to the HTS AC cable or the conventional HVDC cable. Applying both voltage and current simultaneously through the use of a current transformer is impossible due to its AC loss characteristics [21, 22]. Thus, some appropriate new methods for performing reliable and acceptable qualification tests on HTS DC cable are needed. On the other hand, thermal cycling studies of HTS cables is very important for utilities to demonstrate and verify the cable's lifetime since the cables will be subjected to several contractions and expansions during cool-down and warm-up cycles during installation and operation. The thermal cycles are similar to HTS AC cables that are in operation and hence some data exists. The thermochemical issues are design dependent and further work is needed to qualify the designs for 30-40 years of service as required by electric power companies [23, 24]. However, for the planned and installed HVDC based HTS cable projects, the utilities companies have started constructing test centers and establishing specialized tests for the new technology. One of the latest projects that was implemented in South Korea, which has combined two standardized tests, besides the HTS experience. South Korea is one of the earliest countries that adopted the smart and clean energy applications and has been considerably progressing in developing the HTS power cables integration to the real Grid. A project by the Korea electric power corporation (KEPCO) and LS Cable Limited has done the designing and manufacturing part, and started researching the HVDC based HTS cable integration into the Gumak grid in 2011 [25]. The big picture of the project is to connect two conversion stations, from Halim Conversion Station to Gumak Conversion Station, by a 5.3 km long DC line. A 0.5 km long underground HTS DC cable project was meant to connect an existing 4.8 km long DC overhead transmission line to Gumak Conversion Station. Therefore, a ±80 kV, 3.25 kA, 500 MW DC HTS cable was constructed and installed in the Gumak converting station. After passing successfully all the tests by the 101 m cable prototype, the actual cable was installed and started the demonstration in the real grid since October 2014 [26]. For testing, the recommendations were based on Electra 496 and Cigre TB 538, and the HTS experience. In addition, an additional measurement for the DC HTS cable lifetime was performed. That was done through the investigation of the cable response to the mechanical stress due to cool-down and warm-up cycles. The testing period was from 20 Dec. 2012 to 31 Oct. 2013, and the results showed that a combination of HV tests with the HTS experience, which was called the qualification test, is required for the HTS DC cable project. Finally, a test for the cryogenic system is always required to verify the ability of the cooling system to handle the cable performance as designed.

4.2. Cryogenic Systems Challenge to Support Long Distance Superconducting HVDC Systems

One of the major challenges in bringing HVDC based HTS cable systems to the power grid until recently has been the lack of reliable and economic cryogenic refrigerator systems. This challenge has recently been addressed by the industry through the development of large cryo-refrigerators based on Turbo-Brayton cycles with >150 kW cooling power at 70 K [27, 28]. These large cryogenic systems are significantly more efficient and require little maintenance compared to Stirling refrigerators and Gifford-McMahon cryocoolers [29]. Turbo-Brayton cryocoolers have been used successfully for the HTS cable projects in Korea [30, 31]. Further new approaches are being investigated to reduce the complexity and costs of the cryogenic systems for large HTS systems [32, 33]. Further technical advancements are needed to improve the reliability and reduction of capital and operational costs of cryogenic systems and the interfaces between the room temperature components and cryogenic components at the terminals of MT-HVDC systems.

4.3. The MT-HVDC Systems Technology Challenges

The MT-HVDC based HTS cable systems will face the same challenges of the conventional MT-HVDC systems, such as the DC fault protection and the control of the power flow for the meshed DC system with different kinds of converter types.

One major difference between HVDC and HVAC is in the management of faults. While HVAC systems rely on its natural current zero crossings to clear the fault, HVDC requires more complicated circuits to create artificial current zero crossings. It has been reported that the fault current magnitude in HVDC transmission systems will be lower than in HVAC systems for the same amount of the power transmitted [34]. The real problem with the DC systems is the faster rate of the rising fault current than in AC systems [35]. This is due the faster DC system dynamics and transients, which means a very sensitive DC fault relay is needed to detect and remove faults quickly. Unfortunately, these kinds of DC circuit breakers are still not commercially available, but there have been some developments of building and testing of a few prototypes [36]. However, this issue can be dealt with by using the HTS cables, if these cables were designed to operate as fault current limiting (FCL) cables [6]. The Superconducting material loses its superconducting property when the current exceeds its critical value (Ic) and transitions to the normal state of being a high resistance material. However, the HTS cables should be designed in a specific way for that particular purpose, or else they could be damaged by the massive fault currents. Thus, this is a great feature for the HTS power transmission cable systems that give them a distinguished advantage making them more attractive over the conventional systems. To conclude, use of FCL HTS cables can a contributing factor to solve the protection problem for the MT-HVDC systems, which is a big draw back for the conventional cable/line based MT-HVDC systems.

4.4. The Challenge of Establishing Research and Development Centers

The laboratory establishment for research and development of large superconducting systems is still a challenge for the MT-HVDC based HTS cable systems. Unlike the advantages of changing the magnitudes of voltage and current by the transformers in the case of AC system experiments, establishing the required large DC power sources for testing MT-HVDC based HTS cable systems is a major challenge, especially coupled with the need for collocating the large cryogenic systems in the laboratories and research facilities. The conventional cables' laboratories and testing facilities lack for the existence of cryogenic cooling systems. As a result, the experimental research work on MT-HVDC based HTS cables is still an issue that needs to be addressed by making the necessary investments, which will accelerate the development and implementation of this game changing technology. Therefore, for modelling and analysis of the MT-HVDC based HTS cable systems, simulations are needed. The real-time simulation for these system is very important as for any new piece of power equipment that needs to be integrated into the grid. The simulations are also important due to the fact that the DC system stability and dynamics are much faster compared to the corresponding AC systems [35].

5. The System Components

5.1. The Converters

Many types of converters have been utilized in the HVDC systems all over the past years. The load commutated converter (LCC), based on thyristors, was introduced first during the 70s of the past century [37]. It is suitable for bulk power transmission, with low losses (approximately 0.7% of losses per converter), but it lacks for the turn off feature. Thus, it has a drawback regarding to controllability. Besides that, it has disadvantages of needing variable reactive power from the Grid, AC source on both ends for successful commutation, and transformers on AC sides to have a 30-degree phase shift. Then the voltage source converter (VSC), which utilizes insulated gate bipolar transistors (IGBTs) was introduced by ABB and marketed as "HVDC Light" in 1997 [37]. It uses the pulse width modulation (PWM) technique for better voltage waveform and less harmonics. It has higher losses compared to LCC (typically 1.7% of losses per converter). However, it has many other advantages over the LLC, besides the source controllability: First, the conversion station is more compact, less foot print, due to the less of the harmonic filters needed. Second, it can be utilized anywhere in the grid, whether it is a weak or strong network, with no consideration for phase shift. The capability of offering the blackstart and the many other technical features made the VSC to be more preferable over the LLC. In 2010, a new multilevel VSC technology called modular multilevel converter (MMC) was introduced simultaneously by two companies (Siemens and ABB) in two projects [37]. Siemens's HVDC PLUS technology project, the Trans Bay project, was an 88 km subsea cable installed to transfer 400 MW of power from Pittsburg to the city of San Francisco in California [38]. This converter was introduced by Marquardt and Lesnicar in 2002 [39]. It has a better efficiency than the VSC (almost 1% of losses per converter) due the less needed switching frequency, with better control. Thus, it really represents the state of art for the HVDC converters' technology.

5.2. The HVDC HTS Cables

The HTS HVDC cables connect the converter station in one grid to the other, or extend between any two converter stations as in a MT-HVDC radial link. These cables, which are actually pipelines, handle and carry a large DC current between the conversion stations. Since the application in which the HTS cable is going to be used is directly impacted by the design configuration decision, a single pole per envelope design configurations is used for the HV applications [6]. Different cable layouts have been proposed, the Warm, the Cold and the Two Stage Dielectric HTS cable layout. Each one of them has its own advantages and disadvantages. The different aspects of these designs have been investigated and the results showed that the Warm Dielectric Cable design is the right fit for the HVDC HTS cables due the absence of the induced currents on the cryogenic piping [40]. Besides, it is expected to be less costly, because there is no need for a special cryogenic dielectric medium. Additionally, it requires smaller diameter of cryogenic piping. However, it occupies more volume because each DC pole requires an individual separate cryostat [6]. Thus, the total cross section of the Warm Dielectric HTS cable is significantly larger than the Cold Dielectric for a given power rating, which reflects in lower effective power density of the cable. Therefore, the Cold Dielectric cable design is preferable over the Warm dielectric where compact HTS cables are preferred.

Regarding insulation, so far no successful extruded insulation has been reported in the technical literature, except extruding thin layers on the HTS tapes directly [6]. Although the extruded insulation possibilities exist for the Warm Dielectric cable design, but not for the Cold Dielectric designs, where the lapped tape insulation has been used. Here, the insulation tapes wrapped onto the HTS layer and surrounded by the cryogenic coolant, which penetrates the butt gaps in the insulation layers to serve also as dielectric medium, inside the cryostat. However, the dielectric properties vary for the different kinds of the cryogenic coolants. For example, with one sheet of polypropylene laminated paper (PPLP), at 77 K of temperature and within a uniform electric field, liquid nitrogen (LN2) has higher dielectric breakdown strength of values around 270 kV/mm of DC voltage and 115 kV/mm of AC voltage than GHe, at 1.0 MPa, which is around 180 kV/mm and 60 kV/mm of DC and AC voltages, respectively [41]. Nevertheless, it would be more economic and easier in cable manufacturing if the researchers or the manufacturers succeed in developing an extruded cryogenic insulation which can form a thin layer over the HTS layer instead of the lapped tape insulation.

Since the cryogenic coolant submerge the insulation tapes, it is critical to keep the coolant operating temperature under the boiling temperature to avoid having any bubbles which can deteriorate the insulation integrity, and consequently, the performance of the HTS cable. For that purpose, pressurizing and subcooling are used to increase the margin between the operating and the boiling temperatures. For instance, the permissible temperature raise of LN_2 over the operating temperature, which is 75 K, increases from 2 K at 100 kPa pressure to 20 K at 500 kPa of pressure [42]. Increasing the temperature margin will also increase the HVDC cable length limit, with the absence of the charging length limit in HVAC cables, so that the cooling stations can spaced apart farther. Moreover, for long distance HTS cables, the corrugated cryostat pipes, which helps with making the pipes flexible for bending, can be replaced by flat smooth pipes to reduce the pressurized coolant friction and the cost.

5.3. The HTS Cables Terminations

The cable terminations are interface for the room temperature components and the cryogenic components of the system and are the locations of current injection and the cryogenic inlet and outlets. A schematic of a cable termination is shown in Figure 2. The thermal, the mechanical and the electrical stresses are all mostly concentrated in the terminations and hence the MT-HVDC based HTS cables system requires efficient and effective termination designs.

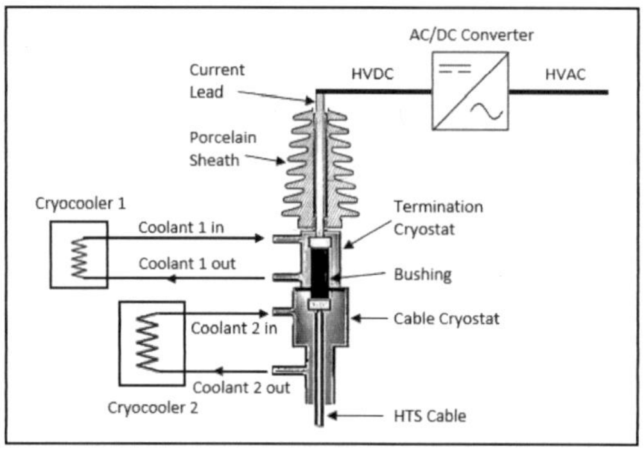

Figure 2. Proposed Two-Stage Cooling HTS Cable's Termination

6. The System Configuration

Several possible configurations exist for MT-HVDC based HTS cables systems. According to the particular application served by the MT-HVDC system, an appropriate configuration has to be designed. If the application is to have a MT-HVDC system that spans long distances, Cross-Country MT-HVDC system, the radial configuration as shown in Figure 3 is more suitable. This Cross-Country radial MT-HVDC scheme is appropriate for transferring the power from a renewable energy source, Terminal 1, all the way to the far away AC grid of the load center, Terminal 4. It also has other taps at certain locations in between to deliver or withdraw power from the cable system, terminals 2 and 3. In addition to the bidirectional power injection that these two terminals offer at their locations close to other generation/load centers, it can serve the purpose as cooling stations for the HTS cable system to maintain the operating temperature and pressure for the cryogenic coolants.

To calculate the distance between any two cooling stations, the following equation is used:

$$Q = \dot{m}\ Cp\ \Delta T$$

For instance, if LN_2 is used with the operating temperature range (ΔT) of 68 to 78 K, a mass flow rate ($\dot{m}$) of 200 (g/s), and specific heat capacity of (Cp) of 2 (kJ/kg/K), then the maximum heat load of the cable (Q) between the cooling stations can be up to 4 kW. Considering 1 W/m of the heat flux for the HTS cable will give us 4 km, which is the maximum length that the HTS cable can go between the cooling stations. However, calculating the distance between the cooling stations involves other factors such as the pressure drop which should be taken into consideration too. An alternating rigid and flexible sections of the cryostat, which was developed by Chevtchenko [43], is a potential solution to decress the friction of the inner surface of the cryostat. Furthermore, smooth cryopipes can be proposed to replace all the corrugated sections since the pipes are going to be buried underground and no flexibility is needed for bending them. This will significantly reduce the total friction of the inner surface of the cryostat. On the other hand, the topography can also add complications to the pressure drop consideration when the cable route involves inclined areas. Moreover, since the heat load of the DC cable is solely from the heat leak into the cryostat, the cryostat design is the key factor to increase the distance that HTS cable can go between refrigeration stations. A team of researchers from EPRI led by Hassenzahl have come with a very interesting result of 20 km distance separating the cooling stations which can be achieved with a cryostat design that uses counter-flow, using go and return streams for the LN_2 [44]. A 3 kg/s of mass flow rate and 3.1 MPa as maximum pressure, with allowable 1.0 MPa of flow pressure drop, were developed for the EPRI study to enable the 20 km distance. Finally, it is possible to add cryogenic cooling stations in the middle of cable sections which adds additional flexibility to the system [2].

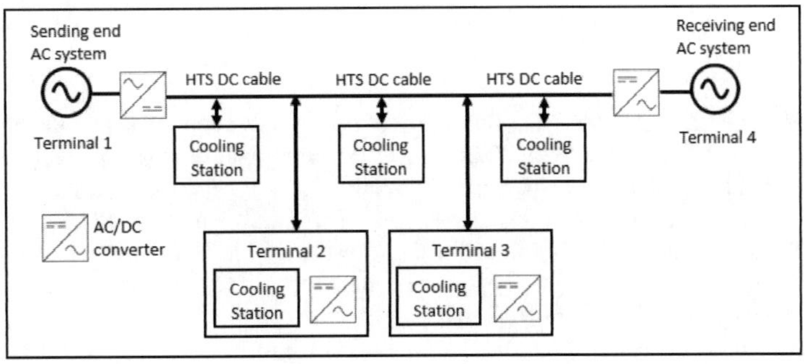

Figure 3. Simplified Four-Terminal Radial MT-HVDC based HTS DC cables with cooling stations scheme.

7. Conclusion

The paper outlined the opportunities for increasing the efficiency of long distance electric power transmission systems by using HTS DC power cables. The multi terminal superconducting HVDC transmission systems allow easier integration of diverse power grids and the insertion of distributed renewable sources into the power network. The integration of the grids and renewable energy sources eliminate the use of fossil fuels for meeting the transient power demands and improve the resiliency of the whole electric power network. There are, however, several technology advancements needed in areas of HTS, cryogenics, and power electronic systems to make the long distance superconducting multi terminal HVDC transmission systems a reality. The technology gaps and possible options for various components of HVDC systems are discussed.

References

[1] Kalair A, Abas N and Khan N 2016 *Renewable and Sustainable Energy Reviews* **59** pp 1653–75

[2] Hassenzahl W, Daneshpooy A, Eckroad S, Grant P, Gregory B and Nilsson S 2009 *A Superconducting DC Cable Report* (Palo Alto) EPRI

[3] Sae-Kok W, Yokoyama A, Verma S C and Ogawa S 2006 Excitation control system design of rotary type frequency converter for performance improvement of power system dynamics *IEEE Trans. on Energy Conversion* **21** 1 pp 210–20

[4] Tiku D 2014 Dc power transmission mercury-arc to thyristor hvdc valves *IEEE Power and Energy* (2014)03/04

[5] Whitaker J C 2006 *AC Power Systems Handbook* (Boca Raton: CRC Press) chapter 1 p 25

[6] Pamidi S, Kim C H and Graber L 2015 *Superconductors in The Power Grid* ed C Rey (Elsevier Science) chapter 7 pp 225–60

[7] Okba M H, Saied M H, Mostafa M Z and Abdel- Moneim T M 2012 High voltage direct current transmission - a review, part I *Energytech 2012 IEEE* (Cleveland) pp 1–7

[8] Kundur P, Balu N J and Lauby M G 1994 *Power System Stability and Control* (New York: McGraw-Hill Companies) chapter 6 pp 199–270

[9] Stemmle M, Marzahn E, West B, Schmidt F and Schippl K 2013 Superconducting hvdc power cables for voltage source converter systems *CIGRE Session*

[10] *Best Paths* http://www.bestpaths-project.eu/

[11] Project demonstration *Best Paths* http://www.bestpaths-project.eu/en/demonstration/demo-5

[12] Chowdhuri P and Laquer H L 1978 Some electrical characteristics of a dc superconducting cable *IEEE Trans. on Power Apparatus and System* **PAS-97** 2 pp 399–408

[13] Kim J H and Pamidi S V 2012 Electrical characteristics of 2g hts tapes under dc current with ac ripple *IEEE Trans. on Applied Superconductivity* **22** 3 p 5801104

[14] Ivanov Y, Vyatkin V, Watanabe H, Chikumoto N, Hamabe M, Sun J, Takano H, Yamaguchi S and Otabe E 2016 Current imbalance and ac losses of long-distance dc hts cable *IEEE Trans. on Applied Superconductivity* **26** 7 pp 1–4

[15] Kim J G, Kim S K, Park M and Yu I K 2015 Hardware-in-the-loop simulation for superconducting dc power transmission system *IEEE Trans. on Applied* Superconductivity **25** 3 pp 1–4

[16] Wang Y, Zheng Y, LiuLiu H, Dai S, Zhang H, Guan X, Teng Y, Zhao L, Xue J and Lin L 2011 A novel approach for design of dc hts cable *IEEE Trans. on Applied Superconductivity* **21** 3 pp 1042–45
[17] Qin J and Saeedifard M 2014 Dc-line current ripple reduction of a parallel hybrid modular multilevel hvdc converter *IEEE PES General Meeting* (National Harbor) pp 1–5
[18] R. L. Garwin and J. Matisoo 1967 Superconducting lines for the transmission of large amounts of electrical power over great distances *Proc. IEEE* **55** no 4 pp 538–48
[19] Grant P M 2007 Superconducting lines for the transmission of large amounts of electrical power over great distances: garwin–matisoo revisited forty years later *IEEE Trans. on Applied Superconductivity* **17** no 2 pp 1641–47.
[20] Gille A 2013 *Insulated Cables Report* (Paris) TB538 Cigre
[21] Dai S *et al* Testing and demonstration of a 10-ka hts dc power cable *IEEE Trans. on Applied Superconductivity* 2014 **24** 2 pp 99–102
[22] Zhang D *et al* 2013 Testing results for the cable core of a 360 m/10 ka hts dc power cable used in the electrolytic aluminum industry *IEEE Trans. on Applied Superconductivity* **23** 3 p 5400504
[23] Hamabe M *et al* 2013 Status of a 200-meter dc superconducting power transmission cable after cooling cycles *IEEE Trans. on Applied Superconductivity* **23** 3 p 5400204
[24] Yumura H *et al* 2013 Update of yokohama hts cable project *IEEE Trans. on Applied Superconductivity* **23** no 3 p 5402306
[25] Lim J H *et al* 2015 Cryogenic system for 80-kv dc hts cable in the kepco power grid *IEEE Trans. on Applied Superconductivity* **25** no 3 pp 1–4
[26] Yang B, Kang J, Lee S, Choi C and Moon Y 2015 Qualification test of a 80 kv 500 mw hts dc cable for applying into real grid *IEEE Trans. on Applied Superconductivity* **25** 3 pp 1–5
[27] Healy R 2016 Taiyo nippon sanso corporation develops cooling neon-based turbo refrigerator *gasworld* https://www.gasworld.com/tnsc-develops-cooling-neon-based-turbo-refrigerator/2010784.articl
[28] Turbo-Brayton cryogenic systems *airliquide* https://advancedtech.airliquide.com/turbo-brayton-cryogenic-systems
[29] Gifford-McMahon cryorefrigerators *cryomech* http://www.cryomech.com/cryorefrigerators/gifford/
[30] Hirai H, Hirokawa M, Yoshida S, Nara N, Ozaki S, Hayashi H, Okamoto H and Shiohara Y 2012 Neon turbo-brayton cycle refrigerator for hts power machines *AIP Conf. Proc*. vol 1434 Issue 1 pp 1672–79
[31] Hirai H, Hirokawa M, Yoshida S, Sano T and Ozaki S 2014 Development of a turbine-compressor for 10 kW class neon turbo-Brayton refrigerator *AIP Conf. Proc*. vol 1573 pp 1236–41
[32] Lee J, Lee C, Jeong S, Park J, Hwang S D, Yang H S, Jung S Y and Lim J H 2015 Investigation on cryogenic refrigerator and cooling schemes for long distance hts cable *IEEE Trans. on Applied Superconductivity* **25** 3 pp 1–4
[33] Demko J A and Hassenzahl W V 2011 Thermal management of long-length hts cable systems *IEEE Trans. on Applied Superconductivity* **21**, Issue 3 pp 957–60
[34] Rahman M M, Rabbi M F, Islam M K and Rahman F M 2014 Hvdc over hvac power transmission system: Fault current analysis and effect comparison *Int. Conf. on Elec.l Eng. and Info. & Comm. Tech.* (Dhaka) pp 1–6
[35] Karthikeyan M, Yeap Y M and Ukil 2014 A Simulation and analysis of faults in high voltage dc (hvdc) power transmission *IECON 2014 - 40th Annual Conf. of the IEEE Industrial Electronics Society* (Dallas) pp 1786–91
[36] Haefner J and Jacobson B 2011 Proactive hybrid hvdc breakers – a key innovation for reliable HVDC grids *Cigre Int. Symp*. (Bologna)
[37] Abildgaard E N and Molinas M 2012 Modelling and control of the modular multilevel converter (mmc) *Energy Procedia* vol 20 pp 227–36
[38] Knaak H 2011 Modular multilevel converters and hvdc/facts: a success story in *Proc. 14th European Conf. on IEEE* pp 1–6
[39] Lesnicar A and Marquardt R 2003 An innovative modular multilevel converter topology suitable for a wide power range *Proc. Power Tech. Conf.* (Bologna) pp 1–6

[40] Morandi A 2015 Hts dc transmission and distribution: Concepts, applications and benefits *Superconductor Science and Technology* **28** 12 p 123001
[41] Kim W J *et al* 2017 Comparative study of cryogenic dielectric and mechanical properties of insulation materials for helium gas cooled hts power devices *IEEE Trans. on Applied Superconductivity* **27** 4 pp 1–5
[42] Chang H M *et al* 2003 Cryogenic cooling system of HTS transformers by natural convection of subcooled liquid nitrogen *Cryogenics* **43** pp 589–96
[43] Chevtchenko O, Zuijderduin R, Smit J, Willén D, Lentge H, Thidemann C and Traeholt C 2012 Low friction cryostat for hts power cable of dutch project *Physics Procedia* **36** pp 1309–12
[44] Hassenzahl W *et al* 2012 *Novel Approaches and Alternative Cryogens for Cooling a Superconducting Cable Report* (Palo Alto) EPRI

IOP Conf. Series: Materials Science and Engineering **279** (2017) 012035 doi:10.1088/1757-899X/279/1/012035

Finite element investigation of the mechanical behaviour of a Twisted Stacked-Tape Cable exposed to large Lorentz loads

Federica Pierro[1], Zijia Zhao[1], Luisa Chiesa[1], Makoto Takayasu[2]

1 Tufts University, Mechanical Engineering Department,
Medford, MA 02155, USA

2 MIT, PSFC, Cambridge, MA 02139, USA

Email: federica.pierro@tufts.edu

Abstract. The mechanical response of Twisted Stacked-Tape Cables (TSTC) experiencing large Lorentz loads generated during its operation in high-current, high-field magnets was investigated using finite element analysis. Two conductor configurations were investigated: a stack of 40 REBCO tapes inside a solid cylindrical copper rod (former) and a solder filled copper tube. Several simulations were conducted to highlight the effect of different parameters in the cable and the differences between the two configurations. A parametric study on the geometrical parameters of the solder filled tube configuration was performed. It was found that increasing the ratio between the thickness of the copper tube and the amount of solder in the cross section lowers the maximum stress experienced by the stack of tapes. Another simulation explored the effect of using different width tapes in the stack and it was found that, for a given Lorentz load, a wider tape reduces the maximum stress experienced by the stack. In certain applications, the cable requires the addition of material for structural support or stability reasons therefore a simulation was performed to understand the effect of copper surroundings. It was found that surrounding the cable with copper as well as using a thick copper tube lowers the stress experienced by the stack and makes the behaviour of the copper core and the solder filled tube configurations almost identical. Finally, the critical current performance of the TSTC conductors as a function of Lorentz load was estimated. A minimal degradation below 2% was predicted for the copper core configuration up to 1000 kN/m and up to 600 kN/m for the solder filled tube configuration.

1. Introduction

Second generation (2G) REBCO (rare-earth-barium-copper-oxide) high temperature superconductors (HTS) have excellent mechanical properties as well as high field and high current capabilities, which make them attractive conductors for high field applications such as fusion reactors and particle accelerators. Due to the flat geometry of HTS tapes, the design and fabrication of multi-tape high current density cables is extremely challenging. Several cabling methods are currently under development such as Roebel assembled coated conductor (RACC) [1], the twisted stacked-tape cable (TSTC) [2], the conductor in round core (CORC) [3] and other cable configurations using stack of tapes [4-5].

In high current and high field applications, the natural occurring Lorentz load can be large enough to cause degradation of the electrical performance of the tapes used in the cables. This large, accumulating transverse electromagnetic compression is due to the interaction between the high currents and the large magnetic field generated in the magnets utilizing those cables.

In previous work [6], the effect of electromagnetic transverse compression on an untwisted 40-tape TSTC was investigated using numerical finite element analysis. Two support methods were investigated: a solid cylindrical copper core and a solder filled copper tube (figure 1). The stress/strain states caused by the accumulating Lorentz load were analyzed in both conditions investigating the influence of the stack orientation and the friction between tapes.

The scope of this work was to provide insights for cable design by investigating how the stress experienced in the stack changes in function of the geometrical parameters of the cable. Therefore, the stress values reported in the result sections should not be interpreted as absolute value for a specific solution, but as a tool to guide the design process and avoid critical situations. The studies on the conductor originally presented in [6] for the solder filled tube configuration and covering electromagnetic loads up to 300 kN/m, were expanded to consider loads up to 1000 kN/m. The effect on the stress distribution in the stack was investigated as a function of the following parameters: thickness and outer diameter dimensions of the copper tube (solder filled configuration), conductor width and structural surrounding material (solid copper core and solder filled configurations). The stress distribution in the stack of tapes as a function of the tape width was investigated to highlight the advantages and disadvantages of different tape widths on the cable design. Furthermore, the effect of having copper structural material surrounding the cable was examined for both solid copper core and solder filled configurations. Finally, the evaluation of the critical current performance of the TSTC is discussed combining literature data of critical current as a function of strain at 4.2 K and 19 T [19] and strain results from the finite element simulations.

2. Finite Element Analysis

Structural finite element analysis was performed using ANSYS® to investigate a full scale three dimensional stacked-tape cable under transverse compression. The cable is studied in the untwisted configuration but the stack was oriented at 45 degrees, which was shown to be the most critical orientation for the tapes in the stack when they experience accumulating electromagnetic loads [6].

A 40-tape stacked-tape cable made with 4 mm wide Superpower tapes [7] was modeled with two cabling support structures. Both configurations are shown in figure 1. In the first case, the stack is completely surrounded by solid copper (figure 1(a)) while in the second method the superconducting stack is enclosed by a copper tube filled with solder (figure 1(b)). Both configurations have the same external diameter of 8.4 mm, and the copper tube of the second method has a thickness of 0.8 mm.

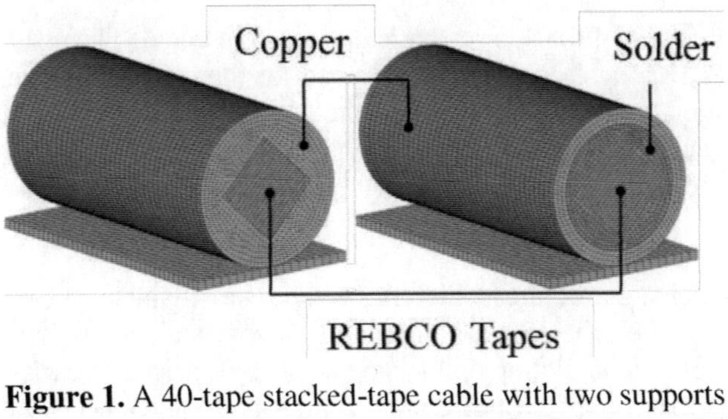

Figure 1. A 40-tape stacked-tape cable with two supports methods: solid copper core (a) and solder filled tube (b) [6].

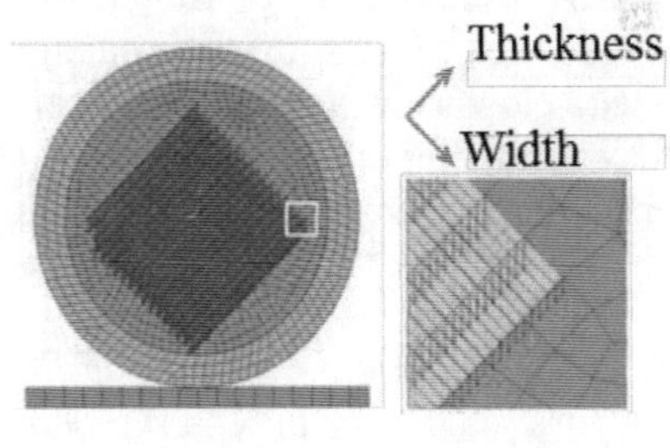

Figure 2. Nodal distribution of the applied electromagnetic Lorentz load for the TSTC solder filled tube model [6].

2.1. *Element Type and Mesh Density*

HTS tapes were modeled as a uniform volume using SOLSH190 structural solid-shell elements using a novel technique developed in [8]. Each tape was modelled with a uniform thickness along the width, neglecting the thickness variation in the copper layers typically observed in real tapes. Details of the modeling of the multi-layer tape as uniform volume are listed in [8]-[9]. The support structure was meshed using SOLID185 homogeneous structural solid elements.

A mesh study was performed in [6] and the optimal mesh was determined to be such that the element size for the HTS tapes was 0.1 mm along its width, resulting in 40 elements for a 4-mm wide tape. Each tape was meshed with only one element through the thickness and with an element size of 0.2 mm along its length, resulting in 50 elements for a 10-mm long stack. The optimal final mesh for the entire cable consisted of about 200,000 elements. 80,000 of those elements are used to mesh the 40-tape stack.

2.2. Material Properties
The materials in the FEA were modeled as bilinear isotropic to capture the nonlinear elastic-plastic deformations that occurs in the HTS tapes when experiencing large deformations. Three parameters describe the bilinear model: the Young modulus, the yield stress, and the tangent modulus. HTS tapes were modelled as a homogeneous material with an average Young modulus of 140 GPa, a yield stress of 825 MPa and a tangent modulus of 13 MPa, obtained experimentally in [9]. Material properties for the support structures were collected from published literature [10]-[16]. Young modulus of 30 GPa and 120 GPa, yield stress of 100 MPa and 400 MPa and tangent modulus of 1 GPa and 4.5 GPa were defined for the solder (Sn60/Pb40) and the copper respectively.

2.3. Boundary Conditions and Loading
A total compressive load of 1000 kN/m with increments of 100 kN/m was applied to the stack to replicate the effect of the electromagnetic Lorentz load for a cable carrying 52.6 kA in a 19 T field.
The load was applied as a nodal force, evenly distributed among the nodes of the 40 tapes of the stack and applied downward in the vertical direction to represent the accumulating effect of Lorentz load in the cross section (current flowing out of the page and horizontal magnetic field pointing left). The results were analysed considering the width and thickness directions of the stack of the tapes (see figure 2). Surface to surface contact pairs were used to described the interaction between tapes and between the stack and the support structure as described in [6]. Following some studies performed in [6], a friction coefficient of 0.2 was used between the tapes of the stack. Symmetry style boundary condition were applied at the two ends of the cable to reduce the computation time. A bottom rigid plate was used to support the structure under the applied load.

3. Support Method Investigations
In this section, the results of the simulations performed for this work are discussed in detail. The results discussed below show the maximum compressive stress as a function of load for the different studies. The maximum stress was found using average elemental stress data in the conductor for 95% of the elements through the cross section of the stack. Results for the remaining 5% of the element were neglected to mitigate the effect of localized stress concentration as described in [6].

3.1. Geometrical analysis of Solder Filled Tube
A parametric study on the characteristic dimensions of the solder filled tube stack-tape cable was performed to identify the effect of the size of the tube (for a constant thickness) and its thickness (maintain a constant outer diameter) on the stress accumulation on the tape-stack under electromagnetic transverse compression. Those studies help identifying the amount of copper and solder needed to reduce the overall stress experienced by the stack.

A study was conducted for four tube diameters: 8.4 mm, 8.9 mm, 9.4 mm and 10.4 mm. In all four cases, the thickness of the copper tube was kept constant at 0.8 mm. Figure 3 shows the effect of the tube diameter on the maximum stress values (for 95% of the elements) as function of load in both width and thickness direction (see figure 2). As shown, increasing the external tube diameter reduces the stress accumulated on the stack. This effect becomes more significant at larger loads. For example, at 1000 kN/m the maximum stress experienced by the stack differ ~60% between the 8.4 mm and the 10.4 mm tube external diameter (width direction, figure 3 (a)).

IOP Conf. Series: Materials Science and Engineering **279** (2017) 012035 doi:10.1088/1757-899X/279/1/012035

A similar study was conducted by varying the tube thicknesses while maintaining a constant outer tube diameter. The copper tube thicknesses were varied between 0.8 mm and 1.4 mm and the results indicated a lower maximum compressive stress in the tape stack for larger thicknesses.

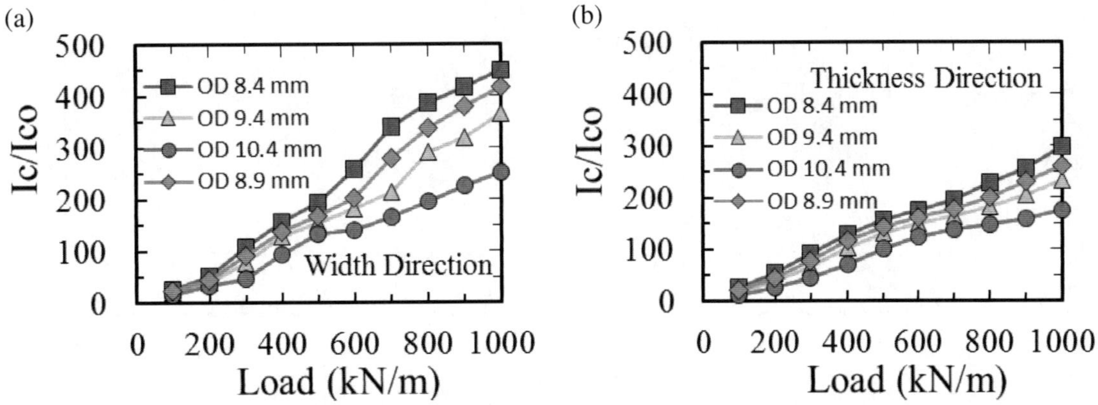

Figure 3. Plot of the maximum compressive stress (95% of elements) in the tape stack as a function of load of the solder filled tube configuration for four copper tube diameters (constant tube thickness).

3.2. Conductor Width Investigation

A conductor width investigation was performed for both cable support methods by considering stacks made with HTS tapes of different widths. The number of tapes in the stack was modified with the width to maintain a square cross section of the TSTC, while the outer diameter of the copper (for both configurations) and the thickness of the copper tube (for the solder filled tube configuration) were scaled linearly. The element size in the FEA analysis was maintained constant while changing the geometrical dimensions (therefore wider tape, larger number of elements). Figure 4 illustrates the compressive stress (95% of elements) in the width direction (most critical for delamination) as function of the engineering current density of the tape-stack (total current over the stack cross section) for different tapes' widths.

The plot is characterized by two regions referred as region one (I) and region two (II). A stress value of 200 MPa delimits both regions horizontally. This value has been used as reference stress in cable for future generation magnets' design [17]. Region I and II are vertically delimited by a current density of 0.45 kA/mm^2 and 3 kA/mm^2, which represent the critical current density of a REBCO tapes for a 19 T field perpendicular and parallel respect to the face of the tape [18]. The two areas on the plot help identifying the maximum achievable cable performance for each configuration based on the limitations of today's tape technology. For example, with a perpendicular background field of 19 T, a 4-mm stack of 40 tapes will carry up to 7.2 kA, while the 8-mm stack of 80 tapes will carry up to 28.8 kA. If we want to carry the same current with a 4-mm tape-stack, a current density of 1.8 kA/ mm^2 would be necessary, which is higher than today's limit of 0.45 kA/ mm^2 and is therefore a concern for applications. As shown in figure 4, the stress experienced in a stack using 4-mm tape would be lower than the 200 MPa limit for a stack current density below 2.5 kA /mm^2, therefore it would be possible to achieve larger cable currents while maintaining mechanical integrity if further improvements in the conductor are obtained in the future.

For the same Lorentz load the use of wider conductor results in smaller compressive stress. For example, for the solder filled tube configuration under a compressive load of 700 kN/m (shaded area in figure 4), the maximum stress is 250 MPa for the 4-mm, 180 MPa for the 6-mm and 140 MPa for the 8-mm tape stack. Finally, while the stress is significantly reduced between a 4-mm and 6-mm stack for a given load (28% reduction at 700 kN/m), a smaller reduction occurs between the 6-mm and the 8-mm wide tape-stack (22% at 700 kN/m), suggesting that the 6-mm tape-stack might be a good candidate for cable design, because it significantly improve the stress distribution compared to the 4 mm, while maintaining a high current density.

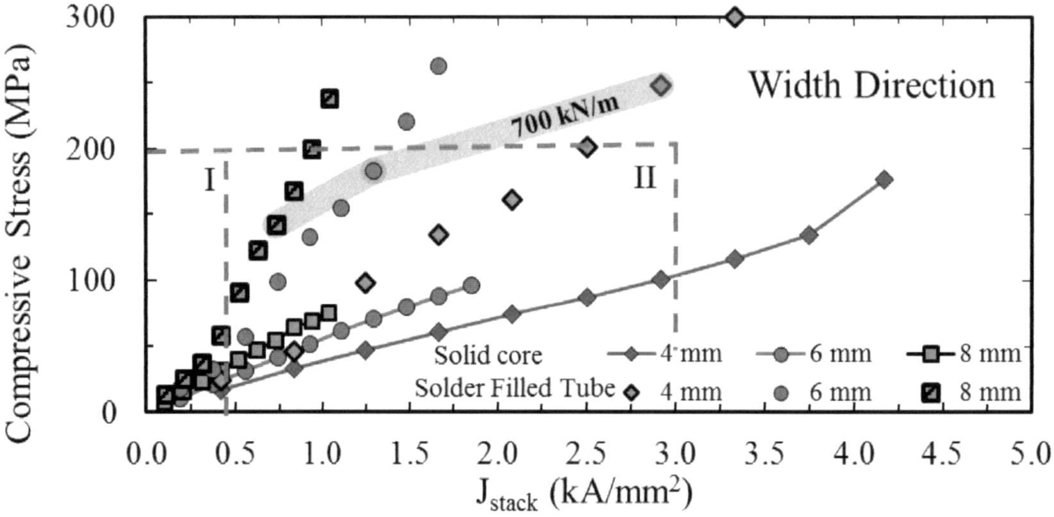

Figure 4. Plot of the maximum compressive stress (95% of elements) as function of the current density of the conductor for both copper core and solder filled tube configurations.

3.3. TSTC Surrounding Conditions

A finite element model was developed to analyse the effect of having an external support on the stress state of a TSTC cable. A structure made of copper was modelled around the cable for both the copper core and the solder filled tube configurations. The copper would mainly serve the function of stabilizer and/or structural material for applications like a cable-in-conduit-conductor for a fusion reactor. As shown in figure 5, two different surroundings conditions were investigated: "infinite" and "half infinite" surrounding. The study was conducted for the 45-degree orientation of the stack and a compressive load of 300 kN/m with the intent to identify the smallest size of the surrounding material for which a variation in the maximum stress experienced by the tapes was not observed. A ratio of two between the height (h) of surrounding copper block and the outer diameter of the copper tube (OD) was chosen. The cross section of the copper was modelled as a square. Bonded contact pair conditions were applied between the cable and the surrounding material to reproduce the effect of soldering. The maximum compressive stress results (95% stress value) in both width and thickness direction for both support methods and surrounding conditions are displayed in figure 6. The "rigid plate" configuration refers to the original configuration in which a rigid plate was used as external support to the cable. Stress results for the "half infinite copper" and "infinite copper" surroundings were significantly lower compared to the original support methods in both width and thickness direction. Although the copper core configuration still experienced the lower stress accumulation compared to the solder filled one, having additional structural material around the conductor reduces the differences between the two configurations from 50% in the "rigid plate" condition to only 2% in the "infinite condition".

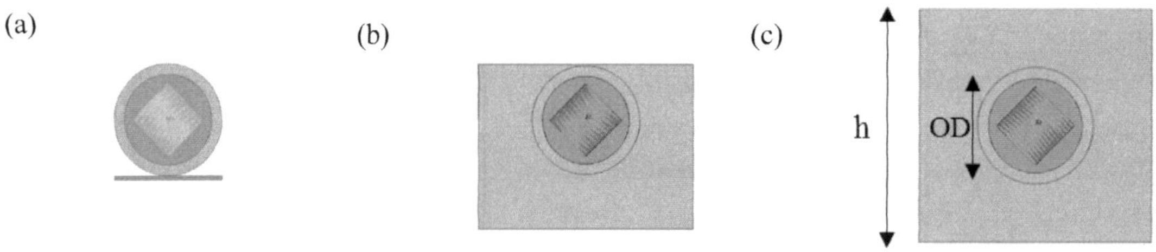

Figure 5. Cable surrounding conditions for the solder filled tube configuration: (a) rigid plate, (b) half infinite surrounding, (c) infinite surrounding. Copper would be used mainly for stabilization and a stronger material could be used for better mechanical support. The study is aimed at providing a guideline for establishing the feasible boundaries for a cable used in large magnets.

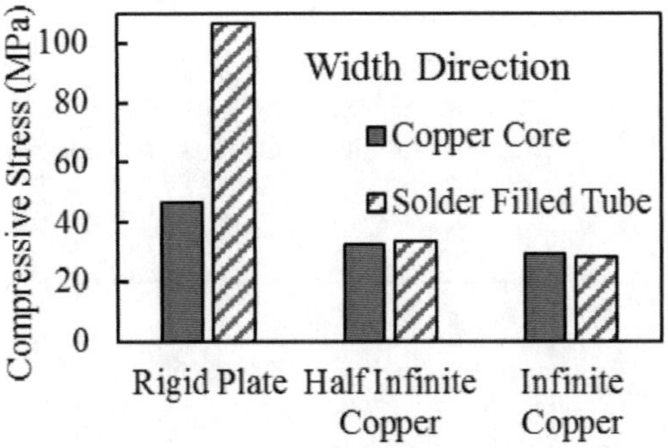

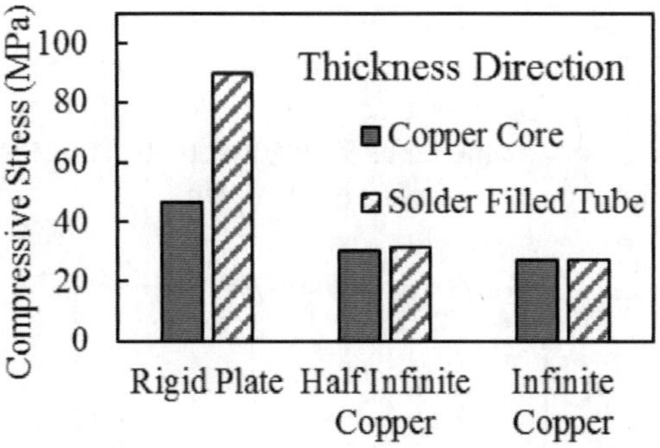

Figure 6. Plot of the compressive stress in both support methods for three surrounding conditions: rigid plate, half infinite surrounding and infinite surrounding.

3.4. Critical Current Performance

It is well known that the electrical behaviour of HTS tapes is highly dependent on the strain state of the conductor. As shown in previous sections of this paper, the stack of tapes experience large stress/strain during operation (accumulating electromagnetic loads). It is therefore important to characterize the critical current performance of the cable keeping in mind the limitations of the model, such as the impossibility to model micro-cracking and delamination. The critical current estimates shown to follow are based on the strain state of the tapes obtained with FEA and fracture and delamination will be taken in consideration by assuming that the tape does not carry any current beyond a certain strain value empirically determined from [19]. Despite those limitations, the analysis discussed below can provide useful insight during the design phase of a cable and a magnet.

Our analysis was performed for the untwisted 40-tape TSTC with an orientation angle of 45 degree (worst condition). The goal of the analysis was to determine the strain map of the tapes under applied transverse compression and associate those strain to a critical current value by utilizing available literature critical current data as a function of axial strain at 4.2 K and 19 T [19]. To represent the effect of the strain experienced in all three directions of the conductors, von Mises strain results were used to calculate the electrical performance of the cable. We believe this approach is justified because the strain state of the tape is rather complex and this is a conservative approach to our analysis.

To accurately predict the strain generated in the brittle superconducting layer, the superconductor was modelled as a layered structure so the strain at the REBCO layer could be easily identified (vs. the homogenized approach described in previous sections). The material properties of each individual layer of the tapes were prescribed using published literature data [20]-[22]. The von Mises strain results were collected for the cross section of each tape at the location of the REBCO layer and at eighty weighted Gaussian points along the width.

The normalized critical current as function of the applied load was calculated for each tape of the stack independently. Results of the 40 tapes were then averaged to estimate the electrical performance of the entire cable. Figure 7 displays the normalized critical current as function of the Lorentz load for both copper core and solder filled tube configurations. As shown, the TSTC supported by the copper core experience minimal critical current degradation up to 1000 kN/m (figure 7, (a)) while the solder filled configuration shows degradation for loads above 500 kN/m (figure 7, (b)). Figure 8 shows the critical current behaviour of some specific tapes in the stack (bottom, middle and top). As it can be seen, in both copper and solder filled cases, the bottom tape experiences the highest degradation, dominating the critical current behaviour of the entire cable. The higher stress experienced in the bottom tape is the result of two factors: the compressive load accumulation from the tapes above and the surrounding conditions of the model. Unlike the majority of the tapes in the stack, which are allowed to slide past one another, the bottom tape is bonded to the solder and therefore the sliding motion is restricted. This results in higher stress in the tape. In the case of a cable supported by a copper core (figure 8, (a)), the analysis suggests that a significant improvement on the critical current performance of the cable can be achieved by replacing the bottom tape with a structural tape. For the solder filled tube configuration (figure 8, (b)), significant degradation is also experienced by the middle tape for loads higher than 600 kN/m. To improve the electrical performance of the conductor in this case one option could be to interlay the superconducting tapes with support structural material. This would come at the expense of reducing the overall current density of the cable.

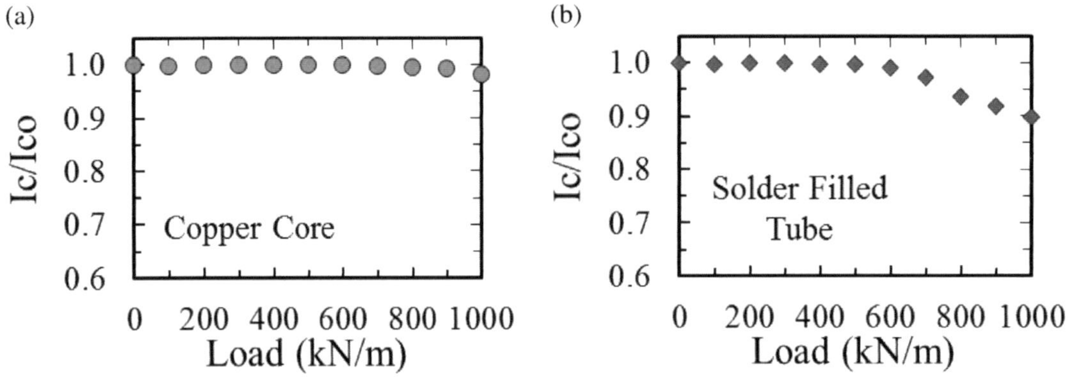

Figure 7. Plot of the critical current performance of the 40-tapes TSTC for both support methods as a function of electromagnetic Lorentz load.

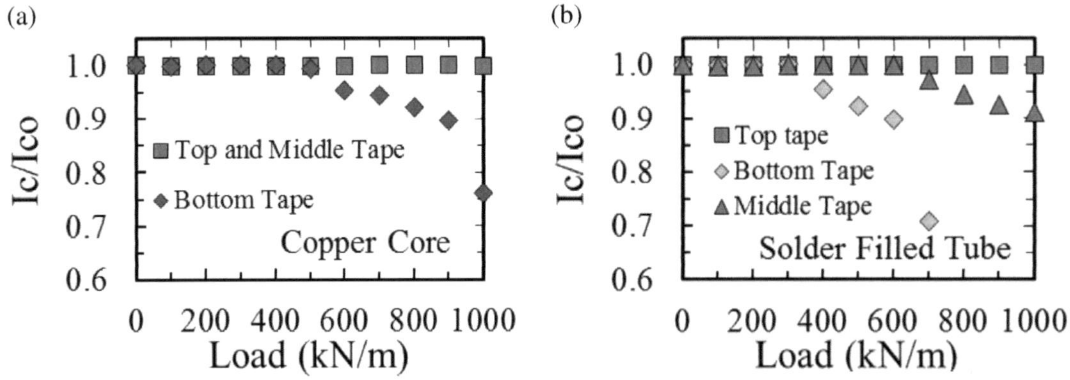

Figure 8. Plot of the critical current for individual tapes of the TSTC for both support methods as a function of electromagnetic Lorentz load.

4. Conclusions

A finite element investigation of a 40-tape untwisted stack cable under electromagnetic Lorentz load up to 1000 kN/m was presented for two cable configurations: copper core and solder filled tube. Several simulations were performed with the intent of guiding the design process of high current cables used in high field magnets and offer insights on the parameters that could be changed to reduce the effects of large accumulating Lorentz loads. Parametric studies were performed for the solder filled tube configuration, highlighting how the maximum compressive stress experienced by the cable can be reduced by increasing the tube external diameter (with a constant thickness) or its thickness (with a constant outer diameter). It was shown that wider tapes reduce the maximum compressive stress experienced by the stack for the same Lorentz load and that having enough material surrounding the cable significantly reduce the stress accumulation in the tape stack. The critical current performance of the cable under Lorentz loads was estimated, indicating a minimal degradation up to 1000 kN/m for the copper core configuration and up to 600 kN/m for the solder filled tube. In both cases the degradation observed was driven by the behavior of the tapes at the bottom of the stack.

Future work will focus on investigating a fully twisted model. A study currently under development and performed for a linearly scaled down model suggests that the twisted and straight models experience similar stress distribution at each corresponding orientation angle of the stack. Additional work will include the analysis of a full scale three-dimensional model for a twisted cable and the results will be compared with available experimental results.

5. References

[1] Goldacker W, *et al.*, "Roebel cables from REBCO coated conductors: a one-century-old concept for the superconductivity of the future," *Supercond. Sci. Technol.*, vol. 27, no. 093001, 2014.
[2] Takayasu M, Chiesa L, Allen NC and Minervini J, "Present status and recent developments of the Twisted Stacked-Tape Cable conductor," *IEEE Trans. Appl. Supercond.*, vol. 26, no. 2, 2016.
[3] Van der Laan DC, et al., "Characterization of a high-temperature superconducting conductor on round core cables in magnetic fields up to 20 T," *Supercond. Sci. Technol.*, vol. 26, no. 045005, 2013.
[4] Celentano G, et al., "Design of an industrially feasible twisted-stacked HTS cable-in-conduit conductor for fusion application," *IEEE Trans. Appl. Supercond.*, vol. 24, no. 4601805, 2014.
[5] Bykovsky N et al, "Performance evolution of 60kA HTS cable prototypes in the EDIPO test facility" *Supercond. Sci. Technol.* 29 (2016) 084002.
[6] Allen NC *et al.* 2016 " Structural finite element evaluation of twisted stacked-tape cables for high-field magnets", *IEEE Trans. Appl. Supercond.,* vol 27, no. 4, 2017
[7] SuperPower, "2G HTS wire", http://www.superpower-inc.com/content/2g-hts-wire
[8] Allen NC, Chiesa L and Takayasu M, "Structural modeling of HTS tapes and cables," *Cryogenics*, doi:10.1016, 2016
[9] N.C. Allen, L. Chiesa and M. Takayasu, "Numerical and experimental investigation of the electromechanical behavior of REBCO tapes," *IOP Conference Series: Materials Science and Engineering (MSE)*, vol. 102, no. 012025, 2015.
[10] Shin HS, et al., "Intrinsic strain effect on critical current in Cu-stabilized GdBCO coated conductor tapes with different substrates," *Supercond. Sci. Technol.,* vol. 26, no. 055005, 2013.
[11] Brechna H, et al., "Cryogenic properties of metallic and non-metallic materials utilized in low temperature and superconducting magnets," *8th Int. Conf. High-Energy Accelerators*, Inst. fur Experim. Kernphysik, C71-09-20, 1971.
[12] Reed R, et al., "Low temperature mechanical properties of copper and selected copper alloys," *National Bureau of Standards*, 1967.
[13] Weiss KP, et al., "Mechanical and thermal properties of central former material for high current superconducting cables," *IEEE Trans. Appl. Supercond.,* vol. 26, no. 4, 2016.
[14] Mruzek MT, "Properties and methods of lead/tin splices for superconductors," TM-944 Fermilab, Batavia, USA, 1980.
[15] Fast RW, et al., "Electrical and mechanical properties of lead/tin solders and splices for superconducting cables," *Cryogenics*, vol. 28, 1988.
[16] Matweb – Material Property Data Website: 60-40 Soft Solder http://matweb.com/search/DataSheet.aspx?MatGUID=06a31d97bb734b509043d81cf131b280
[17] Tommasini D, et al, *"Baseline specifications and assumptions for accelerator magnet: Mileston M5.2",* http://cds.cern.ch/record/2150689
[18] National High Magnetic Field Laboratory, Engineering Critical Current Density vs. Applied Field https://nationalmaglab.org/magnet-development/applied-superconductivity-center/plots
[19] Barth C, Mondonico G, and Senatore C. "Electro-mechanical properties of REBCO coated conductors from various industrial manufacturers at 77 K, self-field and 4.2 K, 19 T." *Superconductor Science and Technology* 28.4 (2015): 045011.
[20] Ekin JW 2006, "Experimental techniques for low-temperature measurements", *Oxford Univ. Press London,* U.K.
[21] Reed RP, et al "Materials at low temperatures American Society for Metals" 1983, Ohio, USA
[22] Matweb – Material Property Data Website: Hastelloy® C-276 alloy http://matweb.com/search/DataSheet.aspx?MatGUID=4361b68268934746ad70542944cdd8cf

Fabrication of long-length cable-in-conduit for superconducting magnets

J Breitschopf[1], D Chavez[2], R Albosta[1], T Elliot[1], R Garrison[1], J Kellams[1], P McIntyre[1] and A Sattarov[1]

[1] Department Physics & Astronomy Texas A&M University, College Station, 77843, TX, USA
[2] Universidad de Guanajuato, Leon, Guanajuato, Mexico.

E-mail: dchavez@fisica.ugto.mx

Abstract. A superconducting cable-in-conduit (CIC)is being developed for the windings of superconducting magnets for a proposed electron-ion collider and for other applications. Short-length samples have been successfully developed and tested for the required performance. The equipment and procedures have been developed for fabrication of long-length conductors. A 10 m CIC has been successfully fabricated and evaluated. The process is now being extended to make 125 m cables in a straight-line process.

1. Introduction

Cable-in-conduit (CIC) conductor has particular benefits for superconducting magnets. It is rugged and provides cable-level stress management; it has internal flow of cryogen; flared ends can be formed readily and are self-stable once formed. CIC conductors have been in use for many years for various applications, including fusion magnets, superconducting magnetic energy storage, electric motors, detector magnets and accelerator magnets [1].

The Accelerator Research Lab has developed a CIC conductor for the windings of the 3 T dipoles and accompanying quadrupoles of the Ion Ring in the Jefferson Lab Electron Ion Collider (JLEIC) [2]. The CIC design was chosen to facilitate cost-effective fabrication of a compact block-coil dipole that can provide collider-homogeneity magnetic fields over a large field range in the large aperture required for the Ion Ring.

The CIC conductor reported in this paper for JLEIC dipoles consists of a single-layer of NbTi/Cu wires which are cabled with a twist pitch around a perforated thin-wall stainless steel center tube (the choice of twist pitch is presented later). The cable is then wrapped with an overwrap of thin-foil 316L stainless steel foil tape (25microns thick, 1.25 cm wide), and the cable is pulled through a copper nickel sheath tube. The sheath tube is drawn down to compress the strands against the center tube and immobilize them. The sheath provides stress management at the cable level, so accumulation of Lorentz stress cannot cause degradation of superconducting performance. It also preserves the benefits of interior helium flow, so that the overall magnet is not required to serve as a He vessel. In windings made from this CIC conductor, liquid helium flows within the center tube and through the perforations in the center tube, so that He bathes all wires to enhance stability against microquenches. Table 1 summarizes the main parameters of the cable and of the JLEIC dipole in which it is used for the windings.

Table 1. Main parameters of the CIC for the JLEIC dipole.

Operating conditions in dipole winding		
Operating temperature (T_0)	5	K
Design dipole field in bore (B_0)	3.0	T
Conductor current @ B_0, T_0	13.7	kA
Maximum field at conductor @ B_0	3.5	T
Bore field at short sample	3.8	T
Specifications of NbTi/Cu wires in CIC		
Copper: SC ratio	1.5	
Cu RRR	140	
# filaments in wire	7000	
Wire diameter	1.2	mm
Specifications of CIC conductor		
# wires	15	
OD of finished CIC conductor	8.128	mm
OD of perforated center tube	4.76	mm
Sheath thickness	0.50	cm
Twist pitch of wires in cable	7.6	cm
316 SS foil tape overwrap on cable		
Tape thickness	25	µm
Tape width	1.7	cm
Overlap wrap? yes/no	No	
Specifications for forming CIC bends		
Bend radius for all 3 T winding ends	5	cm
Overbend angle to produce 180° bend	30°	

The CIC conductor has been developed in three phases. In the first phase, a ~2 m long CIC was fabricated using a small "manual-stranding device" (see figure 1.a), which place the superconductors around the perforated inner tube with the desired twist-pitch. In this process, a uniform and synchronized motion of pulling and twisting the device is vital for providing constant twist-pitch. With this method we were able to control the twist pitch within 0.05% of error in twist-pitch length. That work is summarized in the following section.

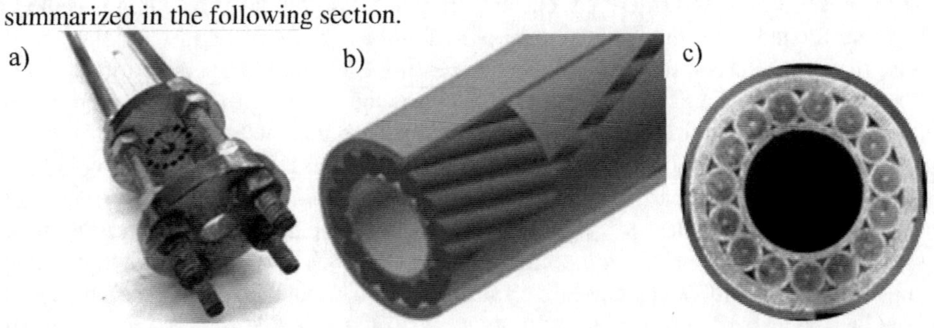

Figure 1. Cable-in-Conduit conductor that is being developed for the windings of the JLEIC dipole: a) Manual-stranding device; b) cutaway showing the cable on the center tube with tape overwrap and sheath tube; c) cross-section photo of a fabricated segment of CIC.

IOP Conf. Series: Materials Science and Engineering **279** (2017) 012036 doi:10.1088/1757-899X/279/1/012036

In the second phase, the equipment and methods were developed for fabricating long-length CIC conductor. A first step was to develop in-line fabrication tooling to make 10 m long CIC samples. In-line fabrication development is complete and is reported in Section 3. From the in-line fabrication development, a complete design has been developed for the 125 m segments of CIC conductor that are required for the windings of the JLEIC dipoles. The complete design is presented in Section 4.

Figure 2. a) wire feed manifold set up on drawbench to cable 10 m CIC segment; b) winding operation in action showing twist pitch of the cable; c) Twist-Pitch uniformity for the 10 m CIC segment.

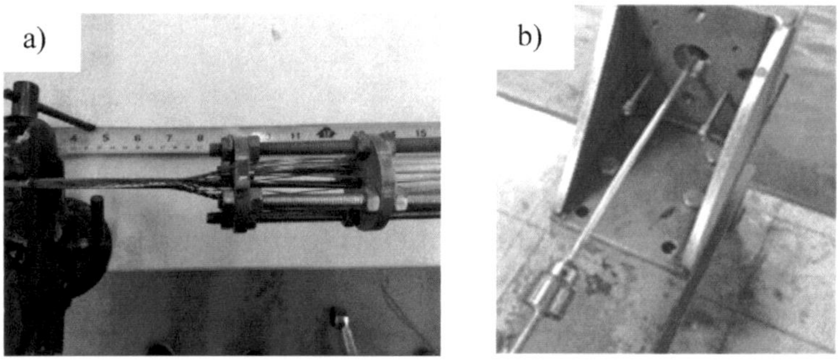

Figure 3. Methods used to hand-fabricate short lengths of CIC cable; a) hand cabling the CIC; b) drawing the outer sheath down onto the core.

2. Fabrication and testing of short segments of CIC cable

The Cable-in-conduit is an arrangement of a number of superconductors (NbTi, MgB2, Nb3Sn depending the desired field and or critical current) in which a single-layer of wires is cabled around a high strength perforated inner tube at a specific twist pitch as shown in figure 2. If there was no twist pitch, a wire that is located on the outside of a bend in the CIC would traverse a longer catenary path after the bend than before, and a wire that is located on the insider of the bend would transverse a shorter catenary path. Therefor the twist pitch of the cabling is adjusted to be equal to the path length of a 180° bend with the bend radius of 5 cm that is used in all bends of the dipole winding. Each wire then has the same path length on the outside as on the inside, so no residual wire strain. The cable is then wrapped with a thin stainless steel foil that acts as sliding plane as seen in figure 1b). Next, the whole assembly is inserted into a high strength outer sheath which is drawn down to compress the wires against the center tube and immobilize them shown in figure 3b). This sheath provides stress management at the cable level, so accumulation of Lorentz stress cannot cause degradation of superconducting performance.

Sample cables were tested to validate that the NbTi/Cu wires were undamaged by the cabling and drawing process. All of the 7000 filaments of the bent NbTi/Cu wires were inspected under a microscope and no broken filaments were found. To confirm that there was no current degradation, Ohio State University performed I_c tests on four strands extracted from a bent CIC and two strands that were directly from the new NbTi/Cu wire spool. As it can be seen in figure 4, the Ic test shows that the cabling, drawing and bending procedures used to reproduce typical conditions as would be experienced by the CIC following JLEIC specifications, do no contribute to the current degradation. The data is presented in figure 4.

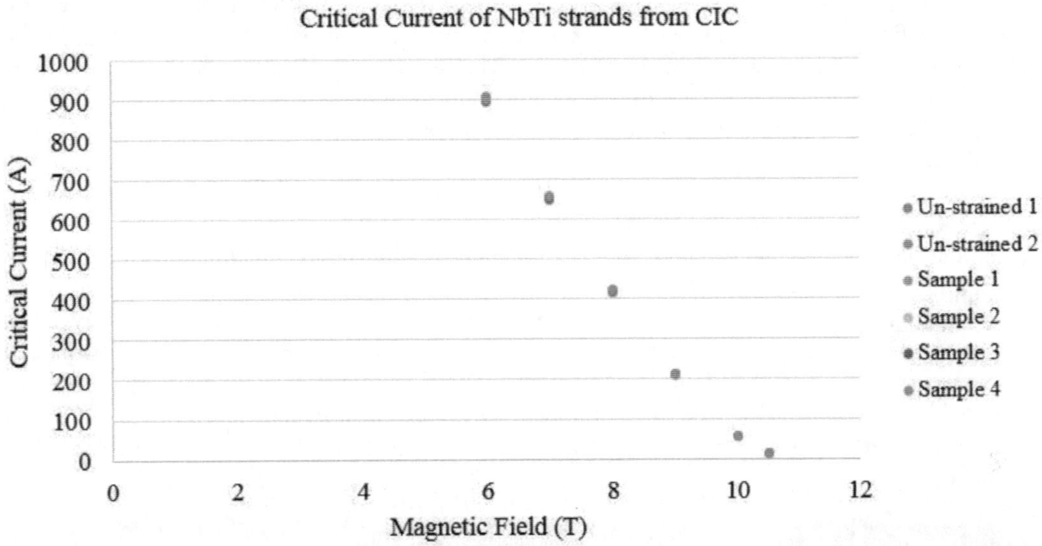

Figure 4. I_c test for NbTi/Cu CIC. Samples 1,2,3 and 4 represent the I_c performance of 4 NbTi/Cu extracted strands taken from a CIC. As seen in the graph, there is no visible difference in the measurements. Samples labeled as "Un-strained 1 and 2" represents the I_c performance of "witness strands." Witness strands are strands that did not experience any stress or strain and were taken directly from the superconductor spool.

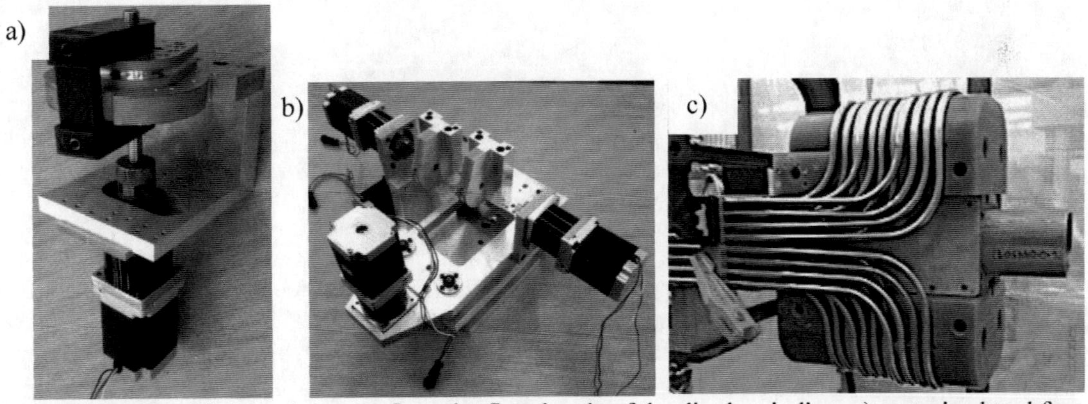

Figure 5. Bending CIC conductor to form the flared ends of the dipole winding: a) motorized tool for forming 180° bend; b) motorized tool for bending the 180 loop to form c) the 90° flare end region of completed dipole winding.

Three highlights from the short-segment studies are noteworthy. First, a procedure for forming bends that preserves the internal registration of the CIC through a 5 cm radius bend, and prevents wire strain from developing in the bend [3].

Second, motorized tooling was developed with which to form the bends required for the flared ends of the JLEIC dipole windings. Figure 5a shows the motorized tool that forms the 180° U-bend with leg spacing appropriate for each layer of the winding. Figure 5b shows the tool that forms the 90° bend of the U-bend to form the flared end and vertical jog for step transitions. Figure 5c shows one flaired end region of a complete mockup winding [3].

Third, a complete mockup winding was fabricated, in which the winding was supported on a precision-machined G-11 structural beam that is integral with the JLEIC dipole design. Metrology was performed to measure the precise location of each of the 24 turns of hollow tube that mimiced the conductor in the geometry of the structural beam. The measured position errors were then fed back into the magnetic simulation of the dipole to determine the multipole field errors that would be produced. The measured locations corresponded to a field distribution that was within tolerance for the collider requirements [3].

3. Fabrication of 10 m segments of CIC cable

An important objective of the second-stage development was to develop methods for the cabling and tape-wrapping, insertion of the cable into the sheath tube, and drawing the sheath to final diameter, that are compatible with extension to the full 125 m length required for the windings of the JLEIC dipole. The cabling, wrapping and pull-through-sheath operations were performed on a drawbench shown in figure 6. The 10 m cables were measured with a micrometer along the length of the cable. The segments showed that the cable diameter and twist pitch were sustained with required precision, and the drawing operation went smoothly. The twist pitch was measured with a set of calipers shown in figure 2c.

a) b) c)

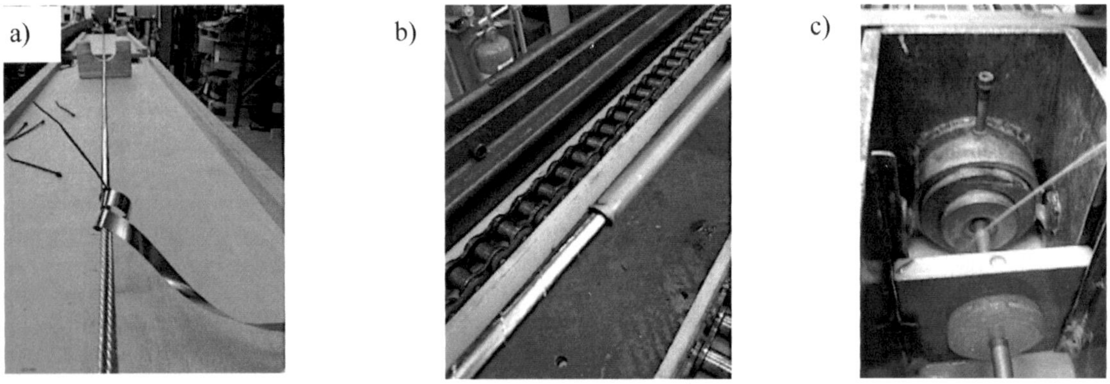

Figure 6. Procedures for fabrication of long-length CIC conductor: a) wrapping the cable; b) pulling the cable through the sheath tube; c) drawing the sheath tube to final size.

The above succession of steps were undertaken with all components – straightened the wires, the perforated center tube, and the sheath tube. We found that the succession steps benefit greatly from the entire process being done as an inline operation, without spooling the cable at intermediate stages. Spooling at intermediate stages would entail a degree of ovalling of the cable and/or tubes that complicate the subsequent operations.

4. Long length CIC fabrication

4.1. In-line cable fabrication

The overall process developed above is now being extended for the construction of 125 m cable CIC segments. A commercial planetary stranding machine is being acquired for the cabling and wrapping operations.

The bed of the 10 m draw bench is being extended to 125 m length. The tape-wrapping will be performed as a last operation on the stranding machine, and the wrapped cable will be pulled through a straightened sheath tube that is mounted on the drawbench. In that way all stages of CIC fabrication can be performed as an in-line cabling and drawing operation.

4.2. Continuous tube forming to make CIC conductor
As opposed to pulling the wrapped cables into an outer sheath, it is possible to use Continuous Tube Forming and Filling (CTFF). Hyper Tech has developed and uses a (CTFF) machine for continuous forming of a cable containing up to 37 filaments of MgB_2 superconducting monofilaments within a Monel sheath superconducting wire [4]. They are collaborating with the Texas A&M group to adapt CTFF processing to continuously form the sheath tube as an in-line process.

In CTFF, the tube is then formed from a foil strip by roll-forming the cusp and curving it around the cable as it passes through the former. The seam is laser-welded to seal the tube. The strip width is chosen to provide a loose fit onto the cable so that there is sufficient gap that the cable can be pulled through the sheath tube. Finally the tube diameter is reduced to the final CIC diameter. Sample cable segments have been prepared, and the sheath tube has been checked along its seam to determine whether there are any vacuum leaks that would compromise the helium containment which is one of the attributes of the CIC approach for dipole magnets.

Sample cable segments were bent using the motorized bend tools and cold-shock tested at liquid nitrogen temperature to evaluate the integrity of the CTFF weld seam. Some early samples showed leaks at the seam. After process development of CTFF for this system, additional 2 m segments were fabricated, and were found to be leak-tight. The segments were then bent to a U-bend and then pressure-tested to 40 bar, under vacuum and at room temperature and at 77 K.

All samples passed the leak test. A dissection was performed on the bend regions and superconducting wires were removed. Half of the wires were etched in a ferric chloride solution and tested for broken filaments. The other half was sent for short-sample measurement of superconducting performance. After etching, there were less than 5 out of 7000 filaments broken filaments in the wires. The superconductivity test revealed no current degradation. Hyper Tech is continuing their development of the CTFF for the CIC application.

5. Conclusion
The construction of a ~10 m sample of Cable-In-Conduit (CIC) has been completed. All challenges regarding the short sample fabrication have been overcome. The next step is scaling to a long length manufacturing process. There are two paths capable of scaling up to long length fabrication. In-Lab production that would easily adopt the short sample fabrication method. CTFF is another option that proved to be promising for long length development of CIC based on a set of integrity tests and current test performed on the samples taken from both manufacturing paths. Both processes have been tested and proven to be viable options of long length production.

6. References
[1] Hamlet K, Egbert F, Alexander K, Gebhard M, Lyudmila P, Alexander S and Gennady V 2005: “Design and test of a hollow superconducting cable based on keystoned NbTi composite wires.” *IEEE Transactions on Applied Superconductivity*, vol. 15, no. 2 p 1529

[2] McIntyre P, Sattarov A, Breitschopf J, Chavez D, Gerity J, Kellams J, O’Quinn K, Elliot T and Garrison R, 2016 Internal Report “*Superferric Magnets for JLEIC,*”, *https://www.jlab.org/indico/event/179/session/13/contribution/32/material/slides/0.pdf*

[3] Breitschopf J, Chavez D, Elliot T, Garrison R, Gerity J, Kellams J, McIntyre P and Sattarov A, October 2016 “*Superferric Arc Dipoles for the Ion Ring and Booster of JLEIC,*” Chicago, IL, USA, , North American Particle Accelerator Conference

[4] Hascicek YS, Aslanoglu Z, Arda L, Akin Y, Sumption MD and Tomsic M 2003: Fabrication and testing of W&R MgB_2 coils using CTFF Cu/MgB_2 wires and the sol-gel insulation process. *Advances in Cryo. Eng.* **711**, 541-5.

IOP Conf. Series: Materials Science and Engineering **279** (2017) 012037 doi:10.1088/1757-899X/279/1/012037

Persistent-current magnetizations of Nb_3Sn Rutherford cables and extracted strands

E W Collings[1], M D Sumption[1], C S Myers[1], X Wang[2], D R Dietderich[2], K. Yagotyntsev[3] and A Nijhuis[3]

1. Center for Superconducting and Magnetic Materials (CSMM), Dept. of Materials Science and Engineering, The Ohio State University, Columbus, OH, USA. Corresponding author e-mail: sumption.3 @osu.edu
2. Superconducting Magnet Group, Lawrence Berkeley National Laboratory (LBNL), University of California, Berkeley, CA, USA
3. Energy, Materials, and Systems Group, the University of Twente, Enschede, NL

Abstract. The magnetizations of eight high-gradient quadrupole cables designated HQ and QXF and a pair of strands, identical in architecture but with different effective strand diameters extracted from an HQ and a related QXF cable, were measured. In the service of field quality assessment, the cable magnetizations and losses were measured by pickup coil magnetometry at 4.2 K in face-on fields, B_m, of ± 400 mT at frequencies, f, of up to 60 mHz. Based on the coupling component of loss, Q_{coup}, the coupling magnetization $M_{coup} = Q_{coup}/4B_m$ was derived for a ramp rate of 7.5 mT/s. Persistent current (shielding) magnetization and loss (M_{sh} and $Q_{h,strand}$) were measured on short pieces of extracted strand by vibrating sample magnetometry at 4.2 K. Unpenetrated *M-B* loops to ± 400 mT and fully penetrated loops to ± 14 T were obtained. M_{coup} can be easily controlled and reduced to relatively small values by introducing cores and adjusting the preparation conditions. But in low fields near injection Nb_3Sn's high J_c and correspondingly high $M_{sh,cable}$ may call for magnetic compensation to preserve field quality. The suitably adjusted cable and strand fully penetrated *M-B* loops were in reasonable accord leading to the conclusion that strand magnetization is a useful measure of cable magnetization, and that when suitably manipulated can provide input to magnet field error calculations.

1. Introduction

1.1 High Field Nb_3Sn Magnets

Rutherford cables wound with Nb_3Sn strands will be used in all the high field superconducting magnets required for ongoing and planned upgrades to the large hadron collider (LHC), viz. the high luminosity LHC (High Lumi LHC, HL-LHC, 11 and 12 T), a higher energy LHC (HE-LHC, 16 T), and a very high energy future circular collider (FCC, 16 T) [1]. Initiated in 2014 and planned for completion around 2024-26 is a set of high field magnets for the HL-LHC upgrade project [2][3] intended to produce 5-10-fold increases in the LHC's luminosity. Final beam focusing for the ATLAS and CMS detectors will be accomplished by four pairs of Nb_3Sn-wound quadrupoles with peak coil fields of 12 T [3]. Also as part of the upgrade, in order to make room for additional collimators in the dispersion-suppressor segments of the ring, some existing 8.33 T 15 m long NbTi dipoles will be replaced by 11 T 11 m long Nb_3Sn dipoles [4][5] [6][7].

Initially suggested in 2001, a higher energy successor to the LHC, the HE-LHC, would collide two proton beams circulating in the LHC tunnel [1][4][8][9][10][11]. A ring of about 1280 14 m long

16 T Nb_3Sn dipoles would support a center-of-mass collision energy of 27 TeV, close to the limit allowed by the tunnel's radius.

In addition to the HE-LHC is a proposed 100 TeV "future circular collider" (FCC). To be housed in a new tunnel 100 km in circumference the FCC is estimated to require 4578 15 m long 16 T Nb_3Sn dipoles [11][12][13]. Accordingly a 16 T Nb_3Sn dipole will be developed to satisfy the requirements of both the FCC and the HE-LHC. In contributing to that development, the US Magnet Development Program will be exploring the limits of applicability of Nb_3Sn for high field magnets [14]. For example in 2015 the Fermi National Accelerator Laboratory (FNAL) reported on the designs of four-layer cosine-theta Nb_3Sn dipoles with maximum bore fields (4.3 K) of 15.7 -16.3 T and a 15 T dipole demonstration magnet for a 100 TeV collider [15], and the Lawrence Berkeley National Laboratory (LBNL) designed a four-layer canted-cosine-theta 16 T Nb_3Sn dipole to satisfy the same need [13].

1.2 Nb_3Sn Strand and Cables for Accelerator Magnets

The numerous planned accelerator applications will demand a continuous supply of Nb_3Sn strand and cables capitalizing on the achievements of conductor development programs in the US (CDP) and Europe (NED) [1]. Studies of Nb_3Sn cable and strand properties are under way at FNAL [16]. Reported elsewhere are the effects of core type, placement, and width and heat treatment condition on interstrand coupling properties of Nb_3Sn cables[17][18]; important properties of strands are the field dependent critical current density, $J_c(B)$, and the effective filament diameter, d_{eff}. Magnetization due to ramp-rate-dependent interstrand coupling currents in cables and persistent currents in strands induce multipoles in the bore fields of dipole and quadrupole magnets [19][20]. Persistent-current magnetizations in Nb_3Sn strands, which being proportional to $J_c(B)d_{eff}$ are much stronger than in the LHCs NbTi, demands special attention. As a contribution to this topic, and indirectly to the US LHC Accelerator Research Program (LARP), we report on the magnetizations of: (i) LARP high gradient quadrupole cables designated HQ and QXF [18], (ii) a pair of strands, identical in architecture but with different d_{eff}s, that had been extracted from a LARP HQ and a related LARP QXF cable.

2. Experimental

2.1. Cable and Strand Samples

Several coils of the stainless-steel-cored HQ- and QXF-type Nb_3Sn Rutherford cables wound at LBNL were supplied for measurement to the Ohio State University's Center for Superconducting and Magnetic Materials (OSU-CSMM). Some cable and strand details are given in Tables 1 and 2, see also [18].

Table 1. Cable details

LBNL name	*	HQ1020 ZB	HQ1021 ZB	QXF 1055z-C	QXF 1055z-K	QXF 1055z-Q	QXF 1055z-O	QXF 1055z-M	QXF 1055z-D
OSU name	**H1**	**H2**		**Q1**	**Q2**	**Q3**	**Q4**	**Q5**	**Q6**
Strand count	35	35	35	40	40	40	40	40	40
pack factor, %	85.54	85.55	85.53	87.04	86.89	87.03	86.98	86.80	87.38
Core width, mm	0	8	--	11.9	15.9	15.4	14.3	13.3	0
Core cover, %	0	60	--	72	96	93	86	80	0

* Mixture of 1020 and 1021with cores extracted

Table 2. Strand details

Cable Type (Table II)	HQ	QXF
Strand source, type	OST-RRP,108/127	OST-RRP,108/127
Strand diam., d_s, mm	0.778	0.852
SC filament count	108	108
Filament OD, d_0, μm	51.5	62.2
Eff. fil. diam., d_{eff}, μm[(b)]	61.8	72.4

Cut pieces (50 cm long) of HQ and QXF cable were first enclosed in s-glass braid. The HQ cables were mounted five-high into bolt-down fixtures designed to apply side constraint as they were uniaxially compressed to 20 MPa at CSMM in preparation for heat treatment (HT) at Brookhaven National Laboratory. Four braid-coated lengths of QXF cable were sent to LBNL for mounting and HT in the same fixture but adjusted to confine the cable stack in a space just large enough to contain it during HT when expansions of 1.5% in width and 4.5% in thickness take place.

After HT the fixtures were returned to CSMM where the cable stacks were wrapped in teflon film, placed in an aluminum mold: (i) under uniaxial pressure of 5 MPa (the HQ stacks) or (ii) under negligible pressure (the QXF stacks) and vacuum impregnated with CTD-101 resin.

2.2 Cable and Strand Magnetization

2.2.1. Cable Measurement: Equipment located at the Energy, Materials, and Systems Laboratory of the University of Twente [21] was used to measure the calorimetric loss [18] and magnetization loss, $Q_t = \int M_t dB$ of the eight cable stacks at 4.2 K in transverse ("face-on", FO) AC fields of amplitude, B_m, = 400 mT at frequencies, f, of up to 60 mHz. In addition, cable stack Q4 was measured in B_m = 0.1- 1.4 T at 10 mHz; an experiment that yielded a set of nested *M-B* loops. In general, the total loss measured is $Q_t(f) = Q_h + Q_{coup}(f)$ where Q_h is the cable's/strand's hysteretic or persistent-current loss and $Q_{coup}(f)$ is the interstrand coupling loss (see below). Calorimetric loss was measured by the He-boil-off technique and calibrated against ohmic loss generated by a submerged 25 Ω resistor. For calibration of the installed pick-up coil magnetometer, the magnetization loss of cable stack H2 was equated to its calorimetric loss around the maximum of $Q_t(f)$. A second such calibration was applied to cable stack Q4's nested loops in terms of the $B_m = \pm 1.4$ T loop.

2.2.2. Strand Measurement: Short pieces of strand extracted from the ends of HT cable stacks were measured using the vibrating sample magnetometer attachment of a Quantum Design Model 6000 Physical Property Measurement System (PPMS). Unpenetrated *M-B* loops to ± 400 mT and fully penetrated loops to ± 14 T were obtained at 4.2 K.

3. Cable Loss and Magnetization

As explained in [18] the coupling loss per cycle per m^3 of cable (width, w, thickness, t, strand count, N, transposition pitch, $2L_p$) exposed to an FO field linearly ramping at a rate dB/dt is given by:

$$Q_{coup(FO)} = \left(\frac{4}{3}\right)\left(\frac{w}{t}\right)L_p B_m \left(\frac{N^2}{20}\right)\left[\frac{1}{R_c} + \frac{20}{N^3 R_a}\right]\left(\frac{dB}{dt}\right) \tag{1}$$

where R_c and R_a are the cable's crossover and adjacent interstrand contact resistances. Then after transforming dB/dt to a sinusoidal frequency, f, as explained in [22] we find:

$$Q_{coup(FO)}(f) = \left(\frac{\pi^2}{30}\right)\left(\frac{w}{t}\right)L_p B_m^2 N^2 \left[\frac{1}{R_c} + \frac{20}{N^3 R_a}\right].f \tag{2}$$

$$= \left(\frac{\pi^2}{30}\right)\left(\frac{w}{t}\right)L_p B_m^2 N^2 \left[\frac{1}{R_{eff}}\right].f \tag{3}$$

Figure 1 displays the total magnetization loss as function of frequency for the H series and QXF series cables. The lower set of curves represents the edge-on total loss measured calorimetrically ([18], Fig.1). The persistent current components, Q_h, are the f = 0 intercepts (FO and EO differ because of demagnetization effects [23]). The lack of a slope is due to the very low coupling currents in the EO orientation.

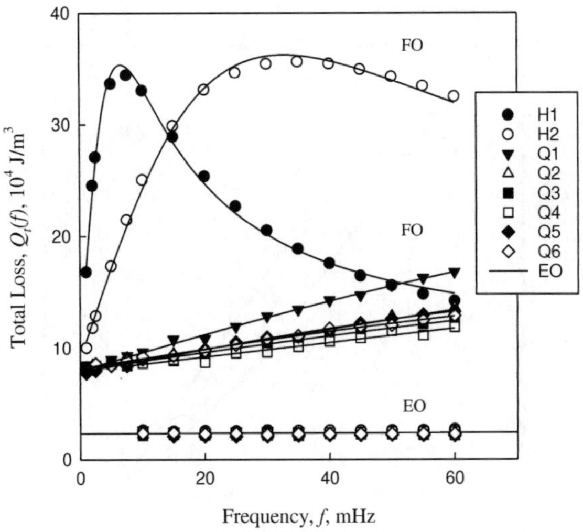

Figure 1. Total face-on (FO) magnetization loss, Q_t, as function of frequency, *f*, for the H series and QXF series cables. The lower set of curves represents the edge-on (EO) total loss measured calorimetrically ([18], Fig.1). The persistent current components, Q_h, are the $f = 0$ intercepts.

After substituting R_{eff} from the experimental dQ_t/df (equation (3), Figure 1, initial slopes) into equation (1) the coupling magnetization $M_{coup} = Q_{coup}/4B_m$ at a typical accelerator charging ramp rate of 7.5 mT/s can be calculated, Table 3.

Table 3. Coupling Magnetizations, $M_{coup} = Q_{coup}/4B_m$, at a ramp rate of 7.5 mT/s and unpenetrated persistent current loss at ± 400 mT

Cable Type	HQ		QXF					
Stack name	H1	H2	Q1	Q2	Q3	Q4	Q5	Q6
M_{coup}, kA/m	206.5	47.1	4.06	2.09	1.73	1.51	2.21	1.84
Q_h, 10^4J/m^3	9.19	8.97	7.95	8.14	8.12	7.92	7.92	8.30

Since the magnetic Q_t data were taken at B_m = ± 400 mT they embody only the unpenetrated Q_h components. It is interesting to note in passing that $\langle Q_h \rangle_{HQ\text{-}set}/\langle Q_h \rangle_{QXF\text{-}set}$ = 1.14 which is very close to the inverse ratios of the strand d_{eff} values (1.17) as expected for unpenetrated magnetizations.

4. Strand Loss and Magnetization

The 4.2 K measured unpenetrated *M-B* loops to ± 400 mT and fully penetrated loops to ± 14 T are presented in Figures 2 and 3, respectively. As indicated in Table 2 the strands extracted from HQ-cable H1 and QXF-cable Q6 were identical in design and differed only in diameter, *d*, and hence effective filament diameter, d_{eff}. We would expect the loop areas, Q_h, to respond to this such that for the fully penetrated loops $Q_{h,\mathrm{H1}}/Q_{h,\mathrm{Q6}} = d_{eff,\mathrm{H1}}/d_{eff,\mathrm{Q6}}$ while for the unpenetrated loops the inverse should hold.

IOP Conf. Series: Materials Science and Engineering **279** (2017) 012037 doi:10.1088/1757-899X/279/1/012037

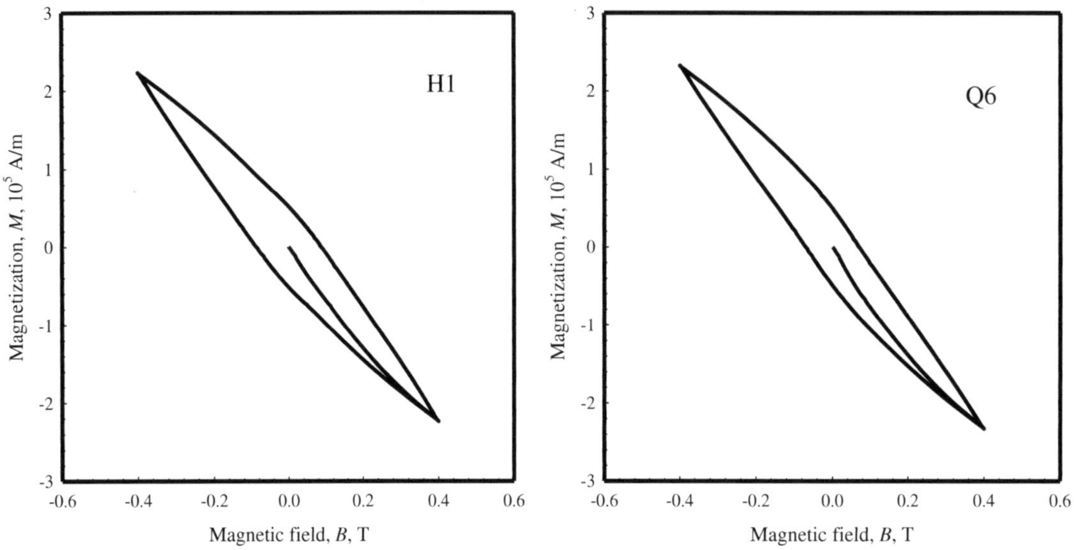

Figure 2. PPMS-measured unpenetrated *M-B* loops at 4.2 K to ± 400 mT

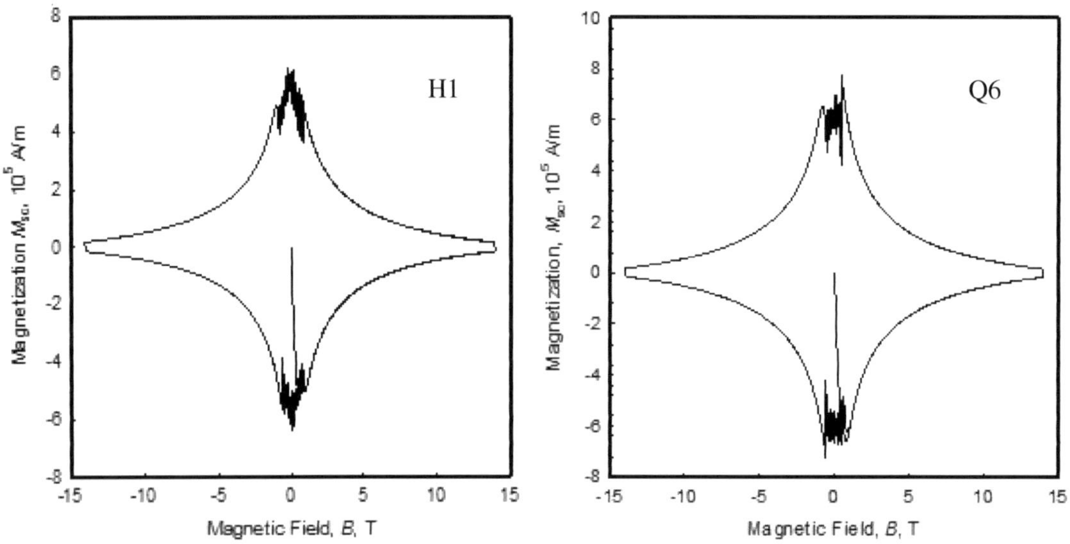

Figure 3. PPMS-measured penetrated *M-B* loops at 4.2 K to ± 14 T.

Both coupling magnetization and persistent current magnetization induce unwanted multipoles in dipole and quadrupole magnets [19][20]. The cable's coupling magnetization can be easily controlled and reduced to relatively small values by introducing cores and/or adjusting the preparation conditions (Table III, [18]). The cable's strand-based persistent current magnetization, which rises to very large values at low fields near injection calls for compensation e.g. by the use of magnetic shims [20]. In preparation for persistent current field error analysis the cable or strand is exposed to a suitable magnetization pre-cycle. In the present study we simply focus on the shielding branches of the *M-B* loops in Figure 3. To improve the relevance of the strand results to future Nb_3Sn accelerator cable applications we have: (i) introduced a cable fill factor of 87% , (ii) dropped the temperature from 4.2 K to 1.9 K and extrapolated the applied field to 16 T [24]. These estimated shielding magnetizations, $M_{sh,cable}$, of cables H1 and Q6 are presented and compared with the coupling results, in Figure 4.

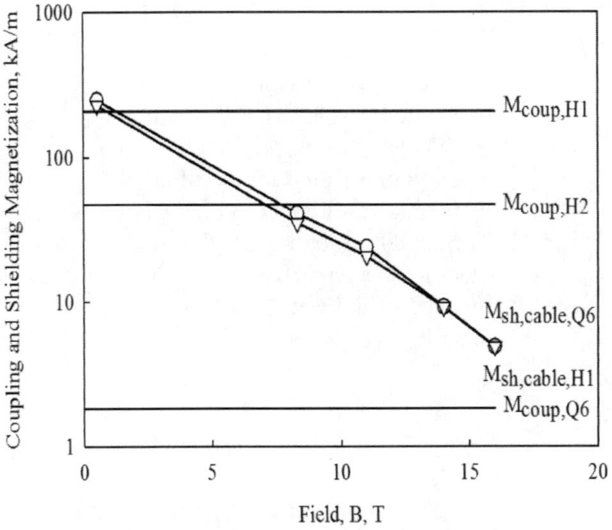

Figure 4. Coupling magnetizations 7.5 mT/s of cables H1, H2, and Q6 ($M_{coup,H1}$, $M_{coup,H2}$, and $M_{coup,Q6}$) and the estimated shielding magnetizations of cables H1 and Q6 ($M_{sh,cable,H1}$ and $M_{sh,cable,Q6}$) at 1.9 K based on strand magnetizations and a cable fill factor of 87%

5. Cable and Strand Magnetizations

5.1. Unpenetrated Persistent Current Strand/Cable Comparisons

Listed in Table 4 are the magnetically measured FO persistent current losses, Q_h, at $B_m = \pm$ 400 mT for all the cable stacks. We note that the averages for types HQ and QXF stacks are 9.078 and 8.057 x10^4J/m^3, respectively; a ratio of 1.14 compared to a d_{eff} inverse ratio (Table 1) of 1.17. Table 4 compares these magnetic FO Q_h values with the calorimetric EO values (Figure 1) and those derived from the *M-B* loops of Figures 2(a) and 2(b). The positioning of the strand Q_hs between the FO and EO cable values is a result of demagnetizations associated with the FO- and EO- oriented highly aspected Rutherford cables (see also [25] Table 4).

Table 4. Unpenetrated persistent current losses, Q_h, 10^4 J/m^3, of cables exposed to FO and EO applied fields and those of corresponding extracted strands

Cable Type	HQ		QXF					
Strand name	H1	H2	Q1	Q2	Q3	Q4	Q5	Q6
Q_h(FO)	9.19	8.97	7.95	8.14	8.12	7.92	7.92	8.30
Q_h(EO)	2.43	2.49	2.63	2.14	2.18	2.34	2.02	2.34
$Q_{h,,cable\text{-}strand}$*	4.38							4.09
$Q_{h,\ strand}$	5.03							4.70

* Based on $Q_{h,strand}$ adjusted for a cable packing factor of 87%

5.2 Penetrated Persistent Current Strand/Cable Comparisons

Figure 5 compares the fully penetrated *M-B* loop for cable stack Q4 (magnetically measured at 10 mHz to ±1.4 T) to that of a corresponding strand, in this case extracted from Cable Q6. Just as in the unpenetrated case strand magnetization is a useful measure of cable magnetization and as has been shown elsewhere [20] can provide input to magnet field error calculations.

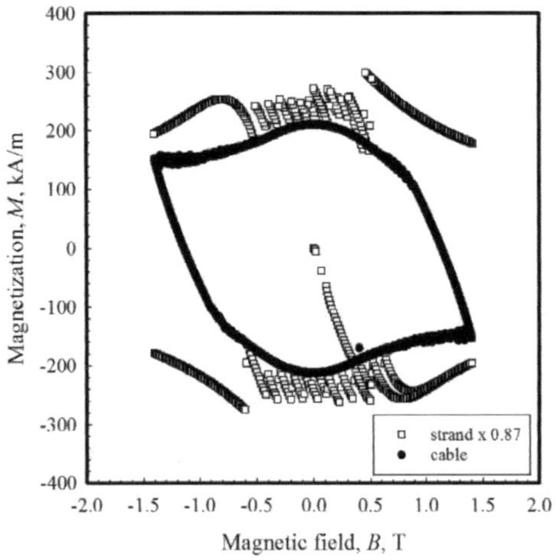

Figure 5. Comparison of penetrated "cable" *M-B* loops to ±1.4 T. For this purpose the strand loop height was modified by 0.87, a cable packing factor. The actual cable loop height measured at 10 mHz was reduced by 7.4% to remove the coupling component.

6. Summary

Magnetizations in the magnet windings induce multipoles in the bore fields of dipole and quadrupole magnets. Strand and cable magnetization data can provide a useful input to field error calculations. As a contribution to this topic, and indirectly to the US LHC Accelerator Research Program (LARP), we have measured the magnetizations of: (i) eight LARP high gradient quadrupole cables designated HQ and QXF, (ii) a pair of strands, identical in architecture but with different d_{eff}s, that had been extracted from a LARP HQ and a related LARP QXF cable.

The magnetization losses, of the eight cable stacks were measured by pickup coil magnetometry at 4.2 K in FO fields, B_m, of ± 400 mT at frequencies, f, of up to 60 mHz. The total loss is $Q_t(f) = Q_h + Q_{coup}(f)$ where Q_h is the cable's/strand's hysteretic or persistent-current loss and $Q_{coup}(f)$ is the interstrand coupling loss. Based on Q_{coup} the coupling magnetization $M_{coup} = Q_{coup}/4B_m$ was derived for a ramp rate of 7.5 mT/s. Persistent current (shielding) magnetization and loss (M_{sh} and $Q_{h,strand}$) were also measured on short pieces of strand extracted from the ends of heat treated cable stacks by vibrating sample magnetometry at 4.2 K. Unpenetrated *M-B* loops to ± 400 mT and fully penetrated loops to ± 14 T were obtained. Thus the cables' M_{coup}s were directly measured and their $M_{sh,cable}$s indirectly measured in terms of the strands' shielding magnetizations, M_{sh}, modified by a cable packing factor of 87%.

M_{coup}, which has no field-dependent components, can be easily controlled and reduced to relatively small values by introducing cores ($M_{coup,H2,cored}/M_{coup,H1,uncored}$ = 0.23) and adjusting the preparation conditions $M_{coup,Q6,prep}/M_{coup,H1,uncored}$ = 0.009). $M_{sh,cable}$ is proportional to $J_c d_{eff}$ in response to which: (1) The ratio $M_{sh,H1}/M_{sh,Q6}$ turned out to be equal to the ratio of the strands' d_{eff}s. (2) In low fields near injection Nb_3Sn's high J_c leads to a correspondingly high $M_{sh,cable}$ Thus at 0.54 T $M_{sh,cable,Q6}$ is over 100 times greater than $M_{coup,Q6}$ which calls for the introduction of some form of magnetic compensation to preserve field quality.

In terms of unpenetrated persistent current loss the directly measured cable Q_h was in reasonable accord with the strand derived $Q_{h,cable\text{-}strand}$. Likewise the suitably adjusted cable and strand fully penetrated *M-B* loops were in reasonable accord. These observations led to the conclusion that strand magnetization is a useful measure of cable magnetization, and that when suitably manipulated can provide input to magnet field error calculations.

Acknowledgments
Funding was provided by the U.S. Department of Energy, Office of High Energy Physics, under Grants No. DE-SC0010312 and DE-SC0011721 (OSU) and DE-AC02-05CH11231 (LBNL). The cables were wound by H.C. Higley (LBNL), heat treated at LBNL (QXF cables) and Brookhaven National Laboratory (A.K, Ghosh, HQ cables). J. Yue and R. Avonce (HyperTech Research) performed the vacuum impregnation.

References

[1] Bottura, L de Rijk G Rossi L and Todesco E 2012 *IEEE Trans. Appl. Supercond.* **22** 4002008
[2] Barachetti A Szeberenyi A and Rossi L 2016 Final Project Report, HiLumi LHC, *CERN-ACC-2016-0007,* 12 January.
[3] Ambrosio G 2015 *IEEE Trans. Appl. Supercond.* **25** 4002107.
[4] Apollinari G 2014 5th International Particle Accelerator Conference (IPAC 2014) 16-20 Jun, Dresden, Germany (4 pp); FERMILAB-CONF-14-326-TD
[5] Apollinari G 2017 HL-LHC and HE-LHC APS April Meeting, January 28–31, Washington, DC
[6] Karppinen M Andreev N Apollinari G 2012 *IEEE Trans. Appl. Supercond.* **22** 4901504.
[7] Karppinen M 2012 11 T Dipole Status May 2012, https://espace.cern.ch/dsdipole/-; see also: hilumilhc.web.cern.ch/hilumilhc/activities/11-T/WP11/
[8] Assmann R Bailey R Brüning O Dominguez O de Rijk G Jimenez JM Myers S Rossi L Tavian L Todesco E and Zimmermann F 2010 *CERN-ATS-2010-177.*
[9] Brüning O Dominguez O Myers S Rossi L Todesco E Zimmermann F 2011 *EuCARD-AccNet-EuroLumi Workshop: The High-Energy Large Hadron Collider*, Malta, 14-16 Oct 2010; CERN Yellow Report CERN-2011-003, pp. 7-14
[10] Zimmermann F 2016 *Challenges and Goals for Accelerators in the XXI Century*, World Scientific, pp. 513-522.
[11] Benedikt M and Zimmermann F 2017 3rd Annual Meeting of the Future Collider Study, Intercontinental Hotel, Berlin, Germany, May 29-June 2.
[12] van Nutgeren J Schoerling D Kirby G Murtomaki J de Rijk G Rossi L Bottura L ten Kate H H J and Dhallé M 2016 *IEEE Trans Appl. Supercond.* **26** 4003206.
[13] Caspi S Arbelaez D Brouwer L Gourlay S Prestemon S and Auchmann B 2017 *IEEE Trans Appl. Supercond.* **27** 4001505.
[14] Goulay S A Prestemon S O Zlobin A V Cooley L and Larbalestier D 2016 The US Magnet Development Program Plan, *Lawrence Berkeley National Laboratory Report No.* LBNL-1006046.
[15] Zlobin A V Andreev N Barzi E Kashikhin V V and Novotski I 2017 *Proc. IPAC 2015,* Richmond, VA 3365-3367 ; also Zlobin A 2017 *US-MDP First General Meeting and Workshop*, Marriott Napa Valley, Feb. 6-8.
[16] Barzi E Li P Turrioni D Xu X and Zlobin A 2017 presentation at the 25TH Magnet Technology Conference, MT25, August 27-31, Amsterdam, NL
[17] Collings E W Sumption M D Majoros M Wang X, et al., 2015 *IEEE Trans. Appl. Supercond.* **25** 4802805.
[18] Collings E W Sumption M D Majoros M Wang X Dietderich D R Yagotyntsev K and Nijhuis A 2017 *IEEE Trans. Appl. Supercond.* **27** 0601305.
[19] Wang X Ambrosio G Borgnolutti F Buehler M Chlachidze G Dietserich D R DiMarco J Felice H Ferracin P Ghosh A Godeke A Marchevsky M Orris D Prestemon S O Sabbi G Sylvester C Tartaglia M Todesco E Velev G and Wanderer P 2014 *IEEE Trans. Appl. Supercond.* **24** 4002607.
[20] Wang X Ambrosio G Chlachidze G DiMarco J Ghosh A K Holik E F Prestemon S O Sabbi G L and Stoynev S E 2017 *IEEE Trans. Appl. Supercond.* **27** 4000805.
[21] Verweij A P 1995 Electrodynamics of Superconducting Cables in Accelerator Magnets *Ph.D. Thesis*, University of Twente Press, p. 94.
[22] Sumption M D Collings E W Scanlan R M Nijhuis A and ten Kate H H J 1999 *Cryogenics* **39** 1-12.
[23] Collings E W Sumption M D Susner M A Dietderich D Kroopshoop E and Nijhuis A, IEEE Trans. Appl. Supercond. 2013 **23** 4702305
[24] Summers L T Guinan M W Miller J R and Hahn P A 1991 *IEEE Trans. Magnetics* **27** 2041-2044.
[25] Collings E W Sumption M D Susner M A Dietderich D R Kroopshoop E and Nijhuis A 2012 *IEEE Trans. Appl. Supercond.* **22** 6000904.